Lecture Notes
in Control and Information Sciences 423

Editors: M. Thoma, F. Allgöwer, M. Morari

Rifat Sipahi, Tomáš Vyhlídal,
Silviu-Iulian Niculescu,
and Pierdomenico Pepe (Eds.)

Time Delay Systems: Methods, Applications and New Trends

 Springer

Editors

Prof. Rifat Sipahi
Northeastern University
Department of Mechanical and
Industrial Engineering
Boston, USA

Prof. Silviu-Iulian Niculescu
L2S (UMR CNRS 8506)
CNRSSupélec
Gif-sur-Yvette
France

Prof. Tomáš Vyhlídal
Czech Technical University
Faculty of Mechanical Engineering
Department of Instrumentation and
   Control Engineering
Prague, Czech Republic

Prof. Pierdomenico Pepe
University of L'Aquila
Department of Electrical and
Information Engineering
L'Aquila, Italy

ISBN 978-3-642-25220-4

e-ISBN 978-3-642-25221-1

DOI 10.1007/978-3-642-25221-1

Lecture Notes in Control and Information Sciences

ISSN 0170-8643

Library of Congress Control Number: 2011940774

Typeset by Scientific Publishing Services Pvt. Ltd., Chennai, India.

Printed on acid-free paper

9 8 7 6 5 4 3 2 1

springer.com

Preface

In many problems in engineering, physics, and biology, various processes take effect only after a certain amount of time elapses after the onset of a stimulus, an input, and any means of cause. This period of time, which is application specific, can arise due to many reasons including, among others, transmitting information across a wireless network in network control systems, shipping products from one location to another in supply chains, producing a decision in car driving upon receiving a stimulus.

Due to the presence of *delays*, instantaneous information cannot be available and therefore many control actions are produced based on the historical evolution of the governing dynamics. This represents a major source of instability, especially in cases when the controllers are designed by neglecting delays. In other words, a controller designed in a conventional way by assuming that the system at hand did not present delays, may not necessarily guarantee the stability of the same system that actually includes delays.

Since 1950s, many studies have been devoted to understanding the effects of delays and exploring control theoretic approaches to circumvent the detrimental effects of delays to dynamical behavior. One of the first tools to deal with delays is the celebrated Smith predictor, dated 1957. On the other hand, an interesting phenomenon discovered in these studies is the so-called *stabilizing effect* of *delays*: inducing delays in the control loop may lead to stable closed-loop system which can be lost when the delay becomes zero. Such a stabilizing feature of delay in dynamical systems works against the human intuition and therefore necessitates analytical studies. Consequently, this feature has become the motivation of many theoretical and experimental studies in the literature, especially in the past two decades.

During this period, many results have been presented at the main control conferences (IEEE CDC, ACC, IFAC symposia), in specialized workshops (IFAC Time-Delay Systems series), and published in the leading journals of control engineering, manufacturing engineering, operations research, mathematical biology, systems and control theory, applied and numerical mathematics. Furthermore, numerous books on the topic appeared, and two review articles were published quite recently in *IEEE Control Systems Magazine*. In the control area, the results are primarily based on the analysis of systems with delays (such as stability, controllability, performance), and on the synthesis of controllers (designing an appropriate control law that achieves certain control objectives). For various specific classes of models, including linear, nonlinear, and stochastic systems, many results have been obtained, while in the numerical analysis area the research has focused on time-stepping, discretization and semi-discretization, computation of stability and corresponding stability charts, and bifurcation analysis.

Although many results are mature, the existing work has so far remained valid for stand-alone systems that have relatively small dimensions and a small number of delays. On the other hand, the challenges faced by the researchers is increasing, in parallel with the rapidly developing technology and higher expectations from researchers and control engineer practitioners. The main challenges are due to handling large-scale systems, nonlinear systems, discrete-continuous type mixed systems, interconnected systems over networks, multi-input multi-output systems with delays of various features such as time-varying, uncertain, and stochastic delays. In the analysis and controller synthesis of such systems, there still remains many uncharted research territories where new results and major impacts are needed.

The arising complexities in the control problems are also in parallel with the versatility of systems aligned with the emergence of multiple disciplines, including engineering, physics, chemistry, biology, operations research, and economics. This trend also calls for researchers from different disciplines to work together, in order to better understand and address control problems. Therefore, it is of no surprise today that control engineers, medical doctors, chemical engineers, and mathematicians work together in defining new frameworks for dealing with the complexity of the systems to be handled. In one sense, the research on time delay systems is expanding and becoming more pervasive, and the publications in this field will grow even further and will be accessed by many researchers including those outside engineering.

This book presents the most recent trends as well as new directions in the field of *control and dynamics of time delay systems*. The field is extremely active and the book captures the most recent snapshot of the research results. The book is collected under five parts: (i) Methodology: From Retarded to Neutral Continuous Delay Models, (ii) Systems, Signals and Applications, (iii) Numerical Methods, (iv) Predictor-based Control and Compensation, and (v) Networked Control Systems and Multi-agent Systems. The themes of these chapters closely tie with the discussions and motivations above. The contributions in the chapters are a well-balanced combination of two main resources; *invited papers*, and the *work presented* at *2009 IFAC Workshop on Time Delay Systems* held in Sinaia, Romania and *2010 IFAC Workshop on Time Delay Systems* held in Prague, Czech Republic. This workshop represents the main specialized meeting venue in the field, and thus captures the most updated research trends. In the selection of the topics and the contributors, the editors have not only aimed at maintaining the highest technical quality of the presentations, but also at achieving an appropriate balance across the chapters (i)-(v). It is worthy to note that the book proposal does not have significant overlap with the contents of 2009-2010 IFAC Workshops on Time Delay Systems (IFAC TDS). Attendees of the workshop contributing to this book proposal also improved their contributions beyond what they presented at the workshop and the editors comfortably claim that at least 50% of the book content-wise comprises *new* contributions that are closely *relevant* to the most recent trends discussed above.

Structure of the Volume

As mentioned earlier, this volume is divided into five parts, where each part is composed of five to eight chapters. They are respectively devoted to Part I: *Methodology: From Retarded to Neutral Continuous Delay Models*, Part II: *Systems, Signals and Applications*, Part III: *Numerical Methods*, Part IV: *Predictor-based Control and Compensation*, and Part V: *Networked Control Systems and Multi-agent Systems*. The first part is concerned with the new results on retarded and neutral type delay differential equations. The second part is devoted to problems with flows, signals and the corresponding real-world applications. The third part is related to numerical methods, which are indispensable part of the research momentum in this field. The fourth and the fifth parts are concerned with the recent trends, which cover respectively predictor-based control/compensation and networked control systems with multi agents. In what follows, we present the different chapters and streams of the different parts.

Part I Methodology: From Retarded to Neutral Continuous Delay Models

Almost all continuous dynamical systems with delays are modeled by either retarded or neutral-type delay differential equations. It is therefore extremely important to develop appropriate analysis tools and methods, and synthesis schemes for such differential equations. In this research direction, many problems are still open, including, among others, improvements on computational efficiency, control of nonlinear systems using observers, new ideas for control system design, connections between frequency and time-domain tools, new ways in constructing Lyapunov quadratic functionals that are not necessarily continuous. In this part of the book, some of these important problems are discussed and addressed.

The first chapter in this part is contributed by VLADIMIR L. KHARITONOV and presents new results in computing quadratic functionals for linear neutral-type systems. Particular emphasis here is on the Lyapunov matrices, which are defined by special matrix valued functions. The focus on such matrices is important as this effort ties with stability analysis and robust stability analysis of time delay systems.

The second chapter is contributed by MICHAEL DI LORETO and JEAN JACQUES LOISEAU and it discusses the stability of positive difference equations, which arise in many stability problems associated with neutral-type delay differential equations. The chapter shows that the stability of linear difference equations with positive coefficients is robust with respect to delays. This result is crucial since otherwise the stability analysis of difference equations with multiple delays can be computationally cumbersome. The result supplements a critical stability analysis step in neutral type systems.

The third chapter contributed by KEQIN GU, YASHUN ZHANG, and MATTHEW PEET concerns the positivity of quadratic functionals, which are key components in time-domain based stability analysis. Under single and double integral positivity conditions, the chapter utilizes appropriate notions from operator theory to show the positivity of quadratic functionals, and studies coupled differential-difference equations using efficient sum-of-squares tools.

The fourth chapter, which is contributed by MILENA ANGUELOVA and MIROSLAV HALÁS, presents a class of retarded-type nonlinear systems that admit an input-output representation of neutral type. Interestingly, the representation becomes possible only with nonlinearities present in the dynamical system, and the authors present conditions under which such a representation exists.

The fifth chapter is contributed by GILBERTO OCHOA, SABINE MONDIÉ, and VLADIMIR L. KHARITONOV. The chapter studies an important problem in linear neutral-type systems, where the computation of critical eigenvalues has direct links with system stability. Different from existing results, the authors consider Lyapunov matrix in order to compute such eigenvalues, which are then used to find the delay intervals in which the system maintains its asymptotic stability.

The last chapter, contributed by ROB H. GIELEN, MIRCEA LAZAR and SORIN OLARU, focusses on the relation between stability of delay difference equations (DDEs) and the existence of $\mathscr{D}$-contractive sets, which provide a region of attraction. First, it is established that a DDE admits a $\mathscr{D}$-contractive set if and only if it admits a Lyapunov-Razumikhin function. This and subsequently derived necessary conditions provide a first step towards the derivation of a notion of asymptotic stability for DDEs which is equivalent to the existence of a $\mathscr{D}$-contractive set.

Part II Systems, Signals and Applications

This part presents an array of contributions that encompass system-level applications to various real-world control problems, in which time delays affect the involved signals. The organization of this chapter is as follows.

The first chapter in this part is contributed by SERGEI AVDONIN and LUCIANO PANDOLFI. The authors focus on the relation between temperature and flux for heat equations with memory. First, it is proved that the relation between temperature and heat governing equations are strict for small times, but these quantities are essentially independent for large times. In the problem analysis, the observation that the *independence* can be interpreted as a kind of *controllability* is utilized.

The second chapter, contributed by LOTFI BELKOURA, investigates the identifiability and algebraic identification of time-delay systems. It is shown that the identifiability property of a general class of systems described by convolution equations can be formulated in terms of approximate controllability or weak controllability, depending on the available models. Consequently, an algebraic method for the identification of delay systems based on both structured inputs and arbitrary input-output trajectories is presented. The proposed on-line algorithms for both parameters and delay estimation are validated on both the simulation and experimental studies with noisy data.

The third chapter is contributed by RUDY CEPEDA-GOMEZ and NEJAT OLGAC. The study addresses the consensus problem for a group of autonomous agents with second order dynamics and time-delayed communications. A consensus protocol for a system of agents with second-order dynamics is proposed, under the assumptions that all agents can communicate with each other and there is a constant communication delay for all. Using the *Cluster Treatment of Characteristic Roots* (CTCR) procedure, a complete stability picture is obtained, taking into account the variations in the control parameters and the communication delay. Case studies and simulations are presented to illustrate the analytical derivations.

The fourth chapter, contributed by ERIK I. VERRIEST, focuses on the analysis of state-space construction for systems with time-varying delay. The fundamental condition for the existence of the state-space is that the delay derivative should be bounded by one. Besides, using a discretization approach, simple derivation of the spectral reachability condition for linear time-invariant (LTI) delay systems is provided. It is also shown that when a system with fixed delay is modeled as one in a class with larger delay, reachability can no longer be preserved.

The fifth chapter of the part is contributed by VLADIMIR RĂSVAN, which discusses some dynamical models in automatic control that are connected with distributed parameters in one dimension. These models are described by boundary value problems for hyperbolic partial differential equations (PDEs). The functional equations associated to these problems are considered by using the integration of the Riemann invariants along the characteristics. The chapter contains some illustrating applications from various fields: nuclear reactors with circulating fuel, canal flows control, overhead crane, drilling devices, without forgetting the standard classical example of the nonhomogeneous transmission lines for distortionless and lossless propagation. Specific features of the control models are discussed in connection with the control approach wherever it applies.

The sixth chapter is contributed by TAMAS INSPERGER, RICHARD WOHLFART, JANOS TURI and GABOR STEPAN. The authors study the dynamics of the stick balancing control where the output for the feedback controller is provided by an accelerometer attached to the stick. In the analysis, different models are considered in the loop with proportional-derivative controller. The stability of the feedback system is studied for the cases with and without feedback delays. Besides, the effect of controller discretization is studied. It is shown that if the accelerometer dynamics is considered in the model with feedback delay, the advanced closed loop dynamics (with infinitely many unstable poles) is obtaiend. Once the controller is discretized, the system can be stabilized despite its advanced nature.

The seventh chapter, contributed by LOUAY SALEH, PHILIPPE CHEVREL and JEAN-FRANÇOIS LAFAY, is dedicated to the study the characteristics of the optimal preview control for lateral steering of a passenger vehicle, provided the *near future* curvature of the road is known. The synthesis is performed in continuous time and leads to a two-degrees of freedom feedback and feedforward controller, whose feedforward part is a finite impulse response filter. Both the theoretical and experimental results show that the preview control action enables the vehicle to track the center of the lane with a smaller tracking error.

In the last, the eighth chapter of this part, HITAY ÖZBAY, CATHERINE BONNET, HOUDA BENJELLOUN and JEAN CLAIRAMBAULT study local asymptotic stability conditions for the positive equilibrium of a system modeling cell dynamics in leukemia. Local asymptotic stability conditions are derived for the positive equilibrium point of this nonlinear model. As the main result, guidelines for development of therapeutic actions are proposed based on the stability conditions derived in terms of inequalities.

Part III Numerical Methods

Roughly speaking, due to the many computational challenges involved in time delay systems, it is of no surprise today that the need and the demand for reliable numerical tools is always a focal point, with more and more versatility expected from these tools. In this line of research, this part presents optimal control design techniques, discretization of solution operators, applications to distributed delays, tools for delay-independent stability test, analysis of non-uniformly sampled systems, numerical methods for constructing Lyapunov matrices and analyzing the stability of systems found in biology. The details of the chapters are as follows.

The first chapter of this part contributed by WIM MICHIELS provides an overview of the eigenvalue-based robust control design methods for linear time delay systems with the fixed-order controllers. The analysis concerns the computation of stability determining characteristic roots and the computation of $\mathcal{H}_2$ and $\mathcal{H}_\infty$ type cost functions. The control synthesis is performed using direct optimization algorithms applied to minimizing either the spectral abscissa (stabilization problem), or $\mathcal{H}_2$, or $\mathcal{H}_\infty$ norms (robust control design), which are, in general, non-convex and even non-smooth objective functions. As mentioned in the chapter, the author and his colleagues have recently implemented the methods into the freely available software tools.

In the second chapter, contributed by DIMITRI BREDA, STEFANO MASET and ROSSANA VERMIGLIO, a numerical scheme to discretize the solution operators of linear time-invariant time-delay systems in Hilbert spaces is proposed and analyzed. Combining polynomial collocation and Fourier projection, a numerical scheme is proposed for discretizing the solution operator of the system under consideration. Next, step-by-step analysis of the discretization approaches results in the matrices necessary for code implementations, and detailed convergence analysis of the method is performed.

The third chapter of this part is contributed by ELIAS JARLEBRING, WIM MICHIELS and KARL MEERBERGEN. The authors show that the Arnoldi method, which is a well-established numerical method for standard and generalized eigenvalue problems, can also be used for infinite-dimensional problems. First, the authors introduce the *infinite Arnoldi method* and provide a step-by-step algorithm for implementation. Consequently, the method is adapted for time-delay system with distributed delays. After outlining the connection with the Fourier cosine transform, the convergence properties in the eigenvalue computation are illustrated on two examples.

The fourth chapter, contributed by ALI FUAT ERGENC, presents an original method for determining delay-independent stability zones of the general LTI dynamics with multiple delays against parametric uncertainties. The method utilizes the extended Kronecker summation and unique properties of self-inversive polynomials. The main result of the chapter is represented by the sufficient condition for delay-independent stability. Several examples demonstrate the applicability of the theoretical results.

In the fifth chapter, LAURENTIU HETEL, ALEXANDRE KRUSZEWSKI and JEAN-PIERRE RICHARD propose a method for computing the Lyapunov exponent of sampled data systems with sampling jitter. The proposed method is hybrid, in the sense that it combines continuous-time models (based on time-delay systems) with polytopic embedding methods (specific to discrete-time approaches). A lower bound of the Lyapunov exponent can be expressed as a generalized eigenvalue problem. Numerical examples complete the presentation and illustrate the improvement in comparison with other classical approaches.

In the sixth chapter, OLGA N. LETYAGINA and ALEXEY P. ZHABKO propose a numerical procedure for the construction of Lyapunov functionals with a prescribed time-derivative for the case of delay systems with periodic coefficients. Similar to the case of time-invariant systems, the functionals are defined by using special Lyapunov matrices. Next, the issues of existence, exponential estimation, robust stability and computational issues are studied.

The seventh chapter by WARODY LOMBARDI, SORIN OLARU and SILVIU-IULIAN NICULESCU proposes a generic approach to obtain polytopic models for systems with inputs affected by time-delays. Three different cases are studied that arise during the discretization process with respect to the structure of the continuous-state transition matrix. The goal is to model the variable input delays as a polytopic uncertainty in order to preserve a linear difference inclusion framework. The proposed approach guarantees a fixed complexity simplex-type global embedding.

Part IV Predictor-Based Control and Compensation

Predictor-based control goes back to 1950s with the results of Smith predictor, which has become one of the major contributions in control systems field in terms of controlling a class of systems with delay. Many other predictive-based approaches currently exist in the literature, including a popular one called model predictive control, which has found broad application in industry. Presence of delay poses major problems in designing predictors and using them as compensation schemes.

The first chapter by MIROSLAV KRSTIC provides a tutorial introduction to methods for stabilization of systems with long input delays, the so-called *predictor feedback* techniques. The methods are based on techniques originally developed for boundary control of partial differential equations using the *backstepping* approach. Several adaptive and compensation schemes are considered for both linear and nonlinear systems with delays of different nature. Primarily, the chapter demonstrates that the construction of backstepping transformations allow one to deal with delays and PDE dynamics at the input, as well as in the main line of applying control action. It is shown that using direct and inverse backstepping transformations, Lyapunov functionals and explicit stability estimates can be constructed.

The second chapter, contributed by ION NECOARA, IOAN DUMITRACHE and JOHAN A.K. SUYKENS, proposes two methods for solving distributively separable convex problems - the proximal center method and the interior-point Lagrangian method. Next, the methods are extended to the case of separable non-convex problems but with a convex objective function using a sequential convex programming framework. It is also proven that some relevant centralized model predictive control (MPC) problems for a network of coupling linear (non-linear) dynamical systems can be recast as separable convex (non-convex) problems for which the proposed distributed methods can be applied.

The third chapter in this part, contributed by SERGIO TRIMBOLI, STEFANO DI CAIRANO, ALBERTO BEMPORAD and ILYA V. KOLMANOVSKY focuses on a model predictive control method with delay compensation for controlling air-to-fuel ratio and oxygen storage in spark ignited engines. The control architecture is based on a delay-free model predictive controller that enforces constraints on the actuators and on the operating range of the variables. Consequently, the architecture comprises a delay compensation strategy based on a state predictor that counteracts the time-varying delay. Simulations of the closed-loop system with a detailed nonlinear model have been shown.

The fourth chapter, contributed by ALFREDO GERMANI, COSTANZO MANES and PIERDOMENICO PEPE, concerns the control of a class of nonlinear retarded systems via an observer-based stabilization scheme. The contributions include both local and global stability, as well as ways to separately design the observer and the controllers, in order to achieve stabilization.

The fifth chapter, contributed by PAVEL ZÍTEK, VLADIMÍR KUČERA and TOMÁŠ VYHLÍDAL, focuses on developing an appropriate cascade control architecture for time-delayed plants based on affine parameterization. The work makes use of quasi-integrating meromorphic functions in order to prescribe the desired open-loop behavior. The arising cascade control has several advantages including its superiority over standard control schemes.

The last chapter, contributed by MARCUS REBLE and FRANK ALLGÖWER, presents results on model predictive control (MPC) for nonlinear time-delay systems. Two schemes for calculating stabilizing design parameters based on the Jacobi linearization of the nonlinear time-delay system are presented. In the first part, an arbitrary stabilizing linear local control law is considered. It is shown that a stabilizable Jacobi linearization implies the existence of a suitable quadratic terminal cost functional and terminal region. In the second part, a simpler terminal region is derived using additional Lyapunov-Razumikhin conditions.

Part V Networked Control Systems and Multi-Agent Systems

Versatile control systems can be created by the synergy of the combination of many subsystems that effectively and synchronously work together to deliver high-performance behavior that cannot be otherwise obtained from individual subsystems. Many engineering as well as biological systems work with this principle, bringing together an ensemble of dynamic systems. In engineering systems, the subsystems can be physically distanced from each other yet can communicate via various communication media, such as internet. The emergence of wireless networks, advancements in robotics, control systems, as well as the need to understand how a group of animals coordinate their group behavior have led to many new trends, especially in the presence of communication delays among the group members. In this part, most recent results in networked control systems and multi-agent systems under the influence of delays are presented.

The first chapter, contributed by ALEXANDRE SEURET and KARL H. JOHANSSON, derives a robust controller for networked control systems with uncertain plant dynamics. The link between the nodes is disturbed by time-varying communication delays, samplings and time-synchronization. A stability criterion for a robust control is presented in terms of LMIs based on Lyapunov-Krasovskii techniques. A second-order system example is considered and the relation between the admissible bounds of the synchronization error and the size of the uncertainties is computed.

The second chapter, contributed by RAFAEL C. MELO, JEAN-MARIE FARINES and JULIO E. NORMEY-RICO, presents the modelling and congestion control of TCP protocols. A nonlinear and a simple linear model are used to represent TCP including comparative results with NS-2 network simulator. Two control schemes are derived based on simple first-order plus dead-time model: i) general predictive controller; ii) proportional-integral plus Smith Predictor. These are compared with a nonlinear based predictive controller. The results obtained using NS-2 demonstrate that the PI plus Smith Predictor offers the best trade-off between complexity and performance to cope with the process dead-time and network disturbances.

The third chapter of the part, contributed by WEI QIAO and RIFAT SIPAHI, studies the indirect relationship between the delay margin of coupled systems and different graphs these systems form via their different topologies. A four-agent linear time-invariant (LTI) consensus dynamics is taken as a benchmark problem with a single delay and second-order agent dynamics. First, *Cluster Treatment of Characteristic Roots* (CTCR) is applied to reveal the delay margin for all possible topologies of the agents. Next, the stability analysis is extended to the case when graphs transform from one to another as the coupling strengths of some links between the agents weaken and vanish.

In the fourth chapter by FATIHCAN M. ATAY, the author studies the consensus problem on directed and weighted networks in the presence of time-delays. The connection structure of the network with information transmission delays is described by a normalized Laplacian matrix. It is shown that consensus is achieved if and only if the underlying graph contains a spanning tree, independently of the value of the delay. Next, the consensus value is calculated and it is shown that the consensus value is determined not just by the initial states of the nodes at time zero, but also on their past history over an interval of time.

The fifth chapter is contributed by IRINEL-CONSTANTIN MORĂRESCU, SILVIU-IULIAN NICULESCU, and ANTOINE GIRARD. It discusses consensus problems for networks of dynamic agents with fixed and switching topologies in presence of delay in the communication channels. The study provides sufficient agreement conditions in terms of delay and the second largest eigenvalue of the Perron matrices defining the collective dynamics. An exact delay bound is determined assuring the preservation of the initial network topology. Besides, the authors present an analysis of the agreement speed when the asymptotic consensus is achieved. Some numerical examples complete the presentation.

The last chapter of this part is contributed by KUN LIU and EMILIA FRIDMAN on the H_∞ control of networked control systems. The chapter presents a new stability and L_2-gain analysis of such systems inspired by discontinuous Lyapunov functions that were introduced in the literature. The novelty in this chapter is represented by the extensions of time-dependent Lyapunov functional approach to networked control systems, where variable sampling intervals, data packet dropouts and variable network-induced delays are taken into account.

Last but not least, we would like to thank the editors of the LNCIS book series for professionally handling the volume and the reviewers for their careful suggestions which improved the overall quality of the volume.

Boston,   Rifat Sipahi
Prague,   Tomáš Vyhlídal
Gif-sur-Yvette,   Silviu-Iulian Niculescu
L'Aquila,   Pierdomenico Pepe
                                                                           August 2011

Contents

Part II Systems, Signals and Applications

Part I
Methodology: From Retarded to Neutral Continuous Delay Models

Lyapunov Functionals and Matrices for Neutral Type Time Delay Systems

Vladimir L. Kharitonov

Abstract. In this chapter we present some basic results concerning the computation of quadratic functionals with prescribed time derivatives for linear neutral type time delay systems. The functionals are defined by special matrix valued functions. These functions are called Lyapunov matrices. Basic results with respect to the existence and uniqueness of the matrices are included. Some important applications of the functionals and matrices are pointed out. A brief historical survey ends the chapter.

1 Introduction

In this chapter we present some advances in the construction of quadratic functionals with prescribed time derivatives for neutral type time delay systems.

There are several issues that will be addressed in this chapter, in particular we focus on the structure of the functionals, we are interested also in lower and upper bounds for them. The functionals are defined by special matrix valued functions known as Lyapunov matrices. The matrices define the functionals. This explains why we dedicate so much attention to the matrices. By definition Lyapunov matrices are solutions of a matrix delay equation which satisfy two additional properties. The delay matrix equation along with the properties form a counterpart of the classical Lyapunov matrix equation.

One of our specific goals is to demonstrate that the computed quadratic functionals can be effectively used in the stability, and robust stability analysis of time delay systems.

Vladimir L. Kharitonov
Faculty of Applied Mathematics
and Control Processes
St.-Petersburg State University
St.-Petersburg, 198504, Russia
e-mail: khar@apmath.spbu.ru

R. Sipahi et al. (Eds.): Time Delay Sys.: Methods, Appli. and New Trends, LNCIS 423, pp. 3–17.
springerlink.com

We restrict our exposition to the case of single delay systems and formally state the main problem in the next section. In Section 3 we define Lyapunov matrices for time delay systems and introduce Lyapunov functionals which time derivative is a given quadratic form. The existence and uniqueness issues for the Lyapunov matrices are discussed in this Section. In Section 4 a new form for the Lyapunov functionals is given. The main feature of the new form is that the functionals in this form do not include terms with time derivative of the independent variable φ. Section 5 is dedicated to some lower and upper bounds for the functionals. Various applications of the Lyapunov matrices and functionals are discussed in Section 6. The last Section 7 with a brief literature review ends the chapter.

2   Preliminaries

2.1   System Description

We consider the class of single delay systems. The main reasons for restricting attention to this class is that dealing with single delay systems simplifies the understanding of basic concepts, and allows to present results without unnecessary cumbersome expressions.

Given a linear neutral type time delay system of the form

$$\frac{d}{dt}\left[x(t) - Dx(t-h)\right] = A_0 x(t) + A_1 x(t-h), \quad t \geq 0, \tag{1}$$

where $h > 0$, and A_0, A_1, D are given constant $n \times n$ matrices.

Let $\varphi : [-h, 0] \to R^n$ be an initial function. We assume that function φ belongs to the space of piece-wise continuously differentiable functions, $PC^1([-h,0],R^n)$. System (1) is time invariant, so without any loss of generality we may assume that the initial time instant $t_0 = 0$. Let $x(t, \varphi)$ stand for the solution of system (1) with an initial function φ,

$$x(\theta, \varphi) = \varphi(\theta), \quad \theta \in [-h, 0],$$

and $x_t(\varphi)$ denotes the restriction of the solution to the segment $[t-h, t]$,

$$x_t(\varphi) : \theta \to x(t+\theta, \varphi), \quad \theta \in [-h, 0].$$

In some cases, when the initial function is not important, or is well defined from the context, we will use notations $x(t)$ and x_t, instead of $x(t, \varphi)$ and $x_t(\varphi)$.

The euclidean norm will be used for vectors, and the induced matrix norm for matrices. For the elements of the space $PC([-h,0],R^n)$ we will use the uniform norm

$$\|\varphi\|_h = \sup_{\theta \in [-h,0]} \|\varphi(\theta)\|.$$

Remark 1. We assume that the following conditions hold:
1. The difference $x(t) - Dx(t - h)$ is continuous for $t \geq t_0$.
2. In (1) the right hand side derivative is assumed in the origin, $t = 0$. By default such agreement remains valid in the situations when only one-sided variation of the independent variable is allowed.

2.2 Exponential Stability

Definition 1. *[4] System (1) is said to be exponentially stable if there exist $\sigma > 0$ and $\gamma \geq 1$, such that every solution of the system satisfies the inequality*

$$\|x(t, \varphi)\| \leq \gamma e^{-\sigma t} \|\varphi\|_h, \quad t \geq 0.$$

Proposition 1. *[4] System (1) is exponentially stable if and only if there exists $\sigma > 0$, such that the real part of any eigenvalue, s_0, of the system satisfies the inequality*

$$Re(s_0) < -\sigma.$$

Corollary 1. *[4] If system (1) is exponentially stable, then any eigenvalue λ_0 of matrix D lies in the open unit disc of the complex plane, $|\lambda_0| < 1$.*

The following statement is a simplified version of the classical Krasovskii-type theorem. It provides sufficient conditions for the exponential stability of system (1).

Theorem 1. *System (1) is exponentially stable if there exists a functional*

$$v : PC^1([-h, 0], R^n) \to R,$$

such that:
1. For some positive α_1, α_2 the functional admits upper and lower bounds of the form

$$\alpha_1 \|\varphi(0) - D\varphi(-h)\|^2 \leq v(\varphi) \leq \alpha_2 \|\varphi\|_h^2.$$

2. For some $\beta > 0$ the inequality

$$\frac{d}{dt} v(x_t) \leq -\beta \|x(t) - Dx(t - h)\|^2, \quad t \geq 0,$$

holds along the solutions of the system.

2.3 Problem Formulation

Motivated by Theorem 1 we address quadratic functionals that satisfy the theorem conditions.

There are two different possibilities to derive such functionals. In the first one we select a particular functional that satisfies the first condition of the theorem, and then check the second condition of the theorem. Usually results obtained in this way are given in the form of special LMIs. In [12] and [13] one can find detailed discussion of results in this direction.

In the other option, following one of the basic ideas of the Lyapunov method, we select first a time derivative, and then compute a functional, which time derivative along the solution of system (1) coincides with the selected one. In this contribution we do not treat the first possibility, but concentrate on the second one.

Since system (1) is linear and time-invariant, it seems natural to start with the case when the selected time derivative is a quadratic form.

Problem 1. Given a symmetric matrix W, we are looking for a functional

$$v_0 : PC^1([-h,0], R^n) \rightarrow R,$$

such that along the solutions of system (1) the following equality holds

$$\frac{d}{dt} v_0(x_t) = -x^T(t) W x(t), \quad t \geq 0.$$

3 Lyapunov Matrices and Functionals

In this subsection we present a solution of Problem 1. But first we introduce Lyapunov matrices for system (1).

Definition 2. [6] We say that $n \times n$ matrix $U(\tau)$ is a Lyapunov matrix of system (1) associated with a symmetric matrix W if it satisfies the properties
1. Dynamic property:

$$\frac{d}{d\tau} [U(\tau) - U(\tau - h)] = U(\tau) A_0 + U(\tau - h) A_1, \quad \tau \geq 0. \tag{2}$$

2. Symmetry property:

$$U(-\tau) = U^T(\tau), \quad \tau \in [0, h]. \tag{3}$$

3. Algebraic property:

$$-W = A_0^T U(0) + U(0) A_0 - A_0^T U(-h) D - D^T U(h) A_0 \tag{4}$$
$$+ A_1^T U(h) + U(-h) A_1 - A_1^T U(0) D - D^T U(0) A_1.$$

We may present property (4) in an alternative form.

Remark 2. The algebraic property (4) of the Lyapunov matrix $U(\tau)$ can be written as

$$-W = \Delta U'(0) - D^T \Delta U'(0) D,$$

where

$$\Delta U'(0) = U'(+0) - U'(-0) = \lim_{\tau \to +0} \frac{dU(\tau)}{d\tau} - \lim_{\tau \to -0} \frac{dU(\tau)}{d\tau}.$$

The following statement sheds light on the role of the Lyapunov matrices in the construction of functionals that solve Problem 1.

Theorem 2. *[2] Let $U(\tau)$ be a Lyapunov matrix associated with a symmetric matrix W. The functional*

$$v_0(\varphi) = \varphi^T(0) \left[U(0) - D^T U(h) - U(-h)D + D^T U(0)D \right] \varphi(0) \tag{5}$$

$$+ 2\varphi^T(0) \int_{-h}^{0} \left[U(-h-\theta) - D^T U(-\theta) \right] \left[D\dot{\varphi}(\theta) + A_1 \varphi(\theta) \right] d\theta$$

$$+ \int_{-h}^{0} \left[D\dot{\varphi}(\theta_1) + A_1 \varphi(\theta_1) \right]^T$$

$$\times \left(\int_{-h}^{0} U(\theta_1 - \theta_2) \left[D\dot{\varphi}(\theta_2) + A_1 \varphi(\theta_2) \right] d\theta_2 \right) d\theta_1.$$

solves Problem 1.

3.1 Existence and Uniqueness Issues

For a given Lyapunov matrix $U(\tau)$ we define two auxiliary matrices

$$Y(\tau) = U(\tau), \quad Z(\tau) = U(\tau - h), \quad \tau \in [0, h]. \tag{6}$$

Lemma 1. *[2], [6] Let $U(\tau)$ be a Lyapunov matrix associated with W, then the auxiliary matrices (6) satisfy the following delay free system of matrix equations*

$$\begin{cases} \frac{d}{d\tau}[Y(\tau) - Z(\tau)D] = Y(\tau)A_0 + Z(\tau)A_1 \\ \frac{d}{d\tau}\left[-D^T Y(\tau) + Z(\tau) \right] = -A_1^T Y(\tau) - A_0^T Z(\tau) \end{cases} \tag{7}$$

and the boundary value conditions

$$\begin{cases} Y(0) = Z(h), \\ -W = Y(0)A_0 + Z(0)A_1 + A_0^T Z(h) + A_1^T Y(h) \\ -D^T \left[Y(h)A_0 + Z(h)A_1 \right] - \left[A_0^T Z(0) + A_1^T Y(0) \right] D. \end{cases} \tag{8}$$

Theorem 3. *[8] Given a symmetric matrix W, if a pair $(Y(\tau), Z(\tau))$ satisfies (7) and (8), then*

$$U(\tau) = \frac{1}{2} \left[Y(\tau) + Z^T(h - \tau) \right], \quad \tau \in [0, h], \tag{9}$$

is a Lyapunov matrix associated with W if we extend it to $[-h, 0)$ by setting $U(-\tau) = U^T(\tau)$, for $\tau \in (0, h]$.

Corollary 2. *If the auxiliary boundary value problem (7)-(8) admits a unique solution $(Y(\tau),Z(\tau))$ then*

$$U(\tau) = Y(\tau), \quad \tau \in [0,h].$$

Theorem 3 motivates us to look for conditions under which the boundary value problem (7)-(8) admits a unique solution. But first we introduce the following definition.

Definition 3. We say that system (1) satisfies Lyapunov condition if there exists $\varepsilon > 0$, such that the sum of any two eigenvalues, s_1, s_2, of the system has module greater than ε,

$$|s_1 + s_2| > \varepsilon.$$

Lemma 2. *[8] System (1) satisfies the Lyapunov condition if and only if the following two conditions hold:*
1. System (1) has no eigenvalue s_0, such that $-s_0$ is also an eigenvalue of the system;
2. Matrix D has no eigenvalue λ_0, such that λ_0^{-1} is also an eigenvalue of the matrix.

Theorem 4. *[8]System (1) admits a unique Lyapunov matrix associated with a given symmetric matrix W if and only if the system satisfies the Lyapunov condition.*

Theorem 5. *[8] Let system (1) do not satisfy the Lyapunov condition. Then there exists a symmetric matrix W for which equation (2) has no solution satisfying properties (3), (4).*

3.2   Computational Issue

It is important to have an efficient numerical procedure for the computation of Lyapunov matrices. In this section we describe such a procedure.

We have already seen that if the boundary value problem (7)-(8) has a solution, then, by Theorem 3, this solution generates a Lyapunov matrix associated with a given W. In the case, when the boundary value problem admits a unique solution, $(Y(\tau),Z(\tau))$, we have

$$U(\tau) = Y(\tau), \quad \tau \in [0,h].$$

According to Theorem 4 the Lyapunov condition guarantees the existence of a unique solution of the boundary value problem (7)-(8) with an arbitrary symmetric matrix W.

3.3   Spectral Properties

The spectrum of system (1) consists of all complex numbers s for which the characteristic matrix of the system,

$$G(s) = sI - se^{-sh}D - A_0 - e^{-sh}A_1,$$

is singular

$$\Lambda = \{ \, s_0 \mid \det G(s_0) = 0 \, \}.$$

The spectrum of the matrix system (7) consists of all complex numbers s for which the following system of algebraic matrix equations

$$\begin{cases} P(sI - A_0) - Q(sD + A_1) = 0_{n \times n} \\ (-sD^T + A_1^T)P + (sI + A_0^T)Q = 0_{n \times n} \end{cases} \tag{10}$$

admits a non-trivial solution (P, Q).

Lemma 3. *[2] The spectrum of the matrix system (7) is symmetric with respect to the origin.*

Theorem 6. *[13] [8] Let system (1) have an eigenvalue s_0, such that $-s_0$ is also an eigenvalue of the system. Then s_0 belongs to the spectrum of system (7).*

4 Lyapunov Functionals: A New Form

Here we are going to present functional (5) in a new form. The main feature of the form is that it does not involve the derivative of function $\varphi(\theta)$. In other words, all terms of the functional that include the derivative are transformed in such a way that the new expressions for them do not include it. There are three such terms. The first one

$$I_1 = 2\varphi^T(0) \int_{-h}^{0} [U(h + \theta) - U(\theta)D]^T D\dot{\varphi}(\theta) d\theta$$

can be transformed as follows

$$\begin{aligned} I_1 &= 2\varphi^T(0) \int_{-h}^{0} [U(h + \theta) - U(\theta)D]^T D\dot{\varphi}(\theta) d\theta \\ &= 2\varphi^T(0) \left[U(-h)D - D^T U(0)D \right] \varphi(0) \\ &\quad - 2\varphi^T(0) \left[U(0)D - D^T U(h)D \right] \varphi(-h) \\ &\quad - 2\varphi^T(0) \int_{-h}^{0} [U(h + \theta)A_0 + U(\theta)A_1]^T D\varphi(\theta) d\theta. \end{aligned}$$

For the second one we have

$$\begin{aligned} I_2 &= 2 \int_{-h}^{0} \left[D\dot{\varphi}(\theta_1) \right]^T \left(\int_{-h}^{0} U(\theta_1 - \theta_2)A_1 \varphi(\theta_2) d\theta_2 \right) d\theta_1 \\ &= 2\varphi^T(0) \int_{-h}^{0} D^T U(-\theta)A_1 \varphi(\theta) d\theta - 2\varphi^T(-h) \int_{-h}^{0} D^T U(-h - \theta)A_1 \varphi(\theta) d\theta \\ &\quad - 2 \int_{-h}^{0} \varphi^T(\theta_1) \left(\int_{-h}^{0} D^T U'(\theta_1 - \theta_2)A_1 \varphi(\theta_2) d\theta_2 \right) d\theta_1. \end{aligned}$$

Finally, the term

$$I_3 = \int_{-h}^{0} \left[D\dot{\varphi}(\theta_1) \right]^T \left(\int_{-h}^{0} U(\theta_1 - \theta_2) D\dot{\varphi}(\theta_2) d\theta_2 \right) d\theta_1$$

can be written as

$$
\begin{aligned}
I_3 = {} & \varphi^T(0) D^T U(0) D\varphi(0) - 2\varphi^T(0) D^T U(h) D\varphi(-h) \\
& + \varphi^T(-h) D^T U(0) D\varphi(-h) + 2\varphi^T(0) D^T \int_{-h}^{0} U'(-\theta) D\varphi(\theta) d\theta \\
& - 2\varphi^T(-h) D^T \int_{-h}^{0} U'(-\theta - h) D\varphi(\theta) d\theta \\
& - \int_{-h}^{0} \varphi^T(\theta_1) D^T \left(\int_{-h}^{\theta_1 - 0} U''(\theta_1 - \theta_2) \varphi(\theta_2) d\theta_2 \right) d\theta_1 \\
& - \int_{-h}^{0} \varphi^T(\theta_1) D^T \left(\int_{\theta_1 + 0}^{0} U''(\theta_1 - \theta_2) \varphi(\theta_2) d\theta_2 \right) d\theta_1 \\
& - \int_{-h}^{0} \varphi^T(\theta) D^T \left[U'(+0) - U'(-0) \right] D\varphi(\theta) d\theta.
\end{aligned}
$$

Substituting these expressions for I_1, I_2, I_3 in $v_0(\varphi)$, and collecting similar terms we arrive at the desired new form of the functional

$$
\begin{aligned}
v_0(\varphi) = {} & [\varphi(0) - D\varphi(-h)]^T U(0) [\varphi(0) - D\varphi(-h)] \qquad\qquad\qquad (11) \\
& + 2 [\varphi(0) - D\varphi(-h)]^T \int_{-h}^{0} \left[U'(-h - \theta) D + U(-h - \theta) A_1 \right] \varphi(\theta) d\theta \\
& + \int_{-h}^{0} \varphi^T(\theta_1) \left(\int_{-h}^{0} A_1^T U(\theta_1 - \theta_2) A_1 \varphi(\theta_2) d\theta_2 \right) d\theta_1 \\
& - \int_{-h}^{0} \varphi^T(\theta_1) \left(\int_{-h}^{0} \left[D^T U'(\theta_1 - \theta_2) A_1 - A_1^T U'(\theta_1 - \theta_2) D \right] \varphi(\theta_2) d\theta_2 \right) d\theta_1 \\
& - \int_{-h}^{0} \varphi^T(\theta_1) D^T \left(\int_{-h}^{\theta_1 - 0} U''(\theta_1 - \theta_2) D\varphi(\theta_2) d\theta_2 \right) d\theta_1 \\
& - \int_{-h}^{0} \varphi^T(\theta_1) D^T \left(\int_{\theta_1 + 0}^{0} U''(\theta_1 - \theta_2) D\varphi(\theta_2) d\theta_2 \right) d\theta_1 \\
& - \int_{-h}^{0} \varphi^T(\theta) D^T P D\varphi(\theta) d\theta.
\end{aligned}
$$

Here P is the solution of the Schur matrix equation

$$P - D^T P D = -W. \qquad\qquad (12)$$

For given symmetric matrices W_j, $j = 0, 1, 2$, one can define the functional

$$
\begin{aligned}
w(\varphi) = {} & \varphi^T(0) W_0 \varphi(0) + \varphi^T(-h) W_1 \varphi(-h) \\
& + \int_{-h}^{0} \varphi^T(\theta) W_2 \varphi(\theta) d\theta, \quad \varphi \in PC^1([-h, 0], R^n).
\end{aligned}
$$

Theorem 7. *Let system (1) satisfy the Lyapunov condition. Then the functional*

$$v(\varphi) = v_0(\varphi) + \int\limits_{-h}^{0} \varphi^T(\theta)\left[W_1 + (h+\theta)W_2\right]\varphi(\theta)d\theta, \qquad (13)$$

where $v_0(\varphi)$ is defined by (11) with Lyapunov matrix $U(\tau)$ associated with $W = W_0 + W_1 + hW_2$, is such that

$$\frac{d}{dt}v(x_t) = -w(x_t), \quad t \geq 0,$$

along the solutions of the system.

Definition 4. We say that functional (13) is of the complete type if matrices W_j, $j = 0, 1, 2$, are positive definite.

5 Quadratic Bounds

Lemma 4. *[6] Let system (1) be exponentially stable. If matrices W_j, $j = 0, 1, 2$, are positive definite, then there exists $\alpha_1 > 0$, such that the complete type functional (13) satisfies the inequality*

$$\alpha_1 \|\varphi(0) - D\varphi(-h)\|^2 \leq v(\varphi), \quad \varphi \in PC^1([-h,0],R^n).$$

Here the value α_1 is such that matrix

$$L(\alpha_1) = \begin{pmatrix} W_0 & 0_{n\times n} \\ 0_{n\times n} & W_1 \end{pmatrix} + \alpha_1 \begin{pmatrix} A_0 + A_0^T & A_1 - A_0^T D \\ A_1^T - D^T A_0 & -D^T A_1 - A_1^T D \end{pmatrix}.$$

is positive definite.

Lemma 5. *[6] Let system (1) satisfy the Lyapunov condition. Given symmetric matrices W_0, W_1 and W_2, then there exists $\alpha_2 > 0$, such that functional (13) satisfies the inequality*

$$v(\varphi) \leq \alpha_2 \|\varphi\|_h^2, \quad \varphi \in PC^1([-h,0],R^n).$$

In order to compute α_2 we introduce the quantities

$$u_0 = \|U(0)\|, \quad u_1 = \sup_{\tau \in (0,h)} \left\| -D^T U'(\tau) + A_1^T U(\tau) \right\|,$$

and

$$u_2 = \sup_{\tau \in (0,h)} \left\| -D^T U''(\tau)D_1 + A_1^T U(\tau)A_1 - D^T U'(\tau)A_1 + A_1^T U'(\tau)D \right\|.$$

Then, the functional admits the following quadratic upper bound

$$v(\varphi) \leq \alpha_2 \|\varphi\|_h^2,$$

where

$$\alpha_2 = u_0 \left(1 + \|D\|\right)^2 + 2hu_1\left(1 + \|D\|\right) + h^2 u_2$$
$$+ h\|W_1\| + h\|W_2\| + h\|D^T PD\|.$$

We present now slightly different upper and lower quadratic bounds for the functionals.

Lemma 6. *[2] [6] Let system (1) be exponentially stable. Given positive definite matrices W_0, W_1 and W_2, then there exist $\beta_j > 0$, $j = 1, 2$, such that the complete type functional (13) satisfies the inequality*

$$\beta_1 \|\varphi(0) - D\varphi(-h)\|^2 + \beta_2 \int_{-h}^{0} \|\varphi(\theta)\|^2 d\theta \le v(\varphi), \quad \varphi \in PC^1([-h, 0], R^n).$$

Here positive values β_1 and β_2 are such that the matrix

$$L(\beta_1, \beta_2) = \begin{pmatrix} W_0 & 0_{n \times n} \\ 0_{n \times n} & W_1 \end{pmatrix} + \beta_1 \begin{pmatrix} A_0 + A_0^T & A_1 - A_0^T D \\ A_1^T - D^T A_0 & -D^T A_1 - A_1^T D \end{pmatrix}$$
$$+ \beta_2 \begin{pmatrix} I & 0_{n \times n} \\ 0_{n \times n} & -I \end{pmatrix}$$

is positive definite.

Lemma 7. *[2] [6] Let system (1) satisfy the Lyapunov condition. Given symmetric matrices W_0, W_1 and W_2, then there exist $\delta_j > 0$, $j = 1, 2$, such that the functional (13) satisfies the inequality*

$$v(\varphi) \le \delta_1 \|\varphi(0) - D\varphi(-h)\|^2 + \delta_2 \int_{-h}^{0} \|\varphi(\theta)\|^2 d\theta, \quad \varphi \in PC^1([-h, 0], R^n).$$

It is shown that positive values δ_1 and δ_2 can be computed as follows

$$\delta_1 = u_0 + hu_1, \text{ and } \delta_2 = u_1 + \|W_1\| + h\|W_2\| + \|D^T PD\| + hu_2.$$

6 Applications

In this section we consider several applications of Lyapunov functionals and matrices.

6.1 Exponential Estimates

In this sub-section we derive exponential estimates for the solutions of system (1).

Lemma 8. *[6] Let system (1) be exponentially stable. Given positive definite matrices W_0, W_1 and W_2, then there exists $\sigma_1 > 0$, such that the complete type functional (13) satisfies the inequality*

$$\frac{d}{dt}v(x_t) + 2\sigma_1 v(x_t) \leq 0, \quad t \geq 0,$$

along the solutions of the system.

Here $\sigma_1 > 0$ is such that matrices

$$L(\sigma_1) = \begin{pmatrix} W_0 & 0_{n \times n} \\ 0_{n \times n} & W_1 \end{pmatrix} - 2\sigma_1 \delta_1 \begin{pmatrix} I & -D \\ -D^T & D^T D \end{pmatrix}$$

and

$$\tilde{L}(\sigma_1) = W_2 - 2\sigma_1 \delta_2 I$$

are positive definite.

We are now able to state the main result of the section.

Theorem 8. *Let system (1) be exponentially stable. Given positive definite matrices* W_0, W_1, W_2, *then the solutions of the system satisfy the inequality*

$$\|x(t, \varphi)\| \leq \gamma \|\varphi\|_h e^{-\sigma t}, \quad t \geq 0.$$

To compute γ and σ we first define

$$\mu = \sqrt{\frac{\alpha_2}{\alpha_1}},$$

see Lemmas 4-5. As matrix D is Schur stable, then there exist $d \geq 1$ and $\rho \in (0, 1)$, such that $\|D^j\| \leq d\rho^j$, $j \geq 0$. So, for some $\sigma_2 > 0$, $\rho = e^{-h\sigma_2}$. Then, we select $\sigma_0 = \min\{\sigma_1, \sigma_2\}$, where σ_1 has been computed in Lemma 8. Finally, the desired values σ and γ are computed as follows

$$\sigma = \sigma_0 - \varepsilon, \text{ where } \varepsilon \in (0, \sigma_0), \text{ and } \gamma = d\left[1 + \mu + \frac{\mu}{he\varepsilon}\right].$$

6.2 Quadratic Performance Index

An interesting application of quadratic functionals to the computation of performance indices for control time delay systems has been studied in [11].

Given a control system of the form

$$\frac{d}{dt}[x(t) - Dx(t-h)] = \tilde{A}_0 x(t) + \tilde{A}_1 x(t-h) + Bu(t), \quad t \geq 0, \qquad (14)$$
$$y(t) = Cx(t),$$

such that the control law

$$\tilde{u}(t) = Mx(t-h), \quad t \geq 0 \qquad (15)$$

makes the closed loop system

$$\frac{d}{dt}[x(t) - Dx(t-h)] = A_0 x(t) + A_1 x(t-h), \quad t \geq 0, \tag{16}$$

where $A_0 = \widetilde{A}_0$, and $A_1 = \widetilde{A}_1 + BM$, exponentially stable.

The value of the quadratic performance index

$$J(\widetilde{u}) = \int_0^\infty \left[y^T(t) P y(t) + u^T(t) Q u(t) \right] dt, \tag{17}$$

where P and Q are given symmetric matrices of the appropriate dimensions, can be written as

$$J(\widetilde{u}) = \int_0^\infty \left[x^T(t, \varphi) W_0 x(t, \varphi) + x^T(t-h, \varphi) W_1 x(t-h, \varphi) \right] dt.$$

Here $\varphi \in PC^1([-h, 0], R^n)$ is an initial function of the solution $x(t, \varphi)$ of the closed loop system (16), and matrices $W_0 = C^T P C$, $W_1 = M^T Q M$.

Theorem 9. *[11] The value of the performance index (17) for the stabilizing control law (15)*

$$J(\widetilde{u}) = v_0(\varphi) + \int_{-h}^0 \varphi^T(\theta) W_1 \varphi(\theta) d\theta,$$

where $v_0(\varphi)$ is the functional (5) computed with Lyapunov matrix $U(\tau)$ associated with matrix $W = W_0 + W_1 = C^T P C + M^T Q M$.

6.3 Robustness Bounds

In this sub-section we demonstrate how the complete type functionals may be used for the robust stability analysis of time delay system. Consider the following perturbed system

$$\frac{d}{dt}[y(t) - Dy(t-h)] = (A_0 + \Delta_0) y(t) + (A_1 + \Delta_1) y(t-h), \quad t \geq 0, \tag{18}$$

where Δ_0, Δ_1 are unknown, but norm bounded matrices

$$\|\Delta_j\| \leq r_j, \ j = 0, 1. \tag{19}$$

Under assumption that system (1) is exponentially stable we would like to derive some bounds for r_0, r_1, under which system (18) remains stable for all perturbation matrices satisfying (19).

We start with the functional (13) computed for the nominal system (1). The time derivative of the functional along the solutions of system (18) is of the form

$$\frac{d}{dt} v(y_t) = -w(y_t) + 2[y(t) - Dy(t-h)]^T U(0) [\Delta_0 y(t) + \Delta_1 y(t-h)]$$
$$+ 2 [\Delta_0 y(t) + \Delta_1 y(t-h)]^T$$

$$\times \int_{-h}^{0} \left[U'(-h-\theta)D + U(-h-\theta)A_1 \right] y(t+\theta)d\theta.$$

Let us introduce the values

$$u_0 = \|U(0)\|, \quad u_1 = \sup_{\tau \in (0,h)} \left\| -D^T U'(\tau) + A_1^T U(\tau) \right\|.$$

Theorem 10. *Let the nominal system (1) be exponentially stable. Given positive definite matrices W_j, $j = 0, 1, 2$, and the Lyapunov matrix $U(\tau)$ associated with $W = W_0 + W_1 + hW_2$, then the perturbed system (18) remains exponentially stable for all perturbations satisfying (19) if r_0, and r_1 satisfy the following inequalities*
1. $\lambda_{\min}(W_0) > u_0 [r_1 + r_0(2 + \|D\|)] + hu_1 r_0,$
2. $\lambda_{\min}(W_1) > u_0 [r_1 + (r_0 + 2r_1) \|D\|] + hu_1 r_1,$
3. $\lambda_{\min}(W_2) > u_1 (r_0 + r_1).$

Corollary 3. *In the previous theorem one may assume that matrices Δ_0, Δ_1 depend continuously on t and y_t.*

6.4 The $\mathcal{H}_2$ Norm of a Transfer Matrix

Assume that control system (14) is exponentially stable. Then, the value of the $\mathcal{H}_2$ norm, see [15], of the transfer matrix

$$F(s) = C \left(sI - se^{-sh}D - \tilde{A}_0 - e^{-hs}\tilde{A}_1 \right)^{-1} B,$$

of the system can be computed.

Theorem 11. *[5]The $\mathcal{H}_2$ norm of the transfer matrix*

$$\|F\|_{\mathcal{H}_2}^2 = Trace\left\{ B^T U_1(0)B \right\} = Trace\left\{ CU_2(0)C^T \right\},$$

where $U_1(\tau)$ is the Lyapunov matrix of the system (14), associated with matrix $W = C^T C$, and $U_2(\tau)$ is the Lyapunov matrix of the system associated with $W = BB^T$.

7 Comments

It seems that the first contribution dedicated to the computation of Lyapunov functionals with a given time derivative in the case of linear neutral type systems has been written by W.B. Castelan and E.F. Infante [2]. In this contribution the authors first compute a quadratic functional for a difference approximation of system (1). Then the desired Lyapunov functional appears as a result of an appropriate limiting procedure. It is shown there that the functional is determined by a special matrix valued function, the Lyapunov matrix in our terminology. A reader can find in this paper the three basic properties of the matrix valued function, the dynamic, the symmetry and the algebraic ones, as well as the fact that the computation of the matrix is reduced to the computation of a special solution of a delay free system of the form (7). The principal aim of the paper was to demonstrate that the presented functionals can be used in the computation of exponential estimates of the solutions of system (1). Unfortunately, this aim fails to be achieved since in the computation of an upper bound for functionals the authors use the exponential estimate that they attempt to compute.

The observation that the critical delay values can be computed on the base of the spectrum of system (7) has been done in [13].

Chapter 5 in [11] contains a detailed analysis of the problem of evaluation of quadratic performance indices for time delay systems, see also references therein. In [5] besides the computation of the H_2 norm of the transfer function of a control time delay systems some interesting comments with respect to the computation of Lyapunov matrices are given.

Several aspects of the problem studied in this chapter are also discussed in the following papers [6], [7], [8], and [14].

The presented bibliography does not pretend to be exhausted. It includes entries that are closely related with the problems discussed in the paper.

References

1. Bellman, R., Cooke, K.L.: Differential-Difference Equations. Academic Press, New York (1963)
2. Castelan, W.B., Infante, E.F.: A Liapunov functional for a matrix neutral difference-differential equation with one delay. Journal of Mathematical Analysis and Applications 71, 105–130 (1979)
3. Gorecki, H., Fuksa, S., Grabowski, P., Korytowski, A.: Analysis and Synthesis of Time-Delay Systems. John Willey and Sons (PWN), Warsaw (1989)
4. Hale, J.K., Verduyn Lunel, S.M.: Introduction to Functional Differential Equations. Springer, New York (1993)
5. Jarlebring, E., Vanbiervliet, J., Michiels, W.: Characterizing and computing the H_2 norm of time delay systems by solving the delay Lyapunov equation. In: Proceedings 49th IEEE Conference on Decision and Control (2010)
6. Kharitonov, V.L.: Lyapunov functionals and Lyapunov matrices for neutral type time-delay systems: a single delay case. International Journal of Control 78, 783–800 (2005)
7. Kharitonov, V.L.: Lyapunov matrices for a class of neutral type time delay systems. International Journal of Control 81, 883–893 (2008)
8. Kharitonov, V.L.: Lyapunov matrices: Existence and uniqueness issues. Automatica 46, 1725–1729 (2010)
9. Kolmanovskii, V.B., Nosov, V.R.: Stability of Functional Differential Equations. Mathematics in Science and Engineering, vol. 180. Academic Press, New York (1986)
10. Louisell, J.: A matrix method for determining the imaginary axis eigenvalues of a delay system. IEEE Transactions on Automatic Control 46, 2008–2012 (2001)
11. Marshall, J.E., Gorecki, H., Korytowski, A., Walton, K.: Time-Delay Systems: Stability and Performance Criteria with Applications. Ellis Horwood, New York (1992)

12. Niculescu, S.-I.: Delay Effects on Stability: A Robust Control Approach. Springer, Heidelberg (2001)
13. Richard, J.-P.: Time-delay systems: an overview of some recent advances and open problems. Automatica 39, 1667–1694 (2003)
14. Velazquez-Velazquez, J., Kharitonov, V.L.: Lyapunov-Krasovskii functionals for scalar neutral type time delay equation. Systems and Control Letters 58, 17–25 (2009)
15. Zhou, K., Doyle, J.C., Glover, K.: Robust and Optimal Control. Prentice-Hall, Upper Saddle River (1996)

On the Stability of Positive Difference Equations

Michael Di Loreto and Jean Jacques Loiseau

Abstract. In this chapter, we are interested with stability of linear continuous-time difference equations. These equations involve delays, which can be non commensurable. Spectrum analysis comes down to the zeros analysis of an exponential polynomial. From previous results on stability dependent or independent of delays, we focus on the particular case of positive difference equations. It is proved that the stability of linear difference equations with positive coefficients is robust with respect to variations of the delays, and are given exponential bounds for the solution.

1 Introduction

We consider a difference equation of the form

$$x(t) = \sum_{i=1}^{n} a_i x(t - \beta_i) \tag{1}$$

where $n \in \mathbb{N}$, $a_i, \beta_i \in \mathbb{R}$, for $i = 1, \ldots, n$, and $0 < \beta_1 < \ldots < \beta_n$. The coefficients β_i are called the delays of the equation. We are interested in studying the stability of the solution of (1), particularly in the case where the coefficients a_i are positive .

Difference equations appear as models in biology, economy, and in propagation phenomena, from the wave equation with particular end points conditions, or from discretization schemes like finite differences methods. See for instance [8]. Another

Michael Di Loreto
AMPERE, UMR CNRS 5005, INSA-Lyon, Avenue Jean Capelle,
69621 Villeurbanne cedex, France
e-mail: michael.di-loreto@insa-lyon.fr

Jean Jacques Loiseau
Institut de Recherche en Communications et Cyberntique de Nantes (IRCCyN), UMR CNRS 6597, Ecole Centrale de Nantes, BP 92101, 44321 Nantes cedex 03, France
e-mail: jean-jacques.loiseau@irccyn.ec-nantes.fr

R. Sipahi et al. (Eds.): Time Delay Sys.: Methods, Appli. and New Trends, LNCIS 423, pp. 19–33.
springerlink.com   © Springer-Verlag Berlin Heidelberg 2012

motivation for studying (1) is the stability of the more general delayed equation of
neutral type

$$\frac{\mathrm{d}}{\mathrm{d}t}\left(x(t) - \sum_{i=1}^{n} a_i x(t - \beta_i)\right) = \sum_{i=1}^{n} b_i x(t - \beta_i), \tag{2}$$

since its poles asympotically converge to those of the corresponding system (1).
Thus, the stability of (1) is necessary for the stability of the neutral equation. Further,
one can see that the stability of system (1) is also necessary for the stabilization of
the controlled system of neutral type

$$\frac{\mathrm{d}}{\mathrm{d}t}\left(x(t) - \sum_{i=1}^{n} a_i x(t - \beta_i)\right) = \sum_{i=1}^{n} b_i x(t - \beta_i) + Bu(t), \tag{3}$$

where $u(t)$ is a control input. Many an author have noticed that the stability of difference equations (and consequently that of differential difference equations of neutral
type) may be lost through arbitrary small perturbations of the delays. This was first
noticed by [3]. A precise explanation of the origin of the phenomena, in terms of
root locus, was presented by [1]. At the contrary of what happens for many linear systems, the pole locations of difference equations are not continuous functions
of the delays β_i. This fact has important consequences in control theory. The first
one is the definition of another stability concept, for practical purpose. A system
is called strongly stable if it is stable, and the stability is kept when the delays are
subject to small perturbations. It is shown that a time delay of neutral type (2) is
strongly stable only if the associated difference equation (1) is also stable. The controlled system (3) is strongly stabilizable only if the open-loop associated difference
equation is strongly stable. These basic results and many variations of them and illustrative examples are provided by [6], and also by [11], [13], [7], [14]. Effective
numerical methods to check the strong stability and the strongly stabilization of a
controlled time-delay system of neutral type are provided by Michiels and Vyhlídal
[15]. It is worth to point out that, in the case of multivariable systems, the method is
based on the calculation of the spectral radius of some matrices.

Difference equations also appear as simple models for networks, when a delay is
associated to each branch. The variables to model the system are the flows along the
branches, and the conservation laws lead to a set of linear difference equations [10].
It is worth remarking that weighted graphs with synchronizations, which arise for instance as flow-shop models in production systems, and in communication systems
(see for instance [9]), also lead to difference equations. The equations are in that
case Max-Plus linear equations, which means that the equations are linear over the
semiring $(\mathbb{R} \cup \{-\infty\}, \max, \text{plus})$, called Max-Plus algebra. To this end, one defines
a weighted graph from the given shop, with nodes $i, j, \cdots$, edges $ij, \cdots$, and weights
m_{ij} and τ_{ij}. The real numbers τ_{ij} and the integers m_{ij} respectively correspond to the
delays and initial materials associated to the tasks. A spectral theory for matrices
over the Max-Plus algebra has been developed (see [2], where a complete account
is given). The asymptotic rate of the production system is given by the so-called

Max-Plus eigenvalue associated to the system, which is equal to the maximal mean weight of the graph circuits, say

$$\lambda_{\max} = \max_{\alpha \in \mathscr{C}} \left(\frac{\sum_{ij \in \alpha} \tau_{ij}}{\sum_{ij \in \alpha} m_{ij}} \right),$$

where $\mathscr{C}$ is the set of (elementary) circuits of the graph. We notice the following facts :

- Max-Plus systems are difference equations,
- they have positive coefficients, $\tau_{ij} > 0$, $m_{ij} > 0$,
- the Max-Plus eigenvalue can be interpreted as a spectral radius,
- the Max-Plus eigenvalue of the system continuously depends on the coefficients of the equation.

These remarks are the starting point of the present work. We shall see that the spectral radius of linear (in the usual sense) difference equations with positive coefficients continuously depends on the delays, and consequently positive linear difference equations are stable if and only if they are strongly stable. Finally, we shall see that the spectral radiusof positive difference equations corresponds the exponential growth rate of the solutions.

2   Stability of Difference Equations

The backgrounds results are presented, concerning the stability and robust stability of linear difference equations with multiple, non commensurable delays. Some hints for the proof are given, since they are important for the proof of our main result in the subsequent section. Two academic examples are provided for the sake of clarity.

2.1   Exponential Stability

Various definitions were proposed for the stability of time delay systems, by Bellman [4], Desoer and Vidyasagar [5], among others. Notice that the definitions are not equivalent for general linear systems, and in particular for the delayed equation of neutral type. Most authors hence deal with exponential stability, which is stated in terms of the zero location of the characteristic function $f(s)$ associated to the system, that is

$$f(s) = 1 - \sum_{k=1}^{n} a_k e^{-\beta_k s}, \tag{4}$$

where s is a complex variable. The system (1) is said to be exponentially stable if there exists a real number $\delta < 0$ such that $f(s) = 0$ implies

$$\operatorname{Re} s \leq \delta.$$

In the sequel, we shall be also interested in the multivariable case, and consider the following multivariable system of difference equations

$$x(t) = \sum_{i=1}^{n} A_i x(t - \beta_i),$$ (5)

where $A_i \in \mathbb{R}^{q \times q}$, for $i = 1$ to n, and $x(t)$ is now a vector function over $\mathbb{R}^q$. This system is said to be (exponentially) stable if all the zeros of the characteristic matrix

$$F(s) = 1 - \sum_{k=1}^{n} A_k e^{-\beta_k s}$$

have their real values less than or equal to a strictly negative real number δ.

2.2 Zeros Location of Difference Equations

2.2.1 Location of the Zeros of the Characteristic Equation

Avellar and Hale [1] have studied the zeros of exponential polynomials and obtained the following result. Let $Z(\beta)$ denote the set of real parts of all the (complex) roots of the characteristic function $f(s)$ of (1), and $\overline{Z}(\beta)$ its closure. We have

$$\overline{Z}(\beta) \subset [\rho_n, \rho_0],$$

where the real numbers ρ_k, for $k = 0$ to n, are defined by

$$|a_k| e^{-\rho_k \beta_k} = \sum_{i \neq k} |a_i| e^{-\rho_k \beta_i},$$

with $a_0 = -1$ and $\beta_0 = 0$. One can show that $\overline{Z}(\beta)$ is the union of a finite number of closed intervals, for which ρ_0 is an upper bound. Further it is remarked that the above inclusion can be strict when the delays are rationally dependent. Note that by construction, ρ_0 is the unique root of the real valued function

$$\varphi(x) = 1 - \sum_{k=1}^{n} a_k e^{-\beta_k x},$$

since $\varphi(x)$ is a continuous strictly increasing function taking its range in $\mathbb{R}$.

2.2.2 Bounds for the Location of the Zeros of Monovariable Systems

To numerically determine the location of the zeros of

$$f(s) = 1 - \sum_{i=1}^{n} a_i e^{-s \beta_i}$$

with $\beta_1 < \beta_2 < \ldots < \beta_n$, we rewrite this equation in the form

$$1 = \sum_{i=1}^{n} |a_i| e^{-x\beta_i} z_i = g(x),$$

with $z_i \in \mathbb{C}$, $|z_i| = 1$, for $i = 1, \ldots, n$, and $x \in \mathbb{R}$, and we determine the interval of variation of $|g(x)|$ when vary the parameters z_i :

$$m(x) \leq |g(x)| \leq M(x).$$

Notice that the upper bound is explicitely given as

$$M(x) = \sum_{i=1}^{n} |a_i| e^{-x\beta_i} .$$

This is illustrated in Fig. 1.

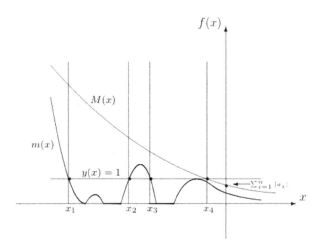

Fig. 1 The region of the zeros corresponds to the condition $1 \in [m(x), M(x)]$.

2.2.3 Multivariable Case

In the multivariable case (5), the solution comes studying the spectral radius —say the maximal value of the module of the eigenvalues— of the matrix

$$\sum_{i=1}^{n} A_i e^{-\beta_i x} \cdot z_i$$

with $|z_i| = 1$. Effective numerical methods for this computation have been developped in [15].

2.3 Stability and Robust Stability of Difference Equations

2.3.1 Preliminary Result

The equation $g(x) = 1$ has only negative solutions if $M(x) < 1$, therefore, the system is stable if $M(0) < 1$, that is

$$\sum_{i=1}^{n} |a_i| < 1.$$

In the multivariable case, the spectral radius should be less than 1 for the system to be stable. Notice that this is a bound, that is not always reached. The condition is not dependent of the delay, and consequently it may be conservative for certain values of the delays. Further, it is observed that the stability is also sensible to the rational dependency of the delays, in general.

2.3.2 Rational Dependence of the Delays

Recall that the delays β_i, $i = 1$ to n, are said to be rationally independent if, for $m_i \in \mathbb{Z}$, $\sum_{i=1}^{k} m_i \beta_i = 0$ implies $m_i = 0$, for $i = 1$ to n.

If $\beta_1, \ldots, \beta_n$ are rationally independent real numbers, and $\mu_1, \ldots, \mu_n$ are real numbers, then there exist integers k_i, $i = 1$ to n, and a real number t such that for all $\varepsilon > 0$, we have

$$|t\beta_i - 2\pi k_i - \mu_i| < \varepsilon, \text{ for } i = 1, \ldots, k.$$

These definitions and result are due to Dedekind and Kronecker. We have by definition of ρ_0:

$$1 = \sum_{i=1}^{n} |a_i| e^{-\rho_0 \beta_i}.$$

Then define the real numbers γ_i, $i = 1, \ldots, n$, such that

$$a_i e^{j\gamma_i} = |a_i|.$$

Applying Kronecker's theorem, we obtain that there exist a real number $q(\varepsilon)$ and integers $k_i(\varepsilon)$, for $i = 1$ to n, such that

$$\forall \varepsilon > 0, \exists q(\varepsilon), k_i(\varepsilon) : |q(\varepsilon)\beta_i - 2\pi k_i(\varepsilon) + \gamma_i| < \varepsilon.$$

Finally, taking

$$s_\varepsilon = \rho_0 + jq(\varepsilon) \, ,$$

we obtain

$$|f(s_\varepsilon)| = |1 - \sum_{i=1}^{n} a_i e^{-\beta_i s_\varepsilon}| < \varepsilon \, .$$

This shows that the set of the real parts of the zeros of system (1) is dense in the intervals where $g(x) = 1$ has a solution. Remark that the rational independence of the delays β_i is required for applying Kronecker's Theorem. This is the base of the following basic result.

2.3.3   Robust Stability of Difference Equations

Theorem 1. [16]
Let the system (1) be given. Then the following claims are equivalent

(i) The system is stable, $\forall \beta_i \geq 0$, $i = 1, \ldots, n$.
(ii) There exists a real number $\varepsilon > 0$, such that the perturbed system

$$x(t) = \sum_{i=1}^{n} a_i x(t - \beta_i - \varepsilon_i)$$

is stable for all $\varepsilon_i < \varepsilon$, $i = 1$ to n.
(iii) $\sum_{i=1}^{n} |a_i| < 1$.

In addition, if the delays β_i, $i = 1, \ldots, n$, are rationally independent, then the claims are equivalent to the stability of system (1).

Claims (i) and (ii) are respectively concerned with the stability and strong stability of system (1). Claim (iii) actually provides a useful test to check stability and strong stability. When the delays are rationally independent, stability and strong stability are equivalent. Furthermore, one can remark that condition (iii) is independent of the delays. This shows that, in that case, the stability is always independent of the delays. Finally notice that the test (iii) is conservative, in general, when the delays are rationally dependent. In this case a system may be stable even when condition (iii) does not hold. However, remark that there are always non rationally dependent delays in any open set of the parameter space with radius ε and centered on the nominal delay. This shows that conditions (ii) and (iii) are always equivalent, hence strong stability is always independent of delays, for difference equations.

2.4 Examples

2.4.1 Example 1 (Unstable)

Consider the following system

$$x(t) = \frac{1}{2}x(t-1) + \frac{1}{2}x(t - \sqrt{2})\,.$$

Its characteristic function is

$$f(s) = 1 - \frac{1}{2}e^{-s} - \frac{1}{2}e^{-s\sqrt{2}} = 0\,.$$

We define, for $x \in \mathbb{R}$:

$$g(x) = 1 - \frac{1}{2}e^{-x} - \frac{1}{2}e^{-x\sqrt{2}}\,.$$

The unique solution of $g(x) = 0$ is $\rho_0 = 0$. We remark that

$$f(\rho_0 + jq) = 1 - \frac{1}{2}e^{-jq} - \frac{1}{2}e^{-jq\sqrt{2}} = 0$$

comes down to the system

$$\begin{cases} q = 2\pi k_1 & ,\ k_1 \in \mathbb{Z} \\ q\sqrt{2} = 2\pi k_2 & ,\ k_2 \in \mathbb{Z} \end{cases}$$

which has no solution, since $\sqrt{2}$ is not rational. For this example, there is no root of the characteristic equation on the imaginary axis, but, following the previous study, there are an infinite number of roots arbitrary close to it, and the system is exponentially unstable.

This is illustrated in Fig. 2. Remark that this system might be considered as stable in the sense of a different definition, since the solution is bounded, which will be shown in section 3.4.

2.4.2 Example 2 (Stable)

We consider now the following system

$$x(t) = \frac{3}{4}x(t-1) - \frac{3}{4}x(t-2)\,,$$

the characteristic equation of which is

$$f(s) = 1 - \frac{3}{4}e^{-s} + \frac{3}{4}e^{-2s} = 0\,.$$

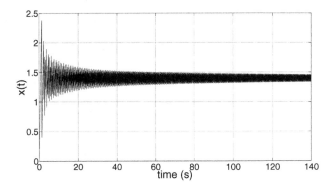

Fig. 2 Solution of the Example 1 with an initial condition that is a sinus of frequency 1 and amplitude 2.

Taking $\lambda = e^s$, we have $\lambda^2 - \frac{3}{4}\lambda + \frac{3}{4} = 0$, thus the zeros are

$$\lambda_{1,2} = \frac{3}{8} \pm j\frac{\sqrt{39}}{8}$$

We deduce from $|\lambda_{1,2}| < 1$ that the system is stable. The following Fig. 3 (obtained under SIMULINK) illustrates the fact. Of course, the system

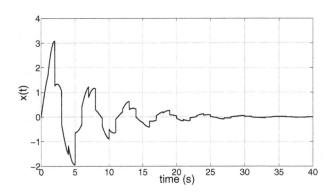

Fig. 3 Stability of the system with $\beta_1 = 1$ and $\beta_2 = 2$.

$$x(t) = \frac{3}{4}x(t-1) - \frac{3}{4}x(t-2)$$

is not robustly stable since $3/4 + 3/4 > 1$. The system stability can be destroyed by an arbitrary small variation of the delays, making them rationally independent. For instance, the system

$$x(t) = \frac{3}{4}x(t-1) - \frac{3}{4}x(t-2-\frac{\pi}{100})$$

is unstable, which is illustrated by the simulation of Fig. 4.

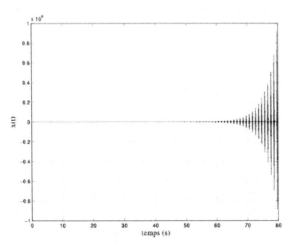

Fig. 4 Unstability of the perturbed system with delays $\beta_1 = 1$ et $\beta_2 = 2 + \frac{\pi}{100}$.

3   Main Results

3.1   Positive Difference Equations

The solution of system (1), and that of system (5), is uniquely defined by its initial conditions $\phi(t)$, such that $x(t) = \phi(t)$, $t \in [-\beta_n, 0[$. The system is said to be positive if the solution $x(t)$ is non negative, for every non negative initial condition.

We say that a vector $x \in \mathbb{R}^q$, and a matrix $P \in \mathbb{R}^{q \times q}$ are non negative, and we write $x \geq 0$, $P \geq 0$, if they have non negative components, say $x_i \geq 0$, $P_{ij} \geq 0$, for $i, j = 1, \cdots, q$. It is readily seen that systems (1) and (5) are positive if and only if they have non negative coefficients, say $a_i \geq 0$, $A_i \geq 0$, for $i = 1, \cdots, n$.

3.2   Stability of Monovariable Positive Difference Equations

For positive systems governed by (1), the coefficients a_i, for $i = 1, \dots, n$, are positive. For this particular case, we are able to analyze more insightful the location of the zeros of the associated characteristic equation. Indeed, let us introduce

$$\alpha(a, \beta) = \sup\{\mathrm{Re}(s) : \sum_{i=1}^{n} a_i \mathrm{e}^{-s\beta_i} = 1\}.$$

For the class of positive difference equations, the above bound $\alpha(a,\beta)$ of real part of the zeros of the characteristic equation coincides with the upper real bound ρ_0. To prove this assertion, we proceed by inequalities. First, by definition of $\alpha(a,\beta)$ and ρ_0, we have $\rho_0 \leq \alpha$. To prove the converse inequality, let $s_r = x + iy$ and consider the characteristic equation

$$1 = \sum_{k=1}^{n} a_k e^{-s_r \beta_k}.$$

Since coefficients a_k are positive, for $k = 1, \ldots, n$, we get

$$1 = \sum_{k=1}^{n} a_k e^{-s_r \beta_k} \leq \sum_{k=1}^{n} a_k e^{-x\beta_k}.$$

The real valued function $M(x) = \sum_{k=1}^{n} a_k e^{-x\beta_k}$ is non increasing. Hence, it follows by definition of ρ_0 that $x \leq \rho_0$. Taking the supremal element x over all $s_r \in \mathbb{C}$ satisfying the characteristic equation gives the result, namely $\alpha \leq \rho_0$. Finally, we obtain the equality $\alpha = \rho_0$.

This result is important, since it relates the supremal bound of the real part of the roots of the characteristic equation with the real number ρ_0. Since this last number is reached, the supremal bound $\alpha(a,\beta) = \rho_0$ is reached, and we know how to compute this bound. Furthermore, this result allows to conclude on the stability of positive systems governed by difference equations.

Theorem 2. A positive system governed by (1) is (exponentially) stable if and only if the condition

$$\sum_{i=1}^{n} a_i < 1$$

is satisfied. In addition, under this condition, system (1) is robustly stable, and stable independent of delays.

Proof. Assume that $\sum_{i=1}^{n} a_i < 1$. Hence, there exists at least one β_k such that

$$e^{-\rho_0 \beta_k} > 1,$$

which in turn implies $\rho_0 < 0$. The converse is immediate, since if $\rho_0 < 0$, we have

$$\sum_{k=1}^{n} a_k < \sum_{k=1}^{n} a_k e^{-\rho_0 \beta_k} = 1.$$

This condition being independant of delays, this ensures the robust stability with respect to the delays.

3.3 Stability of Multivariable Positive Difference Equations

In the multivariable case and in the particular case of positive systems, say $A_i \geq 0$, the matrix

$$\sum_{i=1}^{n} A_i e^{-\beta_i x}$$

is non negative. It has a positive real eigenvalue, say $\lambda(x)$, that is equal to its spectral radius. The matrix is a decreasing function of x, hence its spectral radius is also a decreasing function of x. For this aim, we use a result, stated by Bellman and Cooke [4] (see page 259), that if A and B are real matrices such that $0 \le A \le B$, then the spectral radii verify respectively

$$\rho(A) \le \rho(B).$$

The spectral radius being a decreasing function of x and taking its range into the whole field of real numbers, there is a unique solution $x = \rho_0$ to the equation

$$\lambda(x) = 1.$$

In other words, $x = \rho_0$ is a zero of the characteristic equation of the system. In this case, the system is stable if and only if $\rho_0 < 0$, the delays being rationally dependent or not. Hence, for the multivariable case, we obtain the following result.

Theorem 3. A positive multivariable system governed by (5) is (exponentially) stable if and only if the condition

$$\rho\left(\sum_{i=1}^{n} A_i\right) < 1,$$

is satisfied. In addition, under this condition, system is robustly stable, and stable independent of delays.

3.4 Asymptotic Behavior of Positive Difference Equations

It is interesting to remark that the solution of the difference equation (5) is exponentially growing. More precisely, we can characterize exponential bounds for the behavior, and the exponential growth rate is uniquely defined, given by the number ρ_0 identified above by $\lambda(\rho_0) = 1$.

For every initial condition $\phi(\tau)$, $\tau \in [-\beta_n, 0[$, there is a unique solution satisfying the difference equation (1), for $t \ge 0$, and $x(t) = \phi(t)$, for $\beta_n \le t < 0$. This solution can be iteratively computed, on the intervals $[k\beta_1, (k+1)\beta_1[$, where $k = 0, 1, 2, \cdots$. One can remark that to initial conditions $\phi_1(\tau)$ and $\phi_2(\tau)$ satisfying $\phi_1(\tau) \le \phi_2(\tau)$, for $\tau \in [-\beta_n, 0[$, correspond solutions $x_1(t)$ and $x_2(t)$ that are ordered by $x_1(t) \le x_2(t)$, for $t \ge 0$. Let us define the numbers m_1 and M_1 by

$$m_1 = \inf\{e^{-\rho_0 \tau} \phi(\tau), -\beta_n \le \tau < 0\},$$

and

$$M_1 = \sup\{e^{-\rho_0 \tau} \phi(\tau), -\beta_n \le \tau < 0\}.$$

Both numbers m_1 and M_1 are positive if $\phi(\tau)$ takes positive values on $[-\beta_n, 0[$. From the definition of ρ_0 and previous remarks, we see that the solution $x(t)$ corresponding to the initial condition $\phi(\tau)$ satisfies

$$m_1 e^{\rho_0 t} \leq x(t) \leq M_1 e^{\rho_0 t}, \tag{6}$$

for $t \geq 0$. Consequently, $x(t) = e^{\rho_0 t}, t \geq 0$, is the solution of the difference equation (1) with initial condition $\phi(\tau) = e^{\rho_0 \tau}$, $\tau \in [-\beta_n, 0[$.

This growth rate analysis can be applied on Example 1, for which $\rho_0 = 0$. From (6), we see that for any bounded initial condition, the solution $x(t)$, for $t \geq 0$, is bounded (see Fig. 2, for $\phi(\tau) = 2|\sin(\tau)|$, $\tau \in [-\sqrt{2}, 0[$). More precisely, we have

$$0 \leq x(t) \leq 2\sin(\sqrt{2}), \ t \geq 0.$$

In Example 1, since the initial condition is monotone, we see that the upper and lower bounds m_1 and M_1 will not be reached for any $t \geq 0$.

To extend this remark to the case of a multivariable positive difference equation, we recall another well-known property of positive matrices. There is a unique eigenvector, say $w \in \mathbb{R}^n$ satisfying

$$\sum_{i=1}^{n} A_i e^{-\beta_i \rho_0} w = w,$$

and $\max_i\{w_i\} = 1$. This vector, called Perron eigenvector, is positive, in the sense that all its components are positive. One can see that the solution of (5) corresponding to the initial condition $\phi(\tau) = e^{\rho_0 \tau} w$, $\tau \in [-\beta_n, 0[$, is uniquely defined as $x(t) = e^{\rho_0 t} w$, for $t \geq 0$. Defining now m and M as

$$m = \min_i \inf\{w_i^{-1} e^{-\rho_0 \tau} \phi_i(\tau), -\beta_n \leq \tau < 0\},$$

$$M = \max_i \sup\{w_i^{-1} e^{-\rho_0 \tau} \phi_i(\tau), -\beta_n \leq \tau < 0\},$$

where w_i denotes the ith component of the vector w, and $\phi_i(\tau)$ the ith component of the function $\phi(\tau)$, we obtain the following bounds to the solution $x(t)$,

$$m e^{\rho_0 t} w \leq x(t) \leq M e^{\rho_0 t} w. \tag{7}$$

If the initial condition $\phi(\tau)$ is positive, all the components of $x(t)$ are growing with exponential rate ρ_0. The lower and upper bounds given in (7) are not reached, in general, when t is growing up. However, the bounds are not restrictive, in the sense that there are initial conditions $\phi(\tau)$ for which the bounds (7) are exact, for t greater than an arbitrary large number. For instance, let $\mathbb{Z}[\beta]$ denotes the set of integer combinations of the delays β_i, $i = 1$ to n. From Dedekind-Kronecker's theorem (see Section 2.3.2), it is clear than $\mathbb{Z}[\beta]$ is dense in $\mathbb{R}$. One also remarks that $t \in \mathbb{Z}[\beta]$

if and only if $t - \beta_i \in \mathbb{Z}[\beta]$, for $i = 1$ to n. As a consequence, it appears that the trajectory $x(t)$ defined by

$$x(t) = \begin{cases} m e^{\rho_0 \tau} w, & \text{if } t \in \mathbb{Z}[\beta] \\ M e^{\rho_0 \tau} w, & \text{if } t \notin \mathbb{Z}[\beta] \end{cases} ,$$

for $t \geq 0$, is the unique solution of equation (5) with the initial condition defined by

$$\phi(\tau) = \begin{cases} m e^{\rho_0 \tau} w, & \text{if } t \in \mathbb{Z}[\beta] \\ M e^{\rho_0 \tau} w, & \text{if } t \notin \mathbb{Z}[\beta] \end{cases} ,$$

for $\tau \in [-\beta_n, 0[$. This is an example where the proposed bounds are reached. In this sense, inequalities(7) define the best lower and upper bounds for the solutions of positive difference equations.

4 Conclusions and Final Remarks

We have remarked that in the case of positive difference equations, stability, robust stability and stability independent of the delays are equivalent. It comes from the fact that the supremum of the real parts of the zeros of the characteristic equation is reached, and is a continuous function of the parameters, in particular of the delays.

The starting point of this work was the observation that a similar property holds for another class of difference equations, the so-called Max-Plus linear equations. There are various relationships between Max-Plus equations and positive linear difference equations, in particular they both arise from networks and graph models in production systems or communication systems, and they can be both considered as systems over a semiring (respectively the Max-Plus algebra, and the semiring of positive real numbers). They permit to address different features of the systems. Max-Plus equations are usually associated with synchronization and saturation, and linear equations permit to represent concurrency and routing policies. Another remarkable fact is that both models are designed using the same basic information, that is the set of delays along the graph, and its initial situation. The Max-Plus equations gave rise to a well-developed theory (see [2]). Because of these similarities, extending the theory to take into account difference equations, concurrency and the synthesis of routing policies looks promising. It is a challenging area for future work.

Another perspective of this work concerns the resolution of systems governed by linear partial differential equations. Basically, numerical schemes are proposed to approximate, via a time-space discretization, the exact solution. The existence of this solution over Hilbert spaces is characterized by the Lax-Milgram theorem. Numerically, this theorem can be interpreted through the method of Von Neumann (Strikwerda, 1989), which characterizes the stability property of some discretization scheme using frequency techniques. These schemes, as for instance finite differences method, lead to difference equations, in general. Then, its stability analysis can be done via the tools developed in this paper. Furthermore, for unstable schemes, using stabilization techniques by dynamical state feedback, we are able to ensure robust stabilization, or in other words to develop numerical schemes which converge to the exact solution. The link between the approach proposed by Von Neumann and the one developed in this paper is an open and challenging topic, for both control and numerical analysis communities.

References

1. de Avellar, C.E., Hale, J.K.: On the zeros of exponential polynomials. J. Mathematical Analysis and Applications 73, 434–452 (1980)
2. Baccelli, F., Cohen, G., Olsder, G.J., Quadrat, J.-P.: Synchronization and linearity: An algebra for discrete event systems. John Wiley & Sons, Chichester (1992)
3. Barman, J.F., Callier, F.M., Desoer, C.: L^2-stability and L^2-instability of linear time invariant distributed feedback systems perturbed by a small delay in the loop. IEEE Transactions on Automatic Control 18, 479–484 (1973)
4. Bellman, R., Cooke, K.L.: Differential-difference equations. Academic Press, New York (1963)
5. Desoer, C.A., Vidyasagar, M.: Feedback systems: Input-output properties. Academic Press, New York (1975)
6. Hale, J.K.: Stability, control and small delays. In: Proceedings IFAC Workshop on Time Delay Systems, pp. 37–42. Pergamon (2001)
7. Hale, J.K., Verduyn Lunel, S.M.: Effects of small delays on stability and control. In: Bart, H., Gohberg, I., Ran, A.C.M. (eds.) Operator Theory and Analysis, The M.A. Kaashoek Anniversary Volume. Operator Theory: Advances and Applications 122, pp. 275–301. Birkhauser (2001)
8. Kolmanovski, V., Myshkis, A.: Applied theory of functional differential equations. Springer, New York (1993)
9. Le Boudec, J.-Y., Thiran, P.: Chapter 1: Network calculus. In: Thiran, P., Le Boudec, J.-Y. (eds.) Network Calculus. LNCS, vol. 2050, pp. 3–81. Springer, Heidelberg (2001)
10. Libeaut, L.: Sur l'utilisation des dioïdes pour la commande des systèmes à événements discrets, Ph.D. thesis, Ecole Centrale de Nantes (1996)
11. Logemann, H., Townley, S.: The effect of small delays in the feedback loop on the stability of neutral equations. Systems & Control Letters 27, 267–274 (1996)
12. Loiseau, J.J., Cardelli, M., Dusser, X.: Neutral-type time-delay systems that are not formally stable are not BIBO stabilizable. IMA Journal of Mathematical Control and Information 19, 217–227 (2002)
13. Meinsma, G., Fu, M., Iwasaki, T.: Robustness of the stability of feedback systems with respect to small time delays. Systems & Control Letters 36, 131–134 (1999)
14. Michiels, W., Engelborghs, K., Roose, D., Dochain, D.: Sensitivity to infinitesimal delays in neutral equations. SIAM Journal on Control and Optimization 40(4), 1134–1158 (2002)
15. Michiels, W., Vyhlídal, T.: An eigenvalue based approach for the stabilization of linear time-delay systems of neutral type. Automatica 41, 991–998 (2005)
16. Hale, J.K., Verduyn Lunel, S.M.: Introduction to Functional Differential Equations. In: Applied Mathematical Sciences, vol. 99, Springer, Heidelberg (1993)
17. Strikwerda, J.C.: Finite difference schemes and partial differential equations. Wadworth & Brooks, Cole (1989)

Positivity of Complete Quadratic Lyapunov-Krasovskii Functionals in Time-Delay Systems

Keqin Gu, Yashun Zhang, and Matthew Peet

Abstract. This chapter discusses positivity of quadratic functionals that arise in the stability analysis of time-delay systems. When both the single and double integral terms are positive, a necessary and sufficient condition for positivity is obtained using operator theory. This is applied to the Lyapunov-Krasovskii functional and its derivative. The coupled differential-difference equations are studied using the Sum-of-Squares (SOS) method.

1 Introduction

Stability analysis of time-delay systems represents significant current interest of research. Repin (1965) showed that a complete quadratic Lyapunov-Krasovskii functional is needed for accurate stability analysis for linear time-delay systems. Such a functional takes the form of

$$V(\psi,\phi) = \psi^T P \psi + 2\psi^T \sum_{i=1}^{K} \int_{-r_i}^{0} Q_i(s)\phi_i(s)ds + \sum_{i=1}^{K} \int_{-r_i}^{0} \phi_i^T(s)S_i(s)\phi_i(s)ds$$

$$+ \sum_{i=1}^{K}\sum_{j=1}^{K} \int_{-r_i}^{0}\int_{-r_j}^{0} \phi_i^T(s)R_{ij}(s,\eta)\phi_j(\eta)dsd\eta. \tag{1}$$

Keqin Gu
Department of Mechanical and Industrial Engineering,
Southern Illinois University Edwardsville, Edwardsville, Illinois 62026, USA
e-mail: kgu@siue.edu

Yashun Zhang
School of Automation, Nanjing University of Science and Technology, Nanjing, China
e-mail: yashunzhang@gmail.com

Matthew Peet
Department of Mechanical, Materials, and Aerospace Engineering,
Illinois Institute of Technology, Chicago, Illinois 60616, USA
e-mail: mpeet@iit.edu

R. Sipahi et al. (Eds.): Time Delay Sys.: Methods, Appli. and New Trends, LNCIS 423, pp. 35–47.
springerlink.com

Gu (1997) numerically implemented a stability analysis based on the above functional using piecewise linear discretization. Peet and Papachristodoulou (2006) implemented an alternative discretization using polynomials. These methods achieve asymptotically accurate stability results. Obviously, a more accurate result is obtained at the expense of increased computation. Methods to accelerate the convergence is obviously of great interest.

The quadratic expression given in (1) contains two quadratic terms of ϕ_i: one involves single integration (the third term), and the other involves double integration (the last term). If both terms are positive, then they both have the ability to "offset" the second term to make the complete expression positive. Many existing methods only use one such term to enforce the positivity of (1). In Li and Gu (2010), the Jensen Inequality is used to take advantage of both terms in the discretized Lyapunov-Krasovskii functional formulation. However, this method is not easily applied without undue conservatism for the sum-of-square (SOS) formulation used here. This chapter develops necessary and sufficient conditions for the positivity of the complete expression assuming both the single and double integral terms are positive.

The model this chapter is the coupled differential-difference equations proposed in Gu (2010). Such a model tends to produce a much smaller state space for practical systems. It is numerically implemented using the SOS formulation presented in Peet and Papachristodoulou (2009) and Zhang, et al (2011).

The remaining part of this chapter is organized as follows: Section 2 presents a theorem on quadratic inequalities based on operator theory. Section 3 discusses the corresponding formulation in integral quadratic inequalities. Section 4 introduces the stability problem of time-delay systems formulated in coupled differential-difference equations. Section 5 presents the SOS implementation of stability analysis. Section 6 presents a numerical example to illustrate the effectiveness of the method.

The following notation is used in this chapter: $\mathbb{R}$ denotes the set of reals. $\mathbb{R}^n$ and $\mathbb{R}^{p \times q}$ denote the sets of real n-vectors and $p \times q$ matrices. $\mathbb{S}^n$ denotes the set of real $n \times n$ symmetric matrices. For $X \in \mathbb{S}^n$, $X \geq 0$ ($X > 0$) means that X is positive semidefinite (definite). I denotes the identity matrix. For a given constant $r > 0$ and positive integer n, $\mathbb{PC}(r,n)$ denotes the set of bounded functions $f : [-r,0) \to \mathbb{R}^n$ that are right continuous everywhere with possibly a finite number of discontinuous points. For a given function y defined in an interval $\mathbb{I} \supset [t-r,t)$, $y_{r,t}$ is a function defined on $[-r,0)$ by the relation $y_{r,t}(s) = y(t+s)$. The notation $\phi = (\phi_1, \phi_2, \ldots, \phi_K) \in \mathbb{PC}$ is used to denote $\phi_i \in \mathbb{PC}(r_i, m_i), i = 1, 2, \ldots, K$.

2 Positive Operators

This section presents necessary and sufficient conditions for positivity of a quadratic operator expression that is crucial to evaluating the positivity of (1). The readers are referred to Kato (1966) and Kolmogorov and Fomin (1975) for background in operator theory.

Let $\mathbb{X}$ and $\mathbb{Y}$ be Banach spaces over $\mathbb{R}$. Let $\langle \cdot, \cdot \rangle$ denote the inner products on both $\mathbb{X}$ and $\mathbb{Y}$. However, they are not Hilbert spaces as the norms are not defined from the inner products. The set of all bounded linear operators from $\mathbb{X}$ to $\mathbb{Y}$ is denoted as $\mathbb{L}(\mathbb{X}, \mathbb{Y})$. Let $\mathscr{A} \in \mathbb{L}(\mathbb{X}, \mathbb{X})$ and $\mathscr{B} \in \mathbb{L}(\mathbb{X}, \mathbb{Y})$. Then

1. The *adjoint operator* $\mathscr{B}^*$ of $\mathscr{B}$ is defined as $\mathscr{B}^* \in \mathbb{L}(\mathbb{Y}, \mathbb{X})$ that satisfies $\langle \mathscr{B}^* y, x \rangle = \langle y, \mathscr{B} x \rangle$ for all $x \in \mathbb{X}$ and $y \in \mathbb{Y}$. When $\mathbb{Y} = \mathbb{X}$, then $\mathscr{B}$ is said to be self-adjoint if $\mathscr{B}^* = \mathscr{B}$.
2. If $\mathscr{A} \in \mathbb{L}(\mathbb{X}, \mathbb{X})$ is self-adjoint, then $\mathscr{A}$ is *positive* if $\langle x, \mathscr{A} x \rangle \geq 0$ for all $x \in \mathbb{X}$. It is *coercive* if there exists an $\varepsilon > 0$ such that $\langle x, \mathscr{A} x \rangle \geq \varepsilon \langle x, x \rangle$ for all $x \in \mathbb{X}$. If $\mathscr{A}$ is coercive, then $\mathscr{A}^{-1}$ exists, is bounded, self-adjoint, and coercive.

Theorem 1. *Let $\mathbb{X}$ and $\mathbb{Y}$ be Banach spaces over $\mathbb{R}$, with inner product $\langle \cdot, \cdot \rangle$ defined on both $\mathbb{X}$ and $\mathbb{Y}$. Let $\mathscr{P} \in \mathbb{L}(\mathbb{Y}, \mathbb{Y})$, $\mathscr{Q} \in \mathbb{L}(\mathbb{X}, \mathbb{Y})$, $\mathscr{S} \in \mathbb{L}(\mathbb{X}, \mathbb{X})$, and $\mathscr{R} \in \mathbb{L}(\mathbb{X}, \mathbb{X})$. Suppose $\mathscr{P}, \mathscr{R}$ and $\mathscr{S}$ are self-adjoint, $\mathscr{R}$ is positive, and $\mathscr{S}$ is coercive. Then there exist $\mathscr{Q}^1 \in \mathbb{L}(\mathbb{X}, \mathbb{Y})$ and $\mathscr{P}^1 = \mathscr{P}^{1*} \in \mathbb{L}(\mathbb{Y}, \mathbb{Y})$, such that*

$$\langle u, (\mathscr{P} - \mathscr{P}^1) u \rangle + 2 \langle u, (\mathscr{Q} - \mathscr{Q}^1) \phi \rangle + \langle \phi, \mathscr{S} \phi \rangle \geq 0, \tag{2}$$

$$\langle u', \mathscr{P}^1 u' \rangle + 2 \langle u', \mathscr{Q}^1 \phi' \rangle + \langle \phi', \mathscr{R} \phi' \rangle \geq 0 \tag{3}$$

are satisfied for all $u, u' \in \mathbb{Y}$ and $\phi, \phi' \in \mathbb{X}$ if and only if

$$\langle u, \mathscr{P} u \rangle + 2 \langle u, \mathscr{Q} \phi \rangle + \langle \phi, (\mathscr{S} + \mathscr{R}) \phi \rangle \geq 0 \tag{4}$$

is satisfied for arbitrary $u \in \mathbb{Y}$ and $\phi \in \mathbb{X}$. The conclusion is still valid if we restrict $\mathscr{Q}^1 \in \mathbb{Q} \subset \mathbb{L}(\mathbb{X}, \mathbb{Y})$, as long as $\mathbb{Q}$ contains the element $\mathscr{Q}(\mathscr{S} + \mathscr{R})^{-1} \mathscr{R}$.

Proof. This is a consequence of Lemmas 1 and 2 below. ∎

Lemma 1. *Let $\mathbb{X}$ and $\mathbb{Y}$ be Banach spaces with inner products, and $\mathscr{P} \in \mathbb{L}(\mathbb{Y}, \mathbb{Y})$, $\mathscr{Q}^1, \mathscr{Q}^2 \in \mathbb{L}(\mathbb{X}, \mathbb{Y})$, $\mathscr{S}, \mathscr{R} \in \mathbb{L}(\mathbb{X}, \mathbb{X})$. Suppose $\mathscr{P}$ and $\mathscr{R}$ are self-adjoint, and $\mathscr{S}$ is self-adjoint and coercive. Then there exists a $\mathscr{P}^1 = \mathscr{P}^{1*} \in \mathbb{L}(\mathbb{Y}, \mathbb{Y})$ such that*

$$\langle u_1, (\mathscr{P} - \mathscr{P}^1) u_1 \rangle + 2 \langle u_1, \mathscr{Q}^2 \phi_S \rangle + \langle \phi_S, \mathscr{S} \phi_S \rangle \geq 0, \tag{5}$$

$$\langle u_2, \mathscr{P}^1 u_2 \rangle + 2 \langle u_2, \mathscr{Q}^1 \phi_R \rangle + \langle \phi_R, \mathscr{R} \phi_R \rangle \geq 0 \tag{6}$$

are satisfied for all $u_1, u_2 \in \mathbb{Y}$ and $\phi_S, \phi_R \in \mathbb{X}$, if and only if

$$\langle u, \mathscr{P} u \rangle + 2 \langle u, \mathscr{Q}^2 \phi_S' \rangle + \langle \phi_S', \mathscr{S} \phi_S' \rangle + 2 \langle u, \mathscr{Q}^1 \phi_R' \rangle + \langle \phi_R', \mathscr{R} \phi_R' \rangle \geq 0 \tag{7}$$

is satisfied for all $u \in \mathbb{Y}$ and $\phi_S', \phi_R' \in \mathbb{X}$.

Proof. Necessity is obvious because the left hand side of (7) may be obtained by adding up the left hand side of (5) and (6) and constraining $u_1 = u_2$.

For sufficiency, suppose that (7) is satisfied. Then let

$$\mathscr{P}^1 = \mathscr{P} - \mathscr{Q}^2 \mathscr{S}^{-1} \mathscr{Q}^{2*}. \tag{8}$$

$\mathscr{P}^1$ is well-defined, self-adjoint and bounded. We will show that both (5) and (6) are satisfied with this $\mathscr{P}^1$. The inequality (5) is true because

$$\text{Left hand side of (5)} = \left\langle \left(\phi_S + \mathscr{S}^{-1}\mathscr{Q}^{2*}u_1\right), \mathscr{S}\left(\phi_S + \mathscr{S}^{-1}\mathscr{Q}^{2*}u_1\right)\right\rangle.$$

The left hand side of (6) can be written as

$$\begin{aligned}
&\langle u_2, \mathscr{P}u_2\rangle + 2\left\langle u_2, \mathscr{Q}^2\left(-\mathscr{S}^{-1}\mathscr{Q}^{2*}u_2\right)\right\rangle \\
&+\left\langle\left(-\mathscr{S}^{-1}\mathscr{Q}^{2*}u_2\right), \mathscr{S}\left(-\mathscr{S}^{-1}\mathscr{Q}^{2*}u_2\right)\right\rangle + 2\left\langle u_2, \mathscr{Q}^1\phi_R\right\rangle + \left\langle\phi_R, \mathscr{R}\phi_R\right\rangle, \quad (9)
\end{aligned}$$

which is also nonnegative according to (7).   ∎

Lemma 2. *Let* $\mathscr{P} \in \mathbb{L}(\mathbb{Y},\mathbb{Y})$, $\mathscr{Q} \in \mathbb{L}(\mathbb{X},\mathbb{Y})$, $\mathscr{S},\mathscr{R} \in \mathbb{L}(\mathbb{X},\mathbb{X})$. *Suppose* $\mathscr{S}$ *is coercive, and* $\mathscr{R}$ *is positive. Then there exists a* $\mathscr{Q}^1 \in \mathbb{L}(\mathbb{X},\mathbb{Y})$ *such that*

$$\langle u, \mathscr{P}u\rangle + 2\left\langle u, \left(\mathscr{Q} - \mathscr{Q}^1\right)\phi_S\right\rangle + \langle\phi_S, \mathscr{S}\phi_S\rangle + 2\left\langle u, \mathscr{Q}^1\phi_R\right\rangle + \langle\phi_R, \mathscr{R}\phi_R\rangle \geq 0 \quad (10)$$

is satisfied for all $u \in \mathbb{Y}$ *and* $\phi_S, \phi_R \in \mathbb{X}$, *if and only if*

$$\left\langle u', \mathscr{P}u'\right\rangle + 2\left\langle u', \mathscr{Q}\phi\right\rangle + \left\langle\phi, (\mathscr{S}+\mathscr{R})\phi\right\rangle \geq 0 \quad (11)$$

is satisfied for all $u' \in \mathbb{Y}$ *and* $\phi \in \mathbb{X}$. *The conclusion is still valid if we impose the restriction* $\mathscr{Q}^1 \in \mathbb{Q} \subset \mathbb{L}(\mathbb{X},\mathbb{Y})$, *as long as* $\mathbb{Q}$ *contains the element* $\mathscr{Q}(\mathscr{S}+\mathscr{R})^{-1}\mathscr{R}$.

Proof. Necessity is obvious because (11) may be obtained from (10) with the constraint $\phi_S = \phi_R$.

To show sufficiency, let (11) be satisfied for all $u' \in \mathbb{Y}$ and $\phi \in \mathbb{X}$. It is sufficient to show that (10) is satisfied for

$$\mathscr{Q}^1 = \mathscr{Q}(\mathscr{S}+\mathscr{R})^{-1}\mathscr{R}. \quad (12)$$

This $\mathscr{Q}^1$ is well defined and bounded. Define

$$\mathscr{T} = \mathscr{R}(\mathscr{S}+\mathscr{R})^{-1}\mathscr{S}. \quad (13)$$

It can be easily shown that

$$\mathscr{T} = \mathscr{S} - \mathscr{S}(\mathscr{S}+\mathscr{R})^{-1}\mathscr{S}, \quad (14)$$

$$\mathscr{T} = \mathscr{R} - \mathscr{R}(\mathscr{S}+\mathscr{R})^{-1}\mathscr{R}, \quad (15)$$

$$\begin{aligned}
\mathscr{T} &= \mathscr{S}(\mathscr{S}+\mathscr{R})^{-1}\left[(\mathscr{S}+\mathscr{R})\mathscr{S}^{-1}\mathscr{R}\right](\mathscr{S}+\mathscr{R})^{-1}\mathscr{S} \\
&= \mathscr{S}(\mathscr{S}+\mathscr{R})^{-1}(\mathscr{R}+\mathscr{R}\mathscr{S}^{-1}\mathscr{R})(\mathscr{S}+\mathscr{R})^{-1}\mathscr{S} \geq 0. \quad (16)
\end{aligned}$$

Direct calculation using (12), (13), (14) and (15) yields

Left hand side of (10) $= \langle u, \mathscr{P}u \rangle + 2 \left\langle u, \mathscr{Q}(\mathscr{S}+\mathscr{R})^{-1}\mathscr{S}\phi_S \right\rangle + \langle \phi_S, \mathscr{S}\phi_S \rangle$

$$+ 2 \left\langle u, \mathscr{Q}(\mathscr{S}+\mathscr{R})^{-1}\mathscr{R}\phi_R \right\rangle + \langle \phi_R, \mathscr{R}\phi_R \rangle$$

$$= I_1 + I_2, \tag{17}$$

where,

$$I_1 = \langle u, \mathscr{P}u \rangle + 2\langle u, \mathscr{Q}\phi \rangle + \langle \phi, (\mathscr{S}+\mathscr{R})\phi \rangle, \tag{18}$$

$$I_2 = \langle \phi_S, \mathscr{T}\phi_S \rangle - 2\langle \phi_R, \mathscr{T}\phi_S \rangle + \langle \phi_R, \mathscr{T}\phi_R \rangle = \langle (\phi_S - \phi_R), \mathscr{T}(\phi_S - \phi_R) \rangle, \tag{19}$$

and $\phi = (\mathscr{S}+\mathscr{R})^{-1}(\mathscr{S}\phi_S + \mathscr{R}\phi_R)$. Both I_1 and I_2 are nonnegative in view of (11) and (16). ∎

3 Positive Quadratic Integral Expressions

Theorem 1 may be applied to the quadratic integral expression (1) to obtain conditions for its positivity.

Proposition 1. *Given a matrix $P \in \mathbb{S}^n$, and absolutely continuous matrix functions $Q_i : [-r_i, 0] \to \mathbb{R}^{n \times m_i}, S_i : [-r_i, 0) \to \mathbb{S}^{m_i}$, and $R_{ij} : [-r_i, 0) \times [-r_j, 0) \to \mathbb{R}^{m_i \times m_j}$, $R_{ij}(s, \eta) = R_{ji}^T(\eta, s)$, let S_i and R_{ij} satisfy*

$$S_i(s) > \varepsilon I \text{ for some } \varepsilon > 0 \text{ and all } s \in [-r_i, 0), \tag{20}$$

and

$$\sum_{i=1}^{K} \sum_{j=1}^{K} \int_{-r_i}^{0} \int_{-r_j}^{0} \phi_i^T(s) R_{ij}(s, \eta) \phi_j(\eta) ds d\eta \geq 0, \tag{21}$$

for all $\phi \in \mathbb{PC}$. Then, there exists an $\varepsilon > 0$ such that the functional $V(\psi, \phi)$ defined in (1) satisfies

$$V(\psi, \phi) \geq \varepsilon \psi^T \psi \tag{22}$$

for all $\psi \in \mathbb{R}^n, \phi \in \mathbb{PC}$ if and only if there exist an $\varepsilon > 0$, a matrix $P^1 \in \mathbb{S}^n$, and absolutely continuous matrix functions $Q_i^1 : [-r_i, 0] \to \mathbb{R}^{n \times m_i}, i = 1, 2, \ldots, K$, such that

$$\psi^T(P - P^1)\psi + 2\psi^T \sum_{i=1}^{K} \int_{-r_i}^{0} \left[Q_i(s) - Q_i^1(s) \right] \phi_i(s) ds$$

$$+ \sum_{i=1}^{K} \int_{-r_i}^{0} \phi_i^T(s) S_i(s) \phi_i(s) ds \geq \varepsilon \psi^T \psi \tag{23}$$

and $$\psi^T P^1 \psi + 2\psi^T \sum_{i=1}^{K} \int_{-r_i}^{0} Q_i^1(s) \phi_i(s) ds$$

$$+\sum_{i=1}^{K}\sum_{j=1}^{K}\int_{-r_i}^{0}\int_{-r_j}^{0}\phi_i^T(s)R_{ij}(s,\eta)\phi_j(\eta)dsd\eta \geq 0 \tag{24}$$

are satisfied for all $\psi \in \mathbb{R}^n$ and $\phi \in \mathbb{PC}$.

Proof. Let $\mathbb{X} = \mathbb{PC}$, $\mathbb{Y} = \mathbb{R}^n$. For $\psi, \chi \in \mathbb{R}^n$ and $\phi, \omega \in \mathbb{PC}$. Define

$$\langle \psi, \chi \rangle = \psi^T \chi, \quad \langle \phi, \omega \rangle = \sum_{i=1}^{K}\int_{-r_i}^{0}\phi_i^T(s)\omega_i(s)ds,$$

$$|\psi| = \sqrt{\langle \psi, \psi \rangle}, \|\phi\| = \max_{1\leq i\leq K}\sup_{s\in[-r_i,0)}|\phi_i(s)|.$$

Although the norm $|\cdot|$ in $\mathbb{R}^n$ is defined by the inner product, the norm $\|\cdot\|$ in $\mathbb{PC}$ is not from the inner product. Define the bounded linear operators

$$\mathscr{P}\psi = (P - \varepsilon I)\psi, \mathscr{Q}\phi = \sum_{j=1}^{K}\int_{-r_j}^{0}Q_j(s)\phi_j(s)ds, \tag{25}$$

$$(\mathscr{S}\phi)(s) = \begin{bmatrix} S_1(s)\phi_1(s) \\ S_2(s)\phi_2(s) \\ \vdots \\ S_K(s)\phi_K(s) \end{bmatrix}, \mathscr{R}\phi = \begin{bmatrix} \mathscr{R}_1\phi \\ \mathscr{R}_2\phi \\ \vdots \\ \mathscr{R}_K\phi \end{bmatrix}, \tag{26}$$

$$(\mathscr{R}_i\phi)(s) = \sum_{j=1}^{K}\int_{-r_j}^{0}R_{ij}(s,\eta)\phi_j(\eta)d\eta. \tag{27}$$

Then, the condition (22) may be expressed as (4). This is equivalent to (2) and (3) according to Theorem 1. The operator $\mathscr{P}^1 \in \mathbb{L}(\mathbb{R}^n, \mathbb{R}^n)$ has the expression

$$\mathscr{P}^1\psi = P^1\psi. \tag{28}$$

By Riesz Representation Theorem (Kolmogorov and Fomin, 1975),

$$\mathscr{Q}^1\phi = \sum_{j=1}^{K}\int_{-r_j}^{0}d[T_j(s)]\phi_j(s), \tag{29}$$

where T_j is left continuous and of bounded variation in $[-r_j, 0)$. The proof will be complete if we can show that

$$\mathscr{Q}^1 = \mathscr{Q}(\mathscr{S} + \mathscr{R})^{-1}\mathscr{R} \tag{30}$$

may be expressed in the form of

$$\mathscr{Q}^1\phi = \sum_{j=1}^{K}\int_{-r_j}^{0} Q_j^1(s)\phi_j(s)ds, \tag{31}$$

where $Q_j^1(s)$ is absolutely continuous.

Adding $\mathscr{Q}(\mathscr{S}+\mathscr{R})^{-1}\mathscr{S}$ then right-multiplying $\mathscr{S}^{-1}\mathscr{R}$ on both sides of (30) yields

$$\mathscr{Q}^1 = \mathscr{Q}\mathscr{S}^{-1}\mathscr{R} - \mathscr{Q}^1\mathscr{S}^{-1}\mathscr{R}. \tag{32}$$

Using (26), (27) and (29), after exchanging the order of integrations and summations, we may write

$$\mathscr{Q}^1\mathscr{S}^{-1}\mathscr{R}\phi = \sum_{i=1}^{K}\int_{-r_i}^{0} W_i(\eta)\phi_i(\eta)d\eta, \tag{33}$$

where

$$W_i(\eta) = \sum_{j=1}^{K}\int_{-r_j}^{0} d[T(s)]S_j^{-1}(s)R_{ji}(s,\eta)$$

are absolutely continuous. A similar procedure applied to $\mathscr{Q}\mathscr{S}^{-1}\mathscr{R}$ allows us to conclude that $\mathscr{Q}^1$ may indeed be expressed in the form of (31) in view of (32). ■

4 Stability of Coupled Differential-Difference Equations

Coupled differential-difference equations were initially used to model lossless propagation systems (Răsvan, 2006). The following form was proposed in Gu (2010) to obtain a small state space for practical time-delay systems

$$\dot{x}(t) = Ax(t) + \sum_{j=1}^{K} B_j y_j(t-r_j), \tag{34}$$

$$y_i(t) = C_i x(t) + \sum_{j=1}^{K} D_{ij} y_j(t-r_j), i = 1,2,\ldots K, \tag{35}$$

where $x(t) \in \mathbb{R}^n$, $y_i(t) \in \mathbb{R}^{m_i}$, and K is the number of delay channels. The state of the system at time t is $(x(t),y_t)$, where $y_t := (y_{1_{r_1,t}}, y_{2_{r_2,t}}, \cdots, y_{K_{r_K,t}}) \in \mathbb{PC}$. It is also convenient to write $m = \sum_{i=1}^{K} m_i$.

A necessary and sufficient condition for the stability of the system in the form of quadratic Lyapunov-Krasovskii functional is given in Gu (2010), and is presented below with appropriate adaptation of notation.

Theorem 2. *Suppose there exist $U_i \in \mathbb{S}^{m_i}$, $U_i > 0$, $i = 1,2,\ldots K$ such that*

$$D^T UD - U < 0, \tag{36}$$

where

$$U = \text{diag}(U_1, U_2, \ldots, U_K), \tag{37}$$

$$D = \begin{bmatrix} D_{11} & D_{12} & \cdots & D_{1K} \\ D_{21} & D_{22} & \cdots & D_{2K} \\ \vdots & \vdots & \ddots & \vdots \\ D_{K1} & D_{K2} & \cdots & D_{KK} \end{bmatrix}. \tag{38}$$

Then the system described by (34) and (35) is exponentially stable if and only if there exist a $P \in \mathbb{S}^n$, and absolutely continuous matrix functions $Q_i(s) \in \mathbb{R}^{n \times m_i}$, $R_{ij}(s, \eta) = R_{ji}^T(\eta, s) \in \mathbb{R}^{m_i \times m_j}$, and $S_i(s) \in \mathbb{S}^{m_i}$, such that the functional $V(\psi, \phi)$ given in (1) satisfies

$$V(\psi, \phi) \geq \varepsilon \psi^T \psi, \tag{39}$$

$$\dot{V}(\psi, \phi) \leq -\varepsilon \psi^T \psi, \tag{40}$$

for all $\psi \in \mathbb{R}^n$, $\phi = (\phi_1, \phi_2, \ldots, \phi_K) \in \mathbb{PC}$ and some $\varepsilon > 0$.

From Theorem 2, the stability analysis of the system (34)-(35) amounts to checking the satisfaction of (36), (39) and (40). We will check the satisfaction of (39) by using Proposition 1. This is an improvement over some existing results. For example, in Zhang, et al (2011), (39) is enforced by requiring

$$P^1 = 0, Q_i^1(s) = 0.$$

The derivative of V along the system trajectory has the explicit expression

$$\dot{V}(\psi), \phi$$
$$= \sum_{i=1}^{K} \sum_{j=1}^{K} \int_{-r_i}^{0} \int_{-r_j}^{0} \phi_i^T(s) E_{ij}(s, \eta) \phi_j(\eta) d\eta ds + \sum_{i=1}^{K} \int_{-r_i}^{0} z_i^T(s) F_i(s) z_i(s) ds, \tag{41}$$

where

$$z_i(s) = \begin{bmatrix} \psi^T & \phi_1^T(-r_1) & \cdots & \phi_K^T(-r_K) & \phi_i^T(s) \end{bmatrix}^T,$$

$$E_{ij}(s, \eta) = -\frac{\partial R_{ij}(s, \eta)}{\partial s} - \frac{\partial R_{ij}(s, \eta)}{\partial \eta},$$

$$F_i(s) = \begin{bmatrix} \frac{1}{\sum_{j=1}^{K} r_j} F_{11} & \frac{1}{\sum_{j=1}^{K} r_j} F_{12} & F_{13i}(s) \\ \frac{1}{\sum_{j=1}^{K} r_j} F_{12}^T & \frac{1}{\sum_{j=1}^{K} r_j} F_{22} & F_{23i}(s) \\ F_{13i}^T(s) & F_{23i}^T(s) & F_{33i}^T(s) \end{bmatrix},$$

$$F_{11} = PA + A^T P + \sum_{j=1}^{K} \left[Q_j(0)C_j + C_j^T Q_j^T(0) + C_j^T S_j(0)C_j \right],$$

$$F_{12} = \begin{bmatrix} G_1 & G_2 & \cdots & G_K \end{bmatrix},$$

$$G_j = \sum_{k=1}^{K} [Q_k(0)D_{kj} + C_k^T S_k(0)D_{kj}] + PB_j - Q_j(-r_j),$$

$$F_{22} = D^T \hat{S}_0 D - \hat{S}_r,$$

$$\hat{S}_0 = \text{diag} \left(S_1(0) \; S_2(0) \; \cdots \; S_K(0) \right),$$

$$\hat{S}_r = \text{diag} \left(S_1(-r_1) \; S_2(-r_2) \; \cdots \; S_K(-r_K) \right),$$

$$F_{13i}(s) = A^T Q_i(s) + \sum_{j=1}^{K} C_j^T R_{ij}^T(s,0) - \frac{dQ_i(s)}{ds},$$

$$F_{23i}(s) = \left[H_{i1}^T(s) \; H_{i2}^T(s) \; \cdots \; H_{iK}^T(s) \right]^T,$$

$$H_{ij}(s) = B_j^T Q_i(s) + \sum_{k=1}^{K} D_{kj}^T R_{ik}^T(s,0) - R_{ij}^T(s,-r_j),$$

$$F_{33i}(s) = -\frac{dS_i(s)}{ds}.$$

We may apply Proposition 1 to obtain the following.

Corollary 1. *Let the matrix and matrix functions be defined as in Proposition 1. Let*

$$-F_{33i}(s) > \varepsilon I, \text{ for some } \varepsilon > 0 \text{ and all } s \in [-r_i, 0), \tag{42}$$

and

$$-\sum_{i=1}^{K} \sum_{j=1}^{K} \int_{-r_i}^{0} \int_{-r_j}^{0} \phi_i^T(s) E_{ij}(s,\eta) \phi_j(\eta) ds d\eta \geq 0 \tag{43}$$

be satisfied for all $\phi \in \mathbb{PC}$. Then, there exists an $\varepsilon > 0$ such that (40) is satisfied for all $\psi \in \mathbb{R}^n$ and $\phi \in \mathbb{PC}$ if and only if there exist an $\varepsilon > 0$, matrix $\bar{F}_{aa} \in \mathbb{S}^{n+m}$, and absolutely continuous matrix functions $\bar{F}_{abi} : [-r_i, 0] \to \mathbb{R}^{(n+m) \times m_i}$, $i = 1, 2, \ldots, K$, such that

$$-\sum_{i=1}^{K} \int_{-r_i}^{0} z_i^T(s) [F_i(s) - \bar{F}_i(s)] z_i(s) ds \geq \varepsilon \psi^T \psi \tag{44}$$

and $\quad -\sum_{i=1}^{K} \sum_{j=1}^{K} \int_{-r_i}^{0} \int_{-r_j}^{0} \phi_i^T(s) E_{ij}(s,\eta) \phi_j(\eta) ds d\eta$

$$-\sum_{i=1}^{K} \int_{-r_i}^{0} z_i^T(s) \bar{F}_i(s) z_i(s) ds \geq 0 \tag{45}$$

are satisfied for all $\psi \in \mathbb{R}^n$ and $\phi \in \mathbb{PC}$, where

$$z_i(s) = \left[\psi^T \; \phi_1^T(-r_1) \; \cdots \; \phi_K^T(-r_K) \; \phi_i^T(s) \right]^T, \tag{46}$$

$$\bar{F}_i(s) = \begin{bmatrix} \frac{1}{\sum_{j=1}^{K} r_j} \bar{F}_{aa} & \bar{F}_{abi}(s) \\ \bar{F}_{abi}^T(s) & 0 \end{bmatrix}. \tag{47}$$

5 SOS Formulation

To numerically check the stability conditions, we use the sum-of-square (SOS) method. The basic idea is introduced here. For a given integer $d > 0$, let

$$Z_d(s) := \begin{bmatrix} 1 & s & s^2 & \cdots & s^d \end{bmatrix}^T,$$

and

$$Z_{n,d}(s) = I_{n \times n} \otimes Z_d(s),$$

where $\otimes$ denotes the Kronecker product. A polynomial symmetric matrix $G(s)$ with order not exceeding $2d$ is expressed as a quadratic form of $Z_{n,d}(s)$,

$$G(s) = Z_{n,d}^T(s) J Z_{n,d}(s), J = J^T. \tag{48}$$

A bivariate polynomial matrix $\Pi(s, \eta) = \Pi^T(\eta, s)$ with the order of each variable not exceeding d is expressed as a bilinear form of $Z_{n,d}(s)$,

$$\Pi(s, \eta) = Z_{n,d}^T(s) L Z_{n,d}(\eta), L = L^T. \tag{49}$$

For single variable polynomial matrices, it is useful to define

$$\Sigma_{n,d,\mathbb{I}} = \left\{ G : \mathbb{R} \to \mathbb{S}^n \middle| \begin{array}{l} G(s) = Z_{n,d}^T(s) J Z_{n,d}(s) \\ J = J^T, G(s) \geq 0 \text{ for } s \in \mathbb{I} \end{array} \right\}, \tag{50}$$

where $\mathbb{I}$ is an interval of $\mathbb{R}$. For bivariate polynomial matrices, it is useful to define

$$\Gamma_{n,d} = \left\{ Z_{n,d}^T(s) L Z_{n,d}(\eta) \mid L \in \mathbb{S}^{n(d+1)}, L \geq 0 \right\}. \tag{51}$$

Given the polynomial matrices that depend linearly on some parameters, the software package SOSTOOLS (Prajna, Papachristodoulou and Parrilo, 2002) is available to carry out searches among the parameters for the satisfaction of conditions in the form of $G \in \Sigma_{n,d,\mathbb{I}}$ and $\Pi \in \Gamma_{n,d}$, as well as some other convex constraints.

By restricting the matrix functions to polynomials, the stability conditions can be rendered in a SOS format as stated in the following theorem.

Theorem 3. *The system described by (34) and (35) is exponentially stable if there exist matrices* $P, P^1 \in \mathbb{R}^{n \times n}$, $\bar{F}_{aa} \in \mathbb{S}^{n+m}$, *polynomial matrices* $Q_i(s), Q_i^1(s) \in \mathbb{R}^{n \times m_i}$, $S_i(s) \in \mathbb{S}^{m_i}$, $T_i(s) \in \mathbb{S}^n$, $\bar{F}_{abi}(s) \in \mathbb{R}^{(n+m) \times m_i}$, $W_i(s) \in \mathbb{S}^{n+m}$, $i = 1, 2, \ldots, K$, *and bivariate polynomial matrices* $R_{ij}(\xi, \eta) = R_{ji}^T(\eta, \xi) \in \mathbb{R}^{m_i \times m_j}$, $i, j = 1, 2, \ldots, K$, *such that the following conditions are satisfied,*

$$\begin{bmatrix} P^1 & Q^1(s) \\ Q^{1T}(s) & R(s, \eta) \end{bmatrix} \in \Gamma_{n+m,d}, \tag{52}$$

$$\begin{bmatrix} \frac{-1}{\Sigma_{j=1}^K r_j} \bar{F}_{aa} & -\bar{F}_{ab}(s) \\ -\bar{F}_{ab}^T(s) & -E(s, \eta) \end{bmatrix} \in \Gamma_{n+2m,d}, \tag{53}$$

$$\begin{bmatrix} \frac{P-P^1}{\sum_{i=1}^k r_i} & Q_i(s)-Q_i^1(s) \\ Q_i^T(s)-Q_i^{1T}(s) & S_i(s) \end{bmatrix}$$

$$+ \begin{bmatrix} T_i(s) & 0 \\ 0 & 0 \end{bmatrix} \in \Sigma_{n+m_i,d,[-r_i,0)}, \quad i=1,2,\ldots,K, \quad (54)$$

$$\sum_{i=1}^{K} \int_{-r_i}^{0} T_i(s)ds = 0; \quad (55)$$

$$-F_i(s)+\bar{F}_i(s)+ \begin{bmatrix} W_i(s) & 0 \\ 0 & 0 \end{bmatrix} \in \Sigma_{n+m+m_i,d,[-r_i,0)}, i=1,2,\ldots,K, \quad (56)$$

$$\sum_{i=1}^{K} \int_{-r_i}^{0} W_i(s)ds = 0, \quad (57)$$

where $\bar{F}_i(s)$ is defined in (47) and

$$Q^1(s) = \begin{bmatrix} Q_1^1(r_1 s) & \cdots & Q_K^1(r_K s) \end{bmatrix},$$

$$R(s,\eta) = \begin{bmatrix} R_{11}(r_1 s, r_1 \eta) & \cdots & R_{1K}(r_1 s, r_K \eta) \\ \vdots & \ddots & \vdots \\ R_{K1}(r_K s, r_1 \eta) & \cdots & R_{KK}(r_K s, r_K \eta) \end{bmatrix},$$

$$\bar{F}_{ab}(s) = \begin{bmatrix} \bar{F}_{ab1}(s) & \cdots & \bar{F}_{abK}(s) \end{bmatrix},$$

$$E(s,\eta) = \begin{bmatrix} E_{11}(r_1 s, r_1 \eta) & \cdots & E_{1K}(r_1 s, r_K \eta) \\ \vdots & \ddots & \vdots \\ E_{K1}(r_K s, r_1 \eta) & \cdots & E_{KK}(r_K s, r_K \eta) \end{bmatrix}.$$

Proof. From Theorem 2, Propositions 1 and Corollary 1, it is sufficient to show that (36), (20), (21), (23), (24), (42), (43), (44) and (45) are satisfied. Obviously, (20) is implied by (54). Using a similar proof to that of Proposition 2 in Zhang, Peet and Gu (2011), it can be shown that (23) is implied by (54) and (55). Similar to the proof of Proposition 3 in Zhang, Peet and Gu (2011), we may show that (52) implies (24) by introducing a integral variable transformation and an application of Theorem 7 of Peet and Papachristodoulou (2009). The inequality (21) is implied by (24).

The proofs of (42), (44), (45) and (43) are similar to those of (20), (23), (24) and (21). It remains to be shown that (36) is satisfied. Notice that (56) implies

$$-F_{33i} = \frac{\partial S_i(s)}{ds} > 0, \quad (58)$$

and (56) and (57) together imply

$$F_{22} < 0. \quad (59)$$

As was shown in the first part of the proof of Theorem 5 in Zhang, Peet and Gu (2011), (58) and (59) imply (36). ∎

6 Numerical Example and Observation

The following numerical example is presented to illustrate the effectiveness of the method.

Example 1. Consider the system

$$\dot{x}(t) = \begin{bmatrix} -1 & -1 \\ 0.1 & -0.2 \end{bmatrix} x(t) + \begin{bmatrix} 0 & 1 \\ 1 & 0 \end{bmatrix} y(t-r),$$

$$y(t) = \begin{bmatrix} 1 & 0 \\ 0 & 1 \end{bmatrix} x(t).$$

It can be easily calculated (for example, by using a frequency domain method) that the system is exponentially stable for $r \in [0, r_{max})$, where $r_{max} = 1.69413$. The value of r_{max} is estimated numerically using Theorem 3 (referred to as "joint positivity") and Theorem 5 of Zhang, Peet and Gu (2011) (referred to as individual positivity) through a bisection process. The estimated results corresponding to each method and different order of monomial d are listed in the following table. It is clear that the joint positivity accelerated convergence to the analytical stability limit.

Order of monomial d	0	1	2
Joint positivity	1.6887	1.6934	1.6938
Individual positivity	1.6674	1.6863	1.6869

It is interesting to note that such a quick convergence to analytical stability limit is much more difficult to achieve for systems of neutral type (corresponding to a nonzero D matrix) according to our numerical experiments. Indeed, according to Gu (2010), the theoretical value of R_{ij} may be nonsmooth everywhere in $[-r_i, 0)$, although it is absolutely continuous. It was reported in Li and Gu (2010) that a quick convergence is still possible for discretized Lyapunov-Krasovskii functional method for some systems of neutral type when the theoretical value of R_{ij} contains only a finite number of nonsmooth points.

7 Conclusion

In this chapter, it is shown that the convergence of sum-of-square stability analysis of coupled differential-difference equations can be accelerated through enforcing joint positivity on the entire Lyapunov-Krasovskii functional. The method is less conservative than the previous method with identical order of polynomials.

References

1. Gu, K.: Discretized LMI set in the stability problem of linear uncertain time-delay systems. Int. J. Control 68(4), 923–924 (1997)
2. Gu, K.: Stability problem of systems with multiple delay channels. Automatica 46, 727–735 (2010)

3. Kato, T.: Perturbation Theory of Linear Operators. Springer, New York (1966)
4. Kolmogorov, A.N., Fomin, S.V.: Introductory Real Analysis. Translated and edited by R. A. Silverman. Dover, New York (1975)
5. Li, H., Gu, K.: Discretized Lyapunov Krasovskii functional for coupled differential difference equations with multiple delay channels. Automatica (2010), doi:10.1016/j.automatica.2010.02.007
6. Peet, M.M., Papachristodoulou, A.: Positive forms and the stability of linear time-delay systems. In: Proc. 45th IEEE Conf. Decision Control, San Diego, CA, USA, December 13–15 (2006)
7. Peet, M.M., Papachristodoulou, A.: Positive forms and stability of linear time-delay systems. SIAM J. Control Optim. 47, 3237–3258 (2009)
8. Prajna, S., Papachristodoulou, A., Parrilo, P.A.: Introducing SOSTOOLS: a general purpose sum of squares programming solver. In: Proc. 41th IEEE Conf. Decision Control, Las Vegas, USA, December 10–13 (2002)
9. Răsvan, V.: Functional differential equations of lossless propagation and almost linear behavior. Plenary Lecture. In: 6th IFAC Workshop on Time-Delay Systems, L'Aquila, Italy, July 10–12 (2006)
10. Repin, Y.M.: Quadratic Lyapunov Functionals for Systems With Delay (Russian) Prikl. Mat. Meh. 29, 564–566 (1965)
11. Zhang, Y., Peet, M.M., Gu, K.: Reducing the Complexity of the Sum-of-Squares Test for Stability of Delayed Linear Systems. IEEE Transactions on Automatic Control 56(1), 229–234 (2011)

On Retarded Nonlinear Time-Delay Systems That Generate Neutral Input-Output Equations

Milena Anguelova and Miroslav Halás

Abstract. In this work retarded nonlinear time-delay systems that surprisingly admit an input-output representation of neutral type are discussed. It is shown that such a behaviour is a strictly nonlinear phenomenon, as it cannot happen in the linear time-delay case where the input-output representation of retarded systems is always of retarded type. A necessary and sufficient condition under which nonlinear systems admit a neutral input-output representation is given. Some open problems, like minimality and system transformations, are discussed as well.

1 Introduction

In control theory, systems are usually described either by a set of coupled first-order differential equations, called state-space representation, or by higher order input-output differential equations. In the linear case any control system described by the state-space equations can be equivalently described by higher order input-output differential equations. From that point of view Laplace transforms play a key role. In the nonlinear case the situation is more complicated and several techniques have been developed to find the corresponding input-output equations, see for instance [4, 5]. Considering the algebraic point of view, it was shown in [4] that for a given state-space representation a corresponding set of input-output equations can be, at least locally, always constructed by applying a suitable change of coordinates. Such an idea of the state elimination has recently been carried over in [1] also to nonlinear time-delay systems and it has been shown that even for a state-space system with

Milena Anguelova
Fraunhofer-Chalmers Centre, Chalmers Science Park, SE-412 96 Gothenburg, Sweden
e-mail: milena.anguelova@fcc.chalmers.se

Miroslav Halás
Institute of Control and Industrial Informatics, Fac. of Electrical Engineering and IT,
Slovak University of Technology, Ilkovičova 3, 812 19 Bratislava, Slovakia
e-mail: miroslav.halas@stuba.sk

R. Sipahi et al. (Eds.): Time Delay Sys.: Methods, Appli. and New Trends, LNCIS 423, pp. 49–60.
springerlink.com   © Springer-Verlag Berlin Heidelberg 2012

delays there always exists, at least locally, a set of input-output differential-delay equations. However, as recently pointed out in [7] the state elimination algorithm of [1] might surprisingly result in a set of input-output equations representing a system of neutral type, even when one starts with the state-space equations being of retarded type. Note that by retarded we mean a classical (non-neutral) system and by neutral a system having delays in the highest derivative. This represents an unexpected and strictly nonlinear behaviour, for it cannot happen in the linear time-delay case where retarded systems always admit an input-output representation of a retarded type, and forms the main scope of our interest in this work. In particular, we show why it cannot happen in the linear time-delay case and why and when it happens in the nonlinear time-delay case.

2 Algebraic Setting

An algebraic formalism of differential forms, originally developed for nonlinear systems without delays [2, 4], was recently extended to the case of time-delay systems [8, 10, 13] and was shown to be effective in solving control problems like accessibility and observability, disturbance decoupling, feedback linearization and others. In order to avoid technicalities, we use slightly abbreviated notations and the reader is referred to those works for detailed technical constructions.

The nonlinear time-delay systems considered in this work are objects of the form

$$\dot{x}(t) = f(\{x(t-i), u(t-j); i, j \geq 0\})$$
$$y(t) = h(\{x(t-i), u(t-j); i, j \geq 0\}) \tag{1}$$

where the entries of f and h are meromorphic functions and $x \in \mathbb{R}^n$, $u \in \mathbb{R}^m$ and $y \in \mathbb{R}^p$ denote state, input and output to the system. For the scope of this work, we assume a single output, i.e. $p = 1$, even though the results can be generalized to multi-output systems.

Assuming the system has commensurable delays, it is not restrictive to assume $i, j \in \mathbf{N}$ since all commensurable delays can be considered as multiples of an elementary delay τ, see for instance [8]. Denote by $i_{\max}$ the maximal delay in (1).

The function of initial conditions $\varphi : [-i_{\max}, 0] \rightarrow \mathbb{R}^n$ is assumed to be smooth on an open interval containing $[-i_{\max}, 0]$. The input (control variable) $u : [-i_{\max}, \infty) \rightarrow \mathbb{R}^m$ is smooth for $t > -i_{\max}$. For a given φ, the set of inputs $u(t)$ for which there exists a unique solution to system (1) in the interval $[0, \infty)$ are called admissible inputs. Let $C \subset C^\infty$ denote the open set of φ with a nonempty set of admissible inputs.

Let $\mathcal{K}$ be the field of meromorphic functions of $\{x(t-i), u^{(k)}(t-j); i, j, k \geq 0\}$ and let $\mathcal{E}$ be the formal vector space over $\mathcal{K}$ given by

$$\mathcal{E} = \mathrm{span}_{\mathcal{K}}\{d\xi; \xi \in \mathcal{K}\}.$$

The delay operator δ is defined on $\mathcal{K}$ and $\mathcal{E}$ as

$$\delta(\xi(t)) = \xi(t-1)$$
$$\delta(\alpha(t)\mathrm{d}\xi(t)) = \alpha(t-1)\mathrm{d}\xi(t-1) \qquad (2)$$

where $\xi(t) \in \mathcal{K}$ and $\alpha(t)\mathrm{d}\xi(t) \in \mathcal{E}$.

Remark 1. Note that in case of rational or even algebraic functions the formal vector space of differential one-forms $\mathcal{E}$ coincides with the so-called Kähler differentials that would represent a natural choice in a purely algebraic setting, see [6] for detailed discussion and relevant references.

The delay operator (2) induces the (non-commutative) skew polynomial ring $\mathcal{K}(\delta]$ with the multiplication given by the commutation rule

$$\delta a(t) = a(t-1)\delta$$

for any $a(t) \in \mathcal{K}$. The ring $\mathcal{K}(\delta]$ is a left Ore domain and represents the ring of linear backward shift (delay) operators.

The properties of system (1) can now be analyzed by introducing the machinery of one-forms known from systems without delays [2, 4]. This time, rather than vector spaces we introduce formal modules over $\mathcal{K}(\delta]$, generally

$$\mathcal{M} = \mathrm{span}_{\mathcal{K}(\delta]}\{\mathrm{d}\xi; \xi \in \mathcal{K}\} .$$

The rank of a module over the left Ore domain $\mathcal{K}(\delta]$ is well-defined [3].

Definition 1 ([13]). The closure of a submodule $\mathcal{N}$ in $\mathcal{M}$ is the submodule

$$\overline{\mathcal{N}} = \{w \in \mathcal{M}; \exists a(\delta] \in \mathcal{K}(\delta], a(\delta]w \in \mathcal{N}\} .$$

That is, it is the largest submodule of $\mathcal{M}$ containing $\mathcal{N}$ with rank equal to $\mathrm{rank}_{\mathcal{K}(\delta]}\mathcal{N}$.

The notion of the closure of a submodule will play a key role in showing why and when the systems of the form (1) admit an input-output representation of neutral type.

3   Input-Output Representation

3.1   State Elimination Algorithm

In this subsection we recall the state elimination procedure from [1].

Let f be an r-dimensional vector with entries $f_j \in \mathcal{K}$. Let $\frac{\partial f}{\partial x}$ denote the $r \times n$ matrix with entries

$$\left(\frac{\partial f}{\partial x}\right)_{j,i} = \sum_{\ell \geq 0} \frac{\partial f_j}{\partial x_i(t-\ell)}\delta^\ell \in \mathcal{K}(\delta] .$$

Denote by s the least nonnegative integer such that

$$\text{rank}_{\mathcal{K}(\delta]}\frac{\partial(h,\dots,h^{(s-1)})}{\partial x} = \text{rank}_{\mathcal{K}(\delta]}\frac{\partial(h,\dots,h^{(s)})}{\partial x} \quad . \tag{3}$$

Let $S = \left(h,\dots,h^{(s-1)}\right)$, then

$$\text{rank}_{\mathcal{K}(\delta]}\frac{\partial S}{\partial x} = s \leq n \quad .$$

Hence, $\frac{\partial h^{(s)}}{\partial x}$ is in $\overline{\text{span}_{\mathcal{K}(\delta]}\left\{\frac{\partial(h,\dots,h^{(s-1)})}{\partial x}\right\}}$. Thus, there exists a nonzero polynomial $b(\delta] \in \mathcal{K}(\delta]$ such that

$$b(\delta]\frac{\partial h^{(s)}}{\partial x} \in \text{span}_{\mathcal{K}(\delta]}\left\{\frac{\partial(h,\dots,h^{(s-1)})}{\partial x}\right\} \quad . \tag{4}$$

Therefore

$$b(\delta]\text{d}h^{(s)} + \sum_{r=1}^{m}\sum_{j=0}^{J}c_{j,r}(\delta]\text{d}u_r^{(j)} - \sum_{j=0}^{s-1}b_j(\delta]\text{d}h^{(j)} = 0$$

for some $J \geq 0$, where J is the highest derivative of u appearing in the functions in S and $c_{j,r}(\delta] \in \mathcal{K}(\delta]$. We assume that the polynomials $b(\delta]$, $b_j(\delta]$ and $c_{j,r}(\delta]$ have no common factors other than 1. Since all functions are assumed meromorphic and we have continuous dependence for the output on the input and initial function, the above equality holds on an open and dense subset of C. Applying the Poincaré lemma, we obtain a function $\xi(t) \in \mathcal{K}$ such that

$$\text{d}\xi = b(\delta]\text{d}h^{(s)} + \sum_{r=1}^{m}\sum_{j=0}^{J}c_{j,r}(\delta]\text{d}u_r^{(j)} - \sum_{j=0}^{s-1}b_j(\delta]\text{d}h^{(j)} \tag{5}$$

and

$$\xi_i(\delta,h,\dots,h^{(s)},u,\dots,u^{(J)}) = 0 \tag{6}$$

where we use the notation

$$\psi\left(\delta,y,\dots,y^{(k)},u,\dots,u^{(J)}\right) :=$$
$$\psi(y(t),\dots,y(t-i_0),\dots,y^{(k)}(t),\dots,y^{(k)}(t-i_k),$$
$$u(t),\dots,u(t-j_0),\dots,u^{(J)}(t-j_l),\dots,u^{(J)}(t-j_l))$$

with $i_0,\dots,i_k$ and $j_0,\dots,j_k$ nonnegative. We have obtained the input-output representation of system (1)

$$\xi\left(\delta,y,\dots,y^{(s)},u,\dots,u^{(J)}\right) = 0 \quad . \tag{7}$$

3.2 Neutral Input-Output Equations

Surprisingly, the state elimination algorithm can result in an input-output equation representing a system of neutral type, even if one starts with a classical delay system in state-space form, as demonstrated by the following example.

Example 1. Consider the system

$$\dot{x}(t) = u(t)$$
$$y(t) = x^2(t) + x(t-1) \quad .$$

Note that following the ideas of the state elimination of [4] we can compute

$$\dot{y}(t) = 2x(t)u(t) + u(t-1) \quad . \tag{8}$$

However, contrary to what happens in the case of systems without delays, one cannot go any further, since from $y(t) = x^2(t) + x(t-1)$ one cannot express $x(t)$ and substitute to (8). Technically speaking

$$\frac{\partial h}{\partial x} \notin \text{span}_{\mathscr{K}(\delta]} \left\{ \frac{\partial h}{\partial x} \right\} \quad .$$

Nevertheless, the state elimination algorithm (Subsection 3.1) yields a nonzero polynomial $b(\delta] \in \mathscr{K}(\delta]$ such that

$$b(\delta]\frac{\partial h}{\partial x} \in \text{span}_{\mathscr{K}(\delta]} \left\{ \frac{\partial h}{\partial x} \right\} \quad .$$

That is, an input-output equation can be obtained by also considering shifts of the above equations for $y(t)$ and $\dot{y}(t)$. One can use

$$y(t) = x^2(t) + x(t-1)$$
$$\dot{y}(t-1) = 2x(t-1)u(t-1) + u(t-2)$$

to compute

$$x^2(t) = y(t) - \frac{\dot{y}(t-1) - u(t-2)}{2u(t-1)} \quad .$$

Then, after substituting $x(t)$ to (8), the input-output equation can be found as

$$u(t-1)(\dot{y}(t) - u(t-1))^2 + 2u^2(t)(\dot{y}(t-1) - u(t-2)) - 4y(t)u^2(t)u(t-1) = 0 \tag{9}$$

representing, however, a neutral system.

4   Main Result

The fact that a classical time-delay state space system can admit a neutral input-output equation is a strictly nonlinear phenomenon, for it cannot happen in the linear case as will be shown in this section.

4.1   Input-Output Equations for Linear Delay Systems

Consider linear time-delay systems of the form

$$
\begin{aligned}
\dot{x}(t) &= A[\delta]x(t) + B[\delta]u(t) \\
y(t) &= C[\delta]x(t)
\end{aligned}
\tag{10}
$$

where $A := A[\delta]$, $B := B[\delta]$ and $C := C[\delta]$ are matrices with elements that are polynomials of δ with real coefficients. The latter form a commutative ring, $\mathbb{R}[\delta]$, which is a unique factorization domain and hence also an integrally closed domain [14]. In the case of a single output considered here C is a row-vector. Recall that for nonlinear systems the corresponding algebraic structure is $\mathscr{K}(\delta)$, i.e. a noncommutative ring .

Theorem 1. *The input-output representation of system (10) is of retarded type.*

Proof. Let $\mathscr{M}$ be a module over $\mathbb{R}[\delta]$. We proceed by defining linear recurring sequences over $\mathscr{M}$ in the same way as in [9]. Let $\mathbb{N}_0 = \{0, 1, \ldots\}$. A function $\mu :$ $\mathbb{N}_0 \to \mathscr{M}$ is called a sequence over the module $\mathscr{M}$. The set of all sequences over $\mathscr{M}$ is denoted by $\mathscr{M}^{(l)}$. The multiplication of a polynomial $g(z) = \sum_{i \geq 0}^{r} g_i z^i$ with $g_i \in \mathbb{R}[\delta]$, $i = 0, \ldots, r$ by a sequence $\mu \in \mathscr{M}^{(l)}$ is defined as

$$
g(z)\mu = \nu \in \mathscr{M}^{(l)} \quad \text{such that} \quad \nu(i) = \sum_{s \geq 0}^{r} g_s \mu(i+s)
$$

where $i \in \mathbb{N}_0$. Thus, $\mathscr{M}^{(l)}$ is a module over the ring $\mathbb{R}[\delta][z]$. A sequence $\mu \in \mathscr{M}^{(l)}$ is called a linear recurring sequence (LRS) of the order m over $\mathscr{M}$ if there exists a monic polynomial $p(z) \in \mathbb{R}[\delta][z]$ of the degree m such that $p(z)\mu = 0$. Then $p(z)$ is called a characteristic polynomial of LRS μ and $(\mu(0), \ldots, \mu(m-1))$ is called its initial vector (relative to $p(z)$). The annihilator of a LRS μ in $\mathbb{R}[\delta][z]$ is the ideal $An_{\mathbb{R}[\delta][z]}(\mu)$ of $\mathbb{R}[\delta][z]$ defined by the following equality:

$$
An_{\mathbb{R}[\delta][z]}(\mu) = \{p(z) \in \mathbb{R}[\delta][z] : p(z)\mu = 0\}.
$$

Hence, a sequence $\mu \in \mathscr{M}^{(l)}$ is LRS if and only if $An_{\mathbb{R}[\delta][z]}(\mu)$ is a monic ideal (i.e. contains a monic polynomial). An LRS has a unique minimal polynomial if and only if its annihilator has the form $An_{\mathbb{R}[\delta][z]}(\mu) = \mathbb{R}[\delta][z] \cdot p(z)$, where $p(z)$ is a monic polynomial. In this case, $p(z)$ is the minimal polynomial of μ.

Let now $\mathscr{M}$ be the module generated by $\{C, CA, CA^2, \ldots\}$ over $\mathbb{R}[\delta]$. In the following we use an argument similar to what is used in the proof of Lemma 3.2 in

[11]. Let the sequence $\mu \in \mathscr{M}^{(l)}$ be defined by

$$\mu(i) = CA^i = \frac{\partial h^{(i)}}{\partial x}, \quad i \in \mathbb{N}_0.$$

The commutativity of $\mathbb{R}[\delta]$ allows the use of Cayley-Hamilton's theorem which gives

$$A^n = -\sum_{i=0}^{n-1} g_i A^i$$

for some $g_i \in \mathbb{R}[\delta]$, $i = 0, \ldots, n-1$. Hence,

$$CA^n = -C \sum_{i=0}^{n-1} g_i A^i = -\sum_{i=0}^{n-1} g_i CA^i.$$

With $g(z) = z^n + \sum_{i=0}^{n-1} g_i z^i$ and $v = g(z)\mu$, we have

$$v(k) = \sum_{i=0}^{n} g_i \mu(k+i) = \sum_{i=0}^{n} g_i CA^{k+i} = \left(\sum_{i=0}^{n} g_i CA^i\right) A^k = 0$$

for $k \geq 0$. Therefore, μ is LRS over $\mathscr{M}$ with $g(z)$ as its characteristic polynomial. Consider now μ as a sequence over the field of fractions $\mathbb{R}\langle\delta\rangle$. The ideal $An_{\mathbb{R}\langle\delta\rangle[z]}(\mu)$ in $\mathbb{R}\langle\delta\rangle[z]$ is non-empty since it contains $g(z)$. Since $\mathbb{R}\langle\delta\rangle[z]$ is a principal ideal domain, this ideal has a unique monic generator $p(z)$ and the LRS μ has a minimal polynomial in $\mathbb{R}\langle\delta\rangle$. We must have

$$g(z) = p(z)q(z)$$

for some $q(z) \in \mathbb{R}\langle\delta\rangle[z]$. With $g(z)$ monic in the integrally closed domain $\mathbb{R}[\delta][z]$ and $p(z)$ monic in $\mathbb{R}\langle\delta\rangle[z]$, Theorem 5 in [14], Chapter V, Section 3 can be applied to deduce that $p(z) \in \mathbb{R}[\delta][z]$. On the other hand, since for the linear system (10)

$$\mathrm{rank}_{\mathscr{K}\langle\delta\rangle} \frac{\partial S}{\partial x} = \mathrm{rank}_{\mathbb{R}[\delta]}\{C, CA, \ldots, CA^{n-1}\} = s$$

there exist $b \in \mathbb{R}[\delta]$ and $b_j \in \mathbb{R}[\delta]$, $j = 0, \ldots, s-1$ with no common factors other than 1 such that

$$bCA^s = \sum_{j=0}^{s-1} b_j CA^j.$$

Note that the above is equation (4) for the special case of linear systems. We obtain that $\tilde{p}(z) = z^s - \sum_{j=0}^{s-1} \frac{b_j}{b} z^j$ is a characteristic polynomial for μ in $\mathbb{R}\langle\delta\rangle[z]$ and it is of minimal degree due to the definition of s (3). Hence, $\tilde{p}(z) = p(z)$ and thus, $\frac{b_j}{b} \in \mathbb{R}[\delta]$, $j = 0, \ldots, s-1$. Thus, we conclude that

$$CA^s \in \mathrm{span}_{\mathbb{R}[\delta]}\{C, CA, \ldots, CA^{s-1}\}$$

which means that for linear time-delay systems, the state elimination yields

$$\frac{\partial h^{(s)}}{\partial x} \in \text{span}_{\mathbb{R}[\delta]} \left\{ \frac{\partial(h,\dots,h^{(s-1)})}{\partial x} \right\}. \tag{11}$$

Thus, the input-output representation is of retarded type.

Example 2. Consider the system

$$\dot{x}(t) = x(t) + u(t)$$
$$y(t) = x(t) + x(t-1)$$

for which we have

$$\dot{y}(t) = x(t) + u(t) + x(t-1) + u(t-1).$$

Note that we cannot compute $x(t)$ from $y(t) = x(t) + x(t-1)$ here either. Nevertheless, we have

$$\frac{\partial \dot{h}}{\partial x} \in \text{span}_{\mathbb{R}[\delta]} \left\{ \frac{\partial h}{\partial x} \right\}$$

and thus an input-output representation of retarded type can easily be found as

$$\dot{y}(t) = y(t) + u(t) + u(t-1).$$

4.2 Input-Output Equations for Nonlinear Delay Systems

It is now straightforward to conclude that in the nonlinear time-delay case the existence of a retarded input-output representation is assured by satisfying analogical condition to that given by (11) for the linear time-delay case.

Theorem 2. *The input-output representation of system (1) is of retarded type if and only if*

$$\frac{\partial h^{(s)}}{\partial x} \in \text{span}_{\mathscr{K}(\delta]} \left\{ \frac{\partial(h,\dots,h^{(s-1)})}{\partial x} \right\}.$$

Proof. For the system (1) the state elimination algorithm, recalled in Subsection 3.1, results in the equation

$$b(\delta]\mathrm{d}h^{(s)} + \sum_{r=1}^{m} \sum_{j=0}^{J} c_{j,r}(\delta]\mathrm{d}u_r^{(j)} - \sum_{j=0}^{s-1} b_j(\delta]\mathrm{d}h^{(j)} = 0,$$

where it is assumed that the polynomials $b(\delta]$, $b_j(\delta]$ and $c_{j,r}(\delta]$ have no common factors other than 1. Obviously, the multiplication of $\mathrm{d}h^{(s)}$ by a polynomial $b(\delta]$ of degree at least 1 implies that the Poincaré lemma (see (5) and (6)) will yield an input-output equation of neutral (non-retarded) type. $\qquad\square$

5 Summary

To illustrate the differences between input-output representations of different classes of systems we summarize the results as follows.

The state elimination algorithm yields:

- for linear systems without delays

$$\frac{\partial h^{(s)}}{\partial x} \in \text{span}_{\mathbb{R}} \left\{ \frac{\partial (h, \ldots, h^{(s-1)})}{\partial x} \right\},$$

- for linear delay systems

$$\frac{\partial h^{(s)}}{\partial x} \in \text{span}_{\mathbb{R}[\delta]} \left\{ \frac{\partial (h, \ldots, h^{(s-1)})}{\partial x} \right\},$$

- for nonlinear systems without delays

$$\frac{\partial h^{(s)}}{\partial x} \in \text{span}_{\mathscr{K}} \left\{ \frac{\partial (h, \ldots, h^{(s-1)})}{\partial x} \right\},$$

- for nonlinear delay systems

$$\frac{\partial h^{(s)}}{\partial x} \in \text{span}_{\mathscr{K}(\delta]} \overline{\left\{ \frac{\partial (h, \ldots, h^{(s-1)})}{\partial x} \right\}}$$

or equivalently $b(\delta] \frac{\partial h^{(s)}}{\partial x} \in \text{span}_{\mathscr{K}(\delta]} \left\{ \frac{\partial (h, \ldots, h^{(s-1)})}{\partial x} \right\}$ for some $b(\delta] \in \mathscr{K}(\delta]$ resulting, in general, in an input-output representation of neutral type. Furthermore, it is of retarded (not neutral) type if and only if $b(\delta] = 1$.

6 Retarded Input-Output Representation

Other input-output equations for system (1) can be obtained by considering higher order derivatives of the output and some of these can be of retarded type. Consider once again the system from Example 1 where we have

$$(2x(t)u(t-1) + u(t)\delta) \frac{\partial \dot{h}}{\partial x} - 2u(t)u(t-1) \frac{\partial h}{\partial x} = 0$$

and

$$\frac{\partial \dot{h}}{\partial x} \notin \text{span}_{\mathscr{K}(\delta]} \left\{ \frac{\partial h}{\partial x} \right\}.$$

However, on the other hand, the equations

$$\dot{y}(t) = 2x(t)u(t) + u(t-1)$$
$$\ddot{y}(t) = 2u^2(t) + 2x(t)\dot{u}(t) + \dot{u}(t-1)$$

give us

$$\frac{\partial \ddot{h}}{\partial x} \in \text{span}_{\mathscr{K}(\delta]}\left\{\frac{\partial(h,\dot{h})}{\partial x}\right\}$$

with

$$\frac{\partial \ddot{h}}{\partial x} - \frac{\dot{u}(t)}{u(t)}\frac{\partial \dot{h}}{\partial x} = 0.$$

Thus, in this case the corresponding input-output representation can be found as

$$\ddot{y}(t) = 2u^2(t) + \frac{\dot{y}(t) - u(t-1)}{u(t)}\dot{u}(t) + \dot{u}(t-1) \tag{12}$$

which is of retarded type. However, it involves a second derivative of the output and it is therefore questionable whether it can be considered the *true* (minimal) input-output representation of the scalar time-delay system in the example. In fact, there are solutions to the above input-output equation that do not solve the original delay system. Set, for example, $u(t) = 1$, $t \in [-1,\infty)$. Then, (12) reads $\ddot{y}(t) = 2$ which is solved by, for example, $y(t) = t^2$. However, the pair $(u(t),y(t)) = (1,t^2)$ does not solve the neutral input-output equation (9), since

$$u(t-1)(\dot{y}(t) - u(t-1))^2 + 2u^2(t)(\dot{y}(t-1) - u(t-2)) - 4y(t)u^2(t)u(t-1) =$$
$$(2t-1)^2 + 2(2(t-1) - 1) - 4t^2 = -5 \neq 0.$$

Yet we know that all input-output pairs that satisfy the original state-space system must also satisfy its input-output equation(s) and so we have found a solution to (12) that is extraneous to the original system.

7 Conclusions, Discussion and Open Problems

This work discussed the problem of finding an input-output representation for time-delay systems described by classical state-space equations. Even if such a representation exists for both linear and nonlinear time-delay systems, it was shown that in the linear case the input-output equations involve, as expected, time-derivatives of the output up until the number of observable states of the system. However, in the nonlinear case the corresponding input-output representation might surprisingly be a neutral differential-delay equation which clearly represents unexpected and strictly nonlinear behaviour and has significant consequences, suggests new research directions and creates many open problems. To mention just a few, this implies for instance that, unlike in the linear time-delay case, there exists a class of nonlinear neutral input-output differential-delay equations that can be modelled by retarded (non-neutral) state-space equations. Additionally, the same can be concluded for system transformations, i.e. there exists a class of nonlinear neutral state-space systems that could be transformed, under certain conditions, to state-space systems of retarded type. Other problems arise everywhere the input-output equation is used to provide a solution to control problems. One example is the model matching problem discussed in [7] where the transfer function concept, based on computing an input-output equation, is used. Another example is observer design as suggested in [12], for the computation of an input-output equation is used there as an intermediate step for finding a suitable change of coordinates that transforms a system with delays to the observer canonical form. Clearly, if a system admits a neutral input-output equation this approach cannot be used directly, even if the system might be transformable and observer designed. Finally, the question of minimality of the input-output representations we have found here arises as well. In case a system admits a neutral input-output equation, the problem is to decide whether this is the minimal input-output representation. Note that to find the input-output representation we employed the notion of closure of the module which is in general necessary, unlike in the case of nonlinear system without delays or linear time-delay systems. Thus, the question is whether the use of closure of the module does or does not cause the input-output equation we find will have additional solutions that are not solutions to the original system, forming an open problem. Note also that even if a retarded delay-differential equation can be obtained by taking further derivatives of the output, there will be reasonable doubts about the minimality of such an input-output equation. So, this leaves, in general, an open problem of deciding whether a neutral differential-delay equation or a retarded differential-delay equation of higher order than the number of observable states can be a *true* input-output representation of a retarded time-delay system, or even whether there simply exists a class of nonlinear time-delay systems which do not admit the input-output equation we look for.

Acknowledgements. This work was partially supported by the Swedish Research Council, the Swedish Foundation for Strategic Research via the Gothenburg Mathematical Modeling Center, the Slovak Research and Development Agency, grant No. LPP-0127-06 and the Slovak Grant Agency grants No. VG-1/0656/09 and VG-1/0369/10.

The authors would like to thank prof. B. Wennberg for his valuable contribution to this project and prof. C.H. Moog for helpful discussion and comments on neutral time-delay systems.

References

1. Anguelova, M., Wennberg, B.: State elimination and identifiability of the delay parameter for nonlinear time-delay systems. Automatica 44, 1373–1378 (2008)
2. Aranda-Bricaire, E., Kotta, Ü., Moog, C.H.: Linearization of discrete-time systems. SIAM Journal of Control Optimization 34, 1999–2023 (1996)
3. Cohn, P.M.: Free rings and their relations. Academic Press, London (1985)
4. Conte, G., Moog, C.H., Perdon, A.M.: Algebraic Methods for Nonlinear Control Systems. Theory and Applications. Springer, London (2007)
5. Diop, S.: Elimination in control theory. Math. Contr. Signals Syst. 4, 72–86 (1991)

6. Fu, G., Halás, M., Kotta, Ü., Li, Z.: Some remarks on Kähler differentials and ordinary differentials in nonlinear control theory. Syst. Cont. Letters (provisionally accepted, 2011)
7. Halás, M., Moog, C.H.: A polynomial solution to the model matching problem of nonlinear time-delay systems. In: European Control Conference, Budapest, Hungary (2009)
8. Márquez-Martínez, L.A., Moog, C.H., Velasco-Villa, M.: The structure of nonlinear time-delay systems. Kybernetika 36, 53–62 (2000)
9. Michalev, A.V., Nechaev, A.A.: Linear Recurring Sequences over Modules. Acta Applicandae Mathematicae 42, 161–202 (1996)
10. Moog, C.H., Castro-Linares, R., Velasco-Villa, M., Márquez-Martínez, L.A.: The disturbance decoupling problem for time-delay nonlinear systems. IEEE Transactions on Automatic Control 45, 305–309 (2000)
11. Rouchaleau, Y., Sontag, E.: On the Existence of Minimal Realizations of Linear Dynamical Systems over Noetherian Integral Domains. Journal of Computer and System Sciences 18, 65–75 (1979)
12. Velasco-Villa, M., Márquez-Martínez, L.A., Moog, C.H.: Observability and observers for nonlinear systems with time delays. Kybernetika 38, 445–456 (2002)
13. Xia, X., Márquez-Martínez, L.A., Zagalak, P., Moog, C.H.: Analysis of nonlinear time-delay systems using modules over non-commutative rings. Automatica 38, 1549–1555 (2002)
14. Zariski, O., Samuel, P.: Commutative algebra, vol. I. Springer, Berlin (1991)

Computation of Imaginary Axis Eigenvalues and Critical Parameters for Neutral Time Delay Systems

Gilberto Ochoa, Sabine Mondié, and Vladimir L. Kharitonov

Abstract. In this paper, a procedure for the computation of the Lyapunov matrix of neutral type systems, with delays multiple of a basic one, is recalled: It consists in solving a boundary value problem for a delay free system of matrix equations. The important property that the elements of the spectrum of the delay system that are symmetric with respect to the imaginary axis belong to the spectrum of the delay free system is also established. This property is exploited for the determination of the critical values of the time delay system.

1 Introduction

A number of results on the theoretical analysis of neutral type linear time delays systems have been contributed in the past few years: The form of the Lyapunov functionals with prescribed derivative parametrized in the Lyapunov matrix was presented in [14] for the single delay case. This matrix has been shown to be the solution of a delay free system subject to boundary conditions [5], [10] and its existence and uniqueness were established in [7]. The extension of some of these results to the case of delays multiple of a basic one was recently addressed in [12].

In this paper we present, for the case of neutral type systems with commensurate delays, a methodology that reveals the values of the parameters (called critical parameters) for which the system may cross stability/instability boundaries. This methodology is based on the same concepts as those presented for distributed

G. Ochoa · S. Mondié
Automatic Control Department, CINVESTAV, Av. IPN, C.P. 07360 D.F., México
e-mail: gochoa79@gmail.com, smondie@ctrl.cinvestav.mx

V. L. Kharitonov
Applied Mathematics and Control Process Department, St.-Petersburg State University, St.-Petersburg, Russia
e-mail: khar@apmath.spbu.ru

R. Sipahi et al. (Eds.): Time Delay Sys.: Methods, Appli. and New Trends, LNCIS 423, pp. 61–72.
springerlink.com   © Springer-Verlag Berlin Heidelberg 2012

and point wise delay systems [8]. It exploits the interesting connection between the system spectrum and that of the delay free system used in the computation of the Lyapunov matrix: If s_0 is a root of the delay system such that $-s_0$ is also a root, then s_0 belongs also to the spectrum of the delay free system. This methodology differs from other approaches for the computation of critical values, see [3], [4], [9], [13], [16] and the references therein, by the fact that it is strongly connected to time domain properties.

The structure of the contribution is as follows: In section 1 some basic results and the concept of Lyapunov matrix are introduced. Section 2 is devoted to recall the analytic procedure for the computation of the Lyapunov matrix that is based on the fact that the Lyapunov matrix satisfies a delay free system of matrix equations. The relation between the spectrum of the delay free system and those of the neutral type time delay system is established in section 3. Based on this relation a methodology for the computation of critical parameters is proposed in section 4 and some examples are analyzed in section 5. The contribution ends with some concluding remarks.

2 Preliminary Results

We introduce next some definitions and useful results.

2.1 System Description

In this paper we consider the class of neutral type systems studied in [12]:

$$\sum_{k=0}^{m} B_k \dot{x}(t - hk) = \sum_{k=0}^{m} A_k x(t - hk), \quad t \geq 0. \tag{1}$$

Here $A_k, B_k \in \mathbb{R}^{n \times n}$ are given matrices, and $h > 0$ is the basic delay. Without any loss of generality we assume that $B_0 = I$.

Given an initial condition $\varphi \in C^1([-mh, 0], \mathbb{R})$, there exists a unique solution $x(t, \varphi)$ of system (1). We denote by $x_t(\varphi)$ the segment of the solution $x_t(\varphi) \to x(t + \theta, \varphi), \theta \in [-mh, 0]$.

Definition 1. [1] System (1) is exponentially stable if there exist constants $\sigma > 0$ and $\gamma \geq 1$ such that for every solution $x(t, \varphi)$ of the system the following estimates holds

$$\|x(t, \varphi)\| \leq \gamma e^{-\sigma t} \|\varphi\|_{mh},$$

for $t \geq 0$. Here, $\|\varphi\|_{mh}$ is defined as

$$\|\varphi\|_{mh} = \max_{\theta \in [-mh, 0]} \|\varphi(\theta)\|.$$

2.2 Lyapunov-Krasovskii Functionals

The form of the funtional with prescribed time derivative associated to system (1) is obtained in [12], [14]. This results reads as follows:

Let us assume that system (1) is exponentially stable. Given a functional $w(x_t)$ of the form

$$w(x_t) = \sum_{k=0}^{m} x^\mathsf{T}(t-hk)W_k x(t-hk) + \sum_{k=1}^{m} \int_{-hk}^{0} x^\mathsf{T}(t+\theta)W_{m+k} x(t+\theta)\,d\theta,$$

where W_k, $k = 0,1,...,2m$ are positive definite matrices, there exists a quadratic functional $v(x_t)$, such that

$$\frac{d}{dt}v(x_t) = -w(x_t),$$

along the solutions of system (1).

The functional $v(x_t)$ is of the form

$$v(x_t) = \sum_{j=0}^{m}\sum_{k=0}^{m} x^\mathsf{T}(t)B_j^\mathsf{T}U\left(h(j-k)\right)B_k x(t) + 2\sum_{j=0}^{m}\sum_{k=1}^{m} x^\mathsf{T}(t)B_j^\mathsf{T} \times$$

$$\times \int_{-hk}^{0} U\left(h(j-k)-\theta\right)\left[A_k x(t+\theta) - B_k \dot{x}(t+\theta)\right]d\theta +$$

$$+ \sum_{j=1}^{m}\sum_{k=1}^{m} \int_{-hj}^{0}\int_{-hk}^{0} \left[x^\mathsf{T}(t+\theta_1)A_k^\mathsf{T} - \dot{x}^\mathsf{T}(t+\theta_2)B_k^\mathsf{T}\right] \times$$

$$U\left(h(j-k)+\theta_1-\theta_2\right) \times \left[A_k x(t+\theta) - B_k \dot{x}(t+\theta)\right]d\theta_2 d\theta_1 +$$

$$+ \sum_{k=1}^{m} \int_{-hk}^{0} \dot{x}(t+\theta)\left[W_k + (hk+\theta)W_{m+k}\right]x(t+\theta)\,d\theta.$$

Here

$$U(\tau) = \int_{0}^{\infty} K^\mathsf{T}(t)WK(t+\tau)\,d\tau,$$

is called the Lyapunov matrix of system (1) associated to the matrix $W = W_0 + \sum_{k=1}^{m}[W_k + hkW_{m+k}]$ and matrix $K(t)$ is the fundamental matrix of system (1) [5].

2.3 Properties of the Lyapunov Matrix

The Lyapunov matrix satisfies a set of properties that are the basis for a methodology for its computation, [5]. These properties are:

- The dynamic property

$$\sum_{k=0}^{m} \left[U'(\tau - hk) B_k - U(\tau - hk) A_k \right] = 0, \quad \tau > 0, \tag{2}$$

- The symmetry property

$$U(\tau) = U^{\mathsf{T}}(-\tau), \quad \tau \geq 0, \tag{3}$$

- The algebraic property

$$\sum_{j=0}^{m} \sum_{k=0}^{m} \left[B_j^{\mathsf{T}} U(h(j-k)) A_k + A_k^{\mathsf{T}} U^{\mathsf{T}}(h(j-k)) B_j \right] = -W. \tag{4}$$

3 Computation of the Lyapunov Matrix

In [12] a methodology for the computation of the Lyapunov matrix for system (1) was introduced. This procedure consists in showing that the Lyapunov matrix satisfies a delay free system of matrix equations where the initial conditions are not known. Then, the problem reduces to find the initial conditions via a two point boundary value problem. Due to its importance in the present contribution, we recall this semi-analytic procedure below.

First, $2m$ auxiliary matrices are introduced

$$X_j(\tau) = U(\tau + jh), \quad j = -m, ..., 0, ..., m-1. \tag{5}$$

Let us consider $j \geq 0$ and compute the derivative of matrices $X_j(\tau)$ with respect to τ, that is

$$\frac{d}{d\tau} X_j(\tau) = \frac{d}{d\tau} U(\tau + jh), \quad j = 0, ..., m-1.$$

Since the matrices defined in (5) satisfy the properties (2)-(4), by using the dynamic property (2), the above expression can be rewritten as

$$X_j'(\tau) = -\sum_{k=1}^{m} U'(\tau + h(j-k)) B_k + \sum_{k=0}^{m} U(\tau + h(j-k)) A_k,$$

equivalently,

$$\sum_{k=0}^{m} X_{j-k}'(\tau) B_k = \sum_{k=0}^{m} X_{j-k}(\tau) A_k. \tag{6}$$

Now for $j = -m, ..., -1$ the derivative of the auxiliary matrices is

$$X_j'(\tau) = -\left[U'(-\tau - jh) \right]^{\mathsf{T}}$$

then

$$\sum_{k=0}^{m} B_k^\mathsf{T} X'_{j+k}(\tau) = -\sum_{k=0}^{m} A_k^\mathsf{T} X_{j+k}(\tau). \tag{7}$$

Collecting the matrix equations (6) and (7) we arrive to the following set of delay free matrix equations:

$$\begin{aligned}
\Sigma_{k=0}^{m} X'_{j-k}(\tau) B_k &= \Sigma_{k=0}^{m} X_{j-k}(\tau) A_k, \quad j \geq 0, \\
\sum_{k=0}^{m} B_k^\mathsf{T} X'_{j+k}(\tau) &= -\sum_{k=0}^{m} A_k^\mathsf{T} X_{j+k}(\tau), \quad j < 0.
\end{aligned} \tag{8}$$

The vectorization of the above system can be written as

$$M_1 z'(\tau) = M_2 z(\tau),$$

or

$$z'(\tau) = M z(\tau)$$

with $M = M_1^{-1} M_2$, for details about the existence of M^{-1} see [2]. Here $M_1, M_2 \in \mathbb{R}^{2mn^2 \times 2mn^2}$ are defined as

$$M_1 = \begin{bmatrix}
\bar{B}_0 & \bar{B}_1 & \cdots & \bar{B}_m & 0 & \cdots & 0 & 0 \\
0 & \bar{B}_0 & \cdots & \bar{B}_{m-1} & \bar{B}_m & \cdots & 0 & 0 \\
\vdots & \vdots & \ddots & \vdots & \vdots & \ddots & \vdots & \vdots \\
0 & 0 & \cdots & 0 & 0 & \cdots & \bar{B}_{m-1} & \bar{B}_m \\
\bar{B}_0 & \bar{B}_1 & \cdots & \bar{B}_m & 0 & \cdots & 0 & 0 \\
0 & \bar{B}_0 & \cdots & \bar{B}_{m-1} & \bar{B}_m & \cdots & 0 & 0 \\
\vdots & \vdots & \ddots & \vdots & \vdots & \ddots & \vdots & \vdots \\
0 & 0 & \cdots & 0 & 0 & \cdots & \bar{B}_{m-1} & \bar{B}_m
\end{bmatrix}, M_2 = \begin{bmatrix}
\bar{A}_0 & \bar{A}_1 & \cdots & \bar{A}_m & 0 & \cdots & 0 & 0 \\
0 & \bar{A}_0 & \cdots & \bar{A}_{m-1} & \bar{A}_m & \cdots & 0 & 0 \\
\vdots & \vdots & \ddots & \vdots & \vdots & \ddots & \vdots & \vdots \\
0 & 0 & \cdots & 0 & 0 & \cdots & \bar{A}_{m-1} & \bar{A}_m \\
\bar{A}_0 & \bar{A}_1 & \cdots & \bar{A}_m & 0 & \cdots & 0 & 0 \\
0 & \bar{A}_0 & \cdots & \bar{A}_{m-1} & \bar{A}_m & \cdots & 0 & 0 \\
\vdots & \vdots & \ddots & \vdots & \vdots & \ddots & \vdots & \vdots \\
0 & 0 & \cdots & 0 & 0 & \cdots & \bar{A}_{m-1} & \bar{A}_m
\end{bmatrix}.$$

where $\bar{A}_j = \left(A_j^\mathsf{T} \otimes I\right)$, $\tilde{A}_j = -\left(I \otimes A_j^\mathsf{T}\right)$, $\bar{B}_j = \left(B_j^\mathsf{T} \otimes I\right)$ and $\tilde{B}_j = \left(I \otimes B_j^\mathsf{T}\right)$, $j = 0, 1, ..., m$. The initial conditions of the above system are not known, but we can find them by solving the following boundary value problem: a first set of boundary conditions follows from the definition of the auxiliary matrices (5):

$$X_{j+1}(0) = X_j(h), \quad j = -m, ..., 0, ..., m-2,$$

and an additional condition follows from expressing the algebraic property (4) in terms of the auxiliary matrices (5):

$$-W = \sum_{k=0}^{m} \left[B_k^\mathsf{T} X_0(0) A_k + A_k^\mathsf{T} X_0(0) B_k \right] + \sum_{j=0}^{m} \sum_{k=j+1}^{m} \left[B_k^\mathsf{T} X_{k-j-1}(h) A_j \right.$$

$$\left. + B_j^\mathsf{T} X_{k-j-1}^\mathsf{T}(h) A_k \right] + \sum_{j=0}^{m} \sum_{k=j+1}^{m} \left[A_j^\mathsf{T} X_{k-j-1}^\mathsf{T}(h) B_k + A_k^\mathsf{T} X_{k-j-1}(h) B_j \right].$$

4   Frequency Domain Stability Analysis

In this section the key relationship between the spectrum of the delay free system (8) and the roots of the characteristic function of system (1) is established, [6], [10].

Theorem 1. *Let s_0 be an eigenvalue of the neutral delay system (1), such that $-s_0$ is also an eigenvalue of system (1). Then the value s_0 belongs to the spectrum of the delay free system (8).*

Proof. The characteristic equation of system (1) is given by

$$G(s,h) = \sum_{k=0}^{m} \left[sB_k e^{-shk} - A_k e^{-shk} \right].$$

Since the values s_0 and $-s_0$ belong to the spectrum of the neutral delay system (1), then there exist non zero vectors μ and γ, such that

$$\gamma^{\mathsf{T}} G(s_0, h) = 0 \quad \text{and} \quad G^{\mathsf{T}}(-s_0, h)\mu = 0. \tag{9}$$

If a complex number s belongs to the spectrum of the delay free system (8) then there exists a set of $2m$ non trivial matrices X_j, $j = -m, ..., 0, ...m - 1$ such that the following holds:

$$\sum_{k=0}^{m} \left[sX_{j-k}B_k - X_{j-k}A_k \right] = 0, \quad j \geq 0, \tag{10}$$

$$\sum_{k=0}^{m} \left[sB_k^{\mathsf{T}} X_{j+k} + A_k^{\mathsf{T}} X_{j+k} \right] = 0, \quad j < 0.$$

Computing

$$e^{-s_0 h j} \mu \left[\gamma^{\mathsf{T}} G(s_0, h) \right] = 0, \quad j = 0, ..., m$$

one arrives to the expression

$$\mu \gamma^{\mathsf{T}} \sum_{k=0}^{m} \left[s_0 B_k e^{s_0 h(j-k)} - A_k e^{s_0 h(j-k)} \right] = 0 \tag{11}$$

for $j = 0, ..., m$. Now if we calculate

$$\left[G^{\mathsf{T}}(-s_0, h)\mu \right] \gamma^{\mathsf{T}} e^{s_0 h j} = 0, \quad j = -m, ..., -1$$

it follows that

$$\sum_{k=0}^{m} \left[s_0 B_k^{\mathsf{T}} e^{s_0 h(j+k)} + A_k^{\mathsf{T}} e^{s_0 h(j+k)} \right] \mu \gamma^{\mathsf{T}} = 0. \tag{12}$$

If we define the matrices

$$\bar{X}_l = e^{s_0 h l} \mu \gamma^{\mathsf{T}}, \quad l = -m, ..., 0, ..., m - 1,$$

the matrix equations (11)-(12) take the form

$$\sum_{k=0}^{m} s_0 \bar{X}_{j-k} B_k = \sum_{k=0}^{m} \bar{X}_{j-k} A_k, \quad j \geq 0,$$

$$\sum_{k=0}^{m} s_0 B_k^{\mathsf{T}} \bar{X}_{j+k} = -\sum_{k=0}^{m} A_k^{\mathsf{T}} \bar{X}_{j+k}, \quad j < 0.$$

It is clear that the matrices $\bar{X}_l$ are non trivial, then the value s_0 belongs to the delay free system (8). This procedure also holds for $-s_0$.

Proposition 1. *The spectrum of the delay free system (8) is symmetrical with respect to the imaginary axis.*

Proof. The fact that the spectrum of the delay free system of matrix equations (8) is symmetrical with respect of the imaginary axis follows from the idea that if for a value s there exist non trivial matrices X_i $i = -m, ..., 0, ... m - 1$ that satisfy system (10). By tranposing and multiplying by (-1) system (10), it follows that

$$\sum_{k=0}^{m} \left[-s B_k^{\mathsf{T}} X_{j-k}^{\mathsf{T}} + A_k^{\mathsf{T}} X_{j-k}^{\mathsf{T}} \right] = 0,$$

$$\sum_{k=0}^{m} \left[-s X_{j+k}^{\mathsf{T}} B_k - X_{j+k}^{\mathsf{T}} A_k \right] = 0,$$

and by setting the matrices

$$\tilde{X}_j = X_{-j-1}^{\mathsf{T}}, \quad j = -m, ..., 0, ..., m - 1$$

we conclude that the system (10) is satisfied for $-s$.

5 Proposed Methodology

In this section two different methodologies are proposed: The first one applies to systems with an unknown parameter in the matrices of the system. The second methodology applies to systems where the only unknown parameter is the delay h.

Methodology A

Here we consider the class of systems that are of the form

$$\sum_{k=0}^{m} B_k \dot{x}(t - hk) = \sum_{k=0}^{m} \tilde{A}_k x(t - hk).$$

where $\tilde{A}_k = k\mu A_k$, μ is an unknown parameter that satisfies $\mu > 0$. The methodology for the computation of the critical parameters is as follows:

1. Calculate the corresponding delay free system
2. Compute the characteristic polynomial $p(s,\rho)$ of the delay free system (8). The value ρ represent the delay h or the unknown parameter μ (or both)
3. Set $\lambda = s^2$ and analyze the auxiliary polynomial $p_1(\lambda,\rho)$. Indeed, as the spectrum of the system is symmetrical with respect to the imaginary axis, all the powers of the polynomial are even. Hence, determining the purely imaginary roots of $p(s,\rho)$ reduces to finding the negative real roots of $p_1(\lambda,\rho)$. This is achieved using Sturm's Theorem, see [15]

   - Compute the Sturm sequence of the polynomial $p_1(\lambda,\rho)$
   - Determine with the help of Sturm's Theorem if $p_1(\lambda,\rho)$ has negative real roots

4. If there exist negative real roots compute the corresponding purely imaginary roots and form a set of candidate roots for the delay system
5. For each of these candidate roots, check if there exists a value of the parameter ρ that annihilate the characteristic function. If this is the case then the pair (ω_i^*,ρ_i^*), receives the name of critical frequency and critical parameter
6. Form a table of ρ_i^* in ascending order (consider $\rho_0 = 0$)
7. To determine if an interval $\left[\rho_i^*,\rho_{i+1}^*\right]$ corresponds to a stable or a unstable region, select $\tilde{\rho} \in \left[\rho_i^*,\rho_{i+1}^*\right]$, ω_i^* and using the argument principle determine the number of unstable roots of the characteristic function.

Methodology B

For this methodology we consider the class of neutral type time delay systems of the form (1), the methodology consists of the following steps:

a. The steps (1)-(3) of Methodology A, remain valid
b. Compute the negative real roots and the corresponding imaginary roots
c. Make the change of variable $z = e^{-sh}$
d. Evaluate the characteristic function for $s = i\omega$ obtained in step (b)
e. Compute the roots of polynomial $f(z)$ and select those satisfying $|z| = 1$
f. Compute the critical delays associated to the critical frequency ω^*, that is: $h^* = \frac{-Arg(z)}{\omega^*}$
g. Continue with Steps (8)-(9) of Methodology A for the parameter h.

6 Examples

In this section we illustrate the use of the proposed methodologies for the computation of critical parameters.

Example 1. Let us consider the following system studied in [3], [14]

$$\dot{x}(t) + B_1\dot{x}(t-h) = A_0x(t) + A_1x(t-h)$$

where

$$A_0 = \begin{bmatrix} -2 & 0 \\ 0 & -0.9 \end{bmatrix}, \quad A_1 = \begin{bmatrix} -1 & 0 \\ -1 & -1 \end{bmatrix}, \quad B_1 = \begin{bmatrix} -0.1 & 0 \\ 0 & -0.1 \end{bmatrix}.$$

Since $\rho(B_1) = 0.1 < 1$ the difference operator is stable. In order to find the critical delays, we first compute the characteristic polynomial of the delay free system

$$p(s) = s^8 - 5.6891s^6 + 8.1628s^4 - 0.1955s^2 - 0.3797.$$

By setting $\lambda = s^2$, and analyzing the auxiliary polynomial $p_1(\lambda)$ we can directly check that the only negative real root is $\lambda = -\frac{19}{99}$ which corresponds to the purely imaginary root $s = i\frac{\sqrt{11}\sqrt{19}}{33}$. Now, the characteristic function of the delay system is

$$f(s,h) = \det(sI + sB_1 e^{-sh} - A_0 - A_1 e^{-sh}).$$

Evaluating $f(s,h)$ for $s = i\omega$ and the change of variable $z = e^{-sh}$; $f(s,h)$ takes the form

$$f(z) = z^2(0.998 - 0.0087i) + z(2.9383 + 0.7491i) + (1.608 + 1.2704i),$$

with roots

$$z_1 = -0.8791 - 0.4765i, \quad z_2 = -1.9770 - 0.5246i.$$

Now we are looking for roots $z_{1,2}$ that satisfy $|z| = 1$; the only root that satisfies this condition is z_1. The critical delay is computed as $h^* = 6.0371$. Finally with the help of the argument principle we can conclude that the system is stable for $h \in [0, 6.0371)$ and is unstable for $h \in (6.0371, \infty)$.

Example 2. In this example we consider a system with an unknown parameter μ that satisfies $\mu > 0$. The system is of the form

$$\dot{x}(t) + B_1\dot{x}(t-1) = \mu A_0 x(t) + A_1 x(t-1)$$

where A_0, A_1 and B_1 are the same as in the previous example. The characteristic polynomial of the corresponding delay free system is

$$p(s,\mu) = \frac{1}{99^4} q(s,\mu) r(s,\mu) v(s,\mu)$$

where

$$q(s,\mu) = 99s^2 - 81\mu^2 + 100$$
$$r(s,\mu) = 99s^2 - 400\mu^2 + 100$$
$$v(s,\mu) = 9801s^4 + s^2(-47740\mu^2 + 19800) + (32400\mu^4 - 36000\mu^2 + 10000)$$

With the change of variable $\lambda = s^2$ we can directly check that the auxiliary polynomial $q_1(\lambda, \mu)$ has one negative real root for $\mu \in [0, \frac{10}{9})$, while the auxiliary polynomial $r_1(\lambda, \mu)$ has one negative root for $\mu \in [0, \frac{1}{2})$. With the help of the Sturm array we can conclude that the auxiliary polynomial $v_1(\lambda,\mu)$ does not have negative real roots $\forall \mu > 0$.

The characteristic function of the system is of the form

$$f(s,\mu) = \det\left(sI + sB_1 e^{-s} - \mu A_0 - A_1 e^{-s}\right).$$

We must check if there exist pairs $\{i\omega_k(\mu),\mu\}$ $k = 1,2$ that annihilate the characteristic function of the system. We can proove, by direct calculations that for $\mu_1 = 0.4615$ and $\omega_1^* = 0.3867$, see Figure 1(a), and for $\omega_2^* = 0.3866$ and $\mu_2 = 1.0256$, see Figure 1(b), the real and imaginary parts vanish simultaneously.

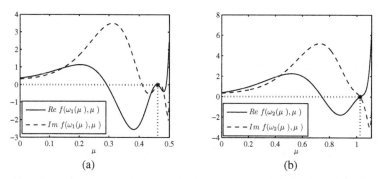

Fig. 1 (a) Real and Imaginary parts of $f(i\omega_1,\mu)$ (b) Real and Imaginary parts of $f(i\omega_2,\mu)$

Example 3. Now let us consider the system introduced in [11], [4],

$$\dot{x}(t) + B_1\dot{x}(t-h) + B_2\dot{x}(t-2h) = A_0 x(t),$$

where

$$A_0 = \begin{bmatrix} -5.5779 & 5.0852 & 3.5478 \\ -1.7151 & 2.2988 & 0.8782 \\ 0.4407 & 3.2284 & -1.3174 \end{bmatrix}, B_1 = \begin{bmatrix} 0 & -0.2 & 0.4 \\ 0.5 & -0.3 & 0 \\ -0.2 & -0.7 & 0 \end{bmatrix}, B_2 = \begin{bmatrix} 0.3 & 0.1 & 0 \\ 0 & -0.2 & 0 \\ -0.1 & 0 & -0.4 \end{bmatrix},$$

The expression of the characteristic polynomial of this delay free system, which is of degree 32, is omitted due to space limitations. With the change of variable $\lambda = s^2$ the polynomial $p_1(\lambda)$ is obtained. We can verify that this polynomial has a pair of negative real roots $\lambda_1 = -0.8807$ and $\lambda_2 = -1.3665 \times 10^{-4}$, with corresponding purely imaginary roots $s_1 = 0.9384i$, $s_2 = 0.0116i$. The characteristic function of the system is of the form

$$f(s,h) = \det\left(sI + sB_1 e^{-sh} + sB_2 e^{-2sh} - A_0\right).$$

With the change of variable $z = e^{-sh}$ and considering $s_1 = 0.9384i$, $f(s,h)$ takes the form

$$f(z) = -(0.0198i)z^6 - (0.0396i)z^5 - (0.6969 - 0.1388i)z^4 + (0.6452 + 0.119i)z^3 + (2.9265 - 0.9274i)z^2 + (0.2915 + 6.5017i)z - (3.9946 + 4.7489i).$$

The only root z_i that satisfies $|z| = 1$ is $z_1 = 0.9839 - 0.1785i$, then the critical delay is $h_1^* = 0.1912$.

If we compute $f(s,z)$ for $i\omega_2 = 0.0116i$, it follows that the polynomial $f(z)$ has the roots

$$z_1 = -9.6132 + 7.6772i, \ z_2 = -39.8206 - 38.2932i, \ z_3 = 0.5582 + 0.8296i,$$
$$z_4 = 36.8117 + 38.2960i, \ z_5 = 5.3012 + 0.07.574i, \quad z_6 = 4.7625 - 8.5854i.$$

The only root that satisfies $|z| = 1$ is z_2 then $h_2^* = 453.7893$. The argument principle shows that the only interval where the system is stable is for $h \in [0, 0.1912)$.

7 Conclusions

A procedure for the computation of the critical parameters of time delay systems of neutral type is presented. The approach takes advantage of the key relation between the roots of the characteristic polynomial of a delay free system of matrix equations and the roots of the characteristic function of the neutral type time delay system. Some illustrative examples show the efectiveness of the method for determining the critical delays as well as other critical parameters.

References

1. Bellman, R., Cooke, K.L.: Differential-Difference Equations. Academic Press, New York (1963)
2. Gohberg, C.I., Lerer, L.E.: Resultants of matrix polynomials. Bull. Amer. Math. Soc. 82, 565–567 (1976)
3. Ivanescu, D., Niculescu, S.I., Dugard, L., Dion, J.M., Verriest, E.I.: On delay-dependent stability for linear neutral systems. Automatica 39, 255–261 (2003)
4. Jarlebring, E.: On critical delays for linear neutral delay systems. In: European Control Conference, Kos, Greece (2007)
5. Kharitonov, V.L.: Lyapunov functionals and lyapunov matrices for neutral type time delay systems: a single delay case. International Journal of Control 78, 783–800 (2005)
6. Kharitonov, V.L.: Lyapunov matrices for a class of time delay systems. Systems and Control Letters 55, 610–617 (2006)
7. Kharitonov, V.L.: Lyapunov matrices: Existence and uniqueness issues. In: 8th Workshop on Time Delay Systems, Sinaia, Romania (2009)
8. Kharitonov, V.L., Mondié, S., Ochoa, G.: Frequency Stability Analysis of Linear Systems With General Distributed Delays. In: Loiseau, J.J., Michiels, W., Niculescu, S.-I., Sipahi, R. (eds.) Topics in Time Delay Systems. LNCIS, vol. 388, pp. 25–36. Springer, Heidelberg (2009)
9. Louisell, J.: Numerics of the Stability Exponent and Eigenvalue Abscissas of a Matrix Delay System. In: Dugard, L., Verriest, E.I. (eds.). LNCIS, vol. 228, pp. 140–157. Springer, Heidelberg (1997)

10. Louisell, J.: A matrix method for determining the imaginary axis eigenvalues of a delay system. IEEE Transactions on Automatic Control 46, 2008–2012 (2001)
11. Michiels, W., Vyhlídal, T.: An eigenvalue based approach for the stabilization of linear time-delay systems of neutral type. Automatica 41, 991–998 (2005)
12. Ochoa, G., Velázquez, J.E., Kharitonov, V.L., Mondié, S.: Lyapunov Matrices for Neutral type Time Delay Systems. In: Loiseau, J.J., Michiels, W., Niculescu, S.-I., Sipahi, R. (eds.) Topics in Time Delay Systems. LNCIS, vol. 388, pp. 61–71. Springer, Heidelberg (2009)
13. Olgac, N., Sipahi, R.: A practical method for analyzing the stability of neutral type lti-time delayed systems. Automatica 40, 847–853 (2004)
14. Rodriguez, S.A., Kharitonov, V.L., Dion, J.M., Dugard, L.: Robust stability of neutral systems: a lyapunov-krasovskii constructive approach. International Journal of Robust and Nonlinear Control 14, 1345–1358 (2004)
15. Uspensky, J.V.: Theory of Equations. McGraw-Hill, New York (1948)
16. Walton, K.E., Marshall, J.E.: Direct method for tds stability analysis. IEEE Proceedings, Pt. D 2, 101–107 (1987)

Set-Induced Stability Results
for Delay Difference Equations

Rob H. Gielen, Mircea Lazar, and Sorin Olaru

Abstract. This chapter focuses on the relation between stability of delay difference equations (DDEs) and the existence of $\mathscr{D}$-contractive sets. Such sets are of importance as they provide a region of attraction, which is difficult to obtain for delay systems. Firstly, it is established that a DDE admits a $\mathscr{D}$-contractive set if and only if it admits a Lyapunov-Razumikhin function. However, it is also shown that there exist stable DDEs that do not admit a $\mathscr{D}$-contractive set. Therefore, secondly, further necessary conditions for the existence of a $\mathscr{D}$-contractive set are established. These necessary conditions provide a first step towards the derivation of a notion of asymptotic stability for DDEs which is equivalent to the existence of a $\mathscr{D}$-contractive set.

1 Introduction

The existence of invariant and contractive sets for discrete-time systems, see, e.g., [3, 4, 9], is an important topic in the field of control theory. For example, invariant sets are crucial [16] in the study of stability for model predictive control (MPC) schemes. As the sub-level sets of a Lyapunov function are contractive sets, Lyapunov theory is frequently used to study the existence of such sets. Indeed, as a dynamical system is asymptotically stable if and only if it admits a Lyapunov function, it follows that any asymptotically stable system admits a non-trivial family of contractive sets. Moreover, it was also shown [2, 15] that a system that admits a contractive set is (under suitable assumptions) asymptotically stable.

R.H. Gielen · M. Lazar
Electrical Engineering Department, Eindhoven University of Technology, Eindhoven, The Netherlands
e-mail: {r.h.gielen,m.lazar}@tue.nl

S. Olaru
SUPELEC Systems Sciences (E3S) - Automatic Control Department and DISCO - INRIA, Gif-sur-Yvette, France
e-mail: sorin.olaru@supelec.fr

R. Sipahi et al. (Eds.): Time Delay Sys.: Methods, Appli. and New Trends, LNCIS 423, pp. 73–84.
springerlink.com   © Springer-Verlag Berlin Heidelberg 2012

While systems affected by delay can be found in many applications in the control field, see, e.g., [11], the existence of contractive sets for such systems remains scarcely studied. As most modern controllers are implemented via a computer and hence delay difference equations (DDEs) form an important modeling class, this chapter focusses on the existence of contractive sets for DDEs. The most straightforward approach [1] to obtain a contractive set for a DDE is to augment the state vector with all delayed states that affect the current state, which yields a standard difference equation of higher dimension. Thus, results on contractive sets for standard difference equations, see, e.g., [4], can be applied directly. However, these techniques do not provide a contractive set in the state space of the original DDE, but rather one in the higher dimensional state space of the augmented system. To obtain a region of attraction for a DDE, which is also useful to design stabilizing MPC schemes [17], a contractive set in the state space of the original DDE, which is called a $\mathscr{D}$-contractive set in this chapter, is required. Sufficient conditions for the existence of a $\mathscr{D}$-contractive set were first obtained in [5, 8]. Then, in [10, 18], a necessary and sufficient characterization for the existence of a $\mathscr{D}$-contractive set was provided. More recently, the construction of $\mathscr{D}$-contractive sets and the verification of $\mathscr{D}$-contractiveness was considered in [13, 14], respectively. However, all of the above results are limited to polytopic sets and none of them relates the existence of a $\mathscr{D}$-contractive set to asymptotic stability of the corresponding DDE.

As such, in this chapter, the relation between asymptotic stability and the existence of a $\mathscr{D}$-contractive set is investigated for DDEs. Firstly, it is established that a DDE admits a $\mathscr{D}$-contractive set if and only if it admits a Lyapunov-Razumikhin function (LRF). Moreover, for linear DDEs, algebraic conditions are derived which are equivalent to the existence of a polytopic $\mathscr{D}$-contractive set and the existence of a polyhedral LRF. Then, an example of an asymptotically stable DDE that does not admit a $\mathscr{D}$-contractive set is provided. Therefore, alternative necessary conditions for the existence of a $\mathscr{D}$-contractive set are derived. However, it is also shown that, while these necessary conditions provide valuable insights, they are not sufficient. Therefore, for linear DDEs, a notion of asymptotic stability, called $\mathscr{D}$-$\mathscr{K}\mathscr{L}$-stability, is proposed and it is shown that if a DDE admits a polytopic $\mathscr{D}$-contractive set then it is $\mathscr{D}$-$\mathscr{K}\mathscr{L}$-stable. This result provides a first step towards the derivation of a notion of asymptotic stability for DDEs that is equivalent to the existence of a $\mathscr{D}$-contractive set.

A limited part of the results presented in this chapter appeared in [6, 7] in the context of Lyapunov theory. These results are reproduced here in terms of contractive sets, which allows us to make several important connections.

2 Preliminaries

Let $\mathbb{R}$, $\mathbb{R}_+$, $\mathbb{Z}$, $\mathbb{Z}_+$ and $\mathbb{C}$ denote the field of real numbers, the set of non-negative reals, the set of integers, the set of non-negative integers and the field of complex numbers, respectively. For every $c \in \mathbb{R}$ and $\Pi \subseteq \mathbb{R}$, define $\Pi_{\geq c} := \{k \in \Pi \mid k \geq c\}$ and similarly $\Pi_{\leq c}$. Furthermore, $\mathbb{R}_\Pi := \Pi$ and $\mathbb{Z}_\Pi := \mathbb{Z} \cap \Pi$. Let $\|\cdot\|$ denote an arbitrary norm and let $|\cdot|$ denote the absolute value. For a vector $x \in \mathbb{R}^n$ let $[x]_i$, $i \in \mathbb{Z}_{[1,n]}$, denote the i-th component of x and let $\|x\|_\infty := \max_{i \in \mathbb{Z}_{[1,n]}} |[x]_i|$. Let $\mathbf{x} := \{x(l)\}_{l \in \mathbb{Z}_+}$ with $x(l) \in \mathbb{R}^n$ for all $l \in \mathbb{Z}_+$ denote an arbitrary sequence and define $\|\mathbf{x}\| := \sup\{\|x(l)\| \mid l \in \mathbb{Z}_+\}$. Furthermore, $\mathbf{x}_{[c_1,c_2]} := \{x(l)\}_{l \in \mathbb{Z}_{[c_1,c_2]}}$, with $c_1, c_2 \in \mathbb{Z}$, denotes a sequence which is ordered monotonically with respect to the index $l \in \mathbb{Z}_{[c_1,c_2]}$. Let $\mathrm{co}(\cdot)$ denote the convex hull. Let $\mathrm{int}(\mathbb{S})$ and $\partial\mathbb{S}$ denote the interior and the boundary of an arbitrary set $\mathbb{S}$, respectively. Moreover, $\mathbb{S}^h := \mathbb{S} \times \ldots \times \mathbb{S}$ for any $h \in \mathbb{Z}_{\geq 1}$ is the h-times cross-product of $\mathbb{S}$. A polyhedron is a set obtained as the intersection of a finite number of half-spaces. A polytope is a compact polyhedron. Throughout this chapter, polytopic sets $\mathscr{S} := \{x \in \mathbb{R}^n \mid \|Fx\|_\infty \leq 1\}$, for some $F \in \mathbb{R}^{p \times n}$ with $p \in \mathbb{Z}_{\geq n}$ and $\mathrm{rank}(F) = n$, are considered. Let I_n denote the n-th dimensional identity matrix and let $1_n \in \mathbb{R}^n$ denote a vector with all elements equal to 1. For a matrix $Z \in \mathbb{R}^{n \times n}$, let $\lambda(Z) \subset \mathbb{C}$ denote the set of all eigenvalues of Z. Let $\rho(Z) := \max_{s \in \lambda(Z)} \sqrt{ss^\dagger}$, with $s^\dagger$ the complex conjugate of s, denote the spectral radius of Z. The matrix Z is called nonnegative if all elements of Z are nonnegative. A function $f : \mathbb{R}_+ \to \mathbb{R}_+$ belongs to class $\mathscr{K}$, i.e., $f \in \mathscr{K}$, if it is continuous, strictly increasing and $f(0) = 0$. Moreover, $f \in \mathscr{K}_\infty$ if $f \in \mathscr{K}$ and $\lim_{s \to \infty} f(s) = \infty$. Furthermore, a function $\beta : \mathbb{R}_+ \times \mathbb{R}_+ \to \mathbb{R}_+$ belongs to class $\mathscr{K}\mathscr{L}$, i.e., $\beta \in \mathscr{K}\mathscr{L}$, if for each $s \in \mathbb{R}_+$, $\beta(r,s) \in \mathscr{K}$ with respect to r and for each $r \in \mathbb{R}_+$, $\beta(r,s)$ is continuous and decreasing with respect to s and $\lim_{s \to \infty} \beta(r,s) = 0$. A function $f : \mathbb{R}^n \to \mathbb{R}^m$ is called homogeneous (positively homogeneous) of order t, $t \in \mathbb{Z}$, if $g(sx) = s^t g(x)$ (if $g(sx) = |s|^t g(x)$) for all $x \in \mathbb{R}^n$, $s \in \mathbb{R}$.

2.1 Delay Difference Equations

Consider the DDE

$$x(k+1) = F(\mathbf{x}_{[k-h,k]}), \quad k \in \mathbb{Z}_+, \tag{1}$$

where $\mathbf{x}_{[k-h,k]} \in (\mathbb{R}^n)^{h+1}$, $h \in \mathbb{Z}_+$ is the maximal delay and $F : (\mathbb{R}^n)^{h+1} \to \mathbb{R}^n$ is an arbitrary nonlinear function. It is assumed that F has the origin as equilibrium point, i.e., $F(\mathbf{0}_{[-h,0]}) = 0$.

Definition 1. *(i)* The DDE (1) is called $\mathscr{D}$-homogeneous of order t, $t \in \mathbb{Z}$, if for any $s \in \mathbb{R}$ it holds that $F(s\tilde{\mathbf{x}}_{[-h,0]}) = s^t F(\tilde{\mathbf{x}}_{[-h,0]})$ for all $\tilde{\mathbf{x}}_{[-h,0]} \in (\mathbb{R}^n)^{h+1}$; *(ii)* The DDE (1) is called linear if $F(\tilde{\mathbf{x}}_{[-h,0]}) = \sum_{\theta=-h}^{0} A_\theta \tilde{x}(\theta)$, for some $A_\theta \in \mathbb{R}^{n \times n}$, $\theta \in \mathbb{Z}_{[-h,0]}$, and for all $\tilde{\mathbf{x}}_{[-h,0]} \in (\mathbb{R}^n)^{h+1}$. $\qquad\square$

In this chapter, $\tilde{\mathbf{x}}_{[-h,0]}$ is used to denote an arbitrary sequence in $(\mathbb{R}^n)^{h+1}$ and to distinguish from the initial condition $\mathbf{x}_{[-h,0]}$ of (1). Moreover, given the initial condition $\mathbf{x}_{[-h,0]}$, let $\{x(k, \mathbf{x}_{[-h,0]})\}_{k \in \mathbb{Z}_{\geq -h}}$ denote a trajectory of (1). Let $x(k, \mathbf{x}_{[-h,0]}) = x(k)$ for all $k \in \mathbb{Z}_{[-h,0]}$, $\mathbf{x}_{[k-h,k]}(\mathbf{x}_{[-h,0]}) := \{x(l, \mathbf{x}_{[-h,0]})\}_{l \in \mathbb{Z}_{[k-h,k]}}$ and $x(k+1, \mathbf{x}_{[-h,0]}) = F(\mathbf{x}_{[k-h,k]}(\mathbf{x}_{[-h,0]})) \; \forall k \in \mathbb{Z}_+$.

Definition 2. Let $\lambda \in \mathbb{R}_{[0,1)}$. A bounded set $\mathbb{V} \subset \mathbb{R}^n$ with $0 \in \mathrm{int}(\mathbb{V})$ is called λ-$\mathscr{D}$-contractive for system (1) if $F(\tilde{\mathbf{x}}_{[-h,0]}) \in \lambda\mathbb{V}$ for all $\tilde{\mathbf{x}}_{[-h,0]} \in \mathbb{V}^{h+1}$. For $\lambda = 1$, a λ-$\mathscr{D}$-contractive set is called a $\mathscr{D}$-invariant set. $\qquad\square$

$\mathscr{D}$-invariant sets were also considered in [13]. $\mathscr{D}$-invariant and $\mathscr{D}$-contractive sets recover the definitions of classical invariant and contractive sets for systems without delay, i.e., with $h = 0$.

Definition 3. Let $\mathbb{X} \subseteq \mathbb{R}^n$. The DDE (1) is called $\mathscr{K}\mathscr{L}$-stable in $\mathbb{X}$ if there exists a function $\beta : \mathbb{R}_+ \times \mathbb{R}_+ \to \mathbb{R}_+$, $\beta \in \mathscr{K}\mathscr{L}$, such that $\|x(k, \mathbf{x}_{[-h,0]})\| \leq \beta(\|\mathbf{x}_{[-h,0]}\|, k)$ for all $\mathbf{x}_{[-h,0]} \in \mathbb{X}^{h+1}$ and all $k \in \mathbb{Z}_+$. $\qquad\square$

Theorem 1 [6], Theorem III.8). *Suppose that* $F(\tilde{\mathbf{x}}_{[-h,0]}) \in \mathbb{X}$ *for all* $\tilde{\mathbf{x}}_{[-h,0]} \in \mathbb{X}^{h+1}$. *Let* $\alpha_1, \alpha_2 \in \mathscr{K}_\infty$ *and* $\rho \in \mathbb{R}_{[0,1)}$. *If there exists a function* $V : \mathbb{R}^n \to \mathbb{R}_+$ *such that*

$$\alpha_1(\|x\|) \leq V(x) \leq \alpha_2(\|x\|), \quad \forall x \in \mathbb{X}, \tag{2a}$$
$$V(F(\tilde{\mathbf{x}}_{[-h,0]})) \leq \rho \max_{\theta \in \mathbb{Z}_{[-h,0]}} V(\tilde{x}(\theta)), \tag{2b}$$

for all $\tilde{\mathbf{x}}_{[-h,0]} \in \mathbb{X}^{h+1}$, *then the DDE* (1) *is* $\mathscr{K}\mathscr{L}$*-stable in* $\mathbb{X}$. $\qquad\square$

A function that satisfies the hypothesis of Theorem 1 is called a Lyapunov-Razumikhin function in $\mathbb{X}$ or shortly, LRF($\mathbb{X}$).

3 $\mathscr{D}$-Contractive Sets and $\mathscr{K}\mathscr{L}$-Stability

For standard difference equations the existence of a contractive set is (under suitable assumptions) equivalent, see, e.g., [2, 12, 15], to the existence of a Lyapunov function. Next, a similar equivalence is established with respect to the existence of a $\mathscr{D}$-contractive set and a LRF.

Theorem 2. *Suppose that the DDE* (1) *is* $\mathscr{D}$*-homogeneous of order* t_1, *for some* $t_1 \in \mathbb{Z}_{\geq 1}$, *and let* $\lambda \in \mathbb{R}_{[0,1)}$. *The following statements are equivalent:*

(i) The DDE (1) *admits a* λ-$\mathscr{D}$*-contractive set;*
(ii) The DDE (1) *admits a LRF($\mathbb{X}$), for some* $\mathbb{X} \subseteq \mathbb{R}^n$, *that is positively homogeneous of order* t_2, *for some* $t_2 \in \mathbb{Z}_{\geq 1}$.

Furthermore, if $t_1 = 1$, *then claim (ii) holds with* $\mathbb{X} = \mathbb{R}^n$.

Proof. First, (i)$\Rightarrow$(ii) is proven. Let $\mathbb{V}$ denote a λ-$\mathscr{D}$-contractive set for the DDE (1) and consider the *Minkowski function*, see, e.g., [4], of $\mathbb{V}$, i.e.,

$$V(x) := \inf\{\mu \in \mathbb{R}_+ \mid x \in \mu \mathbb{V}\}. \tag{3}$$

Obviously, for any $\tilde{\mathbf{x}}_{[-h,0]} \in \mathbb{V}^{h+1}$ it holds that $F(\tilde{\mathbf{x}}_{[-h,0]}) \in \mathbb{V}$. Furthermore, letting $a_1 := \max_{x \in \mathbb{V}} \|x\| > 0$ and $a_2 := \min_{x \in \partial \mathbb{V}} \|x\| > 0$ yields

$$a_1^{-1}\|x\| \leq V(x) \leq a_2^{-1}\|x\|.$$

Next, consider any $\nu \in \mathbb{R}_{(0,1)}$ and let $\tilde{\mathbf{x}}_{[-h,0]} \in (\nu \mathbb{V})^{h+1}$. Then, $\nu^{-1}\tilde{\mathbf{x}}_{[-h,0]} \in \mathbb{V}^{h+1}$ and therefore $F(\nu^{-1}\tilde{\mathbf{x}}_{[-h,0]}) \in \lambda \mathbb{V}$. Moreover, as (1) is $\mathscr{D}$-homogeneous of order t_1

it follows that $F(\tilde{\mathbf{x}}_{[-h,0]}) = v^{t_1}F(v^{-1}\tilde{\mathbf{x}}_{[-h,0]}) \in v^{t_1}\lambda\mathbb{V} \subseteq \lambda(v\mathbb{V})$, where it was used that $v^{t_1} \leq v$. Hence, if the set $\mathbb{V}$ is λ-$\mathcal{D}$-contractive, then $v\mathbb{V}$ is λ-$\mathcal{D}$-contractive for any $v \in \mathbb{R}_{(0,1)}$. Therefore, for all $\tilde{\mathbf{x}}_{[-h,0]} \in \mathbb{V}^{h+1}$

$$V(F(\tilde{\mathbf{x}}_{[-h,0]})) = \inf\{\mu \in \mathbb{R}_+ \mid F(\tilde{\mathbf{x}}_{[-h,0]}) \in \mu\mathbb{V}\}$$
$$\leq \inf\{\mu \mid \max_{\theta \in \mathbb{Z}_{[-h,0]}} \tilde{x}(\theta) \in \mu(\lambda^{-1}\mathbb{V})\} = \lambda \max_{\theta \in \mathbb{Z}_{[-h,0]}} V(\tilde{x}(\theta)).$$

Therefore, the candidate LRF (3) satisfies the hypothesis of Theorem 1 with $\alpha_1(s) := a_1^{-1}s \in \mathcal{K}_\infty$, $\alpha_2(s) := a_2^{-1}s \in \mathcal{K}_\infty$, $\rho := \lambda \in \mathbb{R}_{[0,1)}$ and $\mathbb{X} := \mathbb{V} \subseteq \mathbb{R}^n$. Observing that $V(sx) = sV(x)$ for all $s \in \mathbb{R}_+$ holds, establishes that (i)$\Rightarrow$(ii).

Next, we prove that (ii)$\Rightarrow$(i). Therefore, consider an admissible sublevel set of V, e.g., $\mathbb{V}_\gamma := \{x \in \mathbb{R}^n \mid V(x) \leq \gamma\}$, with $\gamma \in \mathbb{R}_{>0}$ such that $\mathbb{V}_\gamma \subseteq \mathbb{X}$. If $\max_{\theta \in \mathbb{Z}_{[-h,0]}} V(\tilde{x}(\theta)) \leq \gamma$ then it follows from (2b) that $V(F(\tilde{\mathbf{x}}_{[-h,0]})) \leq \rho\gamma$. Hence, as V is positively homogeneous it follows that $V(\rho^{-\frac{1}{t_2}}F(\tilde{\mathbf{x}}_{[-h,0]})) \leq \gamma$, which yields $F(\tilde{\mathbf{x}}_{[-h,0]}) \in \rho^{\frac{1}{t_2}}\mathbb{V}_\gamma$ for all $\tilde{\mathbf{x}}_{[-h,0]} \in \mathbb{V}_\gamma^{h+1}$. Therefore, the set $\mathbb{V}_\gamma$ is λ-$\mathcal{D}$-contractive with $\lambda := \rho^{\frac{1}{t_2}} \in \mathbb{R}_{[0,1)}$ for the DDE (1).

The last claim of the theorem follows from the observation that in the proof of (i)$\Rightarrow$(ii), $v \in \mathbb{R}_{>0}$ instead of $v \in \mathbb{R}_{(0,1)}$ is a valid choice if $t_1 = 1$. $\square$

Moreover, for *linear DDEs* and *polytopic sets*, algebraic conditions can be derived that are equivalent to the existence of a $\mathcal{D}$-contractive set.

Theorem 3. *Let* $\lambda \in \mathbb{R}_{[0,1)}$ *and suppose that the DDE* (1) *is linear. Furthermore, let* $F \in \mathbb{R}^{p\times n}$ *with* $p \in \mathbb{Z}_{\geq n}$ *and such that* $\mathrm{rank}(F) = n$. *The following three statements are equivalent:*

(i) $\mathbb{V} = \{x \in \mathbb{R}^n \mid \|Fx\|_\infty \leq 1\}$ is λ-$\mathcal{D}$-contractive for the DDE (1)*;*
(ii) $V(x) = \|Fx\|_\infty$ is a LRF$(\mathbb{R}^n)$ for the DDE (1)*;*
(iii)   There exist nonnegative matrices $H_\theta \in \mathbb{R}^{p\times p}$*,* $\theta \in \mathbb{Z}_{[-h,0]}$*, such that*

$$\left(\Sigma_{\theta=-h}^0 H_\theta\right)1_p \leq \lambda 1_p, \quad FA_\theta = H_\theta F, \quad \forall\theta \in \mathbb{Z}_{[-h,0]}.$$

Proof. The equivalence (i)$\Leftrightarrow$(iii) can be obtained *mutatis mutandis* from the proof of Theorem 3 in [10]. The equivalence (i)$\Leftrightarrow$(ii) can be proven via the techniques used in the proof of Theorem 2. $\square$

The algebraic conditions in Theorem 3 can be used to verify, using convex optimization algorithms, if a given polyhedral set is $\mathcal{D}$-contractive. Unfortunately, in terms of constructing a $\mathcal{D}$-contractive set they yield a non-convex optimization problem. In fact, to the best of the authors' knowledge, a computationally tractable method to obtain a $\mathcal{D}$-contractive set was only recently proposed in [13]. Therein, an iterative algorithm based on set-operations for computing polytopic $\mathcal{D}$-contractive sets was provided.

Remark 1. For simplicity of exposition, the results in this chapter, for linear systems, are restricted to symmetric sets. However, these results can be extended *mutatis mutandis* to asymmetric sets as done in, e.g., [9].   □

For standard difference equations the existence of a contractive set is equivalent to asymptotic stability. From Theorem 2 and Theorem 1 it follows that if (1) admits a $\mathscr{D}$-contractive set then the DDE (1) is $\mathscr{KL}$-stable. A similar conclusion can be drawn from Theorem 3 and Theorem 1. In what follows, an example is presented which establishes that the converse does not hold, i.e., there exist DDEs that are $\mathscr{KL}$-stable and do not admit a $\mathscr{D}$-contractive set. Therefore, consider the linear DDE

$$x(k+1) = x(k) - 0.5x(k-1), \quad k \in \mathbb{Z}_+, \tag{4}$$

where $\mathbf{x}_{[k-1,k]} \in \mathbb{R} \times \mathbb{R}$ is the system state.

Proposition 1. *Consider any $\lambda \in \mathbb{R}_{[0,1)}$. The following statements hold:*

(i)   *The linear DDE (4) is $\mathscr{KL}$-stable in $\mathbb{R}$;*
(ii) *The linear DDE (4) does not admit a LRF($\mathbb{R}$);*
(iii) *The linear DDE (4) does not admit a λ-$\mathscr{D}$-contractive set.*

Proof. Let $\xi(0) := [x(0), x(-1)]^\top$, which yields

$$\xi(k+1) = \bar{A}\xi(k), \quad k \in \mathbb{Z}_+, \tag{5}$$

where $\bar{A} = \begin{bmatrix} 1 & -0.5 \\ 1 & 0 \end{bmatrix}$. As $\rho(\bar{A}) < 1$, there exists a $\bar{C} \in \mathbb{R}_{>0}$ such that $\|\xi(k, \xi(0))\| \leq \bar{C}\rho(\bar{A})^k \|\xi(0)\|$ for all $\xi(0) \in \mathbb{R}^2$ and all $k \in \mathbb{Z}_+$. As $\beta(r,s) := \bar{C}\rho(\bar{A})^s r \in \mathscr{KL}$, the augmented system (5) is $\mathscr{KL}$-stable in $\mathbb{R}^2$. Hence, it follows from the equivalence of norms that there exists a $C \in \mathbb{R}_{>0}$ such that $\|x(k, \mathbf{x}_{[-1,0]})\| \leq C\rho(\bar{A})^k \|\mathbf{x}_{[-1,0]}\|$ for all $\mathbf{x}_{[-1,0]} \in \mathbb{R} \times \mathbb{R}$ and all $k \in \mathbb{Z}_+$. Thus, as $\beta(r,s) := C\rho(\bar{A})^s r \in \mathscr{KL}$, the DDE (4) is $\mathscr{KL}$-stable in $\mathbb{R}$.

   Next, claim (ii) is proven by contradiction. Therefore, suppose that the DDE (4) admits a LRF($\mathbb{R}$), i.e., $V : \mathbb{R} \to \mathbb{R}_+$. Furthermore, let $x(0) = 1$ and $x(-1) = 0$. From (4) it is obtained that $x(1) = 1$. Hence, (2b) yields

$$V(x(1)) = V(1) \leq \rho \max\{V(x(0)), V(x(-1))\} = \rho V(1). \tag{6}$$

Obviously, as $\rho \in \mathbb{R}_{[0,1)}$, a contradiction was reached. As the function V was chosen arbitrarily, it follows that the DDE (1) does not admit a LRF($\mathbb{R}$).

   To complete the proof it suffices to observe that claim (iii) follows directly from claim (ii) and Theorem 2.   □

Proposition 1 establishes that not every DDE that is $\mathscr{KL}$-stable admits a $\mathscr{D}$-contractive set. Hence, $\mathscr{KL}$-stability is merely a necessary condition for the existence of a $\mathscr{D}$-contractive set and not a sufficient condition. Notice that the above result does not rule out the existence of a $\mathscr{D}$-invariant set, which corresponds to $\lambda = 1$. However, a $\mathscr{D}$-invariant set does not imply $\mathscr{KL}$-stability, nor does it provide a region of a attraction in the absence of a LRF.

4 Necessary Conditions for $\mathscr{D}$-Contractive Sets

In what follows, alternative necessary conditions, in terms of asymptotic stability of
the DDE (1), for the existence of a $\mathscr{D}$-contractive set are established. Moreover, it is
also investigated if these necessary conditions are sufficient as well. Therefore, let
$\mathbb{T} := \{0, 1\}$ and let $\mathbb{S} := \{-1, 0, 1\}$. Then, consider the following family of systems

$$z(k+1) = H_\delta(z(k)), \quad k \in \mathbb{Z}_+, \tag{7}$$

where $z(k) \in \mathbb{R}^n$, $H_\delta(z) := F(\{[\delta]_1 z, \ldots, [\delta]_{h+1} z\})$ and $\delta \in \mathbb{T}^{h+1}$ or $\delta \in \mathbb{S}^{h+1}$. To
illustrate the family of systems (7), consider the DDE (4). Then, $H_{\delta_1}(z) = 1.5z$ and
$H_{\delta_2}(z) = -0.5z$ for $\delta_1 = \begin{bmatrix} -1 & 1 \end{bmatrix}^\top$ and $\delta_2 = \begin{bmatrix} 1 & 0 \end{bmatrix}^\top$, respectively.

Proposition 2. *Let $\lambda \in \mathbb{R}_{[0,1)}$. Suppose that the DDE (1) is $\mathscr{D}$-homogeneous of order t, for some $t \in \mathbb{Z}_{\geq 1}$. Furthermore, suppose that the DDE (1) admits a λ-$\mathscr{D}$-
contractive set, i.e., $\mathbb{V}$. Then:*

(i)   The family of systems (7) is $\mathscr{K}\mathscr{L}$-stable in $\mathbb{V}$ for all $\delta \in \mathbb{T}^{h+1}$;
(ii) If $t = 1$, the family of systems (7) is $\mathscr{K}\mathscr{L}$-stable in $\mathbb{R}^n$ for all $\delta \in \mathbb{S}^{h+1}$.

Proof. To prove claim (i), let $\mathbb{V}$ be a λ-$\mathscr{D}$-contractive set for the DDE (1). Then,
it follows from Theorem 2 that the DDE (1) admits a LRF($\mathbb{V}$). Consider any $\delta \in
\mathbb{T}^{h+1}$. As the conditions of Theorem 1 hold for all $\tilde{\mathbf{x}}_{[-h,0]} \in \mathbb{V}^{h+1}$, they also hold for
$\tilde{\mathbf{x}}_{[-h,0]} := \{[\delta]_1 z, \ldots, [\delta]_{h+1} z\}$ for any $z \in \mathbb{V}$. Hence, as $F(\tilde{\mathbf{x}}_{[-h,0]}) \in \mathbb{V}$, $H_\delta(z) \in \mathbb{V}$ for
all $z \in \mathbb{V}$. Moreover, (2b) also holds for $\tilde{\mathbf{x}}_{[-h,0]} := \{[\delta]_1 z, \ldots, [\delta]_{h+1} z\}$ with $z \in \mathbb{V}$,
which yields

$$V(H_\delta(z)) = V(F(\tilde{\mathbf{x}}_{[-h,0]})) \leq \rho V(z).$$

Next, let $\{z(k, z(0))\}_{k \in \mathbb{Z}_+}$ denote a trajectory of the family (7) from initial condition
$z(0) \in \mathbb{V}$ for some $\delta \in \mathbb{T}^{h+1}$. Applying the above inequality recursively and using
(2a) yields that

$$\|z(k, z(0))\| \leq \alpha_1^{-1}(\rho^k \alpha_2(\|z(0)\|)),$$

for all $z(0) \in \mathbb{V}$ and all $k \in \mathbb{Z}_+$. Letting $\beta(r, s) := \alpha_1^{-1}(\rho^s \alpha_2(r))$ it follows, from the
fact that $\rho(s) < s$ for all $s \in \mathbb{R}_{>0}$, that $\beta \in \mathscr{K}\mathscr{L}$. Therefore, and as $\delta \in \mathbb{T}^{h+1}$ was
chosen arbitrarily, system (7) is $\mathscr{K}\mathscr{L}$-stable in $\mathbb{V}$ for all $\delta \in \mathbb{T}^{h+1}$.

   The proof of claim (ii) follows from the proof of claim (i) with the additional
observation that $V(H_\delta(z)) \leq \rho \max\{V(z), V(-z)\}$ and, as $H_\delta(-z) = -H_\delta(z)$, that
$V(-H_\delta(z)) = V(H_\delta(-z)) \leq \rho \max\{V(z), V(-z)\}$. $\square$

Suppose that the DDE (1) is linear. Then, Proposition 2 yields that (1) admits a $\mathscr{D}$-
contractive set only if $\rho\left(\sum_{\theta=-h}^0 [\delta]_{-\theta+1} A_\theta\right) < 1$, for all $\delta \in \mathbb{S}^{h+1}$. For example, the
DDE (4) does not satisfy the above condition and hence it follows from Proposi-
tion 2 that (4) does not admit a $\mathscr{D}$-contractive set.

   Next, an example is presented of a DDE for which the family of systems (7) is
$\mathscr{K}\mathscr{L}$-stable but which does not admit a $\mathscr{D}$-contractive set. Consider the linear DDE

$$x(k+1) = A_0 x(k) + A_{-d} x(k-d), \quad k \in \mathbb{Z}_+, \tag{8}$$

where $\mathbf{x}_{[k-d,k]} \in (\mathbb{R}^2)^{d+1}$ is the system state, $d \in \mathbb{Z}_+$ is the delay and

$$A_0 = \begin{bmatrix} -0.7 & 0.5 \\ -0.7 & -0.7 \end{bmatrix}, \quad A_{-d} = \begin{bmatrix} 0.2 & 0.1 \\ 0.1 & 0 \end{bmatrix}.$$

Proposition 3. *Let $d = 2$ and let $\lambda \in \mathbb{R}_{[0,1)}$. The following statements hold:*

(i) The family of systems (7) based on the DDE (8) is $\mathcal{KL}$-stable in $\mathbb{R}^2$ for all $\delta \in \mathbb{S}^3$;

(ii) The linear DDE (8) does not admit a λ-$\mathcal{D}$-contractive set.

Proof. To prove claim (i), it suffices to check the spectral radius of the matrices corresponding to the family (7) based on (8) for all $\delta \in \mathbb{S}^3$, i.e.,

$$\begin{aligned} \rho(A_0) = \rho(-A_0) = 0.917, \quad \rho(A_{-2}) = \rho(-A_{-2}) = 0.241, \\ \rho(A_0 + A_{-2}) = \rho(-A_0 - A_{-2}) = 0.843, \\ \rho(A_0 - A_{-2}) = \rho(-A_0 + A_{-2}) = 0.975. \end{aligned}$$

Hence, it can be shown, using similar arguments as used in the proof of Proposition 1, that the family (7) based on (8) is $\mathcal{KL}$-stable in $\mathbb{R}^2$ for all $\delta \in \mathbb{S}^3$.

Next, claim (ii) is proven. Consider, as in Proposition 1, an augmented system similar to (5). Then, $\bar{A} = \begin{bmatrix} A_0 & 0 & A_{-1} \\ I_2 & 0 & 0 \\ 0 & I_2 & 0 \end{bmatrix}$. As $\rho(\bar{A}) = 1.006$, the augmented system is not $\mathcal{KL}$-stable in $\mathbb{R}^6$ and hence the DDE (8) is not $\mathcal{KL}$-stable in $\mathbb{R}^2$. Therefore, it follows from Theorem 1 and Theorem 2 that the DDE (8) does not admit a λ-$\mathcal{D}$-contractive set.   □

Proposition 3 implies that, while they provide a valuable insight for both linear and nonlinear systems, the necessary conditions derived in Proposition 2 are not a sufficient condition for the existence of a $\mathcal{D}$-contractive set. Moreover, it is interesting to observe that for $d = 1$, $\rho(\bar{A}) = 0.915$. Hence, both conditions, i.e., the DDE (1) is $\mathcal{KL}$-stable and the family of systems (7) is $\mathcal{KL}$-stable for all $\delta \in \mathbb{S}^{h+1}$ or $\delta \in \mathbb{T}^{h+1}$, respectively, provide different necessary conditions for the existence of a $\mathcal{D}$-contractive set.

In what follows, *linear DDEs* and *polytopic sets* are considered. Under these assumptions, a stronger set of necessary conditions for the existence of a $\mathcal{D}$-contractive set is derived next.

Definition 4. Suppose that the DDE (1) is linear. The linear DDE (1) is called $\mathcal{D}$-$\mathcal{KL}$-stable in $\mathbb{R}^n$ if the DDE

$$x(k+1) = \sum_{\theta=-h}^{0} A_\theta x(k - d_\theta), \quad k \in \mathbb{Z}_+, \tag{9}$$

is $\mathcal{KL}$-stable in $\mathbb{R}^n$ for all $d_\theta \in \mathbb{Z}_+$, $\theta \in \mathbb{Z}_{[-h,0]}$.   □

It is assumed that the delays d_θ are bounded for all $\theta \in \mathbb{Z}_{[-h,0]}$. Note that, if a DDE is $\mathscr{D}\text{-}\mathscr{KL}$-stable, then it is delay-independently stable, meaning that the DDE (9) is $\mathscr{KL}$-stable for all (bounded) delays.

Proposition 4. *Let $\lambda \in \mathbb{R}_{[0,1)}$ and suppose that the DDE (1) is linear. Furthermore, suppose that (1) admits a polytopic λ-$\mathscr{D}$-contractive set. Then, the linear DDE (1) is $\mathscr{D}\text{-}\mathscr{KL}$-stable in $\mathbb{R}^n$.*

Proof. Let $\mathbb{V} := \{x \in \mathbb{R}^n \mid \|Fx\|_\infty \leq 1\}$, for some $F \in \mathbb{R}^{p \times n}$ with $p \in \mathbb{Z}_{\geq n}$ and such that $\mathrm{rank}(F) = n$, denote a polytopic λ-$\mathscr{D}$-contractive set for (1). As the polytopic set $\mathbb{V}$ is λ-$\mathscr{D}$-contractive, Theorem 3 yields that there exist nonnegative matrices $H_\theta \in \mathbb{R}^{p \times p}$, $\theta \in \mathbb{Z}_{[-h,0]}$, such that

$$\left(\textstyle\sum_{\theta=-h}^{0} H_\theta\right) 1_p \leq \lambda 1_p, \quad FA_\theta = H_\theta F, \quad \forall \theta \in \mathbb{Z}_{[-h,0]}. \tag{10}$$

Note that, (10) is only dependent on A_θ and not on d_θ. Therefore, the same conditions are satisfied for the DDE (9) for any $d_\theta \in \mathbb{Z}_+$, $\theta \in \mathbb{Z}_{[-h,0]}$. Hence, it follows from Theorem 3 that $\mathbb{V}$ is λ-$\mathscr{D}$-contractive for the linear DDE (9) for any $d_\theta \in \mathbb{Z}_+$, $\theta \in \mathbb{Z}_{[-h,0]}$. Thus, it follows from Theorem 2 and Theorem 1 that the linear DDE (9) is $\mathscr{KL}$-stable in $\mathbb{R}^n$ for any $d_\theta \in \mathbb{Z}_+$, $\theta \in \mathbb{Z}_{[-h,0]}$, which completes the proof. $\square$

Proposition 4 implies that a linear DDE admits a $\mathscr{D}$-contractive set only if it is $\mathscr{D}\text{-}\mathscr{KL}$-stable. Indeed, both the DDE (4) and the DDE (8) are not $\mathscr{D}\text{-}\mathscr{KL}$-stable and hence they do not admit a $\mathscr{D}$-contractive set. Furthermore, note that the necessary conditions recently established in [13] can be recovered as particular cases of Proposition 2 and Proposition 4, respectively.

Remark 2. Theorem 1, Theorem 2 and Proposition 2 allow a straightforward extension to delay difference inclusions (DDIs). The extension of Theorem 3 and Proposition 4 to DDIs relies on the extension of Theorem 3 to DDIs which should be made along the lines of the results in [2]. $\square$

Although the necessary conditions obtained so far are not also sufficient, they provide useful insights. Indeed, before pursuing the construction of a $\mathscr{D}$-contractive set, available necessary conditions should be checked. If these tests succeed, it makes sense to employ constructive (sufficient in terms of $\mathscr{KL}$-stability) conditions, such as the ones presented in [13] or in Theorem 3.

5   Illustrative Example

In what follows, the application of the results derived in this chapter is illustrated via an example. Moreover, a $\mathscr{D}$-contractive set is also constructed. Therefore, consider the linear DDE

$$x(k+1) = \begin{bmatrix} 0.9 & -0.34 \\ 0.34 & 0.2 \end{bmatrix} x(k) + \begin{bmatrix} 0.24 & -0.17 \\ 0.17 & 0.24 \end{bmatrix} x(k-1), \quad k \in \mathbb{Z}_+. \tag{11}$$

Firstly, it follows from Theorem 2 and Theorem 1 that (11) admits a $\mathscr{D}$-contractive set only if the DDE (11) is $\mathscr{KL}$-stable. As $\rho\left(\begin{bmatrix} A & A_d \\ I_2 & 0 \end{bmatrix}\right) = 0.9246$ it follows, similarly to the reasoning used in the proof of Proposition 1, that the DDE (11) is $\mathscr{KL}$-stable in $\mathbb{R}^2$. Secondly, it follows from Proposition 2 that (11) admits a $\mathscr{D}$-contractive set only if the family of systems (7) based on the DDE (11) is $\mathscr{KL}$-stable for all $\delta \in \mathbb{S}^2$. As

$$\rho(A) = \rho(-A) = 0.6331, \quad \rho(A + A_d) = \rho(-A - A_d) = 0.8728,$$
$$\rho(A_d) = \rho(-A_d) = 0.2941, \quad \rho(A - A_d) = \rho(-A + A_d) = 0.6159,$$

it follows that the family (7) based on (11) is indeed $\mathscr{KL}$-stable in $\mathbb{R}^2$ for all $\delta \in \mathbb{S}^2$. Thirdly, it follows from Proposition 4 that (11) admits a $\mathscr{D}$-contractive set only if the DDE (11) is $\mathscr{D}$-$\mathscr{KL}$-stable. While it is not clear how to verify if a DDE is $\mathscr{D}$-$\mathscr{KL}$-stable one can easily verify stability of (9) for several values of $d_\theta \in \mathbb{Z}_+$, $\theta \in \mathbb{Z}_{[-1,0]}$, by verifying the stability of the augmented system. For example, let $(d_0, d_{-1}) = (0, 2)$ and $(d_0, d_{-1}) = (2, 0)$, then

$$\rho\left(\begin{bmatrix} A & 0 & A_d \\ I_2 & 0 & 0 \\ 0 & I_2 & 0 \end{bmatrix}\right) = 0.9433, \quad \rho\left(\begin{bmatrix} A_d & 0 & A \\ I_2 & 0 & 0 \\ 0 & I_2 & 0 \end{bmatrix}\right) = 0.9185.$$

As all necessary conditions for the existence of a $\mathscr{D}$-contractive set are satisfied one can now attempt to obtain such a set. For $\lambda = 0.99$, the set $\mathbb{V} = \{x \in \mathbb{R}^2 \mid \|Fx\|_\infty \leq 1\}$ with

$$F = \begin{bmatrix} 1.1729 & 0.1185 & -1.0917 & 0.7283 & -1.1081 & 0.8957 \\ -0.5980 & 0.9406 & 1.4328 & -1.3999 & 0.3542 & 0.0783 \end{bmatrix}^\top,$$

was obtained as a feasible solution to the algebraic conditions of Theorem 3. A plot of $\mathbb{V}$ and several trajectories from its vertices are shown in Figure 1.

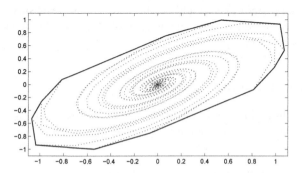

Fig. 1 The polytopic 0.99-$\mathscr{D}$-contractive set $\mathbb{V}$ (——) and several trajectories ($\ldots$).

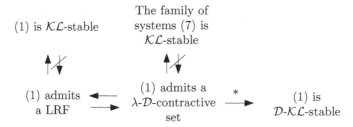

Fig. 2 A schematic overview of all relations established in this chapter. $A \rightarrow B$ means that A implies B, $A \nrightarrow B$ means that A does not necessarily imply B and $A \overset{*}{\rightarrow} B$ means that A implies B under the assumption that the DDE (1) is linear and the set is polytopic.

6 Conclusion

The relation between asymptotic stability and the existence of a $\mathscr{D}$-contractive set was studied for DDEs. Such sets are of importance as they provide a region of attraction which is difficult to obtain for delay systems. Various necessary conditions for the existence of a $\mathscr{D}$-contractive set were derived. A schematic overview of all relations established in this chapter is shown in Figure 2. Establishing if every linear DDE that is $\mathscr{D}$-$\mathscr{K}\mathscr{L}$-stable admits a polytopic $\mathscr{D}$-contractive set makes the subject of future research.

Acknowledgements. The research presented in this chapter is supported by the Veni grant number 10230, awarded by the Dutch organizations STW and NWO.

References

1. Åström, K.J., Wittenmark, B.: Computer controlled systems, theory and design. Prentice Hall International, Inc, Englewood Cliffs, NJ (1990)
2. Barabanov, N.E.: The Lyapunov indicator of discrete inclusions I, II and III. Automation and Remote Control 49, I:152–157, II:283–287, III:558–565 (1988)
3. Bitsoris, G.: Positively invariant polyhedral sets of discrete-time linear systems. International Journal of Control 47(6), 1713–1726 (1988)
4. Blanchini, F., Miani, S.: Set-theoretic methods in control. Birkhäuser, Boston, MA (2008)
5. Dambrine, M., Richard, J.P., Borne, P.: Feedback control of time-delay systems with bounded control and state. Mathematical Problems in Engineering 1, 77–87 (1995)
6. Gielen, R.H., Lazar, M., Kolmanovsky, I.V.: On Lyapunov theory for delay difference inclusions. In: Proceedings of the American Control Conference, Baltimore, MD, pp. 3697–3703 (2010)
7. Gielen, R.H., Lazar, M., Teel, A.R.: On input-to-state stability of delay difference equations. In: Proceedings of the 18th IFAC World Congress, Milano, Italy (2011)
8. Goubet-Bartholoméüs, A., Dambrine, M., Richard, J.P.: Bounded domains and constrained control of linear time-delay systems. Journal Europeen des Systemes Automatises 31(6), 1001–1014 (1997)

9. Hennet, J.-C.: Discrete time constrained linear systems. Control and Dynamic Systems 71, 157–213 (1995)
10. Hennet, J.-C., Tarbouriech, S.: Stability conditions of constrained delay systems via positive invariance. International Journal of Robust and Nonlinear Control 8(3), 265–278 (1998)
11. Kolmanovskii, V., Myshkis, A.: Introduction to the theory and applications of functional differential equations. Kluwer Academic Publishers, Dordrecht (1999)
12. Lazar, M.: On infinity norms as Lyapunov functions: Alternative necessary and sufficient conditions. In: Proceedings of the 49th IEEE Conference on Decision and Control, Atlanta, GA, pp. 5936–5942 (2011)
13. Lombardi, W., Olaru, S., Lazar, M., Niculescu, S.-I.: On positive invariance for delay difference equations. In: Proceedings of the American Control Conference, San Francisco, CA (2011)
14. Lombardi, W., Olaru, S., Lazar, M., Bitsoris, G., Niculescu, S.-I.: On the polyhedral set-invariance conditions for time-delay systems. In: Proceedings of the 18th IFAC World Congress, Milano, Italy (2011)
15. Milani, B.E.A.: Piecewise affine Lyapunov functions for discrete-time linear systems with saturating controls. Automatica 38, 2177–2184 (2002)
16. Rawlings, J.B., Mayne, D.Q.: Model predictive control: Theory and design. Nob Hill Publishing (2009)
17. Reble, M., Allgöwer, F.: General design parameters of model predictive control for nonlinear time-delay systems. In: Proceedings of the 49th IEEE Conference on Decision and Control, Atlanta, GA, pp. 176–181 (2010)
18. Vassilaki, M., Bitsoris, G.: Constrained feedback control of discrete-time systems described by ARMA models. In: Proceedings of the 1999 European Control Conference, Karlsruhe, Germany (1999)

Part II
Systems, Signals and Applications

Temperature and Heat Flux Dependence/Independence for Heat Equations with Memory

Sergei Avdonin and Luciano Pandolfi

Abstract. We present and extend our recent results on the relations between temperature and flux for heat equations with memory. The key observation is that we can interpret "independence" as a kind of "controllability" and this suggests the study of controllability of the pair heat-flux in an appropriate functional space.

1 Introduction

In his famous paper [8] Cattaneo proposed the following heat equation with memory. Here, $t > 0$, $0 < x < l$:

$$\theta'(x,t) = 2\alpha\theta(x,t) + \int_0^t N(t-s)\left(\theta_{xx}(x,s) - b(x)\theta(x,s)\right)\,\mathrm{d}s \qquad (1)$$

(initial and boundary conditions are specified in Section 2). In fact, Cattaneo considered the case $\alpha = 0$, $b(x) = 0$ and $N(t) = e^{-\gamma t}$. Eq. (1) in its full generality was introduced by Gurtin and Pipkin in [13]. The reason for doing this was to overcome the known limitation of the heat equation,

$$\theta_t(x,t) = \theta_{xx}(x,t) - b(x)\theta(x,t), \qquad (2)$$

that a heat signal propagates with infinite speed, a fact especially disturbing Cattaneo, a specialist in relativity theory.

S. Avdonin
Department of Mathematics and Statistics, University of Alaska, Fairbanks,
AK 99775-6660, USA
e-mail: saavdonin@alaska.edu

L. Pandolfi
Politecnico di Torino, Dipartimento di Matematica, Corso Duca degli Abruzzi 24,
10129 Torino — Italy
e-mail: luciano.pandolfi@polito.it

R. Sipahi et al. (Eds.): Time Delay Sys.: Methods, Appli. and New Trends, LNCIS 423, pp. 87–101.
springerlink.com   © Springer-Verlag Berlin Heidelberg 2012

In fact, in the same turn of years other facts suggested that the usual heat equations didn't describe physical facts at extreme temperatures. In particular, we mention Landau theory which implied that heat propagates as a wave at very low temperature, a fact experimentally verified in 1949. In this context, the interest of Eq. (1) stems from the fact that its solutions have certain resemblances with solutions of the wave equation, in particular, they have finite propagation speed. See [26] for an overview of the history and references on Eq. (1).

The limitation of the heat equation (2) is a consequence of the Fourier law stating that the flux $q(x,t)$ instantaneously depends on the gradient of the temperature, at the same place and time,

$$q(x,t) = -\theta_x(x,t) \qquad \text{(Fourier law)}.$$

Instead, the flux associated to Eq. (1) is

$$q(x,t) = -\frac{\partial}{\partial x} \int_0^t N(t-s)\theta(x,s)\,ds, \tag{3}$$

only mildly related to the temperature. So, Cattaneo (see [8, p. 98]) posed the following problem: to understand how strict is the relation between the temperature $\theta(x,t)$, given by Eq. (1), and the corresponding flux, given by (3). *This is the problem that we study in this paper.* First we prove that the relation is very strict for small times. After that we present our results on the regularity of the (heat/flux) pair and finally we prove that flux and temperature are essentially independent for large times, namely for $t \geq T_0$, where $T_0 = 2l$.

The interpretation of the term "independence" is part of the problem, and it is explained in Section 2.

Finally, we state the assumptions in this paper:

- $b(x) \in L^1(0,l)$.
- $N(t) \in C^4(0,\infty)$ and $N(0) > 0$.

Furthermore, for simplicity we assume $N(0) = 1$, a condition which can be achieved using the new variable $\zeta(x,t) = \theta(x,t/N(0))$, which we call again $\theta(x,t)$.

1.1   Further Applications and References

Eq. (1) is encountered in several different applications, most important viscoelasticity and non-fickian diffusion. Early applications to viscoelasticity go back to Maxwell, Kelvin and mostly Boltzmann in the 19-th century. The equation with memory encountered in these applications is of second order in time, Eq. (1) being its integrated form. Another important applications is to non-fickian diffusion, i.e. to diffusion processes in the case Fick law does not hold, mostly due to high concentration of solutes or complex molecular structure. In fact, Eq. (1) was introduced (mid of last century) in order to represent diffusion in polymers. See [25, Sect. 5] for an overview of this application and furthers references. Here we confine ourselves to cite the following book and papers which concern these different applications: [9, 10, 14].

2 The Interpretation of Dependence/Independence of Heat and Flux

Our idea for studying dependence of heat and flux is as follows: we assume that the initial temperature is zero[1]

$$\theta(x,0) = 0$$

and apply a source of heat at the boundary, which controls the temperature at our will. Say, we impose

$$\theta(0,t) = u(t), \qquad \theta(l,t) = 0.$$

The natural class in which u can be chosen is $L^2_{\text{loc}}(0,\infty)$.

Then we study the regularity of the functions $(x,t) \mapsto \theta(x,t)$ and $(x,t) \mapsto q(x,t)$. Let us assume that we can identify Hilbert spaces X and Y such that

$$t \mapsto \theta(\cdot,t) \in C(0,T;X), \qquad t \mapsto q(\cdot,t) \in C(0,T;Y).$$

In this case it makes sense to consider the values at time T of both $\theta(\cdot,t)$ and $q(\cdot,t)$. So, we can consider the reachable set

$$\mathscr{R}(T) = \{(\theta(\cdot,T), q(\cdot,T)) : u \in L^2(0,T)\}.$$

If it turns out that this set is "small" in $X \times Y$ then we interpret this fact as a strict relation between $\theta(\cdot,T)$ and $q(\cdot,T)$; if it is "large", we interpret this fact as a weak relation.

Now we can describe our results. We demonstrate that both $\theta(\cdot,t)$ and $q(\cdot,t)$ are continuous function of time with values in $L^2(0,l)$. Then we prove that for $T_0 = 2l$:

- if $T < T_0$, then $\theta(\cdot,T)$ and $q(\cdot,T)$ are not independent;
- if $T \geq T_0$, then temperature and flux are independent in the sense that $\mathscr{R}(T)$ is as large as possible: it is $L^2(0,l) \times L^2(0,l)$.

These statements can be interpreted as controllability results for the pair (θ,q). Flux and temperature controllability have been first studied, in special cases, in [27]. The results in this paper extend those in [22, 23, 24] and in particular [6].

Finally, we cite the papers [7, 11, 15, 18, 20, 21] which also study different aspects of the controllability of Eq. (1). We mention also the paper [2] where

---

[1] Which is not the absolute zero of thermodynamics: the linear heat equation is an approximation of a nonlinear equation. So, we assume that the initial temperature is an equilibrium temperature of the "real" nonlinear equation.

techniques similar to those in the present paper were used to study controllability of hyperbolic equations with time dependent coefficients.

3   Strict Dependence at Small Times

In order to get a feeling of the problem, in this section we examine a special example and prove that when T is small then the relation between $q(x,T)$ and $\theta(x,T)$ is "strict". The statement in the general case is in Theorem 5. Here we consider the case $\alpha = 0$, $a(x) = 1$, $b(x) = 0$ and $N(t) = 1$. Eq. (1) takes the form

$$\theta_t(x,t) = \int_0^t \theta_{xx}(x,s)\,ds \tag{4}$$

with the conditions

$$\theta(x,0) = 0, \qquad \theta(0,t) = u(t), \qquad \theta(l,t) = 0. \tag{5}$$

Eq. (4) is nothing else then the wave equation

$$\theta_{tt} = \theta_{xx}$$

with the additional initial condition

$$\theta_t(x,0) = 0.$$

First we assume that $u(t)$ is smooth and $u(0) = u'(0) = 0$. For $t \le l = T_0/2$ there is no reflection from the right end of the interval, and the solution of (4), (5) is

$$\theta(x,t) = u(t-x)H(t-x) \tag{6}$$

where $H(t)$ is the Heaviside function. We note that the function $\theta(x,t)$ defined by Eq. (6) has a smooth extension to a left neighborhood of $x = 0$ for every $t > 0$. If $u \in L^2_{loc}(0,+\infty)$, then (6) gives a weak solution of (4)-(5).

It is easy to see using formulas (3), (6) that

$$q(x,T) = \theta(x,T).$$

for $T \le l$. So, if T is "small" the relation between flux and temperature is as strict as possible: flux is uniquely determined by the temperature.

4   Independence at Large Times

The analysis of this problem for large values of T is much more involved and is presented in the next sections. We first present certain asymptotic estimates and give essential information on Riesz bases.

4.1 Preliminary Transformations and Asymptotic Estimates

The transformation $\theta(x,t) \mapsto e^{-N'(0)t}\theta(x,t)$ transforms Eq. (1) to a new equation of the same kind with the kernel $e^{-N'(0)t}N(t)$ whose derivative at $t = 0$ is zero. (We recall that $N(0) = 1$.)

A similar transformation can be applied to the flux, so that we can assume from the outset that we study the original problem (1), (3) with $N'(0) = 0$ (see [22] for details). It turns out that this transformation is very useful in order to simplify the proofs but it changes the value of α. Hence, we cannot confine ourselves to the case $\alpha = 0$.

Now we introduce the operator $\mathscr{A}$ in $L^2(0,l)$,

$$(\mathscr{A}\phi)(x) = -\phi''(x) + b(x)\phi(x), \qquad \operatorname{dom}\mathscr{A} = H^2(0,l) \cap H_0^1(0,l).$$

Let λ_n^2, $n \in \mathbb{N}$, be the eigenvalues of $\mathscr{A}$ and $\phi_n(x)$ be the corresponding eigenfunctions (of norm 1).

Properties of the eigenvalues and eigenfunctions of operator $\mathscr{A}$ are well known (see, e.g. monographs [16, 19, 30]). In this paper we will use the following properties, asymptotic formulas and estimates.

Proposition 1. *1. The spectrum of operator $\mathscr{A}$, $\{\lambda_n^2\}_{n=1}^{\infty}$ is pure discrete, simple, real and $\lim \lambda_n^2 = +\infty$.*
2. Let $\lambda_n = \sqrt{\lambda_n^2}$. The asymptotic representation holds,

$$\lambda_n = \frac{\pi}{l}n + o(1) \text{ as } n \to \infty.$$

3. The sequence $\{\lambda_n\}$ is separated,

$$\inf_{n \neq k} |\lambda_n - \lambda_k| > 0.$$

4. The corresponding (normalized) eigenfunctions $\{\phi_n\}_{n=1}^{\infty}$ form an orthonormal basis in $L^2(0,l)$.
5. We have

$$|\phi_n'(0)| \asymp n \text{ that is } 0 < \inf_n \frac{|\phi_n'(0)|}{n} \leq \sup_n \frac{|\phi_n'(0)|}{n} < \infty.$$

In what follows we need more precise asymptotics for λ_n and ϕ_n. The following asymptotic formulas can be found in [31]. We note that the only assumption $b \in L^1(0,l)$ was used to obtain these formulas:

$$\lambda_n = \frac{\pi}{l}n - \frac{1}{2\pi n}\left[\int_0^l q(x)\left(1 - \cos\left(\frac{2\pi}{l}nx\right)\right) dx\right] + r_n,$$

$$|r_n| \leq 20\frac{lB^2}{(n\pi - 1/4)^2}, \quad B = \int_0^l |q(x)|dx,$$

$$\phi_n(x) = \sqrt{\frac{2}{l}} \left\{ \sin\left(\frac{\pi}{l}nx\right) + \frac{1}{2\pi n}\phi_{n,1}(x) + \Delta\phi_n(x) \right\}.$$

Here

$$\phi_{n,1}(x) = \cos\left(\frac{\pi}{l}nx\right)\left[l\int_0^x q(s)ds - x\int_0^l q(s)ds \right.$$

$$+ x\int_0^l q(s)\cos\left(\frac{2\pi}{l}ns\right)ds - l\int_0^x q(s)\cos\left(\frac{2\pi}{l}ns\right)ds \right]$$

$$- \sin\left(\frac{\pi}{l}nx\right)\left[l\int_0^x q(s)\sin\left(\frac{2\pi}{l}ns\right)ds - \int_0^l (l-s)q(s)\sin\left(\frac{2\pi}{l}ns\right)ds \right],$$

$$|\Delta\phi_n(x)| \le \frac{2.6B^2 + 4.1B}{(n\pi - 1/4)^2}.$$

We have also

$$\phi_n'(x) = \frac{\pi}{l}\sqrt{\frac{2}{l}} \left\{ n\cos\left(\frac{\pi}{l}nx\right) + \frac{1}{2\pi}\zeta_n(x) + \Delta\zeta_n(x) \right\},$$

$$\zeta_n(x) = -\sin\left(\frac{\pi}{l}nx\right)\left[l\int_0^x q(s)ds - x\int_0^l q(s)ds + x\int_0^l q(s)\cos\left(\frac{2\pi}{l}ns\right)ds \right.$$

$$\left. - l\int_0^x q(s)\cos\left(\frac{2\pi}{l}ns\right)ds \right] - \cos\left(\frac{\pi}{l}nx\right)\left[l\int_0^x q(s)\sin\left(\frac{2\pi}{l}ns\right)ds \right.$$

$$\left. - \int_0^l (l-s)q(s)\sin\left(\frac{2\pi}{l}ns\right)ds \right], \quad |\Delta\zeta_n(x)| \le \frac{M}{n}. \tag{7}$$

Remark 1. Only a finite number of λ_n^2 may be negative. In this case, the corresponding λ_n are pure imaginary, $\lambda_n = i\nu_n$, and for definiteness, we assume $\nu_n > 0$.

The eigenfuctions $\phi_n(x)$ are defined and differentiable on $[0, l]$, and zero at $x = 0$ and $x = l$. Let us first take the odd extension of them to $[-l, 0]$. In this way we get continuous functions on $[-l, l]$, which are 0 at $x = \pm l$. Then, we take the periodic extensions to $\mathbb{R}$, which are continuous and a.e. differentiable.

This odd and periodic extension of the eigenfunction $\phi_n(x)$ will be denoted $\phi_n(x)$ without any risk of misunderstanding.

4.2   Riesz Bases and Riesz Sequences

A sequence of vectors $\{e_n\}, n \in \mathbb{N}$, in a Hilbert space H is called a Riesz basis if it is the image of an orthonormal basis under a bounded and boundedly invertible linear transformation. A sequence $\{e_n\}$ in H is called an $\mathscr{L}$-basis (or Riesz sequence) if it is a Riesz basis in its closed linear span. See [3, 12, 17, 32] for results on Riesz bases.

A sequence $\{e_n\}$ is ω-*independent* in a Hilbert space H if the conditions

$$\{\alpha_n\} \in l^2 \quad \text{and} \quad \sum \alpha_n e_n = 0$$

imply that $\alpha_n = 0$ for every n. The convergence of the series is in the norm of H.
We need the following result:

Theorem 1. *Let* $\{\varepsilon_n\}$ *be an* $\mathscr{L}$-*basis in a Hilbert space* H *and let* $\{e_n\}$ *be quadratically close to* $\{\varepsilon_n\}$, *i.e.*

$$\sum \|\varepsilon_n - e_n\|^2 < +\infty.$$

Then, there exists a number N *such that* $\{e_n\}_{n>N}$ *is an* $\mathscr{L}$-*basis.*
If furthermore $\{e_n\}$ *is* ω-*independent, then* span$\{\varepsilon_n\}$ *and* span$\{e_n\}$ *have the same codimension (called the* deficiency *of the* $\mathscr{L}$-*basis). In particular, if* $\{\varepsilon_n\}$ *is a Riesz basis then* $\{e_n\}$ *is also a Riesz basis.*

The statements in the previous theorem combine [12, Theorem 10 p. 32], [32, Theorem 15 p. 38]) with [12, Remarque 2.1, p. 323].
We apply now this result to the sequence $\{\phi_n'(x)/\phi_n'(0)\}$.

Theorem 2. *The family* $\Phi := \{1\} \cup \{\phi_n'(x)/\phi_n'(0)\}$ *forms a Riesz basis in* $L^2(0,l)$.

Proof. It follows from (7) that the family Φ is quadratically close to $\{\cos(\pi n x/l)\}_{n=0}^{\infty}$. The latter family is an orthogonal basis in $L^2(0,l)$, therefore, according to Theorem 1, the family $\{\phi_n'(x)/\phi_n'(0)\}_{n>N}$ is an $\mathscr{L}$-basis in $L^2(0,l)$ for sufficiently large N. To prove the family Φ is ω-independent we suppose that for some $\{\alpha_n\} \in l^2$

$$\alpha_0 + \sum_{n=1}^{+\infty} \alpha_n \frac{\phi_n'(x)}{\phi_n'(0)} = 0.$$

Integrating this series termwise and taking into account that $\phi_n(0) = 0$ we get

$$\alpha_0 x + \sum_{n=1}^{+\infty} \frac{\alpha_n}{\phi_n'(0)} \phi_n(x) = 0.$$

Putting here $x = l$ gives $\alpha_0 = 0$, and so

$$\sum_{n=1}^{+\infty} \frac{\alpha_n}{\phi_n'(0)} \phi_n(x) = 0.$$

It follows that $\alpha_n = 0$ for all n since $\{\phi_n(x)\}$ is an orthonormal basis in $L^2(0,l)$. $\square$

In what follows we shall also use the basis properties of the exponential family $\mathscr{E} = \{\exp(\pm i \lambda_n t)\}$, $n \in \mathbb{N}$. We recall that λ_n^2 are eigenvalues of the operator $\mathscr{A}$, and the properties of λ_n were stated in Proposition 1. If $\lambda_n = 0$ for some $n \in \mathbb{N}$, the corresponding exponential functions $\exp(\pm i \lambda_n t)$ equal identically to one, and we include only one of them into the family $\mathscr{E}$.

Lemma 1. *(i) If $\lambda_n \neq 0$ for all $n \in \mathbb{N}$, then the family $\{1\} \cup \mathcal{E}$ forms a Riesz basis in $L^2(0,2l)$.*

(ii) If for some $n \in \mathbb{N}$, we have $\lambda_n = 0$, the family $\{t\} \cup \{t^2\} \cup \mathcal{E}$ forms a Riesz basis in $L^2(0,2l)$.

Proof. Since $\lambda_n - \pi n/l = o(1)$, the statement (i) follows from the "1/4 in the mean" theorem [1] (see also [3, Proposition II.4.6]).

Using the same asymptotic estimates of λ_n and [1, Theorem 2], one can easily check that the generating function of the family $\{t\} \cup \{t^2\} \cup \mathcal{E}$ is a sine type function (see, e.g. [3, Section II.4]). The statement (ii) follows then from [29] (see also [3, Theorem II.4.22]). $\qquad\square$

4.3 Temperature and Flux Regularity

Let

$$\theta_n(t) = \int_0^l \theta(x,t)\phi_n(x)\,\mathrm{d}x,$$

then

$$\theta(x,t) = \sum_{n=1}^{+\infty} \theta_n(t)\phi_n(x).$$

Assuming u smooth, we multiply both the sides of (1) by $\phi_n(x)$ and integrate over $[0,l]$. Using the assigned boundary conditions, we find that $\theta_n(t)$ solves the following equation

$$\theta_n'(t) = 2\alpha\theta_n(t) - \lambda_n^2\int_0^t N(t-s)\theta_n(s)\,\mathrm{d}s + \phi'(0)\left[\int_0^t N(t-s)u(s)\,\mathrm{d}s\right].$$

We introduce the functions $z_n(t)$ which solve the equations

$$z_n'(t) = 2\alpha z_n(t) - \lambda_n^2\int_0^t N(t-s)z_n(s)\,\mathrm{d}s, \qquad z_n(0) = 1 \tag{8}$$

so that

$$\theta_n(t) = \phi_n'(0)\int_0^t z_n(t-r)\int_0^r N(r-s)u(s)\,\mathrm{d}s\,\mathrm{d}r,$$

$$\theta(x,t) = \sum_{n=1}^{+\infty} \phi_n'(0)\phi_n(x)\int_0^t z_n(t-r)\int_0^r N(r-s)u(s)\,\mathrm{d}s\,\mathrm{d}r. \tag{9}$$

This formula was derived in [23] where it was proved that the series converges in $C(0,T;L^2(0,l))$ for every $T > 0$. Furthermore, it was proved that

$$\{z_n(t)\} \quad \text{and} \quad \left\{\phi_n'(0)\int_0^t N(t-s)z_n(s)\,\mathrm{d}s\right\}$$

are Riesz sequences in $L^2(0,T)$ for every $T \geq l$.

Let us consider now the properties of the flux. Using (9), we see that

$$q(x,t)$$
$$= \sum_{n=1}^{+\infty} \phi_n'(x)\phi_n'(0) \int_0^t u(r) \left\{ \int_0^{t-r} N(t-r-s) \left[\int_0^s N(s-v)z_n(v)\,dv \right]\,ds \right\}\,dr.$$

(10)

This equality holds on $(0,l)$ but it is convenient to read it on $\mathbb{R}$, using the odd and periodic extension of $\phi_n(x)$, as explained in Section 4.1. The series (so extended) converges in $C(0,T;H_{\mathrm{loc}}^{-1}(\mathbb{R}))$. In fact, it is the x-derivative of a series which converges in $C(0,T;L_{\mathrm{loc}}^2(\mathbb{R}))$.

Our goal is to prove that the restriction to $(0,l)$ of the distribution $q(x,t)$ is a regular distribution identified with a square integrable function which, furthermore, as an element of $L^2(0,l)$, depends continuously on t. We need the following asymptotic estimate (see [6, Lemma 22]):

Lemma 2. *Let $\beta_n = \sqrt{\lambda_n^2 - \alpha^2}$, $\mu_n = \lambda_n^2/\beta_n^2$ and let $T > 0$ be fixed. There exists a bounded sequence $\{M_n(t)\}$ of square integrable functions and two continuous functions $F(t)$ and $G(t)$ such that*

$$z_n(t) = \left\{ N''(0)\frac{\mu_n}{2\beta_n}e^{\alpha t}t\sin\beta_n t + \frac{M_n(t)}{\beta_n^3} + \frac{\mu_n}{\beta_n^2}G(t)\cos\beta_n t \right\}$$
$$+ \frac{\mu_n}{\beta_n^2}F(t) + e^{\alpha t}\cos\beta_n t.$$

(11)

Now the idea is to replace $z_n(t)$ in (10) with the right hand side of (11) and study the regularity of the resulting series. All the terms in the brace lead to similar computations. We show the details in the case of the first term. Integrating by parts twice we get

$$\frac{1}{\beta_n} \int_0^t u(r) \int_0^{t-r} N(t-r-s) \left[\int_0^s N(s-v)ve^{\alpha v}\sin\beta_n v\,dv \right]\,ds\,dr$$
$$= \frac{-1}{\beta_n^3} \int_0^t u(r)(t-r)e^{\alpha(t-r)}\sin\beta_n(t-r)\,dr$$

(12)

$$+ \frac{1}{\beta_n^3} \int_0^t u(r) \int_0^{t-r} F_n(t,s,r)\sin\beta_n s\,ds\,dr.$$

(13)

We do not need to specify the form of the functions $F_n(t,s,r)$. Their important properties is that they are differentiable with bounded derivatives, so that a further integration by parts shows that this integral is less then $(\mathrm{const})/\beta_n^4$ and, when substituted in (10), the term (13) gives a series which converges uniformly in $[0,l] \times [0,T]$. When we substitute (12), we get a series which converges in $C(0,T;L^2(0,l))$. In fact, we get

$$\sum_{n=1}^{+\infty} \frac{\phi_n'(0)}{\beta_n} \frac{\phi_n'(x)}{\beta_n} \frac{\mu_n}{\beta_n} \int_0^t [u(t-r)re^r]\sin\beta_n r\,dr.$$

The series converges since $\{\phi_n'(x)/\beta_n\}$ is an $\mathscr{L}$-basis in $L^2(0,l)$, the sequence $\{(\phi_n'(0)\mu_n)/\beta_n^2\}$ is bounded and

$$\sum_{n=1}^{+\infty} \left| \int_0^t [u(t-r)re^r] \sin \beta_n r \, dr \right|^2 \leq C(T) \|u\|_{L^2(0,T)}^2 . \tag{14}$$

The last inequality follows from the fact that the family $\{\sin \beta_n r\}$ is an $\mathscr{L}$-basis in $L^2(0,t)$ for $t \geq l$.

When substituting $(\mu_n/\beta_n^2)F(t)$ in (10), we get

$$\left[\int_0^t u(r) \int_0^{t-r} N(t-r-s) \left(\int_0^s N(s-v)F(v) \, dv \right) ds \, dr \right] \sum_{n=1}^{+\infty} \frac{\phi_n'(0)}{\beta_n} \frac{\phi_n'(x)}{\beta_n} \mu_n .$$

The properties of this term depend on the series

$$\sum_{n=1}^{+\infty} \frac{\phi_n'(0)}{\beta_n} \frac{\phi_n'(x)}{\beta_n} \mu_n \tag{15}$$

which will be studied below.

Now we substitute $e^{\alpha t} \cos \beta_n t$. Integrating by parts we get

$$\int_0^t u(r) \int_0^{t-r} N(t-r-s) \left(\int_0^s N(s-v)e^{\alpha v} \cos \beta_n v \right) ds \, dr = \boxed{1} + \boxed{2}$$

where the term $\boxed{2}$ gives uniformly convergent series, either due to the fact that their terms are of the order $1/\beta_n^4$ or due to the estimate based on $\mathscr{L}$-basis property as in (14). Instead, $\boxed{1}$ gives the series

$$H(t) \sum_{n=1}^{+\infty} \frac{\phi_n'(0)}{\beta_n} \frac{\phi_n'(x)}{\beta_n} , \tag{16}$$

$$H(t) = \int_0^t u(r) \left[N(t-r) - \int_0^{t-r} N(t-r-s) \left(\alpha N(s) - N'(s) \right) ds \right] dr .$$

The series here is the same as (15), with $\mu_n = 1$. So, now we study the series (15). We recall the definition of μ_n and note the following asymptotic estimate obtained using the second item of Proposition 1:

$$\mu_n = \frac{\lambda_n^2}{\beta_n^2} = 1 + \frac{\alpha^2}{\lambda_n^2 - \alpha^2} \qquad \frac{n}{\beta_n} = \frac{l}{\pi} + o\left(\frac{1}{n}\right) . \tag{17}$$

Using these estimates, we analyze the series (15). The computations can be easily adapted to the case $\mu_n = 1$.

The asymptotic estimates (7) combined with (17) imply

$$\frac{\phi_n'(0)}{\beta_n}\mu_n = \sqrt{\frac{2}{l}} + \frac{\sigma_n}{n}, \qquad \{\sigma_n\} \text{ is bounded.}$$

So, the series (15) can be written as

$$\sqrt{\frac{2}{l}} \sum_{n=1}^{+\infty} \frac{\phi_n'(x)}{\beta_n} + \sum_{n=1}^{+\infty} \frac{\sigma_n}{n} \frac{\phi_n'(x)}{\beta_n}.$$

The second series converges in $L^2(0,l)$, and we study now the first one.

Using the asymptotic estimates (7) we see that the first series is equal to

$$\frac{2\pi}{l^2} \sum_{n=1}^{+\infty} \cos\frac{\pi}{l}nx$$

$$+\frac{2\pi}{l^2} \left\{ \sum_{n=1}^{+\infty} \left(\frac{n}{\beta_n} - 1\right) \cos\frac{\pi}{l}nx + \sum_{n=1}^{+\infty} \frac{n}{\beta_n} \left[\frac{1}{2\pi}\zeta_n(x) + \Delta\zeta_n(x)\right]\frac{1}{n} \right\}.$$

Both the series in the brace converge in $L^2(0,l)$, while the first series converges in the sense of distributions:

$$\frac{2\pi}{l^2} \sum_{n=1}^{+\infty} \cos\frac{\pi}{l}nx = \frac{2\pi}{l^2} \sum_{m=-\infty}^{+\infty} \delta\left(\pi x/l + 2m\pi\right) - \frac{\pi}{l^2},$$

see [28, p. 86].

So, the singular part of the series (10) is not supported in $(0,l)$: the restriction of the distribution q to $(0,l)$ is a regular distribution, and we see that it can be identified with an element of $L^2(0,l)$, as we wished to prove.

We can sum up the previous results as follows: the restriction of $q(\cdot,t)$ to $(0,l)$ has the following expansion in $L^2(0,l)$:

$$q(x,t) = \sum_{n=1}^{+\infty} L_n(t,u)\frac{\phi_n'(x)}{\phi_n'(0)} + \mathcal{L}(t;u)\left\{\frac{\pi}{l}\right.$$

$$\left. + \sum_{n=1}^{+\infty} \left(\frac{n}{\beta_n} - 1\right)\cos\frac{\pi}{l}nx + \sum_{n=1}^{+\infty} \frac{1}{\beta_n}\left(\phi_n'(x) - \frac{\sqrt{2}\pi}{l^2}\cos\frac{\pi}{l}nx\right)\right\} \qquad (18)$$

where, for fixed t, $L_n(t;u)$ and $\mathcal{L}(t;u)$ are linear continuous functionals on $L^2(0,t)$ and, for fixed $u \in L^2_{\text{loc}}(0,+\infty)$, the functions $t \mapsto L_n(t;u)$ and $t \mapsto \mathcal{L}(t;u)$ are continuous. Furthermore, the first series converges uniformly. The series in the brace converge in $L^2(0,l)$. Hence, the restriction of the distribution to $(0,l)$ is regular and belongs to $C(0,T;L^2(0,l))$ for every $T > 0$.

Remark 2. The function

$$\sum_{n=1}^{+\infty} \left(\frac{n}{\beta_n} - 1\right)\cos\frac{\pi}{l}nx + \sum_{n=1}^{+\infty} \frac{1}{\beta_n}\left(\phi_n'(x) - \frac{\sqrt{2}\pi}{l^2}\cos\frac{\pi}{l}nx\right) \qquad (19)$$

can be expanded in the series of $\{1\} \cup \{\phi'_n(x)/\phi'_n(0)\}$ and presented in the form

$$\Gamma_0 + \sum_{n=1}^{+\infty} \Gamma_n \frac{\phi'_n(x)}{\phi'_n(0)} . \tag{20}$$

The important point is that $\Gamma_0 = 0$, since the integral over $(0, l)$ of (19) is zero. So, the series expansion of $q(x, T)$ in terms of $\{1, \phi'_n(x)/\phi'_n(0)\}$ contains the constant term due to the presence of the nonzero functional $(\pi/l)\mathscr{L}(T; u)$, see (18).

4.4 The Proof of Independence

Since we have proved continuity in time of both temperature and flux in space $L^2(0, l)$, it makes sense to compute $\theta(\cdot, T)$ and $q(\cdot, T))$ and to investigate their relations. We consider the series representation of $\theta(\cdot, T)$ and $q(\cdot, T)$ on $(0, l)$ in terms of $\phi_n(x)$ and $\phi'_n(x)/\phi'_n(0)$ (see Remark 2) and prove that the coefficients in these expansions can be arbitrary assigned. The extension of $q(\cdot, T)$ to $\mathbb{R}$ is the extension of its regular part (which may be an arbitrary element of $L^2(0, l)$) plus a singular part, not supported on $(0, l)$, which depends on the regular part.

Using (9) and (10), we see that the coefficients can be arbitrarily assigned at a certain time T, if we can impose the equalities

$$\theta(x, T) = \sum_{n=1}^{+\infty} \phi'_n(0)\phi_n(x) \int_0^T z_n(T - r) \int_0^r N(r - s)u(s) \, ds \, dr$$

$$= \sum_{n=1}^{+\infty} \phi_n(x)\xi_n ,$$

$$q(x, T) = \sum_{n=1}^{+\infty} \phi'_n(x)\phi'_n(0)\Gamma_n = \sum_{n=1}^{+\infty} \frac{\phi'_n(x)}{\phi'_n(0)} \eta_n + c \sum_{n=1}^{+\infty} \frac{\phi'_n(x)}{\phi'_n(0)} \tag{21}$$

where both $\{\xi_n\}$ and $\{\eta_n\}$ are arbitrary in l^2 and $c \in \mathbb{R}$ is arbitrary and

$$\Lambda_n = \int_0^T u(r) \left[\int_0^{T-r} N(T - r - s) \left(\int_0^s N(s - v)z_n(v) \, dv \right) ds \right] dr . \tag{22}$$

Note that the last sum in (21) corresponds to the contributions of the series (15), (16) (since $\phi'_n(0) \asymp \beta_n$) and provides both the constant term in the series expansion of $q(\cdot, T)$ and the coefficient of the singular part of the extension of $q(\cdot, T)$ to $\mathbb{R}$.

So, for every $n = 1, 2, \dots$, the input u has to solve two equalities. The first one is

$$\xi_n = \phi'_n(0) \int_0^T u(s) \int_0^{T-s} N(T - s - r)z_n(r) \, dr \, ds . \tag{23}$$

Using the equation for $z_n(t)$ in (8), the expression of ξ_n in (23) and (22), we see that the second equality (21) holds if

$$\int_0^T u(s) \left[\left(\frac{\phi_n'(0)}{\lambda_n} \right)^2 \int_0^{T-s} N(T-s-r)z_n'(r)\, dr \right] ds$$

$$= 2\alpha \left(\frac{\phi_n'(0)}{\lambda_n} \right)^2 \frac{\xi_n}{\phi_n'(0)} - (\eta_n + c). \tag{24}$$

So, the system (1), (3) is exactly controllable in time T, and hence $\theta(\cdot,T)$ and $q(\cdot,T)$ are independent, if we can find $u \in L^2(0,T)$ which solve the moment problem described by the equalities in (23) and (24), for $n \geq 1$.

It is convenient to reduce this problem to a *complex valued moment problem* integrating by parts the left hand side of (24) and introducing

$$d_{\pm n} = \xi_n \pm i \left(2\alpha \frac{\phi_n'(0)}{\lambda_n^2} \xi_n - \eta_n \right), \quad d_0 = -c$$

and

$$R_{\pm n}(t) = \phi_n'(0) \int_0^t N(t-s)z_n(s)\, ds$$

$$\pm i \left(\frac{\phi_n'(0)}{\lambda_n} \right)^2 \left[z_n(t) + \int_0^t N'(t-s)z_n(s)\, ds \right],$$

$$R_0(t) = N(t).$$

The moment problem described by the equalities in (23) and (24) is equivalent to the complex moment problem

$$\int_0^T u(t)R_n(T-t)\, dt = d_n, \qquad n \in \mathbb{Z}.$$

This moment problem is solvable for every $T \geq 2l$ since, as proved in [6, Lemma 19], the family $\{R_n(t)\}_{n\in\mathbb{Z}}$ is an $\mathscr{L}$-basis in $L^2(0,T)$ for $T \geq 2l$. Hence, the following theorem is true.

Theorem 3. *Temperature and flux are independent if $T \geq 2l$ in the sense that it is possible to control the temperature $\theta(\cdot,T)$ and the flux $q(\cdot,T)$ to independent targets, both in $L^2(0,l)$.*

Using the same ideas — quadratical closeness to exponentials and ω−independence — we can also describe properties of the family $\{R_n(t)\}_{n\in\mathbb{Z}}$ in $L^2(0,T)$ for $T < 2l$. The following result is based on Theorem II.4.16 in [3] which states the corresponding properties for exponential families.

Theorem 4. *For any $T \in (0,2l)$ there exists a subfamily of $\{R_n(t)\}_{n\in\mathbb{Z}}$ constituting a Riesz basis in $L^2(0,T)$.*

Using this theorem we can prove the following results about dependence of heat and flux, described by the equations (1) and (3), for small T. It extends the results proved in the special case in Section 3.

Theorem 5. *For $T \leq l$ flux is uniquely determined by temperature and vice versa. More precisely, we can prescribe $\theta(\cdot, T)$ (or $q(\cdot, T)$) to be an arbitrary function from $L^2(0, l)$ supported on $[0, T]$, then $q(\cdot, T)$ (or $\theta(\cdot, T)$) is uniquely determined by this function. For $l < T < 2l$, $\theta(\cdot, T)$ (or $q(\cdot, T)$) may be done an arbitrary function from $L^2(0, l)$, then $q(\cdot, T)$ (or $\theta(\cdot, T)$) is partly determined by this function and cannot be an arbitrary function from $L^2(0, l)$.*

The proof of this theorem is quite similar to the proofs of the corresponding controllability (lack of controllability) results for hyperbolic equations for small time, see, e.g. [3, Ch. V], [4, 5].

Acknowledgements. This paper fits into the research programs of GNAMPA-INDAM. S. Avdonin is supported in part by the National Science Foundation, grant ARC 0724860. L. Pandolfi supported in part by Italian MURST (PRIN 2008 "Analisi Matematica nei Problemi Inversi per le Applicazioni") and by the project "Groupement de Recherche en Contrôle des EDP entre la France et l'Italie (CONEDP)"

References

1. Avdonin, S.A.: On Riesz bases of exponentials in L^2. Vestnik Leningr. Univ., Ser. Mat., Mekh., Astron. 13, 5–12 (1974) (Russian); English transl. Vestnik Leningr. Univ. Math. 7, 203–211 (1979)
2. Avdonin, S., Belinskiy, B., Pandolfi, L.: Controllability of a nonhomogeneous string and ring under time dependent tension. Math. Model. Natur. Phenom. 5, 4–31 (2010)
3. Avdonin, S.A., Ivanov, S.A.: Families of Exponentials. The Method of Moments in Controllability Problems for Distributed Parameter Systems. Cambridge University Press, New York (1995)
4. Avdonin, S.A., Moran, W.: Simultaneous control problems for systems of elastic strings and beams. Systems and Control Letters 44, 147–155 (2001)
5. Avdonin, S., Moran, W.: Ingham type inequalities and Riesz bases of divided differences. Int. J. Appl. Math. Comput. Sci. 11, 101–118 (2001)
6. Avdonin, S., Pandolfi, L.: Simultaneous temperature and flux controllability for heat equations with memory, submitted. Politecnico di Torino, Dipartimento di Matematica, Rapporto Interno (3) (2010)
7. Barbu, V., Iannelli, M.: Controllability of the heat equation with memory. Diff. Integral Eq. 13, 1393–1412 (2000)
8. Cattaneo, C.: Sulla conduzione del calore. Atti del Seminario Matematico e Fisico dell'Università di Modena 3, 83–101 (1948)
9. De Kee, D., Liu, Q., Hinestroza, J.: Viscoelastic (non-fickian) diffusion. The Canada J. of Chemical Engineering 83, 913–929 (2005)
10. Fabrizio, M., Morro, A.: Mathematical problems in linear viscoelasticity. SIAM Studies in Applied Mathematics, vol. (12). Society for Industrial and Applied Mathematics (SIAM), Philadelphia (1992)
11. Fu, X., Yong, J., Zhang, X.: Controllability and observability of the heat equation with hyperbolic memory kernel. J. Diff. Equations 247, 2395–2439 (2009)
12. Gohberg, I.C., Krejn, M.G.: Opèrateurs linèaires non auto-adjoints dans un espace hilbertien, Dunod, Paris (1971)

13. Gurtin, M.E., Pipkin, A.G.: A general theory of heat conduction with finite wave speed. Arch. Rat. Mech. Anal. 31, 113–126 (1968)
14. Jäckle, J.: Heat conduction and relaxation in liquids of high viscosity. Physica A. 162, 377–404 (1990)
15. Leugering, G.: Time optimal boundary controllability of a simple linear viscoelastic liquid. Math. Methods in the Appl. Sci. 9, 413–430 (1987)
16. Levitan, B.M., Sargsjan, I.S.: Introduction to spectral theory: selfadjoint ordinary differential operators. In: Translations of Mathematical Monographs, vol. 39. American Mathematical Society, Providence (1975)
17. Loreti, P., Komornik, V.: Fourier series in control theory. Springer, New-York (2005)
18. Loreti, P., Sforza, D.: Reachability problems for a class of integro-differential equations. J. Differential Equations 248, 1711–1755 (2010)
19. Marchenko, V.A.: Sturm-Liouville operators and applications. Operator Theory: Advances and Applications, vol. 22. Birkhäuser, Basel (1986)
20. Pandolfi, L.: The controllability of the Gurtin-Pipkin equation: a cosine operator approach. Applied Mathematics and Optim. 52, 143–165 (2005)
21. Pandolfi, L.: Controllability of the Gurtin-Pipkin equation. In: SISSA, Proceedings of Science, PoS(CSTNA2005)015
22. Pandolfi, L.: Riesz systems and controllability of heat equations with memory. Int. Eq. Operator Theory 64, 429–453 (2009)
23. Pandolfi, L.: Riesz systems and moment method in the study of heat equations with memory in one space dimension. Discr. Cont. Dynamical Systems, Ser. B 14, 1487–1510 (2010)
24. Pandolfi, L.: Riesz systems and an identification problem for heat equations with memory. Discr. Cont. Dynamical Systems, Ser. S 4, 745–759 (2011)
25. Pandolfi, L.: On–line input identification and application to Active Noise Cancellation. Annual Rev. Control 34, 245–261 (2010)
26. Joseph, D.D., Preziosi, L.: Heat waves. Rev. Modern Phys. 61, 41–73 (1989); Addendum to the paper: Heat waves. Rev. Modern Phys. 62, 375–391 (1990)
27. Renardy, M.: Are viscoelastic flows under control or out of control? Systems Control Lett. 54, 1183–1193 (2005)
28. Richards, J.I., Youn, H.: Theory of distributions: a non technical introduction. Cambridge U.P., Cambridge (1990)
29. Sedletskii, A.M.: Biorthogonal decomposition of functions in exponential series on real intervals. Uspekhi Mathematicheskikh Nauk. 37, 51–95 (1982) (Russian); English transl. Russian Math. Surveys, 37, 57–108 (1983)
30. Tricomi, F.: Differential Equations. Blackie & Sons, Toronto (1961)
31. Vinokurov, V.A., Sadovnichii, V.A.: Asymptotics of any order for the eigenvalues and eigenfunctions of the Sturm–Liouville boundary value problem on a segment with a summable potential. Izvestia: Mathematics 64, 47–108 (2000)
32. Young, R.M.: An Introduction to Nonharmonic Fourier Series. Academic Press, New York (2001)

Identifiability and Algebraic Identification of Time Delay Systems

Lotfi Belkoura

Abstract. Identifiability and algebraic identification of time delay systems are investigated in this paper. Identifiability results are first presented for linear delay systems described by convolution equations. On-line algorithms are next proposed for both parameters and delay estimation. Based on a distributional technique, these algorithms enable an algebraic and simultaneous estimation by solving a generalized eigenvalue problem. Simulation studies with noisy data and experimental results show the performance of the proposed approach.

1 Introduction

The real time delay identification is one of the most crucial open problems in the field of delay systems (see, e.g., [14]), and several on line estimation methods have been suggested in the literature for the identification of delay. While the most popular technique of Pade approximation is limited by the range of validity of the approximation, most of the other approaches (see, e.g., [5, 8] for adaptive techniques or [13] for a modified least squares technique) generally suffer from poor speed performance. Recent developments in [1] in a ditributional framework and in [15, 16] with operational calculus based on Mikusinski's operators, have considered the on line identification of delay systems with particular (structured) inputs, and this paper extends the identification problem to more general input-output trajectories.

In an identification problem, and prior to the design off efficient estimation algorithms, the question arises as to in which sense an input/output description of the process, if any, is unique. An identifiability analysis is then required to ensure uniqueness off the proposed description. This aspect is also addressed in this paper for linear systems, where identifiability conditions, which can be formulated in

Lotfi Belkoura

LAGIS (FRE 3303 CNRS) and INRIA LNE Non-A Project, Université Lille Nord de France

e-mail: lotfi.belkoura@univ-lille1.fr

R. Sipahi et al. (Eds.): Time Delay Sys.: Methods, Appli. and New Trends, LNCIS 423, pp. 103–117.

terms of controllability of time delay systems, are presented for a general class of systems with discrete and distributed delays.

In the proposed continuous time framework, the time-delay is not restricted to be a multiple of some sampling interval, and exact and non asymptotic formulations are obtained for both parameters and delay estimations. Although the parameter estimation technique is still inspired from the fast identification techniques that were proposed [7] for linear, delay free and finite-dimensional models, this paper considers a distributional approach in which the cancelation techniques are reduced to simple multiplication by appropriate functions.

The paper is organized as follows. Section 2 presents sufficient identifiability conditions for time delay systems, and Section 3 presents the estimation methods for parameters and delay identification in case of structured entries. The unstructured case, where the cancelation method can not be applied is considered in Section 4. In this section, the problem of initial conditions that allows experiments to start from an arbitrary starting point is also addressed. Most of our developments are illustrated on a first or second order processes, although extension to higher order systems is generally straightforward. The following subsection presents the general framework and specific tools that are required in the subsequent developments.

1.1 The Distribution Framework

Throughout this paper, functions will be considered through the distributions they define, i.e. as continuous linear functionals on the space $\mathscr{D}$ of C^∞-functions having compact support in $[0, \infty)$. This framework allows the definition of the Dirac distribution $u = \delta$ and its derivative $u = \dot{\delta}$ as $\langle u, \varphi \rangle = \varphi(0)$ and $\langle u, \varphi \rangle = -\dot{\varphi}(0)$, $\varphi \in \mathscr{D}$ respectively. More generally, every distribution is indefinitely differentiable, and if u is a continuous function except at a point a, its distributional derivative writes:

$$u^{(1)} = \frac{du}{dt} + \sigma_a \delta_a. \tag{1}$$

where $\delta_a := \delta(t - a)$, $\sigma_a := u(t_a+) - u(t_a-)$, and $\frac{du}{dt}$ stands for the distribution stemming from the usual derivative (function) of u defined almost everywhere. Note that with $a = 0$, this result, and its extension to higher order derivation, is nothing but the analog part of the familiar Laplace transform $\mathscr{L}(\dot{y}) = sy(s) - y_0$.

We proceed this introductory section with some well-known definitions and results from the convolution products, and as usual, denote $\mathscr{D}'_+$ the space of distributions with support contained in $[0, \infty)$. It is an algebra with respect to convolution with identity δ. For $u, v \in \mathscr{D}'_+$, this product is defined as $\langle u * v, \varphi \rangle = \langle u(x).v(y), \varphi(x + y) \rangle$, and can be identified with the familiar convolution product $(u * v)(t) = \int_0^\infty u(\theta)v(t - \theta)d\theta$ in case of locally bounded functions u and v. Derivation, integration and translation can also be defined from the convolutions

$\dot{u} = \dot{\delta} * u$, $\int u = H * u$, $u(t - \tau) = \delta_\tau * u$, where H is the familiar Heaviside step function. We also recall the following well known property:

$$u(t) * v(t - \tau) = u(t - \tau) * v(t) = \delta_\tau * u * v. \tag{2}$$

As for the supports, one has for $u, v \in \mathscr{D}'_+$:

$$\text{supp } u * v \subset \text{supp } u + \text{supp } v, \tag{3}$$

where the sum in the right hand side is defined by

$$A + B = \{x + y; x \in A, y \in B\} \tag{4}$$

The specific need for the distributional framework also lies in its ability to cancel the singular terms simply by means of multiplication with some appropriate functions. Multiplication of two distributions (say α and u) always make sense when at least one of the two terms (say α) is a smooth function, and the cancelation procedure presented in this paper will be derived from the following general Theorem:

Theorem 1. *[17] If u has a compact support K and is of order r (necessarily finite)*[1], *$\alpha u = 0$ whenever α and its derivatives of order $\leq r$ vanish on K.*

Particularly, one has for the singular Dirac distributions:

$$\alpha \delta_\tau = \alpha(\tau) \delta_\tau, \tag{5}$$

$$\alpha \delta_\tau^{(r)} = 0 \quad \forall \alpha \text{ s.t. } \alpha^{(k)}(\tau) = 0, \quad k = 0, \ldots, r. \tag{6}$$

Finally, with no danger of confusion, we shall sometimes denote $u(s)$, $s \in \mathbb{C}$, the Laplace transform of u.

2 Identifiability Analysis

A standard approach for identification of systems implies that the structure of the system is known and the problem is in finding the values of parameters (including the delays) involved in the set of equations describing the process. The ability to ensure this objective is typically referred to as *parameter* identifiability. First results on identifiability for delay differential equations can be found in [9, 10, 12]. However, these results are limited to the homogeneous case (no forcing term) and use a spectral approach involving infinite dimensional spectrum. The approach used in [2] extends the identifiability analysis to more general systems described by convolution equations of the form:

---

[1] The order of a distribution u is the smallest integer m such that: $\exists C > 0$ such that for any smooth function φ, one has $|\langle u, \varphi \rangle| \leq C \sup_{0 \leq i \leq m} \|\varphi^{(i)}\|_\infty$ (for example, $\delta^{(r)}$ + lower order terms is of order r).

$$R * w = 0, \qquad R = [P, -Q], \qquad w = \begin{bmatrix} y \\ u \end{bmatrix}, \tag{7}$$

where $P(n \times n)$ and $Q(n \times q)$ are matrices with entries in the space $\mathscr{E}'$ of distributions with compact support. Equation (7) correspond to a behavioral approach of systems described by convolutional equations (see [18]). Here, $R(s)$, the Laplace transform of R, provides a kernel representation of the behavior $\mathscr{B}$ which consists in the set $\tilde{w}$ of all admissible trajectories in the space of $C^\infty(\mathbb{R}, \mathbb{R})$ functions, and $\tilde{w} \in \mathscr{B} = \ker_{\mathscr{E}} R(s)$.

The concept of identifiability is based on the comparison of the original system and its associated reference model governed by (7) and in which R, P, Q and y are replaced by $\hat{R}, \hat{P}, \hat{Q}$ and $\hat{y}$ respectively. System (7) is therefore said to be identifiable if there exists a control u such that the output identity $\hat{y} = y$ results in $\hat{R} = R$, which means uniqueness of the matrix coefficients as well as that of the delays. For most practical cases, and provided a sufficiently rich input signal, identifiability of (7) reduces to:

1. $\operatorname{rank} R(s) = n, s \in \mathbb{C}$,
2. $\operatorname{conv} \det P = n \operatorname{conv} R$.

where $\operatorname{conv} R$ denote the smallest closed interval that contains the support of R (i.e. the convex hull of $\operatorname{supp} R$), $\det P$ is the determinant with respect to the convolution product, and $n \operatorname{conv} R$ stands for the sum of n terms as defined in (4). These conditions are closely linked to the property of approximate controllability in the sense that the reachable space is dense in the state space [20]. The following example [2] shows the applicability of the previous result to systems with distributed delays. Consider the multivariable delay system:

$$\begin{aligned} \dot{x}_1(t) &= x_1(t) + \int_{-1}^{0} x_2(t + \tau) d\tau \\ \dot{x}_2(t) &= x_1(t-1) + x_2(t) + \int_{-1}^{0} u(t + \tau) \end{aligned} \tag{8}$$

and denote $\pi(t) = H(t) - H(t-1)$, with H the Heaviside function. Here, $\operatorname{supp} \pi = [0,1]$ and some simple manipulations show that system (8) admits a kernel representation $R * \omega = 0$ with $\omega = (x_1, x_2, u)^T$ and

$$R = [P, -Q] = \begin{bmatrix} \delta' - \delta & -\pi & 0 \\ -\delta_1 & \delta' - \delta & -\pi \end{bmatrix} \tag{9}$$

Clearly, $\operatorname{conv} R = [0,1]$ while $\det P = \delta'' - 2\delta' + \delta - \delta_1 * \pi$, from which one easily gets $\operatorname{conv} \det P = [0,2]$, so condition (2) of is satisfied. On the other hand,

$$R(s) = \begin{bmatrix} s-1 & -(e^{-s}-1)/s & 0 \\ -e^{-s} & s-1 & -(e^{-s}-1)/s \end{bmatrix}, \tag{10}$$

and the determinant formed by the second and third column of $R(s)$ is nonzero for $s \neq 0$, and for $s = 0$, the first and second column of $R(0)$ form a non singular matrix. Hence condition (1) is also satisfied and system (8) is identifiable.

In the case of distributed delays, the major limitation of the previous approach is the need of the largest delay involved in (7). In return for more restrictive models with lumped and commensurate delays of the form

$$\dot{x}(t) = \sum_{i=0}^{r} A_i \, x(t - i.h) + B_i u(t - i.h),\tag{11}$$

a simpler identifiability result which no longer requires the assumption of an a priori known memory length is obtained in [4]. It can be expressed in terms of weak controllability, concept introduced in [11] for systems over rings, through the rank condition (over the ring $R[\nabla]$):

$$\text{rank}\left[B(\nabla),\ldots,A^{n-1}(\nabla)B(\nabla)\right] = n,\tag{12}$$

$$A(\nabla) = \sum_{i=0}^{r} A_i \nabla^i, \quad B(\nabla) = \sum_{i=0}^{r} B_i \nabla^i.\tag{13}$$

Note however that all the previous results are limited to linear and time invariant models. In case of nonlinear delay systems or time dependent delays, general identifiability results are still expected.

2.1 Sufficiently Rich Input

In identification procedures the design of a sufficiently rich input which enforces identifiability is also an important issue. Given a reference model associated to the process under study, one has to know whether equality of the outputs results in that of the transfer functions. Few results are dealing with such issue for time delay systems. In [2, 3] the input design is considered in the time domain rather than the frequency domain and the approaches are mainly based on the non smoothness of the input. More precisely, if

$$\Lambda_u = \{s_0, s_1, \ldots, s_L, \ldots\}\tag{14}$$

denote the singular support of u (i.e. the set of points in $\mathbb{R}$ having no open neighborhood to which the restriction of u is a C^∞ function), the input is required to be sufficiently "discontinuous" in the sense that

$$\text{rank}\left[U_0(D), \ldots, U_L(D)\right] = q\tag{15}$$

where the polynomial matrices $U_l(D)$ are formed with the (possible) jump of $u^{(k)}(t)$ for some $k \geq 0$ at $t = s_l$ by

$$U_l(D) = \sum_{i=0}^{k} \sigma_l^{k-i} D^i,\tag{16}$$

$$\sigma_l^{k-i} = u^{(k-i)}(s_l + 0) - u^{(k-i)}(s_l - 0). \tag{17}$$

On the other hand, "the discontinuity points" $s_0, s_1, ..$ should be sufficiently spaced in the general case of distributed delays, although for lumped delays, this constraint (which may constitute a serious drawback in situations where on line procedures are used) can be relaxed using commensurability considerations. The simplest example consists of a piece-wise constant $\mathbb{R}^q$-valued function with appropriate discontinuities, although inputs of class $\mathscr{C}^r$ for an arbitrary finite integer r can be formed.

While the identifiability analysis of this section is developed for complex systems with possibly distributed delay, most of the subsequent identification algorithms are performed on simple linear and non linear examples for which the present study turn out to be either straightforward or non applicable. The algebraic identification procedures presented in the next sections clearly separate the case of structured entries (that can be annihilated by means of simple algebraic manipulations) from the unstructured ones.

3 Algebraic Identification: The Structured Case

3.1 Structured Signals and Their Annihilation

Structured entries have been introduced (see e.g. [7]) to refer to entities (mainly perturbations) that can be annihilated by means of simple multiplications and derivations. A simple example consists for instance in a load perturbation, modeled by a Heaviside function $\xi(t) = H(t)$, and for which one easily obtains, in the time and operational domain respectively:

$$t \times \frac{d\xi}{dt} = 0, \quad \frac{d}{ds} s\xi(s) = 0. \tag{18}$$

This aspect is generalized in the Schwartz's Theorem of the introductory Section, and the identification procedure is based on the idea that unknown terms to be identified and contained in a structured term can be recovered by the above cancelation procedure. This approach is illustrated in the following subsections for a simulation and an experimental case respectively.

3.2 Application to a Single Delay Identification

Consider a first order nonlinear system subject to a delayed step input $u(t) = H(t - \tau)$. Such systems, where the delays only appear in the control variables, are most common in practice and this simple example emphasizes that linearity w.r.t. input-outputs is not required.

$$\dot{y} + ay^2 = y_0 \delta + bH(t - \tau), \tag{19}$$

where a, b, and τ are constant parameters. Note that in the distribution sense, the initial condition term occurs as an impulsive term. The coefficient a is assumed to be known, and first order derivation results in equation (20) which may be canceled, for instance, and by virtue of Schwartz's Theorem by the polynomial $\alpha = t^3 - \tau t^2$, yielding:

$$\ddot{y} + a\dot{z} = y_0 \dot{\delta} + y_0 \delta + b\delta_\tau, \tag{20}$$

$$(t^3 - \tau t^2)(\ddot{y} + a\dot{z}) = 0, \tag{21}$$

where for ease of notation we have denoted $z = y^2$. As an equality of distributions, equation (21) does not make sense for any t (otherwise we would get from (21) $t = \tau$). However, integration of (21) results in an equality of functions from which the delay becomes available. To ensure causality, $k \geq 2$ integrations can be used, yielding an explicit and non asymptotic formula of the delay:

$$\tau = \frac{\int^{(k)} t^3 (\ddot{y} + a\dot{z})}{\int^{(k)} t^2 (\ddot{y} + a\dot{z})}, \qquad t > \tau, \tag{22}$$

where the symbol $\int^{(k)}$ stands for iterated integration of order k. Note that the integration by part, illustrated in the partial realization scheme of Figure 1 avoid any derivative in the estimation algorithms. Since from (5), $t^q(\ddot{y} + a\dot{z})$, $q = 2,3$ have

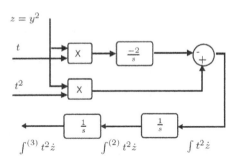

Fig. 1 Realization scheme of $\int^{(k)} t^2 \dot{z}$, $k \geq 2$.

their support reduced to $\{\tau\}$, both numerator and denominator of (22) are with support within (τ, ∞), so the delay is clearly not identifiable for $t < \tau$. Nevertheless, the delay estimation may be achieved in a small time interval $(\tau, \tau + \varepsilon)$. Simulation results in a noisy context are depicted in Figure 2, with $k = 4$ integrations, and

$$(y(0), a, \tau) = (0.3, 2.5, 0.5).$$

In addition to causality requirements, the choice of more than 2 integrations has been used to obtain an additional filtering effect which attenuates the noise from the

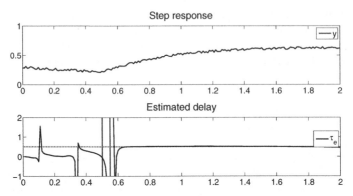

Fig. 2 Trajectory of (19) and delay identification.

measurement y. However, the design of an optimal filter that would contribute to enhance the signal to noise ratio is not a trivial task and is under investigations.

3.3 Application to a Simultaneous Parameters and Delay Identification: Experimental Example

In this section, experiments are carried out on a Feedback process PT37 100. This process consists of heating the air flowing in a tube to the desired temperature level. The physical principle which governs the behavior of the this process is the balance of heat energy. The response of the sensor to a change in the heater power is more-over affected by a pure delay, which depends on the velocity of the process and the distance between the point of change and the sensor. In an open loop configuration, the behavior is approximated by a linear system with delayed input. The process is subject to a step input, and a second order model with both unknown parameters and delay is used to describe it.

$$G(s) = \frac{K e^{-\tau s}}{a_2 s^2 + a_1 s + 1},\qquad(23)$$

The candidate function chosen for the annihilation of the structured part is the complex and bounded function $\alpha = (1 - \lambda e^{-j\omega t})$, with the unknown to be found $\lambda = e^{j\omega \tau} \in \mathbb{C}$, and tunable frequency ω. More explicitly, denoting $e = e^{-j\omega t}$, a derivation of the differential equation derived from (23), followed by the multiplication by α result respectively in:

$$a_2 y^{(3)} + a_1 y^{(2)} + y^{(1)} = K\delta_\tau,\qquad(24)$$

$$(1 - \lambda e)(a_2 y^{(3)} + a_1 y^{(2)} + y^{(1)}) = 0.\qquad(25)$$

We shall focus on the identification of the coefficients $\{\lambda, a_2, a_1\}$, and provided a sufficiently large period $2\pi/\omega$, the delay is deduced from the unique argument

$\tau = \arg(\lambda)/\omega$. Due to the terms λa_i, $i = 1,2$, (25) is not linear w.r.t. the unknown coefficients, but may be written in the following form:

$$[(y^{(3)},\cdots,y^{(1)}) - \lambda(ey^{(3)},\cdots,ey^{(1)})] \begin{pmatrix} a_2 \\ a_1 \\ 1 \end{pmatrix} = 0. \tag{26}$$

As in the previous section, successive integrations transform the equality of singular distributions of (25) into one of continuous functions. Denoting $\Theta = (a_2,a_1,1)^T$ the (normalized) vector of parameters, the specific structure of (26) leads to following generalized eigenvalue problem for possibly non square pencils:

$$(A - \lambda B)\Theta = 0, \tag{27}$$

where, using a Matlab-like notation, the entries of the $m \times 3$ trajectory-dependent matrices A and B are given by

$$A(i,:) = \int^{(i+2)}(y^{(3)},\cdots,y^{(1)}), \quad i = 1,\ldots,m,$$
$$B(i,:) = \int^{(i+2)}(ey^{(3)},\cdots,ey^{(1)}) \quad i = 1,\ldots,m. \tag{28}$$

The implementation of $A(i,j)$ and $B(i,j)$ is performed according to the integration by parts formulas. Therefore, the identification problem has been transformed into the eigenvalue problem (27) in which, at each t, the unknown delay $\tau = \arg(\lambda)/\omega$ is derived from one eigenvalue, while the parameters a_1,a_2 are obtained from the corresponding normalized eigenvector. Solving (27) in a noisy context and in the non-square case (i.e. by considering $m > 3$ lines for A and B) is generally not an easy task, since (27) "has the awkward feature that most matrices have no eigenvalues at all, whilst for those that do, an infinitesimal perturbation will in general remove them" [19]. A possible approach can be based on the pseudo-spectra analysis which consists in introducing the ε−pseudospectra of (27) (see e.g [19] and the references therein). However, this approach is not appropriate for on-line perspectives, and we adopt here a simpler technique, based on the a priori stationarity assumption of the unknown parameters. More precisely, the selected parameters correspond to the eigenpair (λ_i,Θ_i) of the square pencil (27) that minimizes the norm $\|(A + \lambda_i B)\Theta_i\|$ of the rectangular pencil (i.e. for $m > 3$).

As in the single delay estimation problem of the previous section, it can be easily shown that matrices A and B are continuous matrix functions with support within (τ,∞), which means that the delay and parameters are not identifiable for $t < \tau$. Moreover, and although the approach is non asymptotic, this continuity can make the estimation problem sensitive to noise and neglected dynamics in the vicinity of τ. Unlike noise-free contexts or reduced order identification problems such as the delay estimation in the previous section, it is clear a minimum amount of trajectories information is required here to obtain a consistent and relevant eigenvalue problem.

Figure 3 (top) shows the experimental response as well as the simulated trajectories based on the identified delay and parameters (bottom). Although the convergence algorithm is clear, the time history of the identified parameters reflects a singularity of the eigenvalue problem in the vicinity of $t = 0.8s$. The implementation of (28) was converted to discrete-time, assuming a sampling period of 50 msec, and resolution of (27) has been made using the polyeig function of the Matlab software. Taking into account the static gain K estimated by other means, the identified

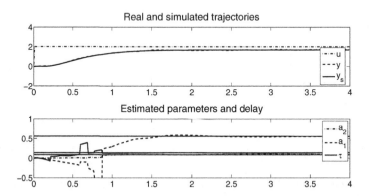

Fig. 3 Experimental and simulated trajectories (top). Estimated delay and parameters (bottom)

second order model for the process reads:

$$G_1(s) \approx \frac{0.84\,e^{-0.13\,s}}{0.09\,s^2 + 0.55\,s + 1}. \tag{29}$$

4 Algebraic Identification: The Unstructured Case

When facing arbitrary inputs, the above annihilation procedure no longer applies, but algebraic estimation results can be still obtained by mean of an approach combining multiplication an cross convolution, as described below. Provided the system is initially at rest (null initial condition), state delay may also be identified.

4.1 The Cross Convolution Approach

We first focus on a single delay identification regardless of any process dynamics. When considered on the whole real line, a delay between two functions $a(t)$ and $b(t)$ reads as in (30) and leads to (31) once multiplied by any deviated known function $\alpha(t - \tau)$.

$$a(t) = b(t - \tau), \tag{30}$$

$$\alpha(t - \tau)a(t) = (\alpha b)(t - \tau). \tag{31}$$

Using (2), a convolution product derived from these two relations results in equation (33) with no deviated argument in the original functions a and b.

$$[\alpha(t - \tau)a(t)] * b(t - \tau) = a(t) * (\alpha b)(t - \tau), \tag{32}$$

$$\Rightarrow [\alpha(t - \tau)a(t)] * b(t) = a(t) * (\alpha b)(t). \tag{33}$$

If the adopted function $\alpha(t - \tau)$ admits an expansion separating its arguments t and τ, i.e.:

$$\alpha(t - \tau) = \sum_{i \text{ finite}} \lambda_i(t)\mu_i(\tau), \tag{34}$$

for some known functions λ and μ, then an algebraic relation is obtained allowing a non asymptotic and explicit delay formulation, as illustrated in the simple following examples:

$$\alpha(t) = t \Rightarrow \tau = \frac{ta * b - a * tb}{b * a} \tag{35}$$

$$\alpha(t) = e^{\gamma t} \Rightarrow e^{\gamma \tau} = \frac{b * e^{\gamma t} a}{a * e^{\gamma t} b} \tag{36}$$

Provided the involved convolution products are well defined, this delay formula holds for all nonzero values of their denominators. More precisely, if the signal b consists in measurements on $(0, \infty)$, then $\operatorname{supp} a \subset (\tau, \infty)$ and hence, by virtue of (3), both numerator and denominator of (35) and (36) have their support within (τ, ∞). Therefore, the delay is not identifiable for $t < \tau$. However, as in the finite dimensional case (see, e.g., [6]), the input signal b being used in this algebraic approach does not necessarily exhibit the classical "persistency of excitation" requirement. Although a local loss of identifiability may occur due to the zero crossing of the denominator, only non trivial trajectories are required.

Once again, when facing derivatives, one of the nice features of multiplication by polynomial or exponential functions lies in the ability to use simple integration by parts formulas to avoid any derivation in the identification algorithm. The next paragraph illustrates the time lag identification for the delayed integrator:

4.2   Application to a Delay Identification

Consider the following linear first order process with delayed input:

$$\dot{y} + y = ku(t - \tau), \tag{37}$$

which correspond to the formulation (30) with $a = \dot{y} + y$ and $b = ku$. In order to avoid multiplications by unbounded functions (polynomials), and hence the amplification of noise and neglected dynamics, a decaying exponential functions is considered and equation (36) reads:

$$\lambda = \frac{u * e^{\gamma t}(\dot{y}+y)}{(\dot{y}+y)*e^{\gamma t}u} \tag{38}$$

where we have denoted $\lambda = e^{-\gamma \tau}$ for some tunable positive parameter γ. Note that the static gain value k is not required nor identified. Denoting $e(t) = e^{-\gamma t}$ and taking into account the integration by parts formula, $\int e\dot{y} = ey + \gamma \int ey$, on gets:

$$\lambda = \frac{ey * u + (1 + \gamma) \int (ey * u)}{eu * y + \int eu * y}, \tag{39}$$

while the delay is obtained from $\tau = \log(\lambda)/\gamma$. For this simple example, and since only a constant delay has to be identified, an additional step considering the integral of the square of equation (39) (*i.e.* $\int (39)^2$) avoids the possible singularities resulting from the zero crossing of the denominator $eu * y$. This finally results in the delay estimation:

$$\lambda = \left[\frac{\int_0^t \left[u * ey + \gamma \int_0^\theta (u * ey) \right]^2 d\theta}{\int_0^t \left[eu * y + \int_0^\theta (eu * y) \right]^2 d\theta} \right]^{\frac{1}{2}}. \tag{40}$$

A simulation result with noisy data is depicted in Figure 4, for an input $u(t) = \sin(t).(0.2 + \sin(0.2t))$, $\gamma = 0.2$, and a delay $\tau = 0.3\,s$. The simulation step size has been fixed to $0.05\,s$, and the integrals involved in the convolutions have been approximated by simple sums.

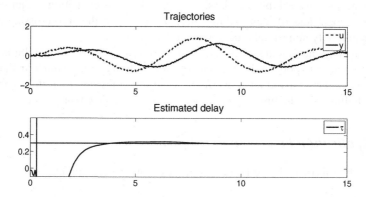

Fig. 4 Trajectories and estimated delay of Eq.(37).

4.3 Experimental Results (Continued)

The above delay estimation procedure is considered in this section for signals a and b describing heating process of Section 3.3. The system is governed by:

$$\sum a_i y^{(i)}(t) = Ku(t - \tau), \tag{41}$$

whose left and right hand sides can be identified with the functions a and b of the previous section (equation 30) as:

$$a = \sum a_i y^{(i)}, \qquad b = K\varepsilon. \tag{42}$$

The exponential function $\alpha(t) = e^{-\gamma t}$, with $\gamma > 0$ is adopted for the multiplication step, leading to a reformulation of equation (33) as:

$$\lambda \sum a_i [u * (e^{-\gamma t} y^{(i)})] = \sum a_i [y^{(i)} * (e^{-\gamma t} u)], \tag{43}$$

where the unknown delay to be identified is contained in the new unknown term $\lambda = e^{-\gamma \tau}$. Note that the process gain K has been removed by this procedure and will not be estimated. Clearly, equation (43) is non causal and non linear with respect to the delay and parameters a_i. The use of successive filters $h_j(s)$ of relative degree > 2 allow one the one hand, to ensure causal relations from (43) and on the other hand, to obtain enough equations for a simultaneous estimation of both parameters and delay. Recalling $\Theta = (a_2, a_1, 1)^T$ the (normalized) vector of parameters, this results in the following estimation problem:

$$(V - \lambda W)\Theta = 0, \tag{44}$$
$$V_{ij} = s^i h_j(s)[y * e^{-\gamma t} u](s), \tag{45}$$
$$W_{ij} = (s + \gamma)^i h_j(s)[e^{-\gamma t} y * u](s), \tag{46}$$

where the realization of each term of the filtering of equation (43), avoiding any measurement's derivative, is based on the familiar property of Laplace transform, $\mathscr{L}[e^{\gamma t} f(t)](s) = F(s - \gamma)$. In the available data V_{ij} and W_{ij}, the notation $[a * b](s)$ emphasizes the measurement based convolution products subject to the filtering procedure. It is worth noticing that in case of a structured input u admitting a simple operational description (for instance a step u(s)=1/s), the realization of the entries of V and W reduces to simple filtering of y and $e^{-\gamma t} y$ as:

$$V_{ij} = v_{ij}(s)y(s), \qquad v_{ij}(s) = s^i h_j(s)u(s + \gamma),$$
$$W_{ij} = w_{ij}(s)y(s + \gamma), \quad w_{ij}(s) = (s + \gamma)^i h_j(s)u(s)$$

Figure 5 shows the convergence of the estimated parameters in case of a step input, where we adopted simple integrations for the filters (i.e. $h_j(s) = 1/s^{j+2}$). Taking

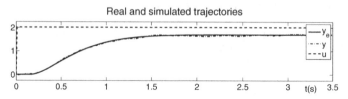

Fig. 5 Estimated delay and parameters of the heat process

into account the static gain K estimated by other means, we recover closed results compared to the transfer function (29) estimated in Section 3.3:

$$G_2(s) \approx \frac{0.84\,e^{-0.136\,s}}{0.1\,s^2 + 0.55\,s + 1}. \tag{47}$$

5   Conclusion

We have presented sufficient conditions for the identifiability of a general class of systems described by convolution equations. More specifically, for delay differential systems, it is shown how the identifiability property can be formulated in terms of approximate controllability or weak controllability, depending on the available models.

The second part of this paper has presented an algebraic method for the identification of delay systems based on both structured inputs and arbitrary input-output trajectories. The ability of identification on bounded sets of measurements also allows us to extend the estimation problem to slowly time varying parameters and delay. Extensions to the identification in the multivariable and multidelay cases, rigorous proofs for non stationary processes, a well as a deeper study of the eigenvalue problem singularities are open problems under investigation.

Acknowledgements. This work was supported by the French Ministry of Higher Education and Research, Nord-Pas de Calais Regional Council and FEDER through the 'Contrat de Projets Etat Region' (CPER) 2007-2013.

References

1. Belkoura, L., Richard, J.P., Fliess, M.: Parameters estimation of systems with delayed and structured entries. Automatica 45(5), 1117–1125 (2009)
2. Belkoura, L.: Identifiability of systems described by convolution equations. Automatica 41, 505–512 (2005)
3. Belkoura, L., Dambrine, M., Orlov, Y., Richard, J.: An Optimized Translation Process and Its Application to ALGOL 68. LNCSE Advances in Time Delay Systems, vol. 38, pp. 132–135. Springer (2004)
4. Belkoura, L., Orlov, Y.: Identifiability analysis of linear delay-differential systems. IMA J. of Mathematical Control and Information 19, 73–81 (2002)

5. Drakunov, S.V., Perruquetti, W., Richard, J.P., Belkoura, L.: Delay identification in time-delay systems using variable structure observers. Annual Reviews in Control, 143–158 (2006)
6. Fliess, M., Sira-Ramirez, H.: Closed-loop parametric identification for continuous-time linear systems via new algebraic techniques. In: Garnier, H., Wang, L. (eds.) Continuous-Time Model Identification from Sampled Data. Springer (2007),
http://hal.inria.fr/inria-00114958
7. Fliess, M., Sira-Ramirez, H.: An algebraic framework for linear identification. ESAIM Control, Optimization and Calculus of Variations 9 (2003)
8. Gomez, O., Orlov, Y., Kolmanovsky, I.V.: On-Line Identification of SISO Linear Time-Delay Systems From Output Measurements. Automatica 43(12), 2060–2069 (2007)
9. Lunel, S.M.V.: Parameter identifiability of differential delay equations. Int J. of Adapt. Control Signal Process. 15, 655–678 (2001)
10. Lunel, S.M.V.: Identification problems in functionnal differential equations. In: Proc. 36th conf. on Decision and Control, pp. 4409–4413 (1997)
11. Morse, A.S.: Ring models for delay differential systems. Automatica 12, 529–531 (1976)
12. Nakagiri, S., Yamamoto, M.: Unique identification of coefficient matrices, time delay and initial functions of functionnal differential equations. Journal of Mathematical Systems, Estimation and Control 5(3), 323–344 (1995)
13. Ren, X.M., Rad, A.B., Chan, P.T., Lo, W.: On line identification of continuous-time systems with unknown time delay. IEEE Tac. 50(9), 1418–1422 (2005)
14. Richard, J.P.: Time-delay systems: an overview of some recent advances and open problems. Automatica 39, 1667–1694 (2003)
15. Rudolph, J., Woittennek, F.: An algebraic approach to parameter identification in linear infinite dimensional systems. In: 16th Mediterranean Conference on Control and Automation, Ajaccio, France, pp. 25–27 (2008)
16. Rudolph, J., Woittennek, F.: Ein algebraischer Zugang zur Parameteridentifikation in linearen unendlichdimensionalen Systemen. Automatisierungstechnik 55, 457–467 (2007)
17. Schwartz, L.: Théorie des distributions. Hermann, Paris (1966)
18. Vettori, P., Zampieri, S.: Controllability of Systems described by convolutionnal or delay-differential equations. SIAM Journal of Control and Optimization 39(3), 728–756 (2000)
19. Wright, T.G., Trefethen, L.N.: Pseudospectra of rectangular matrices. IMA J. of Numer. Anal., 501–519 (2002)
20. Yamamoto, Y.: Reachability of a class of infinite-dimensional linear systems: An external approach with application to general neutral systems. SIAM J. of Control and Optimization 27, 217–234 (1989)

Stability Analysis for a Consensus System of a Group of Autonomous Agents with Time Delays

Rudy Cepeda-Gomez and Nejat Olgac

Abstract. This study addresses the consensus problem for a group of autonomous agents with second order dynamics and time-delayed communications. The constraints on the communication topology can be relatively relaxed to two: it is time invariant, and absolutely connected. This represents a large class of dynamics, of which a small subset is presented here: where all the agents in the group communicate with each other, and that the time delay incurred is constant and equal for all the communication channels. An efficient control structure of PD type is proposed to achieve consensus in the position and velocity of the agents. The proposed control law introduces a particular construction in the characteristic equation of the system, which is first factorized to dramatically simplify the stability analysis in the delay space. Then using cluster treatment of characteristic roots (CTCR) procedure a complete stability picture is obtained, taking into account the variations in the control parameters and the communication delay. Case studies and simulations results are presented to illustrate the analytical derivations.

1 Introduction

Lately, distributed (decentralized) coordination of autonomous systems with multiple agents has been studied in many investigations. This interest is mainly due to the broad spectrum of applications of such systems in many areas, e.g., unmanned explorations, search and rescue operations. The consensus generation is one of the most studied topics. The main objective is to drive all the agents of the group in a

Rudy Cepeda-Gomez
Mechanical Engineering department, University of Connecticut, Storrs, CT, USA
e-mail: rudycepeda@engr.uconn.edu

Nejat Olgac
Mechanical Engineering department, University of Connecticut, Storrs, CT, USA
e-mail: olgac@engr.uconn.edu

R. Sipahi et al. (Eds.): Time Delay Sys.: Methods, Appli. and New Trends, LNCIS 423, pp. 119–133.
springerlink.com   © Springer-Verlag Berlin Heidelberg 2012

way such that they will reach a common behaviour, such as a formation dynamics, rendez-vous operation. The important ground rule in this process is that all agents should be guided under the same autonomous control law although their operating conditions may all be different. For instance, they may be communicating with different number of informant agents, these communications may or may not be directed. The operation is successful when the group of agents reach to consensus in a stable fashion within a selected consensus time. The agents should also be able to reject the unforeseen disturbances, as the control process continues.

The work of Olfati-Saber and Murray [6] is one of the earlier studies published, which introduces the consensus problem for multi agent coordination, focusing on agents with first order dynamics and considering different communication topologies among the agents. They propose a protocol for an average consensus, which is a control strategy that drives the agents to the average value of the group's initial conditions. Under the simplifying features of the first order governing dynamics, they also study the behaviour of their protocol when communication delays are present, keeping the topology fixed. Several other researchers [3, 4, 5, 10, 11, 12] have performed further extensions on this earlier work, proposing different protocols for zero consensus of agents that are driven by second order dynamics. In these studies, the consensus is reached for the position and the velocity of the agents at zero. They also include variations on the topologies and communication delays.

In this chapter, we propose and analyze a different control strategy for consensus over a group of agents with second order dynamics. They operate under a time-delayed communication structure. The main contribution is the treatment of the complete stability picture for such systems, taking into account variations of the control parameters as well as the delay. For simplicity, we consider a fixed topology in which one agent can communicate with all the other members of the set.

One feature of the control logic proposed here is that it introduces an interesting structure for the system dynamics, allowing a factorization of the characteristic equation that simplifies the analysis. This simplification reduces the problem from the analysis of a system whose order depends on the number of agents to the analysis of only two second order systems with no commensuracy (integer multiplicity of delays) in the delay terms.

The text is organized as follows. Section 2 presents the system dynamics and introduces the simplifying construction for the characteristic equation, which is a consequence of the control logic proposed. Section 3 takes advantage of this construction to perform the stability analysis and declares the regions of stability and instability for the system in the parameter space. Section 4 presents some case studies and simulation results. The conclusions and directions for further research are presented in Section 5.

In the rest of the chapter, bold face notation is used for vector quantities, bold capital letters for matrices and italic symbols for scalars.

2 Problem Statement

Consider a group of N autonomous agents, which are driven by second order dynamics:

$$\ddot{x}_j(t) = u_j(t), \qquad j = 1, 2, \ldots, N \tag{1}$$

where $x_j(t)$ is taken as the scalar position and $u_j(t)$ is the control law (also scalar). Here we treat the motion of the agent as one dimensional, but the entire analysis is still valid for higher dimensions. We declare a successful consensus when all N agents are at the same position, i.e. $\lim_{t \to \infty} x_j(t) - x_k(t) = 0$ for any $j, k \in [1, N]$. Notice that this consensus definition does not state if this common position is zero or not.

Assuming there is a communication channel between each pair of agents, and that all the channels have a constant communication delay of τ seconds, i.e. agent j knows the position and velocity of all the other $N - 1$ agents τ seconds earlier, the control logic for the j-th agent is proposed as:

$$\ddot{x}_j(t) = u_j(t) = P \left(\sum_{\substack{k=1 \\ k \neq j}}^{N} \frac{x_k(t - \tau)}{N - 1} - x_j(t) \right) \tag{2}$$

$$+ D \left(\sum_{\substack{k=1 \\ k \neq j}}^{N} \frac{\dot{x}_k(t - \tau)}{N - 1} - \dot{x}_j(t) \right)$$

The controller defined in (3) uses PD logic to bring the position of agent j to the centroid of the rest of the group and consequently its velocity to the mean velocity of the remainder of the group (its informers), using the last known position and velocity of the other agents. Notice that this protocol differs from all previously proposed schemes in the fact that the communication delay affects the information coming from all the other agents, but not the state of the j-th agent, i.e., no self-delay.

With the controller (3) in place, the dynamics of the whole system can be represented in the state space as:

$$\dot{\mathbf{x}}(t) = \mathbf{A}\mathbf{x}(t) + \mathbf{B}\mathbf{x}(t - \tau) \qquad \in \mathbb{R}^{2N} \tag{3}$$

where:

$$\mathbf{x} = \begin{bmatrix} x_1 \\ \dot{x}_1 \\ x_2 \\ \dot{x}_2 \\ \vdots \\ x_N \\ \dot{x}_N \end{bmatrix} \in \mathbb{R}^{2N}, \quad \mathbf{A} = \mathbf{I}_N \otimes \begin{bmatrix} 0 & 1 \\ -P & -D \end{bmatrix}, \quad \mathbf{B} = \begin{bmatrix} 0 & 1 & \cdots & 1 \\ 1 & 0 & \cdots & 1 \\ \vdots & \vdots & \ddots & \vdots \\ 1 & 1 & \cdots & 0 \end{bmatrix} \otimes \begin{bmatrix} 0 & 0 \\ \frac{P}{N-1} & \frac{D}{N-1} \end{bmatrix} \tag{4}$$

and $\otimes$ denotes Kronecker multiplication. The particular construction of the matrices **A** and **B**, which are the consequence of the control law (3), introduces a unique feature to the characteristic equation, stated in the following lemma.

Lemma 1. *The characteristic equation,*$\det(s\mathbf{I} - \mathbf{A} - \mathbf{B}e^{-\tau s}) = 0$, *of the dynamics (3) can be decomposed into two factors of the form:*

$$\left[s^2 + Ds\left(1 - e^{-\tau s}\right) + P\left(1 - e^{-\tau s}\right)\right] \times$$
$$\left[s^2 + Ds\left(1 + \frac{e^{-\tau s}}{N-1}\right) + P\left(1 + \frac{e^{-\tau s}}{N-1}\right)\right]^{N-1} = 0 \qquad (5)$$

Proof. Taking the laplace transform of the dynamics of a single agent in (3), we have N coupled equations of the form:

$$\left(s^2 + Ds + P\right)X_j(s) - \frac{e^{-\tau s}}{N-1}\sum_{\substack{k=1 \\ k \neq j}}^{N}(Ds + P)X_k(s) = 0, \quad j = 1, \ldots, N \qquad (6)$$

It is possible to arrange these N equations as $\mathbf{F}\left[X_1(s)\ X_2(s)\ \ldots\ X_N(s)\right]^T = 0$, with:

$$\mathbf{F} = \begin{bmatrix} a & -b & \cdots & -b \\ -b & a & \cdots & -b \\ \vdots & \vdots & \ddots & \vdots \\ -b & -b & \cdots & a \end{bmatrix} \in \mathbb{R}^{N \times N} \quad , \begin{array}{l} a = s^2 + Ds + P \\ b = \frac{e^{-\tau s}}{N-1}(Ds + P) \end{array} \qquad (7)$$

In order to get a unique and nontrivial equilibrium solution to(6), $\mathbf{F}$ needs to be singular, i.e. $\det(\mathbf{F}) = 0$. Adding all the columns of (7) to the first one, and then subtracting the first row form the others we arrive at:

$$[a - b(N-1)] \begin{vmatrix} 1 & -b & \cdots & -b \\ 0 & a+b & \cdots & 0 \\ \vdots & \vdots & \ddots & \vdots \\ 0 & 0 & \cdots & a+b \end{vmatrix} = [a - b(N-1)](a+b)^{N-1} = 0 \qquad (8)$$

Using the definition of a and b in (7), (8) is transformed into:

$$\left[s^2 + Ds\left(1 - e^{-\tau s}\right) + P\left(1 - e^{-\tau s}\right)\right] \times$$
$$\left[s^2 + Ds\left(1 + \frac{e^{-\tau s}}{N-1}\right) + P\left(1 + \frac{e^{-\tau s}}{N-1}\right)\right]^{N-1} = 0 \qquad (9)$$

$\square$

Remark 1: As it is shown in Appendix 1, the first factor of (9) represents the dynamics of the centroid of the agents, while the second factor is related to the dynamics of the position error between any two agents.

The result of lemma 1 simplifies the problem considerably, transforming it from the stability analysis of a $2N$-order system with time delays into the analysis of two second order systems with single time delay and no commensuracy in the delay terms, i.e. no integer multiples of the delay appear.

According to remark 1, the stability of the first term is related to the stability of the position value the agents will agree on, which forms a part of the consensus. The second factor is related to the stability of the relative motion of the agents with respect to the expected consensus. This means that a stable consensus will be reached if the second factor of (9) is stable, and the consensus value will be stable if the first factor of (9) represents a stable behaviour. Therefore we name the second factor as the governing structure of the "disagreement dynamics". When it is stable, the agents come to a common mode of operation, i.e., the consensus.

3   Stability Analysis

Using the facilitating feature introduced by lemma 1, the stability of each factor of (9) can be assessed separately. Then the stability regions in the parametric space are intersected to find those compositions of (P,D,τ) that bring stable operation for the whole system.

The stability analysis for each factor will be performed following the Cluster Treatment of Characteristic Roots (CTCR) methodology [8]. The first step of CTCR requires an exhaustive determination of the possible imaginary roots of the characteristic equation. For the particular dynamics of concern here, this step becomes extremely simple as the factors in (9) do not have commensurate delays (i.e., integer multiples of the base delay). This simplified procedure is similar to the one presented in the analysis of the Delayed Resonator active vibration absorption system [7, 9], and we describe them in the next section.

3.1   First Factor of (9)

The first factor of (9) can be rearranged and expressed as:

$$s^2 + Ds + P = (Ds + P)e^{-\tau s} \qquad (10)$$

It is clear that $s = 0$ is a stationary root of (10) for any delay value, so this factor will exhibit at best marginally stable behaviour. Furthermore, for $\tau = 0$, this factor creates a double root at the origin, one of which is the stationary zero and the other one moves as the delay increases.

In order to have an imaginary root at $s = \omega i$, both sides of (10) should have the same magnitude and same phase. The first condition implies:

$$\left| -\omega^2 + D\omega i + P \right| = \left| D\omega i + P \right| \qquad (11)$$

which results in:

$$\omega^2 \left(\omega^2 - 2P\right) = 0 \tag{12}$$

The phase equivalency of (10) yields:

$$\angle e^{-\tau\omega i} = \angle \left(P - \omega^2 + D\omega i\right) - \angle \left(P + D\omega i\right) \tag{13}$$

or:

$$\tau\omega = \arctan\left(\frac{D\omega}{P}\right) - \arctan\left(\frac{D\omega}{P - \omega^2}\right) \tag{14}$$

Further developing (14), we arrive at:

$$\tau_k = \frac{1}{\omega} \left(\arctan\left(\frac{-D\omega^3}{P^2 + \omega^2\left(D^2 - P\right)} + 2k\pi\right)\right), \ k = 0, 1, 2, \ldots \tag{15}$$

Equation (12) states that this factor can have only two purely imaginary roots, one at $s = 0$, (which is the stationary root as already mentioned), and the other one is at $s = \pm i\sqrt{2P}$. Equation (15) gives all the possible delay values for which this second root occurs.

The second imaginary root could move to the left as the delay increases, making the system stable, or to the right, making it unstable. The direction of this crossing is found by evaluating the root tendency, which is defined as [8]:

$$RootTendency(RT) = \mathbf{sgn}\left[\Re\left(\frac{ds}{d\tau}\right)\right]\Bigg|_{P, D, \tau = \tau_k, s = \omega i} \tag{16}$$

If RT is -1, the root moves to the left, stabilizing the system, whereas if it is 1, the root moves to the right, introducing instability. The second proposition of CTCR states that RT is invariant for a given imaginary root created by the periodically distributed set of delays determined by (15).

For the characteristic equation (10), the root sensitivity can be determined as:

$$\frac{ds}{d\tau} = \frac{s\left(Ds + P\right)e^{-\tau s}}{2s + D + \left[\left(Ds + P\right)\tau - D\right]e^{-\tau s}} \tag{17}$$

and the corresponding RT is always positive for the crossings at $s = \pm\omega i = \pm i\sqrt{2P}$ and for the infinitely many delay values of (15). This means that each crossing occurs toward the right half of the complex plane, i.e. crossings are all in destabilizing nature.

This direction could also be determined simply by observing that for $\tau = 0$ there are two roots at the origin (one of which is stationary), and the second one has $RT = -1$ from (17) (see Appendix 2). The root moves into the stable left half plane, rendering the system marginally stable for infinitesimally small delays. Since there is no root at the right half plane as delay increases, the first crossing at $s = \pm i\sqrt{2P}$ has to be in the destabilizing direction (i.e., $RT = +1$) and so are the following crossings, as per the RT invariance property of the CTCR paradigm.

In conclusion, the centroid of the group will behave marginally stable only from $\tau = 0$ until the smallest positive τ given by (15), point at which it becomes unstable, and remains unstable for larger delays.

3.2   Second Factor of (9)

This factor of can be expressed as:

$$s^2 + Ds + P = -\frac{1}{N-1}(Ds + P)e^{-\tau s} \tag{18}$$

We follow a similar procedure as in the first factor to determine the imaginary crossings. If there is an imaginary root $s = \omega i$ the magnitude equality condition of both sides of (18), results in the following quadratic equation:

$$\gamma^2 + \left(\frac{D^2 N (N-2)}{(N-1)^2} - 2P\right)\gamma + \frac{P^2 N (N-2)}{(N-1)^2} = 0 \tag{19}$$

where $\gamma = \omega^2$. All the real positive roots of (19) will represent imaginary roots of (18). Equation (19) is quadratic of the form $\gamma^2 + b\gamma + c = 0$, and it will have real roots if its discriminant $b^2 - 4c$ is positive. Since $c > 0$, those roots will both be positive if b is negative. Furthermore $b > 0$ yields no positive real root. The positive-real root generating conditions of $b < 0$ and $b^2 - 4c > 0$ result in the following single relation between P, D and N:

$$D^2 \le \frac{2P(N-1)\left(N-1-\sqrt{N(N-2)}\right)}{N(N-2)} \tag{20}$$

It describes a bounded interval for D. Outside this range the second factor does not have any stability change, so it will remain stable for any τ if P and D are positive. Within the range defined in (20), the system has two crossings of the imaginary axis, at the frequencies given by the solutions of (19).

The delay values causing those crossings can be found again by applying the phase equality condition to (18). This analysis is simple if we consider that (18) and (10) have the same form, except for a negative multiplier $-1/(N-1)$ on the right hand member of (18), which introduces a phase shift of π radians in the phase equality. With this phase shift, the expression for the delays is:

$$\tau_k = \frac{1}{\omega}\left(\arctan\left(\frac{-D\omega^3}{P^2 + \omega^2(D^2 - P)} + (2k+1)\pi\right)\right), \quad k = 0, 1, 2, \ldots \tag{21}$$

The stability analysis of the second factor is completed with the calculation of the root tendency for each crossing. From (18) we obtain:

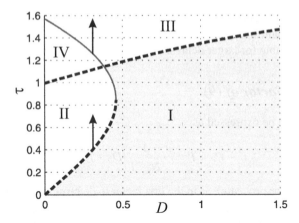

Fig. 1 Stability boundaries for the complete system with $N = 6$ and $P = 5$. Shaded region depicts stable zone.

$$\frac{ds}{d\tau} = \frac{s\left(Ds+P\right)e^{-\tau s}}{(N-1)\left(2s+D\right)+\left[D-\tau\left(Ds+P\right)\right]e^{-\tau s}} \tag{22}$$

The boundaries are given in Figure 1, with different line formats according to the *RT* definition in (16).

3.3 Complete Stability Picture

Using (12), (14), (17), and (19) the stability of the whole system can be determined in 3-D parametric domain for any combination of parameters P, D and time delay τ, by intersecting the stability regions of each factor. Figure 1 shows an example of this construction, for constant $P = 5$ and $N = 6$. The dark dashed lines indicate destabilizing crossings as τ increases (as marked by an arrow), while the light line marks stabilizing crossings to the complex left half-plane.

These stability boundaries divide the parametric space in four regions. In the region I, both factors are stable, so the agents will settle in a common position, while their velocities go to zero. In region II, the second factor is unstable, while the first is still marginally stable. This means that the agents will not reach consensus, but the average position settles in a constant value and the average velocity decays to zero with time. In region III, the second factor is stable, but the first one is not. In this case, the agents reach consensus, although the common position and velocity they reach will increase indefinitely. Finally, in region IV, both factors have roots on the right half-plane, thus they are unstable. There is no consensus and the average velocity will keep increasing.

Since the crossings of the imaginary axis are periodic in τ with period $2\pi/\omega$, the curves in Figure 1 repeat for higher delay values, while keeping the same root tendencies. Those extra crossings create other stable regions for the second factor.

However, since the crossings for the first factor are always destabilizing, in those regions the system behaves unstable, as it does in region III or IV of Figure 1.

The previous analysis gives the information about the absolute stability of the system. In order to have a better understanding of how the variations of the parameters P and D affect the relative stability in the time delay domain, the dominant root of the system is analyzed as D and τ change for different values of P, using the Quasi-Polynomial mapping-based Root-finder (QPmR) algorithm [13]. This routine provides the dominant characteristic root of a quasi-polynomial, as in (9), at a desirable precision.

Figure 2 shows the variations of the real part of the dominant root for $P = 5$, while Figure 3 shows it for $P = 7$. From these plots, it is easy to see that increasing D will make the system more stable, moving the dominant root further to the left. The effect of an increase in P is opposite, but more subtle. Increasing P, however, is not desirable, because it reduces the delay limit of the destabilizing crossing in the first factor, causing a reduction in the tolerance of the system against the communication delay variations.

4 Simulations

In order to validate the analytical results of the previous section, some case studies and simulations results are presented here. Figures 4 to 7 show the time history of the state of the agents for parametric selections corresponding to the four regions marked in Figure 1.

In the upper two panels of Figure 4, which correspond to region I, we see how the average velocity goes to zero while the average position remains constant, as was expected due to the marginal stability of the first factor of the characteristic equation

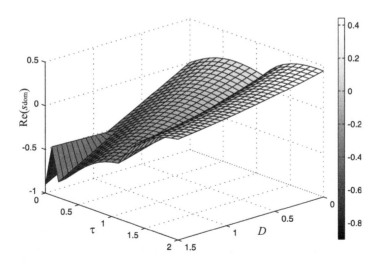

Fig. 2 Real part of the dominant root of the system for $P = 5$ and $N = 6$.

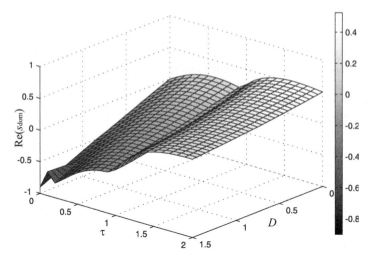

Fig. 3 Real part of the dominant root of the system for $P = 7$ and $N = 6$.

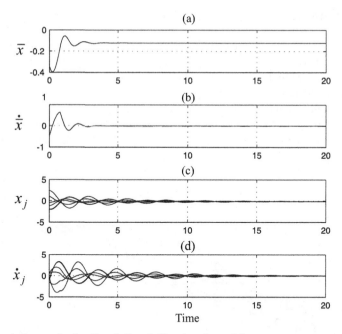

Fig. 4 Simulation results for $N = 6, P = 5, D = 1$ and $\tau = 0.8s$.

(i.e. rigid body mode). Panels (c) and (d) show how the individual agents reach the consensus values. Doubtless to say, these agent behavior is in consensus with the dominant characteristic roots of the system, which are displayed in Figures 2 and 3.

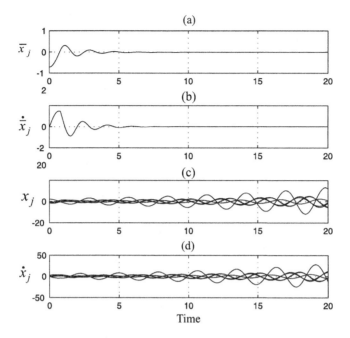

Fig. 5 Simulation results for $N = 6$, $P = 5$, $D = 0.25$ and $\tau = 0.8s$.

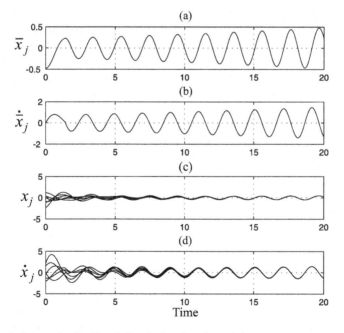

Fig. 6 Simulation results for $N = 6$, $P = 5$, $D = 1$ and $\tau = 1.4s$.

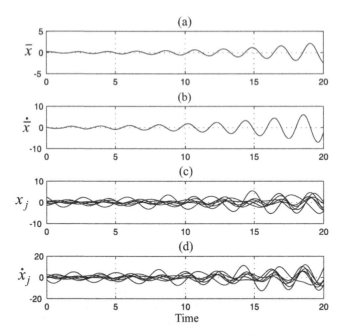

Fig. 7 Simulation results for $N = 6$, $P = 5$, $D = 0.2$ and $\tau = 1.2s$.

In Figure 5 an example behavior for region II is presented. The individual states (panels (c) and (d)) are unstable, while the average values settle after 6 seconds.

In Figure 6, the opposite behavior is shown. In this case, corresponding to region III of Figure 1, the average values are unstable, but the individual agents reach a common position and move with a common velocity. In panels (c) and (d) it looks like the individual states are going to zero, but it is due to the scale of the plot and to the fact that the average values are growing very slowly.

Finally, Figure 7 shows how the system behaves when both factors are unstable. The average values and the individual states are increasing, and there is no consensus. This is the least desirable operation point for the system, corresponding to region IV in Figure 1.

5 Conclusions

This paper proposes a consensus protocol for a system of N agents with second order dynamics, under the assumptions that all agents can communicate with each other and there is a constant communication delay for all.

The stability of the multi agent system is analyzed, taking advantage of a facilitating feature of the characteristic equation of the system. This features yields a factorization that greatly simplifies the problem. The influences of the controller parameters on the relative stability of the system are also studied. The results of this analysis are validated through simulations.

Further steps in this research will include the study of the stability when agents communicate with only a limited number of their neighbors (i.e., the degree number of the topology graph is less than $N-1$, and non-uniform from agent to agent), as well as the introduction of some extra forces, like drag and inter-agent repulsion, to transform the consensus behavior into swarming. Some of these advancements can be followed in the group's most recent publications [1, 2].

Appendix 1: Proof of Remark 1

The dynamics of the centroid of the agents is given by:

$$\ddot{\bar{x}}(t) = \frac{1}{N}\sum_{j=1}^{N}\ddot{x}_j(t)$$

$$= \frac{1}{N}\sum_{j=1}^{N}\left(P\sum_{\substack{k=1\\k\neq j}}^{N}\frac{x_k(t-\tau)}{N-1} - Px_j(t) + D\sum_{\substack{k=1\\k\neq j}}^{N}\frac{\dot{x}_k(t-\tau)}{N-1} - D\dot{x}_j(t)\right) \quad (23)$$

where $\bar{x}(t) = 1/N\sum_{j=1}^{N}x_j(t)$ represents the centroid of the group. The previous equation can be transformed into:

$$\ddot{\bar{x}}(t) = -P\bar{x}(t) - D\dot{\bar{x}}(t) + \frac{P}{N}\sum_{j=1}^{N}\sum_{\substack{k=1\\k\neq j}}^{N}\frac{x_k(t-\tau)}{N-1} + \frac{D}{N}\sum_{j=1}^{N}\sum_{\substack{k=1\\k\neq j}}^{N}\frac{\dot{x}_k(t-\tau)}{N-1} \quad (24)$$

It is not hard to see that each term j inside the double summations in (24) will be added $N-1$ times, and that will cancel the denominator. This leads to:

$$\ddot{\bar{x}}(t) = -P\bar{x}(t) - D\dot{\bar{x}}(t) + \frac{P}{N}\sum_{k=1}^{N}x_k(t-\tau) + \frac{D}{N}\sum_{k=1}^{N}\dot{x}_k(t-\tau) \quad (25)$$

Then, the behavior of the centroid is governed by:

$$\ddot{\bar{x}}(t) = -P\bar{x}(t) - D\dot{\bar{x}}(t) + P\bar{x}(t-\tau) + D\dot{\bar{x}}(t-\tau) \quad (26)$$

The characteristic equation corresponding to (26) is:

$$s^2 + Ds\left(1 - e^{-\tau s}\right) + P\left(1 - e^{-\tau s}\right) = 0 \quad (27)$$

which is the first factor of (9). The origin, $s = 0$, is always a root of this portion. If there is no unstable root of (27), the consensus occurs at zero velocity and at some constant position.

Now we look at the dynamics of the position difference between two agents, $\xi_{jk}(t) = x_j(t) - x_k(t)$ given by:

$$\ddot{\xi}_{jk}(t) = \ddot{x}_j(t) - \ddot{x}_k(t)$$

$$= P\left(x_k(t) - x_j(t)\right) + \frac{P}{N-1}\left(\sum_{\substack{l=1 \\ l \neq j}}^{N} x_l(t-\tau) - \sum_{\substack{l=1 \\ l \neq k}}^{N} x_l(t-\tau)\right) \tag{28}$$

$$+ D\left(\dot{x}_k(t) - \dot{x}_j(t)\right) + \frac{D}{N-1}\left(\sum_{\substack{l=1 \\ l \neq j}}^{N} \dot{x}_l(t-\tau) - \sum_{\substack{l=1 \\ l \neq k}}^{N} \dot{x}_l(t-\tau)\right)$$

In the summations, the only terms that are not canceled out are the j-th and k-th positions and velocities. Then the dynamics of the difference is:

$$\ddot{\xi}_{jk}(t) = -P\xi_{jk}(t) - D\dot{\xi}_{jk}(t) - \frac{P}{N-1}\xi_{jk}(t-\tau) - \frac{D}{N-1}\dot{\xi}_{jk}(t-\tau) \tag{29}$$

with the characteristic equation:

$$s^2 + Ds\left(1 + \frac{e^{-\tau s}}{N-1}\right) + P\left(1 + \frac{e^{-\tau s}}{N-1}\right) = 0 \tag{30}$$

This second factor appears $N-1$ times in the complete characteristic equation (9) because there are at most $N-1$ linearly independent differences between the N agents in the group.

Consequently, the characteristic equation of the complete system (9) is partitioned into that of the centroid of the system and that of the $N-1$-tupled inter-agent dynamics.

Appendix 2: Root Tendency at the Origin

Equation (10), the first factor of (9), has two roots at the origin for $\tau = 0$. When $s = 0$ is replaced in (17), the root sensitivity expression has zero real part, so are the higher order derivatives (i.e., $d^2s/d\tau^2$, $d^3s/d\tau^3$, etc.), which implies one of these roots is stationary. The following steps show that the second root at the origin has negative root tendency.

The first factor is:

$$s^2 + (Ds + P)\left(1 - e^{-\tau s}\right) = 0 \tag{31}$$

Since the product τs is very small, the exponential term can be expanded using Taylor series, transforming (31) into:

$$s^2 + (Ds + P)\left(\tau s - \frac{\tau^2 s^2}{2} + \frac{\tau^3 s^3}{6} - \cdots\right) = 0 \tag{32}$$

Then we divide (32) by s, to remove one of the roots at $s = 0$:

$$s + (Ds + P)\left(\tau - \frac{\tau^2 s}{2} + \frac{\tau^3 s^2}{6} - \cdots\right) = 0 \qquad (33)$$

The root sensitivity, $ds/d\tau$, obtained from (33) for $\tau = 0$ and $s = 0$ is:

$$\left.\frac{ds}{d\tau}\right|_{\substack{s=0\\\tau=0}} = -P \qquad (34)$$

giving negative root tendency for the second root at the origin.

References

1. Cepeda-Gomez, R., Olgac, N.: An exact method for the stability analysis of linear consensus protocols with time delays. IEEE Transactions on Automatic Control (in print, 2011)
2. Cepeda-Gomez, R., Olgac, N.: Exhaustive stability analysis in a consensus system with time delays. International Journal of Control (in print, 2011)
3. Lin, P., Jia, Y.: Consensus of second-order discrete multi-agent systems with nonuniform time delays and dynamically changing topologies. Automatica 45(8), 2154–2158 (2009)
4. Lin, P., Jia, Y.: Further results on decentralized coordination in networks of agents with second order dynamics. IET Control Theory and Applications 3(7), 957–970 (2009)
5. Lin, P., Jia, Y., Du, J., Yuan, S.: Distributed control of multi-agent systems with second order dynamics and delay-dependent communications. Asian Journal of Control 10(2), 254–259 (2008)
6. Olfati-Saber, R., Murray, R.: Consensus problems in networks of agents with switching topology and time delay. IEEE Transactions on Automatic Control 49(8), 1520–1533 (2004)
7. Olgac, N., Holm-Hansen, B.: A novel active vibration absorption technique: delayed resonator. Journal of Sound and Vibration 176(1), 93–104 (1994)
8. Olgac, N., Sipahi, R.: An exact method for the stability analysis of time-delayed linear time invariant systems. IEEE Transactions on Automatic Control 47(5), 793–797 (2002)
9. Olgac, N., Elmali, H., Hosek, M., Renzulli, M.: Active vibration of distributed systems using delayed resonator with acceleration feedback. ASME Transactions, Journal of Dynamic Systems Measurement and Control 119(3), 380–388 (1997)
10. Peng, K., Yang, Y.: Leader-following consensus problem with a varying-velocity leader and time-varying delays. Physica A 388(2-3), 193–208 (2009)
11. Sun, Y., Wang, L.: Consensus problems in networks of agents with double integrator dynamics and time varying delays. International Journal of Control 82(9), 1937–1945 (2009)
12. Sun, Y., Wang, L.: Consensus of multi-agent systems in directed networks with nonuniform time-varying delays. IEEE Transactions on Automatic Control 54(7), 1607–1613 (2009)
13. Vyhlídal, T., Zítek, P.: Mapping based algorithm for large-scale computation of quasipolynomial zeros. IEEE Transactions on Automatic Control 54(1), 171–177 (2009)

State Space for Time Varying Delay

Erik I. Verriest

Abstract. The construction of a state space for systems with time variant delay is analyzed. We show that under causality and consistency constraints a state space can be derived, but fails if the conditions are not satisfied. We rederive a known result on spectral reachability using an discretization approach followed by taking limits. It is also shown that when a system with fixed delay is modeled as one in a class with larger delay, reachability can no longer be preserved. This has repercussions in modeling systems with bounded time varying delay by embedding them in the class of delay systems with fixed delay, equal to the maximum of the delay function $\tau(t)$, or by using lossless causalization.

1 Introduction

One problem with mathematical modeling is that it is easy to loose oneself doing mathematics and forget about the physical reality one tried to model in the first place. For instance connecting two charged capacitors over an Ohmless wire leads to a mathematical model in the realm of impulsive currents. But is this physical reality? A closer inspection would reveal that when a charge packet moves between the capacitors, a magnetic field is created, and inductive effects should not be ignored especially with large di/dt. It becomes appropriate to consider the closed circuit as an LC circuit. This illustrates that mathematical equations may be 'cheap', but insufficient to capture reality. This is especially so if the mathematical equations turn out to lead to inconsistencies. It is easy to write down the equations $x = 1$ and $2x = 3$, but they have no solution. Hadamard's whole idea of well-posedness is an attempt not to stray off-line.

Erik I. Verriest

Georgia Institute of Technology, School of Electrical and Computer Engineering

e-mail: `erik.verriest@ece.gatech.edu`

R. Sipahi et al. (Eds.): Time Delay Sys.: Methods, Appli. and New Trends, LNCIS 423, pp. 135–146.

springerlink.com

Recently, there has been an increased interest in systems with time varying delays. With the condition $\dot{\tau} < 1$, the Lyapunov-Krasovskii method yields simple sufficient conditions for stability [4]. It was emphasized that this is not merely a technical condition to make the proof work, but has some deeper causality meaning associated with it. In particular, we believe that one should be very cautious when models are used for systems where the derivative of the time delay can exceed one [1]. It is not clear what the state space should be in such a case. The same holds for systems where the delay may depend on the solution itself. The term "state-dependent delay" has been used in these cases, but we want to avoid this until the notion of state has been made clear.

One way of avoiding problems with the conceptual definition of state is to envision a system with time varying delay as the limit of one with distributed delay over a fixed interval [5]. In this limit the kernel must be singular, which may already cause a problem. Obviously this embedding into a larger delay can only work if the time variant delay is bounded [1, 2].

2 State

In order to be able to speak of trajectories (the 'state' as function of time) of a dynamical system, it is necessary that what one considers to be a state space is a stationary construct. The space itself cannot depend on time. For a delay systems of the form

$$\dot{x}(t) = Ax(t) + Bx(t - \tau(t)) + bu(t), \tag{1}$$

the 'obvious' choices $C([-\tau(t), 0]$, or $L_2([-\tau(t), 0]$ cannot work. We have shown in [7, 8, 9] that if $d\tau(t)/dt < 1$, then a diffeomeorphism $h : \mathbb{R} \to \mathbb{R}$ may be found such that in the time variable $h(t)$, the delay system is transformed to a timevarying coefficient delay system with fixed time delay. Moreover linearity can be preserved. Hence in the case $\dot{\tau} < 1$, a state space can be well-defined. Also, the limit case, $d\tau(t)/dt = 1$, poses no problem. In this case, the system is equivalent to a finite dimensional one

$$\dot{x}(t) = Ax(t) + Bx(t_0) + bu(t), \tag{2}$$

where t_0 is some fixed time. Causality requires that one should only consider such an equation for $t \geq t_0$. If x is a scalar variable, and $B \neq 0$, the state space is $\mathbb{R}^2$ with the true state at time t equal to (x_0, x). Omitting $x(t_0)$ no longer leaves a sufficient statistic. However, at once we see that this state cannot be reachable. If $\dim x = n$, the state space has dimension $n + \text{rk} B$, as only $Bx(t_0)$ and not all of x_0 is required.

If $\dot{\tau} > 1$, x in the time interval $(t - \tau(t), t)$ no longer is a sufficient statistic to determine future behavior uniquely, and therefore fails to be a state in the sense conceived by Nerode. No diffeomorphism, h, transforming the system into one with constant delay exists. In [7] we introduced *lossless causalization* as an attempt to interpret the state equation with intervals where $\dot{\tau} > 1$ occurs as a causal system. This pertains to embedding the delay, $\tau(t)$, in a bigger delay, i.e., $\hat{\tau}(t)$, such that the graph of $t - \hat{\tau}(t)$ is nondecreasing, but in some minimal sense. See Figure 1.

It was shown in [7, 8] that this only moves the causality problem because of the necessity of an anticipatory information structure. The precise form of $\tau(\cdot)$ must be known at each time, since $t - \hat{\tau}(t) = \inf_{s>t}(s - \tau(s))$. Surely, if the delay depends on system outputs, this cannot be assumed.

We caution that a general time varying system in $\mathbb{R}^n$ does not fall into this class: at time t only the parameters at that time, and not their future values determine $\dot{x}(t)$. Perhaps the information structure is such that only an upper bound to the delay is known [2]. Of course one could be super cautious, and sweep all problems away by letting $\tau = \infty$ from the beginning and cast all systems in a nonparsimonious way as *Volterra systems with infinite aftereffect*[3].

$$\dot{x}(t) = f\left(t, \int_{-\infty}^{t} K(t, \theta, x(\theta)) \, d\theta \right).$$

But there another potential emerges, not related to the information structure: The well-defined state space may not be minimal, just as it wasn't in the border case $\dot{\tau} = 1$. In particular, it may not be possible to reach any arbitrary preassigned state with this system.

We illustrate the difficulty in defining state spaces for some evolutions defined by functional equations. Even if a state space can be defined, we show that reachability may not hold, and thus the quest for a *minimal* state space may remain unsolved.

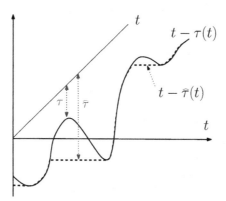

Fig. 1 Lossless Causalization

3 Functional Differential Equation

In this section we consider the scalar evolution system, where the derivative at some future time $t_f(t)$, called *forward time*, is determined by the present state. Thus,

$$\dot{x}(t_f) = f(x(t), u(t)). \tag{3}$$

First, one carefully needs to interpret the left hand side, as it may be ambiguous. Let's introduce an evaluation functional σ_t, mapping functions from a suitable function space to $\mathbb{R}$. For any function f, the number $\sigma_t f = f(t)$. Likewise we have the differentiation operator $\mathbf{D} : x \to \dot{x}$. The left hand side of (3) means $\sigma_{t_f} \mathbf{D}x$, which differs from $\frac{d}{dt} x(t_f(t)) = \sigma_{t_f}(\mathbf{D}x) \sigma_t(\mathbf{D}t_f)$. With $t_f(t) = t + \tau$, this corresponds to a system with delay τ. We favor this representation as causality is built in when $t_f \geq t$ can be guaranteed. Suppose now that at two different times, t_1 ad t_2, it holds that $t_f(t_1) = t_f(t_2)$. This would imply that $x(t_f)$ may have two different values, which is inconsistent. Hence, for consistency, we impose in addition that the forward time function, $t_f(t)$ be monotonically increasing. If $t_f(t)$ is anticipated, i.e., a known function of t, or at least computable at t [8], this system is causal as long as $t_f(t)$ dominates t.

Thus the forward time, $t_f(t)$, has a *causality* constraint

$$\forall t : \quad t_f(t) \geq t. \tag{4}$$

and a *consistency* constraint[1]

$$\forall t : \quad \dot{t}_f(t) > 0. \tag{5}$$

In some systems, the forward time may depend on the actual (partial) state at time t, and therefore only implicitly be time varying [6]. Let thus

$$t_f(t) = t + F(x(t)). \tag{6}$$

Causality imposes $F(x) \geq 0$, and consistency $\frac{d}{dt}[t + F(x(t))] > 0$, or

$$\frac{dF}{dx}\dot{x} > -1. \tag{7}$$

Thus consider a delay system $\dot{x}(t_f) = f(x(t), u(t))$, with a delay dependent on the magnitude of the state, say of the form $F(x) = 1 - e^{-|x|}$. Then the causality and consistency constraints mandate that such a system restricts $\dot{x}$ to $\dot{x} < e^{|x|}$ if $x < 0$ and $\dot{x} > -e^x$ if $x > 0$. When violated such a model cannot accurately represent some physical phenomenon. Clearly the system (3) with $t_f(\cdot)$ as above and with $|f(x, u)| < 1$ will satisfy these conditions.

Likewise, the system with unbounded delay $F(x) = |x|$ imposes the consistency constraint $\dot{x}\,\mathrm{sgn}\,x > -1$. This holds also for a system satisfying $|f(x, u)| < 1$. For a linear system, $f(x, u) = ax + bu$, with $b > 0$, it imposes the control constraints $-(1 + ax)/b < u < -(ax - 1)/b$.

In fact, let us look at the autonomous system

$$\dot{x}(t + |x(t)|) = ax(t) \tag{8}$$

---

[1] $\dot{t}_f = 0$ implies the impossibility of a jump in x at $t_f(t_0)$.

with initial condition $x(0) = x_0$. The first time that something can be discovered about its evolution is at time $t_1 = |x_0|$. Does this mean that $\{\phi(t)|t \in [0,t_1]\}$ is the necessary initial condition? Obviously not all of $C([0,|x_0|],\mathbb{R})$ can be allowed. Causality is obvious, consistency requires that $\dot{\phi}\,\text{sgn}\,\phi > -1$. Strangely enough, the interval length in this space of initial conditions is determined by $\phi(0)$. With this information all subsequent states in the interval (t_1,t_2) are determined, where $t_1 = \phi(0)$ and $t_2 = \phi(t_1)$. More generally, an adaptation of the method of steps allows the computation of $x(t)$ in the k-th step in the interval (t_k,t_{k+1}), where $t_{k+1} = t_k + |x(t_k)|$.

Clearly, if $a > 0$, then $x(t) > 0$, with $\dot{x}(t) > -1$ for all $t < t_1 = x_0$ implies $\dot{x}(t + x(t)) = ax(t) > 0$, and consequently the conditions $x(t) > 0$ and $\dot{x}(t) > -1$ are preserved. The system remains causal and consistent and the solution will diverge, but slower than exponentially. However, the case $a < 0$ can only give consistent solutions for the initial condition $x(0) = x_0 > 0$ if $\dot{x}(t) > -1$ and $x(t) < 1/|a|$ in the interval $[0,x_0]$, or, if $x_0 < 0$, then $\dot{x}(t) < 1$ and $x(t) > -1/|a|$ in the same interval $[0,x_0]$. In these cases the solution converges to zero and is faster than $\exp(at)$. Note that the border line data $\phi(t) = 1/|a| - t$ in the interval $[0,1/|a|]$ is self consistent, except at $t = 1/|a|$, since

$$\dot{x}(t + \phi(t)) = a\phi(t)$$

yields for the left hand side

$$\dot{x}(t + 1/|a| - t) = \dot{x}(1/|a|).$$

and for the right hand side

$$a\phi(t) = a/|a| - at = at - 1.$$

The derivative at $t = 1/|a|$, when x is zero, is multivalued!

If at some time, say t_0, $x(t_0) = 0$, then the evolution equation yields $\dot{x}(t_0) = ax(t_0) = 0$, so that $x(t)$ remains zero. The equilibrium state is $x = 0$. Consequently a trajectory of the autonomous system cannot cross zero. Thus, the above border line case has a deadbeat character and remains zero after $t = 1/|a|$ for the given initial data. In Figure 2 some trajectories are displayed for constant initial data, $\phi(\cdot) = \phi_0$ in the interval $(0,\phi_0)$. Consistency requires that $\phi_0 < 1/|a| = 1$. We plot the nondelay equation solution starting with initial condition ϕ_0 at ϕ_0 as well (exponentials).

The long term behavior of the equation in the asymptotically stable case can be approximated for small $x(t)$. Indeed, using a Taylor expansion, we get ($x_0 > 0$)

$$\dot{x}(t) + \ddot{x}(t)x(t) = ax(t).$$

Upon dividing by $x(t)$, this yields, assuming $x_0 > 0$, (hence $x(t) \geq 0$)

$$\frac{d}{dt}[\dot{x}(t) + \log x(t) - at] = 0.$$

Setting $x(t) = \exp(\psi(t))$ gives, introducing the integration constant, b,

$$\dot{\psi} = (at + b - \psi)e^{-\psi},$$

which has the solution converging asymptotically to

$$\widehat{\psi}(t) = (at + b) - a\exp(at + b).$$

Finally

$$x(t) \approx \exp(at + b - a\exp(at + b)).$$

Note that if $\exp(at + b) < \varepsilon$, then up to first order in ε:

$$\frac{\widehat{\psi}(t)\exp\widehat{\psi}(t)}{at + b - \widehat{\psi}(t)} \approx 1 - 2a\varepsilon.$$

Equation (8) has the remarkable property that first of all the initial data must be restricted for the causality and consistency to hold. In addition, we have seen that the minimal sufficient statistic for the equation is a function in an interval whose length depends on the initial value in that interval. It seems hopeless to try to define an invariant structure as state space in this case. However an invariant structure for the state space is readily constructed. Restricting the solution set of (8) to the set of positive continuous functions for which causality and consistency hold, let us first single out $x(t_0)$, as a partial state at time t_0. Next consider the function

$$\hat{x}_{t_0}(\theta) = \frac{x(t_0 + x(t_0)\theta)}{x(t_0)}\Pi_{[0,1)}(\theta), \tag{9}$$

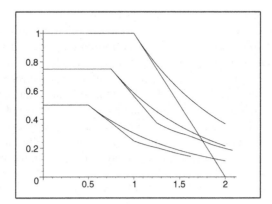

Fig. 2 Solutions to $\dot{x}(t + x(t)) = ax(t)$ for positive initial conditions with $a = -1$, and this in 4 consecutive couplets (not counting the ID). The initial data (minimal sufficient statistic) is chosen to be constant The solutions to this state dependent delay equation decay faster than in the non delay equation. The border case $\phi_0 = 1$ has deadbeat behavior.

where $\Pi_{[0,1)}$ is the indicator function of the interval $[0,1)$. We note that $\hat{x}_{t_0}(0) = 1$ and $\frac{d\hat{x}_{t_0}(\theta)}{d\theta} > -1$ in the interval $[0,1)$. This interval is the same for all t_0. Consistency requires that this function has a derivative larger than -1, and since it starts at 1, its values must be lower bound by $1 - \theta$ in $[0,1)$. A simple characterization of this sufficient statistic is then by the pair $(x(t_0), \psi(\cdot))$ where $\psi = \mathbf{D}\hat{x}_{t_0}$. Hence a suitable state space for (8) (restricted to positive solutions) is the set $\mathbb{R}_+ \times X$, where $X = \{x | x(\theta) = 1 + \int_0^\theta \psi(s)ds, \ \theta \in [0,1), \psi \in L_2([0,1), \mathbb{R}_+ - 1\}$ where $\mathbb{R}_+ - 1 = \{r | r > -1\}$. Then, knowing the state (x, ψ) at time t, readily reproduces the partial state $x(t')$ for $t' \geq t$ as $x(t') = x[1 + \int_0^{(t'-t)/x} \psi(\theta)d\theta]$ in the interval $(t, t+x)$. In the next section we shall explore a somewhat simpler equation with similar behavior.

4 Non-differential Functional Equation

Consider now the simpler functional equation of the form

$$x(t + |x(t)|) = ax(t). \tag{10}$$

One may think of this as a continuous form of a discrete system, prescribing the value of the state at some future time t_f in terms of the value of x at the present time, t. In addition, we let this future time depend on the present value of x. Here, the forward time jump is precisely equal to $|x(t)|$.

What are equilibria for this equation? Clearly if it has a constant solution, say x_∞, then it must hold that $x_\infty = ax_\infty$. Hence, if $a = 1$, all constant solutions are possible, while if $a \neq 1$, the only equilibrium solution is $x(t) \equiv 0$. However, note that if at some time t_0, we have $x(t_0) = 0$, then it follows that the 'next' step is also t_0, so that the solution does not evolve any further. Thus also, the initial condition $x(0) = 0$ does not lead to a propagating solution if $a \neq 1$.

Taking $V(x) = |x|$, then between successive jumps, we get the forward increment

$$\Delta V(t) = |x(t + |x(t)|)| - |x(t)| = (|a| - 1)|x|(t). \tag{11}$$

Clearly, then $\Delta V(t)$ is decreasing if $|a| < 1$, but this does not yet imply asymptotic stability as the time sequence $t_0 = t, t_1 = t_0 + |x(t_0)|, \cdots, t_{k+1} = t_k + |x(t_k)|$, may be clustering. However, we find

$$\frac{\Delta V(t)}{\Delta t} = (|a| - 1). \tag{12}$$

Not only does this imply asymptotic stability, but also that the equilibrium must be found in *finite time*. Necessarily, this autonomous system has time jumps that cluster. Consider thus the behavior for small differentiable initial data. Thus $x(t_1) = \varepsilon > 0$ implies $x(t_1 + \varepsilon) = a\varepsilon$. The left hand side can be approximated by

$$x(t_1 + \varepsilon) = x(t_1) + \dot{x}(t_1)\varepsilon.$$

Hence, $\dot{x}(t_1) = (a-1)$ approximately, and $x(t)$ must decrease for $t > t_1$ if $a < 1$. So, let us check the decay towards zero: If t_F is the finite time when the differentiable solution hits 0, then $x(t) = (1-a)(t_F - t)$. It is now readily verified that this solution is consistent for all $t < t_F$. (It seems also to imply that this would hold even if $|a| > 1$, but in this case the assumption of a finite t_F is false.

What is in this case a minimal statistic that allows to evolve a solution to either t_F (in the asymptotically stable case) or indefinitely (if unstable)? Let $x_0 > 0$ (we exclude zero for there is no propagation, and the case $x_0 < 0$ can be dealt with in a symmetrical fashion.) Just as in the case of the differential functional equation, causality imposes the constraint that the forward time exceeds present time, $t_f(t) > t$, and the consistency constraint imposes that as function of time t, the forward time cannot trace it steps back, i.e., $\dot{t}_f(t) > 0$. In the example this means that the initial data, $\phi(t)$, must be restricted to the class of functions satisfying $\dot{\phi}(t) > -1$, for $\phi(t) > 0$. But what is the minimal interval? With $t_0 = 0$, we get $t_1 = t_f(t_0) = x_0$, the initial condition. Let $x_1 = ax_0$. Is any differentiable curve from $(0, x_0)$ to (x_0, x_1), satisfying the causality and consistency constraint allowed?

Suppose that at time θ, it holds that $x(\theta) = \theta$. Define the iterates as $t_{k+1} = t_k + x(t_k)$, and $x_{k+1} = ax_k$ with $x_0 = t_0 = \theta$. It is easily shown by induction that, if $|a| < 1$, then $t_k = (1 + \frac{a^k-1}{a-1})\theta$ and $x_k = x(t_k) = a^k\theta$. It follows from the above that t_k and x_k respectively converge to $t_k \to t_F = \frac{(a-2)}{(a-1)}\theta$ and $x_k \to x_F = 0$. If $a = 1$, $t_k = (k+1)\theta$ and $x_k = \theta$, and the solution periodically revisits its initial value θ. It should not go unnoticed that $x(t)$ in each interval $(t_k, t_k + 1)$ determines a minimal statistic for evolving the solution. These intervals have in general a nonuniform length. Moreover, this length is dictated by the data itself.

5   Reachability of Systems with Fixed Point Delay

We give in this section a simple proof for the spectral reachability condition for sytems with fixed delay, and use it to illustrate the loss of reachability when the system is embedded in one with a larger delay. Consider the system (1) with constant delay. Assume that u is a locally integrable function, and that the initial data, ϕ, is integrable. Then the solution $x(t)$ is continuous for $t > 0$. We are interested in conditions that guarantee that any arbitrary continuous function x over $(T - \tau, T)$ is attainable, by proper choice of $u(t)$ for $0 < t < T$.

5.1   PBH Test for Delay Systems

As a first approach of the problem, we discretize the delay model by letting $N\Delta = \tau$, and setting

$$\dot{x}(t) \approx \frac{1}{\Delta}(x(t+\Delta) - x(t))$$

from which a forward difference model follows

$$x(t+\Delta) = (I+\Delta A)x(t) + \Delta B x(t-N\Delta) + \Delta b u(t).$$

In turn, this gives the approximate dynamics for the delay system in terms of the approximating state $\xi_N(t)^\top = [x(t)^\top, x(t-\Delta)^\top, \dots, x(t-N\Delta)^\top]$.

$$\xi_N(t+\Delta) = \begin{bmatrix} I+\Delta A & 0 & \cdots & 0 & \Delta B \\ I & & & & 0 \\ 0 & I & & \vdots & \vdots \\ \vdots & & \ddots & & \vdots \\ 0 & \cdots & \cdots & I & 0 \end{bmatrix} \xi_N(t) + \begin{bmatrix} \Delta b \\ 0 \\ \vdots \\ 0 \\ 0 \end{bmatrix} u(t). \tag{13}$$

or, $\xi_N(t+\Delta) = \Phi_N(A,B)\xi_N(t) + \beta_N u(t)$. By the PBH-test, this discrete time system is reachable if $\operatorname{rank}[zI - \Phi_N : \beta_N] = (N+1)n$. Using elementary column and row operations, this reduces to

$$\operatorname{rank} \begin{bmatrix} [(z-1)I - \Delta A] & 0 & \cdots & 0 & -\Delta B & -\Delta b \\ -I & zI & & & & \\ & -I & zI & & & \\ & & & \ddots & & \\ & & & -I & zI & 0 \end{bmatrix}$$

$$= \operatorname{rank} \begin{bmatrix} 0 & & 0 & [z^N[(z-1)I - \Delta A] - \Delta B] & -\Delta b \\ I & & & & \\ & I & & & \\ & & \ddots & & \\ & I & & 0 & 0 \end{bmatrix}$$

$$= Nn + \operatorname{rank}[z^N[(z-1)I - \Delta A] - \Delta B : \Delta b]$$

It follows that the discretized model is reachable iff

$$\operatorname{rank}[z^N[(z-1)I - \Delta A] - \Delta B : \Delta b] = n, \tag{14}$$

In the limit for $N \to \infty$ and $\Delta \to 0$ such that $N\Delta = \tau$, we get for $z \neq 0$ the reachability condition

$$\operatorname{rank} \lim \left[\frac{(z-1)}{\Delta} I - A - z^{-N} B : b \right] = n, \tag{15}$$

When $z = 0$, an additional condition is obtained:

$$\operatorname{rank}[B : b] = n. \tag{16}$$

Set now $z = e^{s\Delta}$, to get

$$\operatorname{rank}[sI - A - e^{-s\tau} B : b] = n, \tag{17}$$

which nicely generalizes the PBH test for finite dimensional LTI systems.

Also note that the condition (16) is easily understood. If it is desired to obtain the state ψ at time t, then it follows from the dynamics that

$$\frac{d}{dt}\psi - A\psi = Bx(T-\tau) - bu(T)$$

Hence since $\frac{d}{dt}\psi - A\psi$ can be arbitrary, it follows that the range space of $[B:b]$ must be the $\mathbb{R}^n$. This corresponds to the limit $\Re s \to -\infty$ in (17).

For a reachable finite dimensional discrete time system of order n, it is known that at most n steps are required to accomplish the requisite state transfer. Hence the approximating discretized system with state $\xi_N(t)$ will require at most $(N+1)n$ steps. Since with each step a time Δ elapses, the total elapsed time is $n(N+1)\Delta$. In the limit for $N \to \infty$ and $\Delta \to 0$ with constraint $N\Delta = \tau$, we obtain a lower bound on the requisite time for the reachability problem for the delay system. This lower limit is $n\tau$.

This gives an alternative proof to the spectral reachability theorem.

Theorem 1. *The state of the constant delay system (1) can be made arbitrary iff*

$$\text{rank}[sI - A - e^{-s\tau}B : b] = n, \quad \forall s,$$

provided that $n\tau$ time units are available to achieve this state transfer.

5.2   State Augmentation

It has been argued that for systems with time variant delay, the state should be taken as the fixed interval corresponding to the maximum of the delay. Likewise, loss-less causalization embeds the system into one with larger delay. We show that this necessarily compromises reachability. In order to streamline the ideas we analyze the case for a linear time invariant delay system with delay τ. It will be shown that if one embeds this system into the class of delay systems with state space $C^1([-(\tau+\varepsilon),0),\mathbb{R}^n)$, where $\varepsilon > 0$, then reachability will not hold in this bigger space.

Theorem 2. Failure of reachability by embedding.
Let the LTI delay system (1) be reachable. Then in general, it is not possible to make $\{x(t+\theta)|\theta \in (-\tau-\varepsilon,0])\}$ for any $t > 0$ coincide with an arbitrary preassigned continuous function, no matter what value of t is chosen, for any particular choice of the input $u(\cdot)$.

Proof. Proceed with the discretization as in the reachability problem. Assume first that $\tau = N\Delta$ and $\varepsilon = p\Delta$. We now get, letting again $\xi_{N+p}(t)^\top = [x(t)^\top,$ $x(t-\Delta)^\top,\ldots,x(t-N\Delta)^\top,\ldots,x(t-(N+p)\Delta))^\top]$.

$$
\xi_{N+p}(t+\Delta) =
\begin{bmatrix}
I+\Delta A & 0 & \cdots & \Delta B & \cdots & 0 \\
 & I & & & & \\
 & & \ddots & \ddots & & \\
 & 0 & I & 0 & & \\
 & & & \ddots & \ddots & \\
 & & & & I & 0
\end{bmatrix}
\xi_{N+p}(t) +
\begin{bmatrix}
\Delta b \\
0 \\
\vdots \\
0 \\
\vdots \\
0
\end{bmatrix}
u(t). \qquad (18)
$$

The PBH test applied to this approximation with $(N+p+1)$ steps yields now

$$
\mathrm{rank}
\begin{bmatrix}
[(z-1)I - \Delta A] & 0 & \cdots & -\Delta B & 0 & \cdots & 0 & -\Delta b \\
-I & zI & & & & & & \\
 & & \ddots & \ddots & & & & \\
 & & -I & zI & & & & \\
 & & & & \ddots & \ddots & & \\
 & & & & & -I & zI &
\end{bmatrix}
$$

$$
= \mathrm{rank}
\begin{bmatrix}
0 & 0 & \cdots & 0 & z^p[z^N[(z-1)I - \Delta A] - \Delta B] & -\Delta b \\
I & & & & & \\
 & I & & & & \\
 & & \ddots & & & \\
 & & & I & & 0
\end{bmatrix}
$$

$$
= (N+p)n + \mathrm{rank}[z^p[z^N[(z-1)I - \Delta A] - \Delta B] : \ \Delta b]
$$

Clearly, for $z=0$, $\mathrm{rank}[z^p[z^N[(z-1)I-\Delta A]-\Delta B] : \ \Delta b] = \mathrm{rank}[b]$. Unless $n=1$, and $b \neq 0$, the embedded delay system cannot be reachable. □

If the embedding does not yield a reachable realization, the lossless causalized systems cannot be minimal. Consequently, pole placement based on the maximal delay may not be possible.

Remark: The reachability condition is of the form $\mathrm{rank}[\Delta(s), b] = n$. where $\Delta(s) = sI - A - Be^{-s\tau}$. The one-sided Laplace transform of $x(t)$ satisfies $\Delta(s)X(s) = E(s)$ for some entire function $E(s)$. Hence $\Delta(s)$ is the denominator matrix in a left fraction description of the system. See [10] for further generalizations.

6 Conclusions

We discussed some issues regarding the existence of an invariant state space for systems with time varying delay. It was shown that the condition $\dot{\tau} < 1$ is fundamental for its existence. We have given a simple derivation of the spectral reachability condition for LTI delay systems. We also discussed embedding the system into one with larger delay when the delay derivative condition is not met. In this case we have shown that reachability may fail. Physical reality should constrain the mathematical models. [11]

References

1. Banks, H.T.: Necessary conditions for control problems with variable time lags. SIAM J. Contr. 6(1), 9–47 (1968)
2. Fridman, E., Shaked, U.: An Improved Stabilization Method for Linear Time-Delay Systems. IEEE Transactions on Automatic Control 47(11), 1931–1937 (2002)
3. Kolmanovskii, V., Myshkis, A.: Applied Theory of Functional Differential Equations. Kluwer Academic Publishers (1992)
4. Verriest, E.I.: Robust Stability of Time-Varying Systems with Unknown Bounded Delays. In: Proceedings of the 33rd IEEE Conference on Decision and Control, Orlando, FL, pp. 417–422 (1994)
5. Verriest, E.I.: Stability of Systems with Distributed Delays. In: Proceedings IFAC Workshop on System Structure and Control, Nantes, France, pp. 294–299 (1995)
6. Verriest, E.I.: Stability of Systems with State-Dependent and Random Delays. IMA Journal of Mathematical Control and Information 19, 103–114 (2002)
7. Verriest, E.I.: Causal Behavior of Switched Delay Systems as Multi-Mode Multi-Dimensional Systems. In: Proceedings of the 8-th IFAC International Symposium on Time-Delay Systems, Sinaia, Romania (2009)
8. Verriest, E.I.: Well-Posedness of Problems involving Time-Varying Delays. In: Proceedings of the 18-th International Symposium on Mathematical Theory of Networks and Systems, Budapest, Hungary, pp. 1985–1988 (2010)
9. Verriest, E.I.: Inconsistencies in systems with time varying delays and their resolution. To appear: IMA Journal of Mathematical Control and Information (2011)
10. Yamamoto, Y.: Minimal representations for delay systems. In: Proc. 17-th IFAC World Congress, Seoul, KR, pp. 1249–1254 (2008)
11. Willems, J.C.: The behavioral approach to open and interconnected system. IEEE Control Systems Magazine 27(6), 46–99 (2007)

Delays. Propagation. Conservation Laws.

Vladimir Răsvan

Abstract. Since the very first paper of J. Bernoulli in 1728, a connection exists between initial boundary value problems for hyperbolic Partial Differential Equations (PDE) in the plane (with a single space coordinate accounting for wave propagation) and some associated Functional Equations (FE). The functional equations may be difference equations (in continuous time), delay-differential (mostly of neutral type) or even integral/integro-differential. It is possible to discuss dynamics and control either for PDE or FE since both may be viewed as self contained mathematical objects. A more recent topic is control of systems displaying conservation laws. Conservation laws are described by *nonlinear* hyperbolic PDE belonging to the class "lossless" (conservative). It is not without interest to discuss association of some FE. Lossless implies usually distortionless propagation hence one would expect here also lumped time delays. The paper contains some illustrating applications from various fields: nuclear reactors with circulating fuel, canal flows control, overhead crane, without forgetting the standard classical example of the nonhomogeneous transmission lines for distortionless and lossless propagation. Specific features of the control models are discussed in connection with the control approach wherever it applies.

1 Introduction and Basics

We shall start from two elementary facts. First, any electrical or control engineer has dealt with mathematical models where either a complex domain term like $e^{-\tau s}$ with $\tau > 0$, $s \in \mathbb{C}$, or a time domain term like $u(t - \tau)$, where u was some signal, were present. Such models were called *time delay* or *time lag systems*. A more involved interest to such systems would inevitably have sent to some reference about

Vladimir Răsvan
Dept. of Automatic Control, University of Craiova, A.I.Cuza, 13,
Craiova, RO-200585, Romania
e-mail: vrasvan@automation.ucv.ro

R. Sipahi et al. (Eds.): Time Delay Sys.: Methods, Appli. and New Trends, LNCIS 423, pp. 147–159.
springerlink.com   © Springer-Verlag Berlin Heidelberg 2012

the underlying equations of these models - the *equations with deviating argument*. A still more involved interest would concern origins of these equations: the first differential equation with deviating argument, reported in [16], was published by Johann (Jean) Bernoulli in 1728 [2] and reads as

$$y'(t) = y(t-1) \tag{1}$$

As the title of this paper shows, this equation appears to be associated to a partial differential equation of hyperbolic type - the string equation; it thus sends to the second elementary fact, less known, that propagation is associated to time delay. In order to explain this, we shall discuss a special case of propagation - the *lossless propagation*. By lossless propagation it is understood the phenomenon associated with long (in a definite sense) transmission lines for physical signals. In electrical and electronic engineering there are considered in various applications circuit structures consisting of multipoles connected through LC transmission lines (A long list of references may be provided, starting with [3] and going up to a quite recent book [13]). The lossless propagation occurs also for non-electric signals as water, steam or gas flows and pressures - see e.g. the pioneering papers of [10], [11] on steam pipes for combined heat-electricity generation, the papers dealing with waterhammer and many other. In order to illustrate these assertions, we shall consider one of the early benchmark problems, the nonlinear circuit containing a tunnel diode and a lossless transmission - the so called Nagumo-Shimura circuit described by

$$L\frac{\partial i}{\partial t} = -\frac{\partial v}{\partial \lambda} \ , \ C\frac{\partial v}{\partial t} = -\frac{\partial i}{\partial \lambda}, 0 \leq \lambda \leq 1$$
$$E = v(0,t) + R_0 i(0,t) \ , \ -C_0\frac{\mathrm{d}}{\mathrm{d}t} v(1,t) = -i(1,t) + \psi(v(1,t)) \tag{2}$$

Proceeding in "an engineering way" we may apply formally the Laplace transform to compute the solution of the boundary value problem viewed as independent of the differential equation but being nevertheless controlled by it. A similar approach of applying formally the Laplace transform and deducing a characteristic equation accounting for time delays (deviating arguments) was used in the pioneering papers [10, 11, 21] dealing with steam pipes; for water pipes a pioneering paper is [22] where the same approach is applied.

Continuing the investigation of the above benchmark system one may observe that the aggregate

$$v(t) + \sqrt{\frac{L}{C}}i(1,t) \equiv u(1,t) + \sqrt{\frac{L}{C}}i(1,t)$$

represents the so called progressive (forward) wave of the system at the boundary $\lambda = 1$. Both the voltage $u(\lambda,t)$ and the current $i(\lambda,t)$ are linear combinations of the progressive (forward) and reflected (backward) waves hence it is useful to express (2) in terms of these waves

$$\frac{\partial u_1}{\partial t} + \frac{1}{\sqrt{LC}} \frac{\partial u_1}{\partial \lambda} = 0 \, , \, \frac{\partial u_2}{\partial t} - \frac{1}{\sqrt{LC}} \frac{\partial u_2}{\partial \lambda} = 0$$

$$(1 + R_0\sqrt{C/L})u_1(0,t) + (1 - R_0\sqrt{C/L})u_2(0,t) = 2E(t) \tag{3}$$

$$u_1(1,t) + u_2(1,t) = 2v(t) \, , \, C_0\frac{dv}{dt} + \psi(v) = \frac{1}{2}\sqrt{\frac{C}{L}}[u_1(\lambda,t) - u_2(\lambda,t)]$$

The propagation (partial differential) equations of the two waves are decoupled; the two waves are exactly the Riemann invariants of the problem. We may consider now the standard version of the d'Alembert method i.e. of integrating along the two families of characteristics; there is a family of increasing characteristics and one of decreasing; as (3) shows, the forward wave should be considered along the increasing characteristics while the backward wave along the decreasing ones. If we perform this integration we shall find

$$u_1(0,t) = u_1(1,t + \sqrt{LC}) \, , \, u_2(1,t) = u_2(0,t + \sqrt{LC}) \tag{4}$$

By denoting $\eta_1(t) = u_1(1,t)$, $\eta_2(t) = u_1(1,t)$ we associate to (2) the following system of equations with delayed argument

$$C_0\frac{dv}{dt} + \psi(v) = \frac{1}{2}\sqrt{\frac{C}{L}}(\eta_1(t) - \eta_2(t - \sqrt{LC})),$$

$$\eta_2(t) = -\rho_0\eta_1(t - \sqrt{LC}) + (1 + \rho_0)E(t) \, , \, \eta_1(t) = -\eta_2(t - \sqrt{LC}) + 2v(t) \tag{5}$$

associated in a rigorous way, starting from the solutions of (2); even the initial conditions may be associated in this way. Moreover, the converse association is also possible. Using the representation formulae for the two waves

$$u_1(\lambda,t) = \eta_1(t + (1 - \lambda)\sqrt{LC}) \, , \, u_2(\lambda,t) = \eta_2(t + \lambda\sqrt{LC}) \tag{6}$$

we may construct the solutions of (2) starting from the solutions of (5).

To end this introductory discussion we just mention that (4) and (6) define what is usually known as *lossless propagation*. Since the two waves propagate from one boundary to the other in finite time $\tau = \sqrt{LC}$ but without changing their waveform (just with a pure - lumped - time delay) this propagation is also distortionless. In the following we shall discuss both these aspects.

2 Lossless and Distortionless Propagation

A. We shall consider again the Nagumo-Shimura circuit but with a lossy transmission line

$$\frac{\partial u}{\partial \lambda} + L\frac{\partial i}{\partial t} + Ri = 0 \ , \ \frac{\partial i}{\partial \lambda} + C\frac{\partial u}{\partial t} + Gu = 0$$

$$ (7) $$

$$R_0 i(0,t) + u(0,t) = E(t) \ , \ u_1(t) = v(t) \ , \ C_0\frac{dv}{dt} + \psi(v) = i(1,t)$$

Introducing the forward and backward waves as previously we find their propagation equations which are no longer decoupled unless the "matching" condition of Heaviside is met i.e. $RC = LG$ which "destroys" the coupling terms. Introducing the new "waves"

$$u_1(\lambda,t) = e^{-\delta\lambda}w_1(\lambda,t) \ , \ u_2(\lambda,t) = e^{\delta\lambda}w_2(\lambda,t) \ , \ \delta = R\sqrt{C/L} \qquad (8)$$

we obtain a lossless-like system. Moreover, denoting

$$\eta_1(t) = w_1(1,t) \ , \ \eta_2(t) = w_2(0,t) \qquad (9)$$

we associate the system

$$C_0\frac{dv}{dt} + \psi(v) = \frac{1}{2}\sqrt{\frac{C}{L}}[e^{-\delta}\eta_1(t) - e^{\delta}\eta_2(t - \sqrt{LC})]$$

$$\eta_2(t) = -\rho_0\eta_1(t - \sqrt{LC}) + (1 + \rho_0)E(t) \qquad (10)$$

$$\eta_1(t) = -e^{-2\delta}\eta_2(t - \sqrt{LC}) + 2e^{-\delta}v(t)$$

and an additional damping is introduced in the second difference equation. Adapting (6) to the new case, we have

$$u_1(\lambda,t) = e^{-\delta\lambda}\eta_1(t + (1-\lambda)\sqrt{LC}) \ , \ u_2(\lambda,t) = e^{\delta\lambda}\eta_2(t + \lambda\sqrt{LC}) \qquad (11)$$

and it is easily seen that the progressive wave propagates forwards from $\lambda = 0$ to $\lambda = 1$ being retarded and damped along the propagation while the reflected wave propagates backwards from $\lambda = 1$ to $\lambda = 0$ being also retarded and damped. Since the basic waveforms $\eta_i(\cdot)$ are not modified but just retarded during propagation, the propagation is also distortionless.

B. The natural development of the distortionless propagation is to consider the so called inhomogeneous media and transmission lines. The theory of the waveguides is their most straightforward application. The mathematical model of the inhomogeneous transmission line is given by the space varying telegraph equations [4]

$$-\frac{\partial v}{\partial \lambda} = r(\lambda)i(\lambda,t) + l(\lambda)\frac{\partial i}{\partial t} \ , \ -\frac{\partial i}{\partial \lambda} = g(\lambda)v(\lambda,t) + c(\lambda)\frac{\partial v}{\partial t} \qquad (12)$$

with the standard notations, the line having length L. Here $l(\lambda) > 0$, $c(\lambda) > 0$ for standard physical reasons. The distortionless definition (*op. cit.*) states that

$$v(\lambda,t) = f(\lambda)\phi(t - \tau(\lambda)) \qquad (13)$$

where $f(\cdot)$ is called *attenuation* and $\tau(\cdot)$ is called *propagation delay* while $\phi(\cdot)$ is the waveform. There exist also other cases of interest, for instance the time independent voltage/curent ratio i.e. when the line is resistive. Our approach includes these cases in the general setting of the distortionless propagation. Since (12) are exactly like (7), we introduce the Riemann invariants by

$$u^{\pm}(\lambda,t) = v(\lambda,t) \pm a(\lambda)i(\lambda,t) \tag{14}$$

or by the converse equalities; with the choice $a(\lambda) = \sqrt{l(\lambda)/c(\lambda)}$ which is similar to (8) the cross derivative terms are "destroyed" and the equations for the forward and backward waves are obtained. It appears in these equations that the coupling terms cannot be canceled by a unique choice of the line coefficients. This explains the option in [4] for the distortionless propagation forwards: such choice requires decoupling of the progressive wave $u^+(\lambda,t)$. Therefore

$$a'(\lambda) = r(\lambda) - g(\lambda)a^2(\lambda) \tag{15}$$

which is a condition on line's parameters. Remark that *this is a Riccati differential equation*. Consequently the equations of the waves become

$$-\frac{\partial u^+}{\partial \lambda} = \sqrt{l(\lambda)c(\lambda)}\frac{\partial u^+}{\partial t} + a(\lambda)g(\lambda)u^+(\lambda,t)$$

$$-\frac{\partial u^-}{\partial \lambda} = -\sqrt{l(\lambda)c(\lambda)}\frac{\partial u^-}{\partial t} - \frac{r(\lambda)}{a(\lambda)}u^-(\lambda,t)+ \tag{16}$$

$$+(r(\lambda)/a(\lambda) - g(\lambda)a(\lambda))u^+(\lambda,t)$$

Having in mind (8) we introduce the new "waves" by

$$u^+(\lambda,t) = \exp\left(-\int_0^{\lambda} g(\sigma)a(\sigma)d\sigma\right)w^+(\lambda,t)$$

$$u^-(\lambda,t) = \exp\left(-\int_{\lambda}^1 (r(\sigma)/a(\sigma)d\sigma\right)w^-(\lambda,t)$$

to obtain with the corresponding notation

$$-\frac{\partial w^+}{\partial \lambda} = \sqrt{l(\lambda)c(\lambda)}\frac{\partial w^+}{\partial t} , \; \frac{\partial w^-}{\partial \lambda} = \sqrt{l(\lambda)c(\lambda)}\frac{\partial w^-}{\partial t} + \beta(\lambda)w^+(\lambda,t) \tag{17}$$

We perform now integration along the characteristics to find

$$w^+(0,t) = w^+(L,t+\tau) \tag{18}$$

Denoting $\eta^+(t) = w^+(L,t)$ the following representation formula is obtained

$$w^+(\lambda,t) = \eta^+ \left(t + \int_\lambda^L \sqrt{l(\mu)c(\mu)}d\mu \right) \tag{19}$$

obviously accounting for distortionless propagation of the forward wave. For the backward wave we obtain, by integrating along the decreasing characteristics but taking also into account (19)

$$w^-(L,t) = w^-(0,t+\tau) + \int_0^L \beta(\sigma)\eta^+ \left(t + 2\int_\sigma^L \sqrt{l(\mu)c(\mu)}d\mu \right) d\sigma \tag{20}$$

Denoting $\eta^-(t) = w^-(0,t+\tau)$ the following representation formula is obtained

$$w^-(\lambda,t) = \eta^- \left(t + \int_0^\lambda \sqrt{l(\mu)c(\mu)}d\mu \right) +$$
$$+ \int_0^\lambda \beta(\sigma)\eta^+ \left(t + \int_\sigma^\lambda \sqrt{l(\mu)c(\mu)}d\mu + \int_\sigma^L \sqrt{l(\mu)c(\mu)}d\mu \right) d\sigma \tag{21}$$

and the propagation is clearly associated with the distortions introduced by the integral term. To obtain distortionless of the backward wave, it is necessary to have $\beta(\lambda) = 0$ a.e. This will give finally $a'(\lambda) = 0$ hence the ratio $l(\lambda)/c(\lambda)$ *has to be piecewise constant on* $(0,L)$. *Not only constant coefficients can ensure distortionless propagation for both forward and backward waves!*

3   The Multi-wave Case. Application to the Circulating Fuel Nuclear Reactors

When several transmission lines (channels) are included in the system, several couples of waves are present, leading to the model of e.g. [20]

$$\frac{\partial u}{\partial t} + A(\lambda)\frac{\partial u}{\partial \lambda} = B(\lambda)u \, , \, t > 0 \, , \, 0 \le \lambda \le L \tag{22}$$

where u is a m-dimensional vector and $A(\lambda)$, $B(\lambda)$ are $m \times m$ matrices. Also A is supposed diagonal, having distinct diagonal elements, of which k are strictly positive (corresponding to the forward waves) and $m - k$ are strictly negative (corresponding to the backward waves). If $B(\lambda)$ could be also diagonal then propagation would be distortionless, otherwise it is not.

There exist situations when this diagonal structure is inherent to the basic equations, for instance, the model of the circulating fuel nuclear reactor [7, 8, 9]

$$\begin{cases} \dfrac{d}{dt}n(t) = \rho n(t) + \sum_{i=1}^m \beta_i(\bar{c}_i(t) - n(t)) \, , \, t \ge t_0 \, , \, 0 \le \eta \le h \\ \bar{c}_i(t) = \int_0^h \phi(\eta)c_i(\eta,t)d\eta \, , \, \dfrac{\partial c_i}{\partial t} + \dfrac{\partial c_i}{\partial \eta} + \sigma_i c_i = \sigma_i\phi(\eta)n(t) \, , \, i = \overline{1,m} \\ c_i(0,t) = c_i(h,t) \, , \, c_i(\eta,t_0) = q_i^0(\eta) \, , \, n(t_0) = n_0 \, ; \, i = \overline{1,m}. \end{cases} \tag{23}$$

The boundary conditions are of periodic type. The PDE (partial differential equations) are completely decoupled, the coupling taking place at the level of the differential equation. All eigenvalues of $A(\lambda) = I$ are equal and positive hence there exist m forward waves. Integration along the characteristics and computation of the integral of (23) will give again a system of FDE (functional differential equations) of neutral type [14], where it may be seen that propagation is distortionless.

4   A Control Problem

We shall discuss here the control model of an overhead crane with a flexible cable, given by [6]

$$y_{tt} - (a(s)y_s)_s = 0 \, , \, t > 0 \, , \, 0 < s < L \, ; \, a(s) = g(s + \frac{m}{\rho})$$

$$y_{tt}(0,t) = gy_s(0,t) \, , \, y(L,t) = X_p(t) \tag{24}$$

$$\ddot{X}_p = K(a(s)y(s,t))(L,t) + u(t) \, , \, K = \frac{m + \rho L}{ma(L)} = \frac{\rho}{Mg}$$

It will appear in what follows that the nonhomogeneous material properties account for propagation with distortions. But we have to mention first that the basic model of (op.cit.) contained the boundary condition $y_s(0,t) = 0$, explained by the physical assumption that the acceleration of the load mass is negligible with respect to the gravitational acceleration g i.e. $y_{tt}(0,t)/g \approx 0$; in fact this is not rigorous and definitely cannot be ascertained for all t; the only valid argument is connected to singular perturbations. For this reason we shall deal with the complete model (24).

If the rated cable length variable $\sigma = s/L$ is introduced, then, with a slight abuse of notation, the following model containing possible small parameters is obtained

$$\frac{L}{g} \cdot \frac{\rho L}{m} y_{tt} - \left(\left(1 + \frac{\rho L}{m} \sigma \right) y_\sigma \right) = 0 \, , \, 0 \leq \sigma \leq 1 \, , \, t > 0$$

$$\frac{L}{g} y_{tt}(0,t) = y_\sigma(0,t) \, , \, y(1,t) = X_p(t) \, , \, \frac{L}{g} \ddot{X}_p = \frac{m}{M} \left(1 + \frac{\rho L}{M} \right) y_\sigma(1,t) + \frac{L}{g} u(t) \tag{25}$$

A preliminary comment is useful: supposing we would like to neglect nonuniformity of the cable parameters, this would require the assumption that the cable mass is negligible with respect to the carried mass i.e. $\rho L/m \approx 0$. However, this will destroy the entire distributed dynamics since (25) would become

$$y_{\sigma\sigma} = 0 \, ; \, \frac{L}{g} y_{tt}(0,t) = y_\sigma(0,t) \, , \, y(1,t) = X_p$$

$$\frac{L}{g} \ddot{X}_p = \frac{m}{M} y_\sigma(1,t) + \frac{L}{g} u(t) \tag{26}$$

We shall then have $y(\sigma,t) = \phi_1(t)\sigma + \phi_0(t)$ which is substituted in the boundary conditions. Therefore

$$\frac{L}{g}\ddot{\phi}_0 + \phi_0 = X_p \; ; \; \phi_1 = X_p - \phi_0 \; , \; \frac{L}{g}\ddot{X}_p = \frac{m}{M}(X_p - \phi_0) + \frac{L}{g}u(t) \qquad (27)$$

Its uncontrolled dynamics is given by the roots of the characteristic equation

$$\frac{L}{g}s^2\left(\frac{L}{g}s^2 + 1 - \frac{m}{M}\right) = 0 \qquad (28)$$

i.e. by two purely imaginary modes and a double zero mode; this is but well known. Instead of this approach, we start by introducing new functions and by making some other notations

$$v(\sigma,t) := y_t(\sigma,t) \; , \; w(\sigma,t) := (1+\gamma_0\sigma)y_\sigma(\sigma,t) \; , \; \gamma_0 = \frac{\rho L}{m} \; , \; T^2 = \frac{L}{g} \; , \; \delta_0 = \frac{m}{M} \tag{29}$$

Define further the forward and backward waves as below

$$v(\sigma,t) = u^+(\sigma,t) + u^-(\sigma,t) \; , \; w(\sigma,t) = T\sqrt{\gamma_0(1+\gamma_0\sigma)}(u^-(\sigma,t) - u^+(\sigma,t))$$

The following equations are then obtained

$$\frac{\partial u^+}{\partial t} + \frac{1}{T\sqrt{\gamma_0}}\sqrt{(1+\gamma_0\sigma)}\frac{\partial u^+}{\partial t} = \frac{1}{T\sqrt{\gamma_0}} \cdot \frac{\gamma_0}{\sqrt{(1+\gamma_0\sigma)}}(u^- - u^+)$$

$$\frac{\partial u^-}{\partial t} - \frac{1}{T\sqrt{\gamma_0}}\sqrt{(1+\gamma_0\sigma)}\frac{\partial u^-}{\partial t} = \frac{1}{T\sqrt{\gamma_0}} \cdot \frac{\gamma_0}{\sqrt{(1+\gamma_0\sigma)}}(u^- - u^+) \qquad (30)$$

$$T(u_t^- + u_t^+)(0,t) = \sqrt{\gamma_0}(u^-(0,t) - u^+(0,t)) \quad u^-(1,t) + u^+(1,t) = \dot{X}_p$$

$$T\ddot{X}_p = \delta_0\sqrt{\gamma_0(1+\gamma_0)}(u^-(1,t) - u^+(1,t)) + Tu(t)$$

It is clear that *under no conditions can be made this system distortionless*. We may however try to replace this system by an approximation which would be such. To find such an approximation, we turn back to the basic equation of (24) where $a(\cdot)$ is a sufficiently smooth function. With the new variables

$$v(s,t) = y_t(s,t) \; , \; w(s,t) = a(s)y_s(s,t)$$

there are obtained the first order equations of the propagation; the forward and backward waves are defined by

$$v(s,t) = u^-(s,t) + u^+(s,t) \; , \; w(s,t) = \sqrt{a(s)}(u^-(s,t) - u^+(s,t))$$

and satisfy

$$u_t^{\pm} \pm \sqrt{a(s)}u_s^{\pm} = \frac{a'(s)}{4\sqrt{a(s)}}(u^- - u^+) \tag{31}$$

Obviously the distortionless condition is $a'(s) = 0$ a.e., but in our case $a'(s) = g \neq 0$. The piecewise constant approximation is thus the only suitable. This means approximation of $a(s)$ piecewise constantly in order that e.g. the propagation time should remain constant

$$\int_0^L \frac{d\lambda}{\sqrt{a(\lambda)}} = \sum_1^N \frac{l_i}{\sqrt{a_i}} \,,\, \sum_1^N l_i = L \tag{32}$$

We shall not discuss here specific approximation problems such as concatenation conditions and convergence but just take $N = 1$ and write down the associated system. In this simplest case we find

$$T\sqrt{\gamma_0} \int_0^1 \frac{d\sigma}{\sqrt{1 + \gamma_0\sigma}} = T\sqrt{\gamma_0} \frac{1}{\sqrt{1 + \gamma_1}} \tag{33}$$

If $T_d = T\sqrt{\gamma_0/(1+\gamma_1)}$ - the propagation time - is introduced and the cyclic variable X_p is eliminated, we obtain a genuine system of neutral type

$$T_d \frac{d}{dt}(y^+(t) + y^-(t - T_d)) = -\gamma_0(y^+(t) - y^-(t - T_d))$$

$$T_d \frac{d}{dt}(y^-(t) + y^+(t - T_d)) = \gamma_0\delta_0(y^-(t) - y^+(t - T_d)) + T_d u(t) \tag{34}$$

$$\dot{X}_p = y^-(t) + y^+(t - T_d)$$

with adequate notations [18]. For $u(t) \equiv 0$ the inherent stability of (34) has been studied [18]. Its characteristic equation is

$$(T_d s + \gamma_0)(T_d s - \gamma_0\delta_0) - (T_d s - \gamma_0)(T_d s + \gamma_0\delta_0)e^{-2sT_d} = 0 \tag{35}$$

and obviously has a zero root. The output $\dot{X}_p$ being a cyclic variable, we have here a double zero root of the controlled configuration with $u(t)$ as input and X_p as measurable control output. Since (28) has also a pair of of purely imaginary roots, we may check for purely imaginary roots of (35) and find them to be of the form $\pm \iota x_k/Td$ where x_k are the positive roots of

$$\tan x = \frac{\gamma_0(1 - \delta_0)x}{\gamma_0\delta_0 + x^2} \tag{36}$$

The equation is well studied [19]: it has real roots of the form $k\pi + \delta_k$ where $\{\delta_k\}_k$ is a positive bounded sequence approaching 0 for $k \to \infty$.

We have thus discovered an infinity of purely imaginary roots; this infinity of oscillating modes is well known in the theory of the elastic rods; mathematically, its presence can be explained by the fact that the difference operator of (34) has its roots on $\iota\mathbb{R}$ - the imaginary axis. Other details may be found in [18].

5   Dynamics and Control for Systems of Conservation Laws

A more contemporary trend in the field of control for systems with distributed parameters consists in applying control theory to general structures that may be considered as benchmark problems. Due to their broad applications, the systems of
conservation laws which describe various physical phenomena with a single space
parameter distribution are very suitable for such applications [12]. The systems of
conservation laws are interesting also for their nonlinear character; when linearized
they reduce to the quite well propagation equations - see [17] or the previous sections - and, therefore, a comparison to some known results is also available.

We shall consider in this section a system of two conservation laws on $\mathbb{R}^2$ (one
space variable) which reads

$$Y_t + f(Y)_x = 0 \tag{37}$$

where $Y : [0, \infty) \times [0, L] \mapsto \Omega \subseteq \mathbb{R}^2$ is the vector of the two dependent variables and
$f : \Omega \subseteq \mathbb{R}^2 \mapsto \mathbb{R}^2$ is the *flux density*. The solution is defined by the initial conditions

$$Y(x, 0) = Y_0(x) , \ 0 \leq x \leq L \tag{38}$$

and by some boundary conditions of Dirichlet type that may contain some control
input variables

$$g_0(Y(0, t), u_0(t)) = 0 , \ g_L(Y(L, t), u_L(t)) = 0 , \ t > 0 \tag{39}$$

The standard problem we are approaching reads as follows

For constant control actions $u_i(t) \equiv \bar{u}_i$, $i = 0, L$, a steady state solution is a constant solution $\overline{Y}$ satisfying (37) and (39). Depending on the form of the boundary
conditions this steady state solution may be stable or unstable. Accordingly it may
be stated the *boundary control problem* - that of defining the control inputs $u_i(t)$
from a feedback structure such that for any smooth enough initial condition in (38)
the unique smooth solution should converge to a desired steady solution defined by
$\bar{u}_i$ - the controllers' set points.

A. Consider the first application - the control of the flows in open canals [1, 5,
15] By choosing the flow velocity $V(x, t)$ and the cross section $A(x, t)$ (instead of
the liquid level) as variables, the standard Saint Venant equations are conservation
laws. In the simplest case of the *prismatic level canal* whose geometric parameters
are independent of the coordinate x and whose bed is lying at the same constant
elevation Y_b they are given by

$$\frac{\partial}{\partial t} \begin{pmatrix} A \\ V \end{pmatrix} + \frac{\partial}{\partial x} \begin{pmatrix} AV \\ \frac{1}{2}V^2 + g\psi(A) \end{pmatrix} = 0 \tag{40}$$

where $h = \psi(A)$ is the liquid level and ψ is the inverse of the monotone mapping
defining the cross section in a prismatic canal

$$\Phi(h) = \int_0^h \sigma(y)dy$$

(σ is the canal width corresponding to the liquid elevation y). To these equations we may add the boundary conditions which arise from the canal conditions; for constant flow at $x = 0$ and constant level (area) at $x = L$

$$A(0,t)V(0,t) = Q_0 \ , \ A(L,t) = A_L \tag{41}$$

In [15] the existence of physically significant invariant sets was proved. First

$$-F(A_0) < V(x,t) < F(A_0) \ , \ 0 < F(A(x,t)) < 2F(A_0) \ , \ 0 \le x \le L \ , \ t > 0 \tag{42}$$

This shows both limited flow reversals as well as some limitations of the liquid level. It is not quite clear if these conditions may ensure *the invariance of the Froude number* i.e. $\mathfrak{Fr}(A(x,t),V(x,t)) < 1$ provided $\mathfrak{Fr}(A(x,0),V(x,0)) < 1$ thus ensuring sub-criticality of the flow; actually one can hope that initial conditions that are sufficiently far away from the critical limit will generate sub-critical evolutions.

B. We shall address now to a problem that has been considered much earlier (see [17] but also its references). In the technology of combined heat electricity generation there are steam pipes whose dynamics affect the stability of the control systems for basic operating parameters. The traditional approach of pipe dynamics started from the equations of the hydrodynamic flow

$$\frac{\partial w}{\partial t} + w\frac{\partial w}{\partial l} + \frac{1}{\rho}\frac{\partial p}{\partial l} = 0 \ , \ \frac{\partial \rho}{\partial t} + \rho\frac{\partial w}{\partial l} + w\frac{\partial \rho}{\partial l} = 0 \tag{43}$$

where the flow characteristics (velocity w, mass density ρ and steam pressure p) are also related by the polytropic equation

$$p/p_\infty = (\rho/\rho_\infty)^\kappa \tag{44}$$

with the subscript ∞ accounting for steady state values and $\kappa > 1$ being the polytropic exponent. Instead of the linearization we introduce the rated (per cross section area) mass flow $\phi = \rho w$ and eliminate the pressure p to obtain a system of conservation laws

$$\frac{\partial}{\partial t}\begin{pmatrix} \rho \\ \phi \end{pmatrix} + \frac{\partial}{\partial l}\begin{pmatrix} \phi \\ \phi^2/\rho + \gamma_\infty\rho^\kappa \end{pmatrix} = 0 \tag{45}$$

where $\gamma_\infty = p_\infty(\rho_\infty)^{-\kappa}$ is the polytropic steady state constant. The boundary conditions are defined by the controlled admission of the steam into the pipe at $l = 0$ and the steam consumption from the pipe at $l = L$, thus being analogous to those of the previous application

$$\phi(0,t) = \sqrt{2}\,\gamma_\infty\eta(\rho(0,t))^{\frac{\kappa+1}{2}} \quad , \quad \phi(L,t) = \sqrt{2}\,\gamma_\infty\frac{f(t)}{F}(\rho(L,t))^{\frac{\kappa+1}{2}} \tag{46}$$

Here $f(t)$ - the admission cross section of the steam to the consumer acts as a disturbance; since the steam has to be supplied at constant pressure, this pressure has to be
the measured output. Taking into account that the controller acts using the control
error, the controller equations might be as follows

$$\psi_m^2 \ddot{\zeta} + \psi_D \dot{\zeta} + \zeta = \gamma_\infty (\rho(0,t))^\kappa - p_\infty , \ \dot{\eta} = -\varphi(\zeta + \gamma_0(\eta - \eta_\infty)) \qquad (47)$$

where η_∞ - the steady state of the actuator - may be computed using the steady state
equations of (45)-(47). Summarizing there was obtained a boundary value problem
for a system of conservation laws.

For this system various problems may be stated, some of them being already
mentioned at the previous application: discussion of the hyperbolicity, associated
functional equations and basic theory, invariant sets, control synthesis, stability.

In all, the control of the systems of conservation laws is at its beginnings (at least
in the nonlinear case). Our point of view is that *the most suitable approach would be
to use the energy integral as a Liapunov functional* (possibly for synthesis purposes
also: a nonlinear counterpart of the standard results e.g.[17] may be obtained. Unlike
[17] here even the partial differential equations are nonlinear. Finding the associated
functional equations is still a challenge [15].

6 Conclusions

This survey is an attempt to discuss some dynamical models in automatic control
that are connected with distributed parameters in one dimension. These models are
described by boundary value problems for hyperbolic partial differential equations.
We considered here the functional equations associated to these problems using the
integration of the Riemann invariants along the characteristics. Such quite known
models correspond to linear partial differential equations with possible nonlinear
boundary conditions. The most interesting and significant fact is that these equations
arise from the linearization of the equations of the conservation laws. The control
of the nonlinear systems of conservation laws is one of the most recent challenges
in engineering. It is felt that the energy integral combined with integration along
the characteristics could produce new advancement. And, last but not least, the so
called *model validation* (basic theory, invariant sets) may turn helpful for better
control issues.

References

1. Bastin, G., Coron, J.B., d'Andréa Novel, B.: Using hyperbolic system of balance laws
   for modeling, control and stability analysis of physical networks. In: Conf. on Contr. of
   Phys. Syst. and Partial Diff. Eqs (Lect. Notes). Inst. Henri Poincaré, 16 p (2008)
2. Bernoulli, J.: Comm. Acad. Sci. Imp. Petropolitanae 3, 13–28 (1728)
3. Brayton, R.K.: IBM. Journ. Res. Develop. 12, 431–440 (1968)
4. Burke, V., Duffin, R.J., Hazony, D.: Quart. Appl. Math. XXXIV, pp. 183–194 (1976)

5. Coron, J.B., d'Andréa Novel, B., Bastin, G.: IEEE Trans. on Aut. Control 52, 2–11 (2007)
6. d'Andréa Novel, B., Boustany, F., Conrad, F., Rao, B.P.: Math. Contr. Signals Systems 7, 1–22 (1994)
7. Gorjachenko, V.D.: Methods of stability theory in the dynamics of nuclear reactors (in Russian). Atomizdat, Moscow (1971)
8. Gorjachenko, V.D.: Methods for nuclear reactors stability study (in Russian). Atomizdat, Moscow (1977)
9. Gorjachenko, V.D., Zolotarev, S.L., Kolchin, V.A.: Qualitative methods in nuclear reactor dynamics (in Russian). Energoatomizdat, Moscow (1988)
10. Kabakov, I.P.: Inzh. Sbornik. 2, 27–60 (1946)
11. Kabakov, I.P., Sokolov, A.A.: Inzh. Sbornik. 2, 61–76 (1946)
12. Lax, P.D.: Hyperbolic Partial Differential Equations, Courant Lecture Notes in mathematics 14. AMS & Courant Inst. of Math. Sci., Providence (2006)
13. Marinov, C., Neittaanmäki, P.: Mathematical Models in Electrical Circuits:Theory and Applications. Kluwer Academic, Dordrecht (1991)
14. Niculescu, S.I., Răsvan, V.l.: Stability of some models of circulating fuel nuclear reactors - a Liapunov approach. In: Agarwal, R.P., Perera, K. (eds.) Differential and Difference Equations and Applications (Proceedings), pp. 861–870. Hindawi Publ. Corp., New York (2005)
15. Petre, E., Răsvan, V.l.: Rev. Roum. Sci. Techn.-Électrotechn. et Énerg. 54, 311–320 (2009)
16. Pinney, E.: Ordinary Difference-Differential Equations. Univ. of California Press, Berkeley (1958)
17. Răsvan, V.l.: Absolute stability of time lag control systems (in Romanian). Editura Academiei, Bucharest (1975)
18. Răsvan, V.l.: Contr. Engineering and Appl. Informatics 10(3), 11–17 (2008)
19. Răsvan, V.l.: Mathem. Reports 9(59), 99–110 (2007)
20. Smirnova, V.B.: Diff. Uravnenya 9, 149–157 (1973)
21. Sokolov, A.A.: Inzh. Sbornik. 2, 4–26 (1946)
22. Solodovnikov, V.V.: Avtomat. i Telemekh 6(1), 5–20 (1941)

Equations with Advanced Arguments in Stick Balancing Models

Tamas Insperger, Richard Wohlfart, Janos Turi, and Gabor Stepan

Abstract. A stick balancing problem is considered, where the output for the feedback controller is provided by an accelerometer attached to the stick. This output is a linear combination of the stick's angular displacement and its angular acceleration. If the output is fed back in a PD controller with feedback delay, then the governing equation of motion is an advanced functional differential equation, since the third derivative of the angular displacement (the angular jerk) appears with a delayed argument through the derivative term. Equations with advanced arguments are typically non-causal and are unstable with infinitely many unstable poles. It is shown that the sampling of the controller may still stabilize the system in spite of its advanced nature. In the paper, different models for stick balancing are considered and discussed by analyzing the corresponding stability diagrams.

1 Introduction

Systems, where the rate of change of the state depends on the state at deviating arguments are described by functional differential equations (FDEs). FDEs can be categorized into retarded, neutral and advanced types (see, e.g., [1, 6]). If the rate of change of the state depends on the past states of the system, then the corresponding mathematical model is a retarded functional differential equation (RFDE). If the rate of change of the state depends on its own past values as well, then the corresponding equation is called neutral functional differential equation (NFDE). If the rate of change of the state depends on the past values of higher derivatives of the

Tamas Insperger · Richard Wohlfart · Gabor Stepan
Budapest University of Technology and Economics, Department of Applied Mechanics, Budapest, Hungary
e-mail: inspi@mm.bme.hu, wohl@mm.bme.hu, stepan@mm.bme.hu

Janos Turi
Programs in Mathematical Sciences, University of Texas at Dallas, Richardson, TX, USA
e-mail: turi@utdallas.edu

R. Sipahi et al. (Eds.): Time Delay Sys.: Methods, Appli. and New Trends, LNCIS 423, pp. 161–172.
springerlink.com   © Springer-Verlag Berlin Heidelberg 2012

state, then the system is described by an advanced functional differential equation (AFDE). Note that these equations are also referred to as FDEs of retarded, neutral or advanced type [7].

The reason for the phrase "advanced" can be demonstrated by the following example. Consider the simple AFDE

$$\dot{x}(t) = \ddot{x}(t - \tau) \,. \tag{1}$$

By a τ-shift transformation in time, and by using the new variable $z = \dot{x}$, this equation can be written in the form

$$\dot{z}(t) = z(t + \tau) \,. \tag{2}$$

Here, the rate of change of state is determined by future values of the state.

As opposed to RFDEs and NFDEs, AFDEs are rarely used in practical applications due to their inverted causality explained by (2). While linear autonomous RFDEs have infinitely many poles on the left half of the complex plane, linear autonomous AFDEs have infinitely many poles on the right half of the complex plane [3, 9]. In this sense, linear autonomous AFDEs are always strongly unstable.

Control systems with feedback delay are usually described by RFDEs or NFDEs. For instance, displacement and velocity feedback in a second order system results in an RFDE, while acceleration feedback induces an NFDE (see, e.g., [12]). However, if the jerk (the third derivative of the displacement) is fed back in a second order system with feedback delay, then the governing equation is an AFDE, which is always unstable independently on the system and the control parameters. Although, jerk is practically never intended to be fed back in a control system, it may still appear in the input signal due to measuring error or noise.

In this paper, a stick balancing problem is considered where the angular displacement is to be measured by an accelerometer attached to the stick. Different types of models are considered with respect to the arguments of the feedback term. The corresponding model equations are RFDEs, NFDEs, and AFDEs. It is shown that the sampling effect stabilizes the system even if an advanced term (the angular jerk with feedback delay) appears in the equation.

2 Different Mechanical Models

Figure 1 presents the mechanical model of a simple stick balancing process using the signal of an accelerometer to measure the angular position of the stick compared to the vertical direction. The equation of motion reads

$$J\ddot{\varphi}(t) - Hmg\sin\varphi(t) = Q(t) \,, \tag{3}$$

where φ is the angular position of the pinned stick, m is the mass, H is the distance between the suspension point O and the centre of gravity C, J is the mass moment of inertia with respect to the axis normal to the plane of the figure through point O, and Q is the control torque.

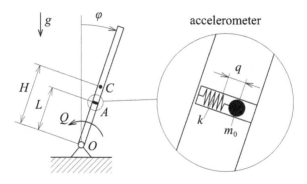

Fig. 1 One-DoF model of stick balancing, and the mechanical model of the accelerometer.

The angular position of the stick is measured by the piezo-accelerometer attached to the stick at point A. This accelerometer operates normal to the stick as a mass m_0 attached to a spring of stiffness k modeling the piezo crystal (see Fig. 1). The accelerometer's output y is proportional to the displacement q of the mass: $y = K_A q$ with K_A [V/m] being the characteristic constant of the accelerometer. The displacement q of the mass depends on the angular position φ of the rod. If the rod is standing still in an oblique position, then

$$q = \frac{m_0 g}{k} \sin \varphi \,. \tag{4}$$

Assuming small angular displacements, the output can be approximated linearly as

$$y = K\varphi \,, \tag{5}$$

where $K = K_A m_0 g / k$ [V/rad].

If the rod is in motion, then the dynamic effects should also be taken into account. In this case the displacement of the accelerometer reads

$$q = \frac{m_0 g}{k} \sin \varphi - \frac{m_0 L}{k} \ddot{\varphi} \,, \tag{6}$$

where $\ddot{\varphi}$ is the angular acceleration of the rod. Here, the linearized output is

$$y = K\varphi - K_1 \ddot{\varphi} \,, \tag{7}$$

where $K_1 = K_A m_0 L / k$ [Vs^2/rad] and K is the same as above. As it can be seen, the second derivative of the angular position φ appears in the output.

Based on the measurement technique of the output, three different models are distinguished:

(a) real-time continuous measurement: $y(t)$;
(b) continuous measurement with feedback delay: $y(t - \tau)$;
(c) sampled measurement (digital control) with feedback delay: $y(t_{j-r})$ with $t \in [t_j, t_{j+1}]$, $t_j = jh$, $j \in \mathbb{Z}$, here h is the sampling period, r is the delay parameter and $\tau = rh$ is the feedback delay.

Table 1 Different models for stick balancing using accelerometer

Model	output	control torque
1.0	$y(t) = K\varphi(t)$	$Q(t) = -PK\varphi(t) - DK\dot{\varphi}(t)$
1.1	$y(t) = K\varphi(t-\tau)$	$Q(t) = -PK\varphi(t-\tau) - DK\dot{\varphi}(t-\tau)$
1.2	$y(t) = K\varphi(t_{j-r})$	$Q(t) = -PK\varphi(t_{j-r}) - DK\dot{\varphi}(t_{j-r})$
1.3	$y(t) = K\varphi(t_{j-r})$	$Q(t) = -PK\varphi(t_{j-r}) - DK\frac{\varphi(t_{j-r}) - \varphi(t_{j-r-1})}{h}$
2.0	$y(t) = K\varphi(t) - K_1\ddot{\varphi}(t)$	$Q(t) = -PK\varphi(t) + PK_1\ddot{\varphi}(t) - DK\dot{\varphi}(t) + DK_1\varphi^{(3)}(t)$
2.1	$y(t) = K\varphi(t-\tau) - K_1\ddot{\varphi}(t-\tau)$	$Q(t) = -PK\varphi(t-\tau) + PK_1\ddot{\varphi}(t-\tau)$
		$\qquad -DK\dot{\varphi}(t-\tau) + DK_1\varphi^{(3)}(t-\tau)$
2.2	$y(t) = K\varphi(t_{j-r}) - K_1\ddot{\varphi}(t_{j-r})$	$Q(t) = -PK\varphi(t_{j-r}) + PK_1\ddot{\varphi}(t_{j-r})$
		$\qquad -DK\dot{\varphi}(t_{j-r}) + DK_1\varphi^{(3)}(t_{j-r})$
2.3	$y(t) = K\varphi(t_{j-r}) - K_1\ddot{\varphi}(t_{j-r})$	$Q(t) = -PK\varphi(t_{j-r}) + PK_1\ddot{\varphi}(t_{j-r})$
		$\qquad -DK\frac{\varphi(t_{j-r}) - \varphi(t_{j-r-1})}{h} + DK_1\frac{\ddot{\varphi}(t_{j-r}) - \ddot{\varphi}(t_{j-r-1})}{h}$

It is assumed that the output is fed back in a PD controller . In case of sampled systems (digital controller), two derivative models can be defined:

(I) continuous differentiation: $\dot{y}(t_{j-r})$, $t \in [t_j, t_{j+1}]$;

(II) discrete differentiation: $\frac{y(t_{j-r}) - y(t_{j-r-1})}{h}$, $t \in [t_j, t_{j+1}]$.

Equations (5) and (7) and cases (a), (b), (c) and (I), (II) rises up 8 different models listed in Table 1. In the next section, these models are analyzed in details.

3   Analysis of the Different Models

In this section, the models listed in Table 1 are considered and their stability properties are described using stability charts.

Model 1.0

In this model, it is assumed that the exact angular position is measured continuously real-time and is fed back without any delay. The corresponding linearized equation of motion reads

$$\ddot{\varphi}(t) - \frac{Hmg}{J}\varphi(t) = -\frac{PK}{J}\varphi(t) - \frac{DK}{J}\dot{\varphi}(t) , \qquad (8)$$

which implies

$$\ddot{\varphi}(t) + d\dot{\varphi}(t) + (p-a)\varphi(t) = 0 , \qquad (9)$$

with

$$a = \frac{Hmg}{J}, \qquad p = \frac{PK}{J}, \qquad d = \frac{DK}{J}. \tag{10}$$

This system is asymptotically stable if $d > 0$ and $p > a$.

Model 1.1

Here, it is assumed that the feedback loop contains a delay τ. The corresponding linearized equation of motion reads

$$\ddot{\varphi}(t) - a\varphi(t) = -p\varphi(t - \tau) - d\dot{\varphi}(t - \tau), \tag{11}$$

where a, p and d are defined in (10). This equation is a basic equation for balancing with feedback delay [11]. Although complete pole placement is not possible for this system [8], it can be stabilized for certain system and delay parameters. The corresponding stability boundaries are the line $p = a$ and the parametric curve

$$p = (\omega^2 + a)\cos(\omega\tau), \qquad d = \frac{\omega^2 + a}{\omega}\sin(\omega\tau), \tag{12}$$

with $\omega \in \mathbb{R}^+$. The corresponding stability chart can be seen in Fig. 2 for $a = 0.2$ and $\tau = 1$. Note that this system is always unstable if $a > a_{\text{crit}} = 2/\tau^2$ (see, for instance, [10] or [11]).

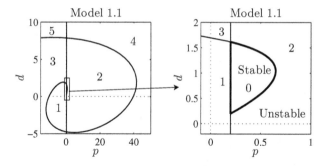

Fig. 2 Stability boundaries and the number of unstable poles for (11) with $\tau = 1$ and $a = 0.2$.

Model 1.2

In this model, it is assumed that sampling effect (or zero-order hold) also arises in addition to the feedback delay. The governing equation reads

$$\ddot{\varphi}(t) - a\varphi(t) = -p\varphi(t_{j-r}) - d\dot{\varphi}(t_{j-r}), \qquad t \in [t_j, t_{j+1}], \tag{13}$$

where $t_j = jh$, $j \in \mathbb{Z}$ denotes the discrete instants of sampling, h is the sampling period and r is the delay parameter. The feedback delay in this case is $\tau = rh$. The parameters a, p and d are defined in (10).

Equation (13) can also be written in the form

$$\ddot{\varphi}(t) - a\varphi(t) = -p\varphi(t - \sigma(t)) - d\dot{\varphi}(t - \sigma(t)), \tag{14}$$

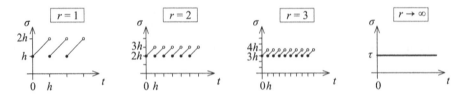

Fig. 3 Sampling effect as time-periodic delay.

where

$$\sigma(t) = \tau - t_j + t, \quad t \in [t_j, t_{j+1}) \tag{15}$$

is a time-periodic delay shown in Fig. 3. In fact, the sampling effect introduces a periodic parametric excitation at the time delay according to (15), thus (13) is an non-autonomous RFDE. It can be seen that the time-periodic delay $\sigma(t)$ tends to the constant delay τ as $h \to 0$ and $r \to \infty$ by keeping $rh = \tau$. This limit is basically the key point of the semi-discetization method of delayed systems [4].

Equation (13) can be transformed to the form

$$\dot{x}(t) = Ax(t) + Bu(t_{j-r}), \quad t \in [t_j, t_{j+1}], \tag{16}$$
$$u(t_j) = Dx(t_j) \tag{17}$$

with

$$x(t) = \begin{pmatrix} \varphi(t) \\ \dot{\varphi}(t) \end{pmatrix}, \quad A = \begin{pmatrix} 0 & 1 \\ a & 0 \end{pmatrix}, \quad B = \begin{pmatrix} 0 \\ 1 \end{pmatrix}, \quad D = (-p \ -d). \tag{18}$$

If $r \geq 1$, then piecewise solution of the system over a sampling period and state augmentation gives the finite dimensional discrete map

$$\begin{pmatrix} x_{j+1} \\ u_j \\ u_{j-1} \\ \vdots \\ u_{j-r+1} \end{pmatrix} = \underbrace{\begin{pmatrix} P & 0 & \dots & 0 & RB \\ D & 0 & \dots & 0 & 0 \\ 0 & 1 & \dots & 0 & 0 \\ \vdots & & \ddots & & \vdots \\ 0 & 0 & \dots & 1 & 0 \end{pmatrix}}_{= \Phi_{1.2}} \begin{pmatrix} x_j \\ u_{j-1} \\ u_{j-2} \\ \vdots \\ u_{j-r} \end{pmatrix}, \tag{19}$$

where $x_j = x(t_j)$, $u_j = u(t_j)$, and

$$P = e^{Ah}, \quad R = \int_0^h e^{A(h-s)} ds. \tag{20}$$

The system is asymptotically stable if the eigenvalues of the monodromy matrix $\Phi_{1.2}$ in (19) are in modulus less than one. The corresponding stability chart can be seen in Fig. 4 for $a = 0.2$ and for different delay parameters r such that $rh = \tau = 1$.

If the delay $\tau = rh$ is fixed to 1, then the system is always unstable if

$$a > a_{\text{crit}} = \left(r \ln \left(\frac{r(r+1) + 1 + \sqrt{2r(r+1) + 1}}{r(r+1)} \right) \right)^2. \tag{21}$$

For $r = 1$, this formula reduces to

$$a_{\mathrm{crit}} = \ln^2\left(\frac{3+\sqrt{5}}{2}\right) = 0.926\,, \qquad (22)$$

that was derived for the digitally controlled inverted pendulum in [2]. The critical values to different delay parameters r with $rh = \tau = 1$ are summarized in Table 2. It can be seen that as $r \to \infty$ and $h \to 0$ by keeping $\tau = rh = 1$ constant, the critical system parameter tends to that of Model 1.1, that is, to $a_{\mathrm{crit}} = 2$ in case of $\tau = 1$.

Table 2 Some critical system parameters for Model 1.2 with $\tau = rh = 1$

r	1	2	5	10	100
a_{crit}	0.926	1.297	1.658	1.815	1.980

Model 1.3

The new feature of this model compared to the previous one is that here the angular velocity for the derivative term of the feedback is determined by discrete differentiation. The corresponding equation reads

$$\ddot{\varphi}(t) - a\varphi(t) = -p\varphi(t_{j-r}) - d\frac{\varphi(t_{j-r}) - \varphi(t_{j-r-1})}{h}\,, \qquad t \in [t_j, t_{j+1}]\,. \qquad (23)$$

This system can be transformed to a system with two delays of the form

$$\ddot{\varphi}(t) - a\varphi(t) = -p_1\varphi(t_{j-r}) + p_2\varphi(t_{j-r-1})\,, \qquad t \in [t_j, t_{j+1}]\,, \qquad (24)$$

where $p_1 = p + d/h$ and $p_2 = d/h$. The first-order representation reads

$$\dot{x}(t) = Ax(t) + BD_1y(t_{j-r}) + BD_2y(t_{j-r-1})\,, \qquad t \in [t_j, t_{j+1}]\,, \qquad (25)$$
$$y(t_j) = Cx(t_j)\,, \qquad (26)$$

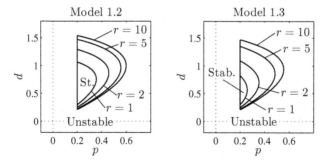

Fig. 4 Stability boundaries for (13) (left) and for (23) (right) with $a = 0.2$ for different delay parameters r such that $rh = \tau = 1$.

with $x(t)$, A, and B defined in (18) and

$$C = \begin{pmatrix} 1 & 0 \end{pmatrix}, \quad D_1 = (-p_1), \quad D_2 = (p_2). \tag{27}$$

If $r \geq 1$, then this semi-discrete system implies the discrete map

$$\begin{pmatrix} x_{j+1} \\ y_j \\ y_{j-1} \\ \vdots \\ y_{j-r} \end{pmatrix} = \underbrace{\begin{pmatrix} P & 0 & \dots & 0 & RBD_1 & RBD_2 \\ C & 0 & \dots & 0 & 0 & 0 \\ 0 & 1 & \dots & 0 & 0 & 0 \\ \vdots & & \ddots & & & \vdots \\ 0 & 0 & \dots & & 1 & 0 \end{pmatrix}}_{= \Phi_{1.3}} \begin{pmatrix} x_j \\ y_{j-1} \\ y_{j-2} \\ \vdots \\ y_{j-r-1} \end{pmatrix}, \tag{28}$$

where P and R are defined in (20). The system is asymptotically stable if the eigenvalues of the monodromy matrix $\Phi_{1.3}$ in (28) are in modulus less than one. The associated stability chart is presented in Fig. 4 for $a = 0.2$.

Model 2.0

In this model, it is assumed that the angular position is measured by accelerometers continuously without any feedback delay. In this case the output of the system is affected by the angular acceleration of the stick, and the equation of motion reads

$$\ddot{\varphi}(t) - \frac{Hmg}{J}\varphi(t) = -\frac{PK}{J}\varphi(t) + \frac{PK_1}{J}\ddot{\varphi}(t) - \frac{DK}{J}\dot{\varphi}(t) + \frac{DK_1}{J}\dddot{\varphi}(t), \tag{29}$$

where $\dddot{\varphi}(t)$ is the angular jerk. This equation can be written in the form

$$-\varepsilon d\dddot{\varphi}(t) + (1 - \varepsilon p)\ddot{\varphi}(t) + d\dot{\varphi}(t) + (p - a)\varphi(t) = 0, \tag{30}$$

where $\varepsilon = K_1/K$ describes the weight of the angular acceleration in the output, and the parameters a, p and d are defined in (10). The condition for asymptotic stability for this ordinary differential equation is

$$d > 0 \quad \text{and} \quad a < p < 1/\varepsilon < 0. \tag{31}$$

Clearly, this system is unstable for any parameters p and d, if $a > 0$.

Model 2.1

This model assumes that the output of the system is affected by the angular acceleration as in the previous model, but the feedback loop involves a delay τ. The corresponding equation of motion reads

$$\ddot{\varphi}(t) - a\varphi(t) = -p\varphi(t - \tau) + \varepsilon p\ddot{\varphi}(t - \tau) - d\dot{\varphi}(t - \tau) + \varepsilon d\dddot{\varphi}(t - \tau). \tag{32}$$

where, again, a, p and d are defined in (10), and $\varepsilon = K_1/K$. As it can be seen, the highest derivative, the angular jerk, appears with a delayed argument, thus this equation is an autonomous AFDE. Consequently, this system is always unstable with infinitely many unstable poles if $\varepsilon d \neq 0$.

Model 2.2

Here, the sampling effect is also modeled, and the governing equation is shaped as

$$\dddot{\varphi}(t) - a\varphi(t) = -p\varphi(t_{j-r}) + \varepsilon p\ddot{\varphi}(t_{j-r}) - d\dot{\varphi}(t_{j-r}) + \varepsilon d\dddot{\varphi}(t_{j-r}), \quad t \in [t_j, t_{j+1}]. \tag{33}$$

This equation can also be considered as an AFDE, since the highest derivative appears with a delayed argument. Note that the term $\dddot{\varphi}(t_{j-r})$ with $t \in [t_j, t_{j+1}]$ can be written in the form $\dddot{\varphi}(t - \sigma(t))$ with $\sigma(t)$ defined in (15) (see also Fig. 3). Still, this equation does not face the non-causality of (32), since this equation can simply be solved by the method of steps if the initial function and its appropriate derivatives are given. The corresponding first-order representation reads

$$\dot{x}(t) = Ax(t) + Bu(t_{j-r}), \quad t \in [t_j, t_{j+1}], \tag{34}$$
$$u(t_j) = Dx(t_j) - \varepsilon D\ddot{x}(t_j) \tag{35}$$

with $x(t)$, A, B, and D defined in (18). Piecewise solution of the system over a sampling period and state augmentation gives the finite dimensional discrete map

$$\begin{pmatrix} x_{j+1} \\ \ddot{x}_{j+1} \\ u_j \\ u_{j-1} \\ \vdots \\ u_{j-r+1} \end{pmatrix} = \underbrace{\begin{pmatrix} P & 0 & \cdots & 0 & RB \\ A^2 P & 0 & \cdots & 0 & QB \\ D & -\varepsilon D & & 0 & 0 \\ 0 & 1 & & 0 & 0 \\ \vdots & & \ddots & & \vdots \\ 0 & 0 & \cdots & 1 & 0 \end{pmatrix}}_{= \Phi_{2.2}} \begin{pmatrix} x_j \\ \ddot{x}_j \\ u_{j-1} \\ u_{j-2} \\ \vdots \\ u_{j-r} \end{pmatrix}, \tag{36}$$

for the case $r \geq 1$, where P and R are the same as in (20), and

$$Q = \int_0^h A^2 e^{A(h-s)} ds + A. \tag{37}$$

The system is asymptotically stable if the eigenvalues of the monodromy matrix $\Phi_{2.2}$ in (36) are in modulus less than one. The stability boundaries can be seen in Fig. 5 for $a = 0.2$ with different r and ε values. It can be seen that there exist some domains of the control parameters, where the system is stable. Note that the case

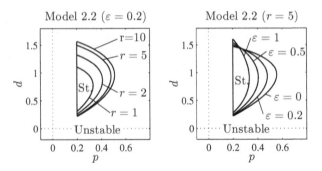

Fig. 5 Stability boundaries for (33) for different delay parameters r (left) and for different ε (right) with $rh = \tau = 1$ and $a = 0.2$.

$\varepsilon = 0$ gives Model 1.2, and the corresponding stability boundaries are identical to that of Model 1.2.

Model 2.3

In this model, the angular velocity for the derivative term of the feedback is determined by digital differentiation as it was done for Model 1.3. The corresponding equation reads

$$\ddot{\varphi}(t) - a\varphi(t) = -p\varphi(t_{j-r}) + \varepsilon p\dot{\varphi}(t_{j-r}) - d\frac{\varphi(t_{j-r}) - \varphi(t_{j-r-1})}{h}$$
$$+ \varepsilon d\frac{\dot{\varphi}(t_{j-r}) - \dot{\varphi}(t_{j-r-1})}{h}, \qquad t \in [t_j, t_{j+1}]. \quad (38)$$

In this case, the highest derivative, the angular acceleration, appears both with delayed and with non-delayed arguments, thus, this equation is an NFDE. Similarly to (38), this system can be transformed to a system with two delays of the form

$$\ddot{\varphi}(t) - a\varphi(t) = -p_1\varphi(t_{j-r}) + p_2\varphi(t_{j-r-1})$$
$$+ \varepsilon p_1\dot{\varphi}(t_{j-r}) - \varepsilon p_2\dot{\varphi}(t_{j-r-1}), \qquad t \in [t_j, t_{j+1}], \quad (39)$$

where $p_1 = p + d/h$ and $p_2 = d/h$. The first-order representation reads

$$\dot{x}(t) = Ax(t) + BD_1 y(t_{j-r}) + BD_2 y(t_{j-r-1}), \qquad t \in [t_j, t_{j+1}] \quad (40)$$
$$y(t_j) = C_1 x(t_j) + C_2\dot{x}(t_j), \quad (41)$$

where $x(t)$, A, and B are defined in (18) and

$$C_1 = \begin{pmatrix} 1 & 0 \end{pmatrix}, \quad C_2 = \begin{pmatrix} -\varepsilon & 0 \end{pmatrix}, \quad D_1 = \begin{pmatrix} -p_1 \end{pmatrix}, \quad D_2 = \begin{pmatrix} -p_2 \end{pmatrix}. \quad (42)$$

The corresponding discrete map for $r \geq 1$ reads

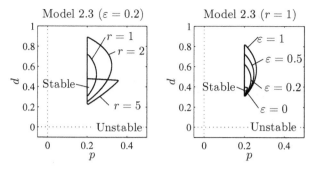

Fig. 6 Stability boundaries for (38) for different delay parameters r (left) and for different ε (right) with $rh = \tau = 1$ and $a = 0.2$.

$$\begin{pmatrix} x_{j+1} \\ \ddot{x}_{j+1} \\ y_j \\ y_{j-1} \\ \vdots \\ y_{j-r} \end{pmatrix} = \underbrace{\begin{pmatrix} P & 0 & \dots & 0 & RBD_1 & RBD_2 \\ A^2P & 0 & \dots & 0 & QBD_1 & QBD_2 \\ C_1 & C_2 & \dots & 0 & 0 & 0 \\ 0 & 1 & \dots & 0 & 0 & 0 \\ \vdots & & \ddots & & \vdots & \\ 0 & 0 & \dots & 1 & 0 \end{pmatrix}}_{= \, \Phi_{2.3}} \begin{pmatrix} x_j \\ \ddot{x}_j \\ y_{j-1} \\ y_{j-2} \\ \vdots \\ y_{j-r-1} \end{pmatrix}, \qquad (43)$$

where P, R and Q are defined in (20) and in (37). The system is asymptotically stable if the eigenvalues of the monodromy matrix $\Phi_{2.3}$ in (43) are in modulus less than one. The corresponding stability diagram is presented in Fig. 6 for $a = 0.2$ with different r and ε. Note that the case $\varepsilon = 0$ gives Model 1.3, and the corresponding stability boundaries are identical to that of Model 1.3.

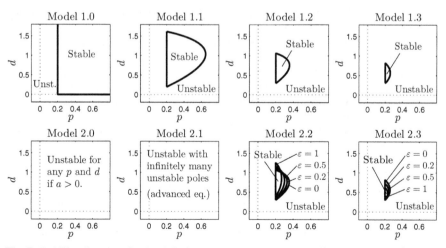

Fig. 7 Stability diagrams for the different models. Parameters: $a = 0.2$, $\tau = 1$ (for models 1.1 and 2.1) and $r = 1$ (for models 1.2, 1.3, 2.2 and 2.3).

4 Results and Conclusions

Stability diagrams to the different models listed in Table 1 are summarized in Fig. 7. The systems described by Models 1.0, 1.1, 1.2 and 1.3, where the output contains only the angular position of the stick, can all be stabilized if the system parameter a and the feedback delay τ satisfies certain conditions. The system described by Model 2.0, where the linear combination of the angular position and the angular acceleration is fed back continuously without delay, cannot be stabilized if the open loop system is unstable (i.e., if $a < 0$). Model 2.1 shows that if this output signal is fed back continuously with feedback delay, then the third derivative of the angular position (the angular jerk) appears in the equation with a delayed argument resulting in an AFDE that is unstable for any system parameters with infinitely many unstable poles. This strong instability is however reversed by the sampling effect of the (digital) controller, as it was shown by Model 2.2. Finally, Model 2.3 shows that digital differentiation may also help in the stabilization of the process. A detailed study about this phenomenon for first order delayed, neutral and advanced scalar equations was presented in [5].

Acknowledgements. This work was supported by the János Bolyai Research Scholarship of the HAS and by the grants OTKA K72911, OTKA K068910 and NSF DMS-0705247.

References

1. Èl'sgol'c, L.È.: Qualitative methods in mathematical analysis. AMS, Providence (1964)
2. Enikov, E., Stepan, G.: Micro-chaotic motion of digitally controlled machines. J. Vib. Control 4, 427–443 (1998)
3. Hale, J.K., Lunel, S.M.V.: Introduction to functional differential equations. Springer, New York (1993)
4. Insperger, T., Stepan, G., Turi, J.: On the higher-order semi-discretizations for periodic delayed systems. J. Sound Vib. 313, 334–341 (2008)
5. Insperger, T., Stepan, G., Turi, J.: Delayed feedback of sampled higher derivatives. Philos. T. R. Soc. A. 368, 469–482 (2010)
6. Kolmanovskii, V.B., Myshkis, A.D.: Introduction to the theory and applications of functional differential equations. Kluwer Academic Publishers, Dordrecht (1999)
7. Kolmanovskii, V.B., Nosov, V.R.: Stability of functional differential equations. Academic Press, London (1986)
8. Michiels, W., Engelborghs, K., Vansevenant, P., Roose, D.: Continuous pole placement for delay equations. Automatica 38, 747–761 (2002)
9. Niculescu, S.-I.: Delay effects on stability – A robust control approach. Springer, London (2001)
10. Stepan, G.: Retarded dynamical systems. Longman, Harlow (1989)
11. Stepan, G.: Delay effects in the human sensory system during balancing. Philos. T. R. Soc. A. 367, 1195–1212 (2009)
12. Vyhlídal, T., Michiels, W., Zítek, P., McGahan, P.: Stability impact of small delays in proportional-derivative state feedback. Control Eng. Pract. 17, 382–393 (2009)

Optimal Control with Preview for Lateral Steering of a Passenger Car: Design and Test on a Driving Simulator

Louay Saleh, Philippe Chevrel, and Jean-François Lafay

Abstract. This paper is dedicated to studying the characteristics of the optimal preview control for lateral steering of a passenger vehicle. Such control is known to guarantee improved performance when the near future of the exogenous signal, here the road curvature, is known. The synthesis is performed in continuous time and leads to a two-degrees of freedom feedback and feedforward controller, whose feedforward part is a finite impulse response filter. The controller has been implemented on the SCANeR^{TM} Driving Simulator available at IRCCyN, whose steering column is electrically powered. A methodology for choosing the weighting matrices in the quadratic index and the preview time are finally proposed. The obtained experimental results are discussed as well.

1 Introduction

Many works have been dedicated recently to the design of control laws for the lateral assistance of vehicles, e.g synthesis of optimal and robust control LQ, LQG, H2 and H∞ with constraints, [7] and [12]. These syntheses have proven their effectiveness in tracking and correcting the excessive deviations from the centerline of the road, but their level of anticipation is often quite low, particularly due to the absence of prediction for the coming trajectory curvature.

Recently, an approach with predictive aspect has appeared in the literature. A so-called LQ-Preview control law is proposed in [6] to exploit the information known in advance about the trajectory curvature. Such a control law is used, when the exogenous signals (disturbances or references) are known in advance, to anticipate

Louay Saleh · Philippe Chevrel · Jean-François Lafay
IRCCyN, Institut de Recherche en Communications et Cyberntique de Nantes, UMR CNRS 6597, Ecole Centrale de Nantes,1, rue de la No - B.P. 92101 - F-44321 Nantes, France
e-mail: Louay.Saleh@irccyn.ec-nantes.fr,
    Philippe.Chevrel@irccyn.ec-nantes.fr,
    Jean-Francois.Lafay@irccyn.ec-nantes.fr

R. Sipahi et al. (Eds.): Time Delay Sys.: Methods, Appli. and New Trends, LNCIS 423, pp. 173–185.
springerlink.com   © Springer-Verlag Berlin Heidelberg 2012

and then improve the performance of the controlled scheme in terms of tracking or disturbance rejection. This is obtained in an optimal way by introducing an infinite dimensional feedforward term to the control law. The application of such optimal preview control to the lateral guidance of a vehicle has been considered also in [11] and [4]. Recent theoretic progress in preview control can be found in [5] and [2].

This article focuses on an improved way to use preview control for an automated steering of vehicles, with the implicit goal to consider it later on for developing assistance to the driver. It defines the H2/LQ-Preview problem and proposes a solution for the more general case when one input disturbance model is used to anticipate the trajectory. The method is then applied to the lateral guidance of a passenger vehicle, and the controller performance is evaluated both in simulation (see [9]) and on a driving simulator. The paper proposes a methodology for choosing the preview horizon and the weighting matrices in the quadratic index.

After presenting the generic problem in §2, its solution is given in §3 and analyzed in §4. Its application to the unmanned lateral vehicle control using a driving simulator and the experimental results are discussed in §5 and §6.

2   Problem Statement

Consider the LTI system Σ:

$$\dot{x} = Ax + B_1 u + B_2 w$$
$$z = Cx + D_1 u + D_2 w \tag{1}$$

$$A \in \mathfrak{R}^{n \times n}, B_1 \in \mathfrak{R}^{n \times m}, B_2 \in \mathfrak{R}^{n \times r}, C \in \mathfrak{R}^{p \times n}, D_1 \in \mathfrak{R}^{p \times m}, D_2 \in \mathfrak{R}^{p \times r}$$

where x is the state vector, u the control input , w the disturbance input and z the performance vector output. Let's now define the "optimal preview controller problem" as the problem of finding a controller Σ_c which rejects the effect of the input disturbances w (known in advance with a finite preview horizon T) on the output z (see figure 1). The state is assumed to be known and the state-feedback controller Σ_c calculates $u(t) = f(x(t), w(\sigma))$, where $\sigma \in [t, t+T]$. The controller has to stabilize the closed-loop system of figure 1 and to minimize the performance index $J = \|z\|_2^2 = \int_0^{+\infty} z^T(t)z(t)dt$ under the assumption that, beyond the preview horizon, the previewed exogenous input $w_p(t) = w(t+T)$ is generated through the generator model Σ_w described by (2), where w' is an unpredictable signal.

Let $R = D_1^T D_1$, $Q = C^T C$, $S = C^T D_1$. To solve the problem, we will consider first in the proof that the feedthrough matrix D_2 is zero before showing how to generalize the solution to the case where $D_2 \neq 0$.

$$\dot{x}_w = A_w x_w + B_w w'$$
$$w_p = C_w x_w \tag{2}$$

$$A_w \in \mathfrak{R}^{q \times q}, C_w \in \mathfrak{R}^{r \times q}, B_w \in \mathfrak{R}^{q \times q}$$

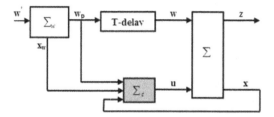

Fig. 1 H2/LQ-Preview controller problem

Assuming $D_2 = 0$, the the performance index J can be recast as the LQR problem (3).

$$J = \|z\|_2^2 = \int_0^{+\infty} \left[x(t)^T \ u(t)^T \right] \begin{bmatrix} Q & S \\ S & R \end{bmatrix} \begin{bmatrix} x(t)^T \\ u(t)^T \end{bmatrix} dt \qquad (3)$$

Since the transfer function matrix $T_{zw'}(s)$ between w' and z can be specified as $T_{zw'}(s) = T_{zw}(s) \times e^{Ts} \times \Sigma_w(s)$, the performance index J can be written also as $J = \|T_{zw'}\|_2^2$ (see [1]). As the cost criterion represents both the LQR and the H2 problem, the problem is named the "H2/LQ-Preview problem".

3 The H2/LQ-Preview Problem Solution

Assume that:
(A-1) The pair (A, B_1) is stabilisable
(A-2) The quadruple (A, B_1, C, D_1) has no invariant zeros on $i\Re$
(A-3) D_1 is a full-column rank matrix (which ensures that $R = R^T > 0$)
(A-4) $Im D_2 \subseteq Im D_1$, which guarantees the existence of the solution when $D_2 \neq 0$
(A-5) A_w is Hurwitz
The main result of this paper is given in the following Theorem (considering the general case of the problem where $D_2 \neq 0$):

Theorem 3.1: let a system (Σ, Σ_c) defined by (1) and (2), the solution of the H2/LQ-preview problem is given by the controller Σ_c defined through the following relation:

$$u(t) = -K_+ x(t) + \int_0^T \phi(\tau) w_p(t - \tau) d\tau - R^{-1} B_1^T e^{A_+^T T} M x_w(t) - D_1^+ D_2 w(t) \qquad (4)$$

where:

- $K_+ = R^{-1}(S^T + B_1^T P_+)$ is the gain feedback matrix
- $\phi(\tau) = -R^{-1} B_1^T e^{A_+^T(T-\tau)} P_+ B_2$,
  P_+ is the stabilizing solution of the algebric Riccati equation:
  $PA + A^T P - (S + PB_1)R^{-1}(S^T + B_1^T P) + Q = 0$

- $A_+ = A - BR^{-1}(S^T + B_1^T P_+)$ is the closed loop matrix
- M is the solution of the Sylvester equation:
  $A_+^T M + MA_w + P_+B_2C_w = 0$
- D_1^+ is defined by $D_1D_1^+D_2 = D_2$

Proof: Suppose first that $D_2 = 0$. At the time t, the signal w is known over the time interval $[t, t+T]$. Assuming w' to be zero over the interval $[t+T, \infty]$, the signal w may be predicted by using the generator model (2). So, it is possible to apply the solution of the deterministic nonhomogeneous LQR problem [3], (which considers that w is perfectly known in the future) to get the optimal solution from the infinite-horizon LQR problem by introducing the additional term given below:

$$u^*(t) = -K_+x(t) - R^{-1}B_1^T \int_t^\infty e^{A_+^T(\tau-t)}P_+B_2w(\tau)d\tau \tag{5}$$

Here, K_+, P_+ and A_+ are defined as in (4). The existence of P_+ is guaranteed by the assumptions (A-1) and (A-2). Considering the signal w as defined by $w_p(t) = w(t+T)$ and equation (2), equation (5) is equivalent to:

$$u^*(t) = -K_+x(t) - R^{-1}B_1^T \underbrace{\int_t^{t+T} e^{A_+^T(\tau-t)}P_+B_2w(\tau)d\tau}_{*}$$

$$-R^{-1}B_1^T \underbrace{\int_{t+T}^\infty e^{A_+^T(\tau-t)}P_+B_2w(\tau)d\tau}_{**} \tag{6}$$

The integral $(*)$ can be rewritten as:

$\int_t^{t+T} e^{A_+^T(\tau-t)}P_+B_2w(\tau)d\tau = \int_0^T e^{A_+^T(T-\theta)}P_+B_2w_p(t-\theta)d\theta$

Using (2), the integral $(**)$ can be written as:

$\int_{t+T}^\infty e^{A_+^T(\tau-t)}P_+B_2w(\tau)d\tau = \int_{t+T}^\infty e^{A_+^T(\tau-t)}P_+B_2C_wx_w(\tau-T)d\tau$

$= \int_t^\infty e^{A_+^T(\theta-t+T)}P_+B_2C_wx_w(\theta)d\theta$

$= \int_t^\infty e^{A_+^T(\theta-t+T)}P_+B_2C_we^{A_w(\theta-t)}x_w(t)d\theta$

$= e^{A_+^T T}e^{-A_+^T t}(\int_t^\infty e^{A_+^T \theta}P_+B_2C_we^{A_w\theta}d\theta)e^{-A_wt}x_w(t)$

$= e^{A_+^T T}Mx_w(t)$

where:

$$M = e^{-A_+^T t}(\int_t^\infty e^{A_+^T \theta}P_+B_2C_we^{A_w\theta}d\theta)e^{-A_wt} \tag{7}$$

The substitution of terms $(*)$ and $(**)$ into (6) leads to (4).

Note that M is independent over the time and verifies the Sylvester equation $A_+^T M + M A_w = -P_+ B_2 C_w$, which can be proved by realizing the integration by parts in (7) as follows:

$$M = e^{-A_+^T t} \left((A_+^T)^{-1} [e^{A_+^T \tau} P_+ B_2 C_w e^{A_w \tau}]_t^\infty \right.$$
$$\left. - (A_+^T)^{-1} \int_t^\infty e^{A_+^T \tau} P_+ B_2 C_w A_w e^{A_w \tau} d\tau \right) e^{-A_w t}$$

So,

$$A_+^T M = e^{-A_+^T t} \left(-e^{A_+^T t} P_+ B_2 C_w e^{A_w t} - \left(\int_t^\infty e^{A_+^T \tau} P_+ B_2 C_w e^{A_w \tau} d\tau \right) A_w \right) e^{-A_w t}$$
and then: $A_+^T M = -P_+ B_2 C_w - M A_w$

Finally, when D_2 differs from zero, the problem can still be solved thanks to the assumption (A-4) which allows us to introduce a compensating term $-D_1^+ D_2 w(t)$ in the control $u(t)$ to perfectly reject $D_2 w(t)$ in the performance vector $z(t)$, with D_1^+ defined such that $D_1 D_1^+ D_2 = D_2$.

4 Solution Analysis

The optimal preview controller consists of 4 terms (figure 2):

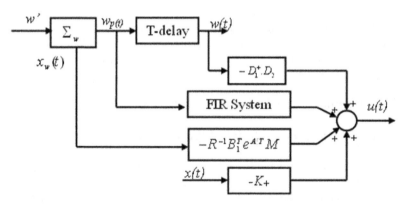

Fig. 2 Optimal preview controller

- the state-feedback term $-K_+ x$,
- an anticipation term represented by the finite impulse response filter (FIR), $\int_0^T \Phi(t) w_p(t - \tau) d\tau$, which reflects the exploitation of the previewed signal w on the preview horizon,

- a pre-compensation term, $-R^{-1}B_1^T e^{A_+^T T} M x_w(t)$, which copes with the predicted signal w beyond the preview horizon T,
- a direct compensation term, $-D_1^+ D_2 w(t)$, instantaneously compensating the direct link from $w(t)$ to the criterion.

This result encompasses other one already known in the context of optimal preview control. Two particular cases are considered in the following Corollaries:

Corollary 4.1: The result of [6] is a particular case of Theorem 3.1 when $D_2 = 0$ and the system Σ undergoes only one disturbance signal with a first order generator model ($A_w = a_w$ is scalar),

$$u(t) = -K_+ x(t) + \int_0^T \Phi(\tau) w_p(t - \tau) d\tau + F_p w_p(t)$$

where $F_p = R^{-1} B_1^T (A_+^T + a_w I)^{-1} e^{A_+^T T} P_+ B_2$

The proof comes directly from the explicit solution in that case of the Sylvester equation $A_+^T M + M a_w = -P_+ B_2 C_w$ (note that M is an n × 1 vector) which induces that $M = -(A_+^T + a_w I)^{-1} P_+ B_2 C_w$, hence the control $u(t)$ given in the Corollary.

Corollary 4.2: The result of [2] is got in the absence of the dynamical generator model Σ_w,

$$u(t) = -K_+ x(t) + \int_0^T \Phi(\tau) w_p(t - \tau) d\tau - D_1^+ D_2 w(t)$$

The proof is trivial by considering $x_w(t) = 0$ in Theorem (3.1).

Next, the optimal preview control of figure (2) will be applied for the lateral steering of a vehicle in order to minimise the offset from the centerline of the lane.

5   Application to Car Lateral Steering Control

5.1   Simplified Lateral Model of the Vehicle

The general model of the lateral steering system consists of three embedded submodels: the vehicle dynamics, the lane keeping dynamics and the steering column.

Decoupling the longitudinal and lateral dynamics and retaining only lateral and yaw dynamics, a linear model is got. The so-called bicycle model is used in what follows, by considering small steering angles and a linear tire model. This model is parameterized by the current longitudinal velocity. Coupling the two front wheels together, and the two rear wheels as well, then considering the vehicle side slip angle β and yaw rate r as state variables, (figure 3), the model is described by equation (8) [8]:

$$\begin{bmatrix} \dot{\beta} \\ \dot{r} \end{bmatrix} = \begin{bmatrix} -\frac{2(c_f + c_r)}{mV_x} & \frac{2(c_r l_r - c_f l_f)}{mV_x^2} - 1 \\ \frac{2(c_r l_r - c_f l_f)}{I_\psi} & -\frac{2(c_f l_f^2 + c_r l_r^2)}{I_\psi V_x} \end{bmatrix} \begin{bmatrix} \beta \\ r \end{bmatrix} + \begin{bmatrix} \frac{2c_f}{mV_x} \\ \frac{2c_f l_f}{I_\psi} \end{bmatrix} \delta_f \qquad (8)$$

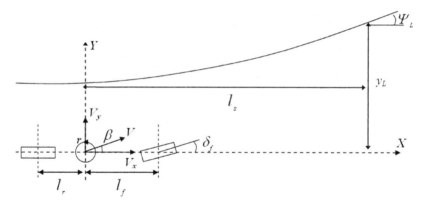

Fig. 3 Bicycle model and vision system

in which δ_f is the front wheels steering angle, V_x is the vehicle longitudinal speed, m is the vehicle mass, I_ψ is the yaw inertia, c_f and c_r are the stiffness of the front and rear tires. l_f and l_r are the distance from the vehicle centre of gravity to the front and rear axles, respectively. To model the lane keeping dynamics, the vision system provides two additional measurements [8](see figure(3)): the offset y_L from the centreline of the lane projected forward a distance said the lookahead distance l_s, and the angle Ψ_L between the tangent to the road and the vehicle orientation(eq(9)). ρ_{ref} represents the curvature of the road.

$$\begin{bmatrix} \dot{\Psi}_L \\ \dot{y}_L \end{bmatrix} = \begin{bmatrix} 0 & 1 & 0 & 0 \\ V_x & l_s & V_x & 0 \end{bmatrix} \begin{bmatrix} \beta \\ r \\ \Psi_L \\ y_L \end{bmatrix} + \begin{bmatrix} -V_x \\ 0 \end{bmatrix} \rho_{ref} \tag{9}$$

Let us consider an electronic power steering mechanism through the following model [8]:

$$\begin{bmatrix} \dot{\delta}_d \\ \ddot{\delta}_d \end{bmatrix} = \begin{bmatrix} 0 & 0 & 0 & 1 \\ \frac{2K_p c_f \eta_t}{R_s I_s} & \frac{2K_p c_f \eta_t}{R_s I_s} \frac{l_f}{V_x} & -\frac{2K_p c_f \eta_t}{R_s^2 I_s} & \frac{B_s}{I_s} \end{bmatrix} \begin{bmatrix} \beta \\ r \\ \delta_d \\ \dot{\delta}_d \end{bmatrix} + \begin{bmatrix} 0 \\ \frac{1}{I_s} \end{bmatrix} T_v \tag{10}$$

I_s is the inertia moment of the steering column, R_s the reduction ratio of the column, η_t the width of the tire contact, B_s damping factor of the column and K_p is the manual steering column coefficient. Concerning the input and output, T_v is the motorized steering torque (the control input) while δ_d is the steering angle.

The three subsystems (8), (9) and (10) can be combined in the single global dynamical system:

$$\begin{bmatrix} \dot{\beta} \\ \dot{r} \\ \dot{\Psi}_L \\ \dot{y}_L \\ \dot{\delta}_d \\ \ddot{\delta}_d \end{bmatrix} = \begin{bmatrix} a_{11} & a_{12} & 0 & 0 & \frac{b_1}{R_s} & 0 \\ a_{21} & a_{22} & 0 & 0 & \frac{b_2}{R_s} & 0 \\ 0 & 1 & 0 & 0 & 0 & 0 \\ V_x & l_s & V_x & 0 & 0 & 0 \\ 0 & 0 & 0 & 0 & 0 & 1 \\ \frac{T_{S\beta}}{I_s} & \frac{T_{Sr}}{I_s} & 0 & 0 & -\frac{T_{S\beta}}{R_s I_s} & -\frac{B_s}{I_s} \end{bmatrix} \begin{bmatrix} \beta \\ r \\ \Psi_L \\ y_L \\ \delta_d \\ \dot{\delta}_d \end{bmatrix} + \begin{bmatrix} 0 \\ 0 \\ 0 \\ 0 \\ 0 \\ \frac{1}{I_s} \end{bmatrix} T_v + \begin{bmatrix} 0 \\ 0 \\ -V_x \\ 0 \\ 0 \\ 0 \end{bmatrix} \rho_{ref}$$

$$\begin{bmatrix} \Psi_L \\ y_L \\ a \\ T_v \end{bmatrix} = \begin{bmatrix} 0 & 0 & 1 & 0 & 0 & 0 \\ 0 & 0 & -l_s & 1 & 0 & 0 \\ V_x a_{11} & V_x a_{12} & 0 & 0 & \frac{V_x b_1}{R_s} & 0 \\ 0 & 0 & 0 & 0 & 0 & 0 \end{bmatrix} \begin{bmatrix} \beta \\ r \\ \Psi_L \\ y_L \\ \delta_d \\ \dot{\delta}_d \end{bmatrix} + \begin{bmatrix} 0 \\ 0 \\ 0 \\ 1 \end{bmatrix} T_v \qquad (11)$$

$$a_{11} = -\frac{2(c_f + c_r)}{mV_x}, \quad a_{12} = \frac{2(c_r l_r - c_f l_f)}{mV_x^2} - 1, \quad a_{21} = \frac{2(c_r l_r - c_f l_f)}{I_\Psi} - 1, \quad a_{22} = -\frac{2(c_f l_f^2 + c_r l_r^2)}{I_\Psi V_x},$$

$$b_1 = \frac{2c_f}{mV_x}, \quad b_2 = \frac{2c_f l_f}{I_\Psi}, \quad T_{S\beta} = \frac{2K_p c_f \eta_t}{R_s}, \quad T_{Sr} = \frac{2K_p c_f \eta_t}{R_s} \frac{l_f}{V_x}.$$

where $x = \begin{bmatrix} \beta, r, \Psi_L, y_L, \delta_d, \dot{\delta}_d \end{bmatrix}^T$ is the state vector, $u = T_v$ is the control input and $y = [\Psi_L, y_{act}, a, T_v]^T$ is the output vector; y_{act} is the offset of the vehicle gravity centerfrom the centerline and a is the lateral acceleration tracking error (offset from the lateral acceleration imposed by the centerline). The road curvature ρ_{ref} enters the model as an exogenous disturbance signal.

5.2 Experimental Setup and Results

This section presents the experimental setup of the lateral control and the performance comparison with and without preview. The test was carried out using a fixed-base driving simulator (SCANeR^{TM}). The road track considered is about 2.5km long, and consists of several curved sections including tight bends (radius up to 70m) (figure 4). The optimum controller was calculated considering the linearized 'bicycle' model matching the parameters of a Peugeot 307 given in Table(1). The longitudinal speed was fixed at 18m/s (the maximum safe speed regarding to some tight bens of the track). The road curvature generator model allows considering different profiles of road curvature beyond the preview horizon. What seems to be more consistent with the human anticipation approach is to consider that the road will maintain its curvature. Thus, the generator model (12), with $\tau = 0.05s$, was designed assuming that the variation of the raod curvature, as an exogenous disturbance to the system, does not exceed $2(m.s)^{-1}$.

$$\begin{bmatrix} \dot{x}_w \\ w_p \end{bmatrix} = \begin{bmatrix} -\frac{1}{\tau} & \frac{1}{\tau} & 0 \\ 0 & -\frac{1}{\tau} & \frac{1}{\tau} \\ \hline 1 & 0 & 0 \end{bmatrix} \begin{bmatrix} x_w \\ w' \end{bmatrix} \qquad (12)$$

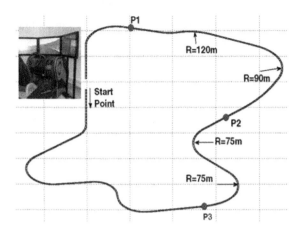

Fig. 4 Test track

Table 1 Vehicle parameters

Parameter	Value
μ:adhesion	0.8
l_f:distance form GC to front axle	$1.127m$
l_r:distance from GC to rear axle	$1.485m$
M:total mass	$1476kg$
J:vehicle yaw moment of inertia	$1810kg.m^2$
c_{f_0}:front cornering stiffness	$65000N/rad$
c_{r_0}:rear cornering stiffness	$57000N/rad$
η_t: tire length contact	$0.185m$
K_p:manual steering column coefficient	0.038
l_s: look-ahead distance	$5m$
B_s: Steering system damping coefficient	5.7
I_s: inertial moment of steering system	$0.05\ kg.m^2$
R_s: steering gear ratio	16

The generator model is useful when there is a lack of preview information as in cases of vision disturbance during driving, but it will no more improve the performance if a sufficient preview was available (see next paragraph).

Furthermore, to implement numerically the preview control law, the infinite dimensional term in (4) which corresponds to the FIR filter is approximated by a finite one using Simpson's rule which is as follows with n=8:

$$\int_0^T f(x)dx \approx \frac{T}{3n}\left[f(0)+2\sum_{i=1}^{n-1}\alpha f(i*\frac{T}{3})+f(T)\right]$$

with $\alpha=1$ if i is even and $\alpha=2$ when i is odd.

The performance vector was considered through a weighting matrix $z = W.y$ (y is the output vector defined as in eq(11)) (see figure (5)), with $c1 = 100$, $c2 = 70$ and $c3 = 10$. The choice of the weighting matrix and its influence on the performance are analyzed in the next paragraph.

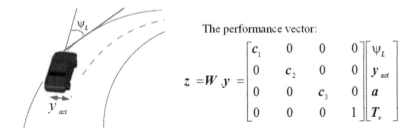

The performance vector:

$$z = W.y = \begin{bmatrix} c_1 & 0 & 0 & 0 \\ 0 & c_2 & 0 & 0 \\ 0 & 0 & c_3 & 0 \\ 0 & 0 & 0 & 1 \end{bmatrix} \begin{bmatrix} \psi_L \\ y_{act} \\ a \\ T_v \end{bmatrix}$$

Fig. 5 Weighting matrix

Figure (6) presents the lateral deviation from the centreline, for preview horizons: $T = 0$ (LQ controller, i.e., when the curvature of the road is accessible for measurement but not previewed), $T = 0.1s$, $T = 0.5s$ and $T = 0.8s$. The maximum of the lateral position error decreases from about $60cm$ for $T = 0$ to $10cm$ for $T > 0.5s$. As expected in the simulation study (see [9]), the tracking of the centerline is considerably improved as the preview time increases. Beyond a certain value however, increasing T has no more impact, as it leads to negligible improvement in the performance index. For $T > 0.5s$, the controller maintains a vehicle position of $\pm 10cm$ around the lane centre.

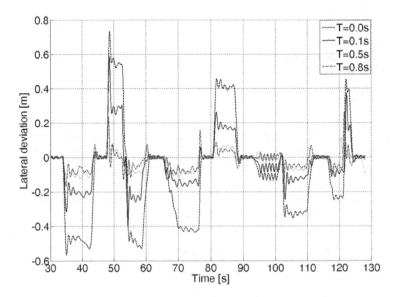

Fig. 6 Lateral position error

6   Preview Horizon and Weighting Matrix Impact

This section shows the impact of the feedforward term (FIR) on the control performance and presents a methodology to choose the optimal preview horizon. The influence of the weighting matrix W is studied as well.

The anticipation feedforward term (FIR), helps to initiate the steering action before the curvature starts to change. This term can be written as $\int_0^T R^{-1}B_1^T e^{A_+^T \tau}P_+B_2 w(t + \tau)d\tau$ where $w(t)$ is the road curvature at the current instant t. The exponential coefficient $\phi(t) = R^{-1}B_1^T e^{A_+^T \tau}P_+B_2$ decays to zero with time as A_+ is stable, so the integration beyond a certain horizon is no more necessary (figure 7).

$\phi(t)$ shows how much current steering action depends on what road curvature will be in the future especially over the near future due to its exponential form that decays to 0, reflecting that the steering command depends less and less on remote road curvature.

The weighting matrix W (figure 5) represents the trade off between the steering command T_v, and the lateral position which is represented by three variables: the heading angle error Ψ_L (anticipated position error), the actual position y_{act} and the error of lateral acceleration a (pre-sensed lateral deviation). The preview horizon depends on this trade off as well as on the dynamics of the system. The FIR coefficient $\phi(t)$ enters definitely in its 5% of its max value for time horizon 3τ where τ is the dominant time constant of $e^{A_+^T}$, this time constant can be determined by the minimum real part of the closed loop matrix A_+ eigenvalues: $\tau = \frac{1}{min(|Re(\lambda_i)|)}$ where $\lambda_{i=1..n}$ are the A_+ eigenvalues.

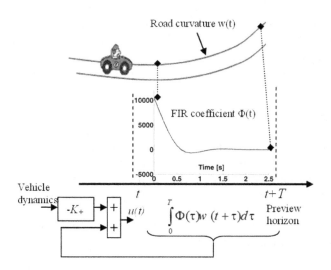

Fig. 7 Preview horizon

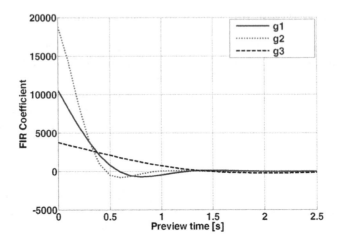

Fig. 8 FIR coefficient as function of preview time

Figure(8) shows the exponential form of $\phi(t)$ for different weighting choices, g1 represents the considered weighting ($c1 = 100$, $c2 = 70$ and $c3 = 10$, see figure 5). This weighting was chosen to be moderate between an exigent position error and the steering effort. for the sake of comparison, g2 represents a strict punishment of position error (high values for $c1$ and $c2$), as shown, this requires more steering effort and faster dynamics, but lower preview information. g3 represents a strict punishment of lateral acceleration error (high value for $c3$), this requires calm steering actions with further ahead preview information.

Practically, it is sufficient to consider a preview horizon in the interval $[\tau, 3\tau]$, especially when using a numerical method to approximate the integral of the FIR, here the approximation error increases with time horizon. In the previous setup, the dominant time constant of the matrix A_+ is $0.42s$ (see g1 in figure 8), this time is found sufficient to get close of the maximum improvement (figure 6).

7   Conclusion

In this paper, a generalized H2/LQ preview control problem has been studied, in which the previewed disturbance type is defined beyond the preview horizon through a generator model giving some insight on the type of disturbance to be encountered. The preview controller has been applied to the lateral guidance of a vehicle. The results show that the preview control action enables the vehicle to track the center of the lane with a smaller tracking error. The controller performance is improved by increasing the preview horizon until certain limit (depending on the system dynamics in the closed loop). The perspective of this work is to study the utility of using the preview controller as a co-pilot in an assist driving system using a driver model as proposed in [10], the controller will have to act as a soft electronic copilot (see http://www.projet-partage.fr/).

Acknowledgements. This work was supported by PARTAGE research program, funded by the ANR "Agence Nationale de la Recherche" (n 0866C0222).

References

1. Chevrel, P.: Méthodologie de la commande par approche d'état. In: Commande des Systèmes Linéaires, ch.5, Hermès, Paris, France (2002)
2. Ferrante, A., Marro, G., Ntogramatzidis, L.: A Hamiltonian approach to the H2 decoupling of previewed input signal. In: Proceedings of the European Control Conference 2007, Greece, July 2-5 (2007)
3. Hampton, R.D., Knospe, C.R., Towensend, M.A.: A practical solution to the deterministic nonhomogeneous LQR problem. Journal of Dynamic Systems, Measurement and Control 118, 354–360 (1996)
4. Hazell, A.: Discrete-time optimal preview control. Phd thesis. University of London (2008)
5. Moelja, A., Meinsma, G.: H2 control of preview systems. Automatica 42(6), 945–952 (2006)
6. Peng, H., Tomizuka, M.: Optimal preview control for vehicle lateral guidance. PATH research report, UCB-ITS-PRR-91-16 (1991)
7. Raharijaona, T.: Commande robuste pour l'assistance au contrôle latéral d'un véhicule routier. Phd thesis, Paris XI et Supélec (2004)
8. Rajamani, R.: Vehicle Dynamics and Control, 471 p. Springer, US (2006) ISBN 978-0-387-28823-9
9. Saleh, L., Chevrel, P., Lafay, J.F.: Generalized H2-preview control and its application to car lateral steering. In: IFAC Time Delay Systems 2010, Prague, Czech, p. LS796 (2010)
10. Sentouh, C., Chevrel, P., Mars, F., Claveau, F.: A human-centred Approach of Steering Control Modelling. In: Proceedings of the 21st IAVSD Symposium on Dynamics of Vehicles on Roads and Tracks (2009)
11. Sharp, R.S., Valtetsiotis, V.: Optimal preview car steering control. Vehicle System Dynamics Supplement 35, 101–117 (2001)
12. Switkes, J.P.: Hand wheel force feedback with lane keeping assistance: combined Dynamics, Stability and Bounding. Phd thesis. Stanford University (2006)

Local Asymptotic Stability Conditions for the Positive Equilibrium of a System Modeling Cell Dynamics in Leukemia

Hitay Özbay, Catherine Bonnet, Houda Benjelloun, and Jean Clairambault

Abstract. A distributed delay system with static nonlinearity has been considered in the literature to study the cell dynamics in leukemia. In this chapter local asymptotic stability conditions are derived for the positive equilibrium point of this nonlinear system. The stability conditions are expressed in terms of inequalities involving parameters of the system. These inequality conditions give guidelines for development of therapeutic actions.

1 Introduction

Starting with the early works of Mackey and his colleagues, [9, 10] there has been a growing interest in the development of mathematical models for cell dynamics in hematological processes. Over the last ten years, significant improvements have been made in this direction and, in particular, models for cell dynamics in leukemia (blood cancer) have been refined, see e.g. [1, 5, 6, 8, 11, 13, 20] and their references. In this chapter, the model of [1] will be considered. This is a cascade connection

Hitay Özbay
Dept. of Electrical and Electronics Eng., Bilkent Univ., Ankara, 06800, Turkey
e-mail: hitay@bilkent.edu.tr

Catherine Bonnet
INRIA Saclay - Île-de-France, Parc Orsay Université, 4 rue J. Monod, 91893,
Orsay Cedex, France
e-mail: catherine.bonnet@inria.fr

Houda Benjelloun
Ecole Centrale Paris, France
e-mail: houdabenjelloun@gmail.com

Jean Clairambault
INRIA Paris-Rocquencourt, Domaine de Voluceau, B.P. 105,
78153 Le Chesney, Cedex, France
e-mail: jean.clairambault@inria.fr

R. Sipahi et al. (Eds.): Time Delay Sys.: Methods, Appli. and New Trends, LNCIS 423, pp. 187–197.
springerlink.com

of a series of systems (compartments) containing distributed delays and a static nonlinear feedback. There are several possible equilibrium points for the system, the origin is being one of them. Here, local asymptotic stability conditions are studied for the "positive equilibrium" where the equilibrium states of all the compartments (sub-systems) are positive.

In [2] a global stability condition is obtained for the case where the only equilibrium is the origin. Some of the works mentioned above consider the "point delay" version of the problem; a recent one is [20], where conditions for global asymptotic stability of the origin and instability of the positive equilibrium are obtained in terms of the delay values.

Rest of the chapter is organized as follows. Details of the mathematical model are given in the next section. Then, the main results are derived and concluding remarks are made. Preliminary versions of the results of this chapter have been already presented in various meetings, [14, 15, 16, 17].

2   Mathematical Model of Cell Dynamics in Leukemia

Since the identification of leukemic stem cells (LSCs) in humans, [4], many studies have been conducted to characterize the process of formation of leukemic cells. It is now well understood that LSCs can self-renew and they can differentiate to generate leukemic progenitors which can also self-renew and differentiate. There are many stages of differentiation (compartments of progenitors between LSCs and leukemic cells) until leukemic cells are released into the blood, [7]. At each stage, there is a compartment (population) of cells of a certain biological property, characterized by specific cluster definition (CD) molecules, such as CD34, CD38, CD123, CD90, CD117, CD135 and CD33. For example, in a certain type of acute myelogenous leukemia (AML), cells with the concentration of molecules CD34+CD38-CD33- can be identified as LSCs, i.e. the first compartment, (respectively, CD34+CD38+CD33- for progenitors and CD34+CD38+CD33+ for leukemic cells, i.e., second and third compartments in a 3 compartment model), [12]. Recently, it has been shown that for mathematical modeling purposes, 4 to 8 compartment models are sufficient to diagnose chronic myelogenous leukemia in humans, [19].

At each compartment, the cells can be grouped into two: the ones in growth phase (proliferation) and the quiescent (non-proliferating) ones. At the end of growth phase, each cell is divided into two. Some of the new cells stay in the same compartment (having the same biological property as the mother cell - self renewal) and some go to the next compartment (differentiation). The dynamical behavior of cell populations in the quiescent and proliferating phases can be characterized as shown in Figure 1, where δ and γ represent the death rates of the quiescent and proliferating cells respectively, $\beta(\cdot)$ is the re-introduction function, τ is the maximal time spent in the growth phase before cell division occurs and $L = 1 - K \in (0\,,\,1)$ is the rate of proliferating cells that divide without differentiation. Note that each of these parameters can be different for different compartments, i.e. δ_i, γ_i, τ_i, L_i and $\beta_i(\cdot)$ are

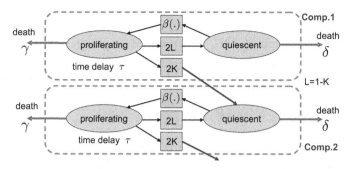

Fig. 1 Cell population dynamics in compartmental modeling.

the parameters of the ith compartment. The notation $x_i(t)$ and $y_i(t)$ will be used to denote the cell population in the quiescent and proliferation phases, respectively, in compartment i at time t.

With the above definitions, dynamical equations for x_i and y_i can be given as follows, see e.g. [1],

$$\dot{x}_i(t) = -\delta_i x_i(t) - w_i(t) + 2L_i \int_0^{\tau_i} e^{-\gamma_i a} f_i(a) w_i(t-a)da + u_{i-1}(t) \tag{1}$$

$$\dot{y}_i(t) = -\gamma_i y_i(t) + w_i(t) - 2\int_0^{\tau_i} e^{-\gamma_i a} f_i(a) w_i(t-a)da \tag{2}$$

$$u_i(t) = 2K_i \int_0^{\tau_i} e^{-\gamma_i a} f_i(a) w_i(t-a)da \tag{3}$$

$$w_i(t) := \beta_i(x_i(t))x_i(t) , \tag{4}$$

$$f_i(a) \geq 0 \quad \text{for all } a \in [0, \tau_i] \text{ and } \int_0^{\tau_i} f_i(a)da = 1 \tag{5}$$

with $K_0 = 0$. Here f_i is the cell division probability and we consider the form

$$f_i(a) = \frac{m_i}{e^{m_i \tau_i} - 1} e^{m_i a}, \quad a \in [0 , \tau_i] \quad m_i > \gamma_i \tag{6}$$

which is originally proposed in [14]. Define $g_i(a) := e^{-\gamma_i a} f_i(a)$ for $0 \leq a \leq \tau_i$ and $g_i(a) = 0$ otherwise. Then, the Laplace transform $G_i(s)$ of $g_i(t)$ is

$$G_i(s) = q_i \frac{1 - e^{-\tau_i(s-r_i)}}{(s-r_i)} \tag{7}$$

where $q_i = m_i/(e^{m_i \tau_i} - 1) > 0$ and $r_i = m_i - \gamma_i > 0$.

In [1], the above system is analyzed for the choice of $G_i(s) = e^{-\tau_i(s+\gamma_i)}$, which is a system with "point delay". We feel that the choice (7) is more natural, it corresponds to a distributed delay system, [14].

Dynamical equations given above for the ith compartment can be combined into a single block diagram as shown in Figure 2. Note that the sub-system Σ_{yi} is a stable system, i.e. when its input $(I - 2G_i)w_i$ is bounded we get a bounded y_i. Therefore, we will be interested in the analysis of the system represented by the equations (1), (3) and (4), with the distributed delay term (7) and nonlinearity β_i specified as

$$\beta_i(x) = \frac{\beta_i(0)}{1 + b_i x^{N_i}} \tag{8}$$

where $\beta_i(0) > 0$, $b_i > 0$ and N_i is an integer greater or equal to 2, see [5, 6, 9] for biological justifications of this selection.

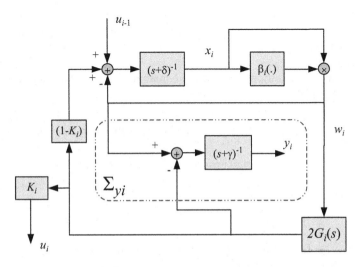

Fig. 2 Block diagram representation of the ith compartment cell dynamics.

3 Stability Analysis for the Positive Equilibrium

In this section local asymptotic stability conditions are obtained for the "positive equilibrium" point $\bar{\mathbf{x}} = [\bar{x}_1, \ldots, \bar{x}_n]^{\mathrm{T}}$ where all $\bar{x}_i$ are strictly positive. Existence of such an equilibrium point depends on certain conditions derived as follows. First define

$$\alpha_i := 2L \int_0^{\tau_i} g_i(t) dt - 1 = 2L_i G_i(0) - 1 \tag{9}$$

and make the following assumption.

Assumption. We have $\alpha_i > 0$ for all $i = 1, \ldots, n$, and $\beta_1(0) > \delta_1 / \alpha_1$. ☐

Then, a unique positive equilibrium exists, see e.g. [1]. It can be computed from the following equations: $\bar{x}_1$ is such that

$$\beta(\bar{x}_1) = \delta_1/\alpha_1 \; ; \tag{10}$$

and for $i \geq 2$, the equilibrium points $\bar{x}_i$ are the unique solutions of

$$\beta_i(\bar{x}_i) = \frac{1}{\alpha_i}\left(\delta_i - \frac{1}{\bar{x}_i}\left(\frac{\bar{x}_{i-1}K_{i-1}(\beta(\bar{x}_{i-1}) + \delta_{i-1})}{L_{i-1}}\right)\right). \tag{11}$$

Since $G_i(s)$ is strictly proper, the system is locally asymptotically stable around the positive equilibrium if and only if all the roots of

$$s + \delta_i + \mu_i - 2L_i\mu_iG_i(s) = 0 \tag{12}$$

are in $\mathbb{C}_-$ for all i, where

$$\mu_i := \frac{\mathrm{d}}{\mathrm{d}x}x\,\beta_i(x)\,|_{\bar{x}_i}\,. \tag{13}$$

As noted in [3] depending on the parameters of the system, μ_i can be positive, negative or zero. Clearly, when $\mu_i = 0$ the the equation (12) has its roots at $-\delta_i < 0$. Therefore, the most interesting case is $\mu_i \neq 0$.

Since the analysis has to be done individually for each compartment, in the rest of the paper the subscript i is dropped whenever it is clear from the context that ith characteristic equation (12) is considered.

3.1   Local Asymptotic Stability for $\mu > 0$

Consider the characteristic equation (12) with $\mu > 0$. Figure 3 shows that under different parameter selections one may have a common equilibrium point with different positive μ values.

When $\mu > 0$, the system is locally asymptotically stable if and only if

$$\mu < \frac{\delta}{\alpha} \quad \text{which is equivalent to} \quad 2LG(0) < \frac{\delta + \mu}{\mu}. \tag{14}$$

For the proof, see [1, 14]. Also, it has been recently shown, [18], that he condition (14) holds true for all β in the form (8). So, whenever we have a unique positive equilibrium with $\mu_i > 0$ for all i, we have local asymptotic stability.

3.2   Local Asymptotic Stability for $\mu < 0$

Consider the system whose characteristic equation is in the form (12) with $\mu < 0$. In this case (12) can be re-written as

$$1 + |\mu|\frac{(2LG(s) - 1)}{(s + \delta)} = 0 \;. \tag{15}$$

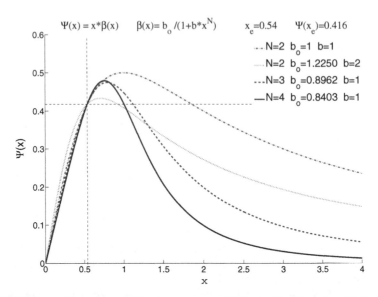

Fig. 3 Different parameters resulting in a same equilibrium with different positive μ.

The equation (15) can be seen as a characteristic equation of a feedback system composed of stable open loop transfer functions $(2LG(s) - 1)$ and $|\mu|/(s + \delta)$. Hence the small gain condition

$$|\mu| \, \|(s + \delta)^{-1}\|_\infty \, \|2LG(s) - 1\|_\infty < 1 \tag{16}$$

implies stability.

Clearly, a sufficient condition for (16) is $|\mu| \, (2LG(0) + 1) < \delta$ (see also [1, 3]), i.e.,

$$2LG(0) < \frac{\delta - |\mu|}{|\mu|}, \tag{17}$$

which is valid only when $\delta > |\mu|$.

A weaker condition for stability, again by the small gain on (15), is

$$|\mu| < 1/\|H\|_\infty \tag{18}$$

where

$$H(s) = \frac{2LG(s) - 1}{(s + \delta)}. \tag{19}$$

Note that $H(0) = \alpha/\delta$. Thus (18) is equivalent to

$$|\mu| < \frac{1}{K_H} \frac{\delta}{\alpha} \tag{20}$$

where

$$K_H := \|\ \frac{1}{H(0)}\ H(s)\ \|_\infty. \tag{21}$$

We now investigate K_H for G in the form (7).

Proposition 1. *Consider the function $G(s)$ in the form (7) and define*

$$\kappa := \frac{(\alpha+1)(\tau r+1)+0.28}{\alpha\ \sqrt{1+r^2/\delta^2}}. \tag{22}$$

The feedback system represented by the characteristic equation (15) is stable if one of the following two conditions are satisfied:

(i) $\kappa \le 1$ *and* $|\mu| < (\delta/\alpha)$;
(ii) $\kappa > 1$ *and* $|\mu| < \kappa^{-1}\ (\delta/\alpha)$.

Proof. We claim that (i) when $\kappa \le 1$ we have $K_H = 1$, and (ii) when $\kappa > 1$ we have $K_H \le \kappa$. Recall that

$$H(s) = \left(\frac{1}{s+\delta}\right)\left(q\left(\frac{1-e^{-\tau(s-r)}}{(s-r)}\right)-1\right) ; \quad H(0) = \frac{\alpha}{\delta}.$$

Then, scaling the frequency by r and using simple algebra it can be shown that

$$K_H = \max_{\omega \in \mathbb{R}}\left|\frac{1+j\omega\frac{\delta}{r\alpha}+j\omega\frac{qe^{\tau r}\delta\tau}{r\alpha}\left(\frac{e^{-j\tau r\omega}-1}{j\tau r\omega}\right)}{(1+j\omega)(1-j\omega\frac{\delta}{r})}\right|.$$

Expanding the numerator of the above expression into its real and imaginary parts, we get

$$1 \le K_H^2 \le \max_{\omega \in \mathbb{R}}\frac{1+\omega^2\frac{\delta^2}{r^2\alpha^2}\left((1-q\tau e^{\tau r}\frac{\sin(\tau r\omega)}{\tau r\omega})^2+(q\tau e^{\tau r}(\frac{1-\cos(\tau r\omega)}{\tau r\omega}))^2\right)}{(1+\omega^2)(1+\omega^2\frac{\delta^2}{r^2})}$$

Since

$$q = 2LG(0)\frac{r}{e^{\tau r}-1} = \frac{r\ (\alpha+1)}{e^{\tau r}-1}$$

we have

$$1 \le q\tau e^{\tau r} = \frac{(\alpha+1)\ \tau r}{1-e^{-\tau r}} \le (\alpha+1)(\tau r+1). \tag{23}$$

Also note that for all $q\tau e^{\tau r} \ge 1$ we have

$$\max_{\omega \in \mathbb{R}}\sqrt{(1-q\tau e^{\tau r}\frac{\sin(\tau r\omega)}{\tau r\omega})^2+(q\tau e^{\tau r}\frac{1-\cos(\tau r\omega)}{\tau r\omega})^2} \le q\tau e^{\tau r}+0.28.$$

Thus

$$1 \leq K_H^2 \leq \max_{\omega \in \mathbb{R}} \frac{1 + \omega^2 A^2 \delta^2 / r^2}{1 + (1 + \delta^2 / r^2) \omega^2 + (\delta^2 / r^2) \omega^4} \tag{24}$$

where

$$A := \alpha^{-1} ((\alpha + 1)(\tau r + 1) + 0.28).$$

By studying the maximum condition on the right hand side of (24) we see that $K_H = 1$ if $A^2 \leq (1 + \frac{r^2}{\delta^2})$. Note that $\kappa = A / \sqrt{1 + (r^2/\delta^2)}$. Hence part (i) of the proposition is proven. For the second part, when $\kappa > 1$, it can be shown that the maximum on the right hand side of (24) gives

$$K_H^2 \leq \left(1 - \frac{r^2}{A^4 \delta^2} (\sqrt{1 + \varpi^2} - 1)^2\right)^{-1} \quad \text{where } \varpi^2 = \frac{A^4 \delta^2}{r^2} \left(1 - \frac{1}{\kappa^2}\right). \tag{25}$$

Now using the fact

$$\sqrt{1 + \varpi^2} - 1 = \frac{\varpi^2}{\sqrt{1 + \varpi^2} + 1} \leq \varpi$$

a new bound can be found from (25)

$$K_H^2 \leq \left(1 - \frac{r^2}{A^4 \delta^2} \varpi^2\right)^{-1} = (1 - (1 - \frac{1}{\kappa^2}))^{-1} = \kappa^2.$$

In conclusion, if $\kappa > 1$ then $K_H \leq \kappa$.   □

The inequality conditions expressed in Proposition 1 can be easily checked once the parameters of the system are given. The first stability condition is equivalent to

$$2LG(0) < \frac{\delta + |\mu|}{|\mu|} \tag{26}$$

when $\kappa \leq 1$, and the second condition means

$$2LG(0) < \kappa^{-1} \frac{\delta + |\mu|}{|\mu|} \tag{27}$$

when $\kappa > 1$. In both cases there is a lower bound for $2LG(0)$ given by

$$\frac{(1 - e^{-\tau r})}{(\tau r + 1)} < 2LG(0), \tag{28}$$

which is derived from (23) by recalling that $2LG(0) = \alpha + 1$.

Proposition 1 gives the above sufficient conditions, (26) and (27), which are valid for $\delta > |\mu|$ as well as $\delta < |\mu|$. Necessary and sufficient conditions for these two

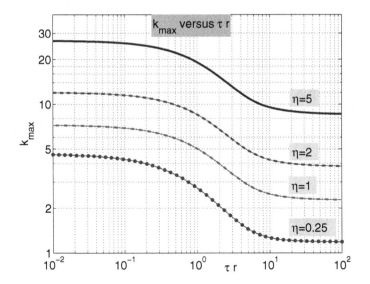

Fig. 4 Gain $k_{\max}$ versus τr for different values of η.

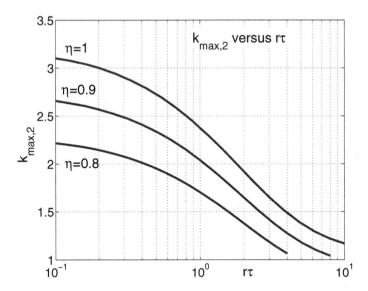

Fig. 5 Gain $k_{\max,2}$ versus τr for different values of η and τr pairs satisfying (30).

different cases are obtained in [15] as inequalities in the following forms. For $\delta > |\mu|$, the system is locally asymptotically stable if and only if

$$2LG(0) < \frac{\delta - |\mu|}{|\mu|} k_{\max} \qquad (29)$$

where $k_{\max} > 1$ depends on τr, and $\eta := \tau^{-1}(\delta - |\mu|)^{-1}$, as shown in Figure 4.

Similarly, for $\delta < |\mu|$, the system is locally asymptotically stable if and only if

$$\eta > (1 - e^{-\tau r})^{-1} - (\tau r)^{-1} \qquad (30)$$

and

$$\frac{|\mu| - \delta}{|\mu|} < 2LG(0) < \frac{|\mu| - \delta}{|\mu|} k_{\max,2}, \qquad (31)$$

where $k_{\max,2} > 1$ depends on τr, and η, as shown in Figure 5.

4 Conclusions

In this chapter, local asymptotic stability conditions are studied for a distributed delay system modeling cell dynamics in leukemia. Proposition 1 gives a simple sufficient condition which is valid for the case $\mu < 0$, independent of the relative size of δ with respect to $|\mu|$. Necessary and sufficient conditions for local asymptotic stability are obtained in [15, 18] and they can be checked graphically (there are no analytic expressions for the functions $k_{\max}$ and $k_{\max,2}$). The conditions derived here can be easily checked in terms of the parameters of the dynamical equation δ, τ, μ and the product $2LG(0) = (\alpha + 1)$ which depend on the mitosis function f, the death rate γ as well as the gain L. Some of these parameters can be adjusted by therapeutic actions, that may be useful in achieving stability.

For global asymptotic stability, a nonlinear small gain argument is used in [18] and an inequality condition is obtained. However the level of conservatism in this inequality has not been established yet. In particular, checking whether the following conjecture holds is an interesting open problem: if the positive equilibrium of the system represented by the equations (1)–(4) is locally asymptotically stable, then it is globally asymptotically stable. Recently, for the case where the origin is the only equilibrium point for the point delay version of the system, the conjecture has been proven to hold [20], see also for a related result [2].

References

1. Adimy, M., Crauste, F., El Abdllaoui, A.: Discrete maturity-structured model of cell differentiation with applications to acute myelogenous leukemia. J. Biological Systems 16(3), 395–424 (2008)
2. Adimy, M., Crauste, F., El Abdllaoui, A.: Boundedness and Lyapunov function for a nonlinear system of hematopoietic stem cell dynamics. C. R. Acad. Sci. Paris, Ser. I 348, 373–377 (2010)

3. Adimy, M., Crauste, F., Ruan, S.: A Mathematical Study of the Hematopoiesis Process with Applications to Chronic Myelogenous Leukemia. SIAM J. Appl. Math. 65, 1328–1352 (2005)
4. Bonnet, D., Dick, J.E.: Human acute myeloid leukemia is organized as a hierarchy that originates from a primitive hematopoietic cell. Nature Medicine 3, 730–737 (1997)
5. Colijn, C., Mackey, M.C.: A mathematical model of hematopoiesis: I. Periodic chronic myelogenous leukemia. J. Theoretical Biology 237, 117–132 (2005)
6. Foley, C., Mackey, M.C.: Dynamic hematological disease: a review. J. Mathematical Biology 58, 285–322 (2009)
7. Huntly, B.J.P., Gilliland, D.G.: Leukemia stem cells and the evolution of cancer-stem-cell research. Nature Reviews: Cancer 5, 311–321 (2005)
8. Kold-Andersen, L., Mackey, M.C.: Resonance in periodic chemotherapy: A case study of acute myelogenous leukemia. J. Theoretical Biology 209, 113–130 (2001)
9. Mackey, M.C.: Unified hypothesis for the origin of aplastic anaemia and periodic hematopoiesis. Blood 51, 941–956 (1978)
10. Mackey, M.C., Glass, L.: Oscillation and chaos in physiological control systems. Science 197(4300), 287–289 (1977)
11. Mackey, M.C., Ou, C., Pujo-Menjouet, L., Wu, J.: Periodic Oscillations of Blood Cell Populations in Chronic Myelogenous Leukemia. SIAM J. Appl. Math. 38, 166–187 (2006)
12. Marie, J.P.: Private communication, Hôpital St. Antoine, Paris, France (July 2010)
13. Niculescu, S.-I., Kim, P.S., Gu, K., Lee, P.P., Levy, D.: Stability Crossing Boundaries of Delay Systems Modeling Immune Dynamics in Leukemia. Discrete and Continuous Dynamical Systems. Series B 13, 129–156 (2010)
14. Özbay, H., Bonnet, C., Clairambault, J.: Stability Analysis of Systems with Distributed Delays and Application to Hematopoietic Cell Maturation Dynamics. In: Proc. of the 47th IEEE Conference on Decision and Control, Cancun, Mexico, pp. 2050–2055 (December 2008)
15. Özbay, H., Benjelloun, H., Bonnet, C., Clairambault, J.: Stability Conditions for a System Modeling Cell Dynamics in Leukemia.In: preprints of IFAC Workshop on Time Delay Systems, TDS 2010, Prague, Czech Republic (June 2010)
16. Özbay, H., Bonnet, C., Benjelloun, H., Clairambault, J.: Global Stability Analysis of a System Modeling Cell Dynamics in AML. In: Abstracts of the 3rd Conference on Computational and Mathematical Population Dynamics (CMPD3), Bordeaux, France, p. 186 (June 2010)
17. Özbay, H., Bonnet, C., Benjelloun, H., Clairambault, J.: Absolute Stability of a System with Distributed Delays Modeling Cell Dynamics in Leukemia. In: Proc. of the 19th International Symposium on Mathematical Theory of Networks and Systems (MTNS 2010), Budapest, Hungary, pp. 989–992 (July 2010)
18. Özbay, H., Bonnet, C., Benjelloun, H., Clairambault, J.: Stability Analysis of a Distributed Delay System Modeling Cell Dynamics in Leukemia (March 2010) (submitted for publication) (revised January 2011)
19. Peixoto, D., Dingli, D., Pacheco, J.M.: Modelling hematopoiesis in health and disease. Mathematical and Computer Modelling (2010), doi:10.1016/j.mcm.2010.04.013
20. Qu, Y., Wei, J., Ruan, S.: Stability and bifurcation analysis in hematopoietic stem cell dynamics with multiple delays. Physica D 239, 2011–2024 (2010)

Part III
Numerical Methods

Design of Fixed-Order Stabilizing and $\mathcal{H}_2$ - $\mathcal{H}_\infty$ Optimal Controllers: An Eigenvalue Optimization Approach

Wim Michiels

Abstract. An overview is presented of control design methods for linear time-delay systems, which are grounded in numerical linear algebra techniques such as large-scale eigenvalue computations, solving Lyapunov equations and eigenvalue optimization. The methods are particularly suitable for the design of controllers with a prescribed structure or order. The analysis problems concern the computation of stability determining characteristic roots and the computation of $\mathcal{H}_2$ and $\mathcal{H}_\infty$ type cost functions. The corresponding synthesis problems are solved by a direct optimization of stability, robustness and performance measures as a function of the controller parameters.

1 Introduction

We consider the system

$$\begin{cases} \dot{x}(t) &= A_0 x(t) + \sum_{i=1}^{m} A_i x(t - \tau_i) + B\zeta(t), \\ \eta(t) &= Cx(t) + D\zeta(t), \end{cases} \tag{1}$$

where $x(t) \in \mathbb{C}^n$ is the state variable at time t, $\zeta \in \mathbb{C}^{n_\zeta}$ is the input, $\eta \in \mathbb{C}^{n_\eta}$ is the output, and τ_i, $i = 1, \dots, m$, represent time-delays. We assume that

$$0 < \tau_1 < \cdots < \tau_m.$$

It is well known that the solutions of (1), with $\zeta \equiv 0$, satisfy a spectrum determined growth property, in the sense that the asymptotic behavior and stability properties are determined by the location of the characteristic roots, i.e., the solutions of the nonlinear eigenvalue problem

Wim Michiels
Department of Computer Science, K.U. Leuven, Celestijnenlaan 200A,
3001 Heverlee, Belgium
e-mail: Wim.Michiels@cs.kuleuven.be

R. Sipahi et al. (Eds.): Time Delay Sys.: Methods, Appli. and New Trends, LNCIS 423, pp. 201–216.
springerlink.com

$$\left(\lambda I - A_0 - \sum_{i=1}^{m} A_i e^{-\lambda \tau_i} \right) v = 0, \ v \in \mathbb{C}^n, \ v \neq 0. \tag{2}$$

For example the zero solution of (1) is asymptotically stable if and only if all characteristic roots are confined to the open left half plane [9, 22]. As commonly done in robust control, we assume that ζ and η are defined in such a way that performance and robustness requirements for (1) can be expressed in terms of norms of the corresponding transfer function

$$G(\lambda) := C \left(\lambda I - A_0 - \sum_{i=1}^{m} A_i e^{-\lambda \tau_i} \right)^{-1} B + D. \tag{3}$$

In Section 2 we present an overview of numerical methods for the computation of the rightmost characteristic roots of (1) and for the computation of the $\mathcal{H}_2$ / $\mathcal{H}_\infty$ norm of the transfer function (3). These analysis tools lay at the basis of the controller synthesis methods discussed in Section 3. Here we assume that the system matrices in (1) depend on a finite number of parameters, which may originate from the parametrization of a controller (hence, (1) corresponds to the closed-loop system). The stabilization problem and the optimization of the $\mathcal{H}_2$ / $\mathcal{H}_\infty$ norm of (3) are solved. The approach is inspired by recently proposed controller synthesis methods for LTI systems which rely on eigenvalue optimization, as, e.g., implemented in the package HIFOO [5]. These methods have proven very useful for synthesis problems where the order (dimension) of the controller is much smaller than the dimension of the plant. This indicates their potential for time-delay systems, where any design problem involving the determination of a finite number of parameters is in fact a reduced-order control design problem (due to the infinite dimension of the system), while a successful implementation has become possible due to the advances on the level of fast analysis tools. In Section 5 we present some concluding remarks.

The presented approach is grounded in the eigenvalue based framework developed in [20]. Surveys on time-domain methods include [9, 14, 18] and the references therein.

2   Solving Analysis Problems

We start with the reformulation of (1) as a standard infinite-dimensional linear system, based on [7], because the interplay between the two representations has played an important role in the development of computational tools.

Consider the Hilbert space $X := \mathbb{C}^n \times \mathcal{L}_2([-\tau_m, 0], \mathbb{C}^n)$, equipped with the inner product

$$< (y_0, y_1), (z_0, z_1) >_X = < y_0, z_0 >_{\mathbb{C}^n} + < y_1, z_1 >_{\mathcal{L}_2}.$$

Let $\mathscr{A} : X \to X$ be the derivative operator defined by

$$\mathscr{D}(\mathscr{A}) = \{z = (z_0, z_1) \in X : z_1 \in \mathscr{C}([-\tau_m, 0], \mathbb{C}^n),$$
$$z_1' \in \mathscr{C}([-\tau_m, 0], \mathbb{C}^n), \; z_0 = z_1(0)\},$$

(4)

$$\mathscr{A} z = \left(A_0 z_0 + \textstyle\sum_{i=1}^{m} A_i z_1(-\tau_i), \; z_1' \right), \; z \in \mathscr{D}(\mathscr{A})$$

and let the operators $\mathscr{B} : \mathbb{C}^{n\zeta} \to X$ and $\mathscr{C} : X \to \mathbb{C}^{n\eta}$ be given by

$$\mathscr{B}\zeta = \left(B\zeta, \, 0 \right), \; \zeta \in \mathbb{C}^{n\zeta},$$
$$\mathscr{C} z = C z_0, \qquad z = (z_0, z_1) \in X.$$

We can now equivalently rewrite (1) as

$$\begin{cases} \dot{\Xi}(t) &= \mathscr{A}\,\Xi(t) + \mathscr{B}\zeta(t), \\ \eta(t) &= \mathscr{C}\,\Xi(t) + D\zeta(t), \end{cases}$$

(5)

where $\Xi(t) \in \mathscr{D}(A) \subset X$. The connection between corresponding solutions of (5) and (1) is given by $\Xi_0(t) = x(t)$, $\Xi_1(t) \equiv x(t+\theta)$, $\theta \in [-\tau_m, 0]$.

2.1 Computation of Characteristic Roots and the Spectral Abscissa

The spectral properties of the operator $\mathscr{A}$ in (5) are described in detail in [20, Chapter 1]. The operator only features a point spectrum. Hence, its spectrum, $\sigma(\mathscr{A})$, is fully determined by the eigenvalue problem

$$\mathscr{A} \, z = \lambda z, \; z \in X, \; z \neq 0.$$

(6)

The connections with the characteristic roots are as follows. The characteristic roots are the eigenvalues of the operator $\mathscr{A}$. Moreover, if $\lambda \in \sigma(\mathscr{A})$, then the corresponding eigenfunction takes the form

$$z(\theta) = v e^{\lambda \theta}, \; \theta \in [-\tau_m, 0],$$

(7)

where $v \in \mathbb{C}^n$ and (λ, v) satisfies (2). Conversely, if a pair (λ, v) satisfies (2), then (7) is an eigenfunction of $\mathscr{A}$ corresponding to the eigenvalue λ. We conclude that, as a heritage of the equivalent representation of (1) as (5), the characteristic roots can be equivalently expressed as

1. the solutions of the finite-dimensional nonlinear eigenvalue problem (2);
2. the solutions of the infinite-dimensional linear eigenvalue problem (6).

This dual viewpoint lies at the basis of the available tools to the compute the rightmost characteristic roots. On the one hand, discretizing (6) and solving the resulting standard eigenvalue problems allows to obtain global information, for example, estimates of *all* characteristic roots in a given compact set or in a given right half plane. On the other hand, the (finitely many) nonlinear equations (2) allow to make *local corrections* on characteristic root approximations up to the desired accuracy, e.g., using Newton's method or inverse residual iteration.

There are several possibilities to discretize (6). Given a positive integer N and a mesh Ω_N of $N+1$ distinct points in the interval $[-\tau_m, 0]$,

$$\Omega_N = \{\theta_{N,i}, \ i = 1, \ldots, N+1\}, \tag{8}$$

with

$$-\tau_m \leq \theta_{N,1} < \ldots < \theta_{N,N} < \theta_{N,N+1} = 0,$$

a spectral discretization as in [2] leads for example to the eigenvalue problem

$$\mathscr{A}_N \mathbf{x}_N = \lambda \mathbf{x}_N, \ \mathbf{x}_N \in \mathbb{C}^{n(N+1)}, \ \mathbf{x}_N \neq 0, \tag{9}$$

where

$$\mathscr{A}_N = \begin{pmatrix} d_{1,1} & \cdots & d_{1,N+1} \\ \vdots & & \vdots \\ d_{N,1} & \cdots & d_{N,N+1} \\ a_1 & \cdots & a_{N+1} \end{pmatrix} \in \mathbb{R}^{(N+1)n \times (N+1)n} \tag{10}$$

and

$$\begin{aligned}
d_{i,k} &= l'_{N,k}(\theta_{N,i})I_n, & i &= 1, \ldots, N, \ k = 1, \ldots, N+1, \\
a_k &= A_0 l_{N,k}(0) + \sum_{i=1}^m A_i l_{N,k}(-\tau_i), & k &= 1, \ldots, N+1.
\end{aligned}$$

The functions $l_{N,k}$ represent the Lagrange polynomials relative to the mesh Ω_N, i.e. polynomials of degree N such that, $l_{N,k}(\theta_{N,i}) = 1$ if $i = k$ and $l_{N,k}(\theta_{N,i}) = 0$ if $i \neq k$, In [2] it is proved that spectral accuracy on the individual characteristic root approximations (approximation error $O(N^{-N})$) is obtained with a grid consisting of (scaled and shifted) Chebyshev extremal points, that is,

$$\theta_{N,i} = -\cos \frac{\pi(i-1)}{N}, \ i = 1, \ldots, N+1.$$

The discretization of (6) into (9) lays at the basis of the software tool TRACE-DDE [3]. The stability routine for equilibria of the package DDE-BIFTOOL [8] fully exploits the dual representation of the eigenvalue problem since it is based on discretizing the solution operator associated with (5), whose infinitesimal generator is $\mathscr{A}$, followed by Newton corrections on (2). Both tools rely on computing all eigenvalues of the discretized systems, which heavily restricts the size of the problem form a computational point of view. In [13] an iterative method is described for computing selected eigenvalues of large-scale systems. This method has an interpretation as the Arnoldi method (see, e.g., [24]) in a function setting, applied to the infinite-dimensional operator $\mathscr{A}$, and, therefore, it does not explicitly rely on a discretization.

Complementary to the above approach, methods and software have been designed, based on computing zeros of quasipolynomials, see, e.g. [30] and the references therein.

2.2 Computation of $\mathscr{H}_\infty$ Norms

For systems without delay, level set methods are standard methods for computing $\mathscr{H}_\infty$ norms and related problems, see, e.g., [1, 4] and the references therein. These methods originate from the property that all intersections of the singular value curves, corresponding to the transfer function, and a constant function (the level) can be directly computed as the solutions of a structured eigenvalue problem. This makes a fast two-directional search for the dominant peak value in the singular value plot possible.

In [19] an extension of this approach for computing the $\mathscr{H}_\infty / \mathscr{L}_\infty$ norm of the transfer function (3) is described. The theoretical foundation is contained in the following proposition [19, Lemma 2.1 and Proposition 2.2].

Proposition 1. *Let $\xi > 0$ be such that the matrix $D_\xi := D^T D - \xi^2 I$ is non-singular. For $\omega \geq 0$, matrix $G(j\omega)$ has a singular value equal to ξ if and only if $\lambda = j\omega$ is a solution of the equation*

$$\det H(\lambda; \, \xi) = 0, \tag{11}$$

where

$$H(\lambda; \, \xi) := \lambda I - M_0 - \sum_{i=1}^{m} \left(M_i e^{-\lambda \tau_i} + M_{-i} e^{\lambda \tau_i} \right) - \left(N_1 e^{-\lambda \tau_0} + N_{-1} e^{\lambda \tau_0} \right),$$

with

$$M_0 = \begin{bmatrix} A_0 & -BD_\xi^{-1}B^T \\ -C^T C + C^T DD_\xi^{-1} D^T C & -A_0^T \end{bmatrix},$$

$$M_i = \begin{bmatrix} A_i & 0 \\ 0 & 0 \end{bmatrix}, \quad M_{-i} = \begin{bmatrix} 0 & 0 \\ 0 & -A_i^T \end{bmatrix}, \quad 1 \leq i \leq m,$$

$$N_1 = \begin{bmatrix} 0 & 0 \\ 0 & C^T DD_\xi^{-1} B^T \end{bmatrix}, \quad N_{-1} = \begin{bmatrix} -BD_\xi^{-1} D^T C & 0 \\ 0 & 0 \end{bmatrix}.$$

Moreover, (11) holds if and only if λ is an eigenvalue of the operator $\mathscr{L}_\xi$, defined by

$$\mathscr{D}(\mathscr{L}_\xi) = \left\{ \phi \in Z : \, \phi' \in Z, \, \phi'(0) = M_0 \phi(0) + \sum_{i=1}^{m} (M_i \phi(-\tau_i) + M_{-i} \phi(\tau_i)) \right.$$
$$\left. + N_1 \phi(-\tau_0) + N_{-1} \phi(\tau_0) \right\},$$

$$\mathscr{L}_\xi \, \phi = \phi', \quad \phi \in \mathscr{D}(\mathscr{L}_\xi),$$

where $Z := \mathscr{C}([-\tau_m, \, \tau_m], \, \mathbb{C}^{2n})$.

According to Proposition 1, the intersections of the constant function $\mathbb{R} \ni \omega \mapsto \xi$, with $\xi > 0$ given, and the curves

$$\mathbb{R} \ni \omega \mapsto \sigma_i(G(j\omega)), \ 1 \le i \le \min(n_u, n_y),$$

where $\sigma_i(\cdot)$ denotes the i-th singular value, can be found by computing the solutions on the imaginary axis of

1. either, the finite-dimensional nonlinear eigenvalue problem

$$H(\lambda; \xi)v = 0, v \in \mathbb{C}^{2n}, \ v \ne 0, \tag{12}$$

2. or, the infinite-dimensional linear eigenvalue problem

$$\mathscr{L}_\xi \phi = \lambda \phi, \ \phi \in Z, \ \phi \ne 0.$$

These two characterizations are similar to the representations of characteristic roots as eigenvalues. As a consequence, the methods outlined in §2.1 can be adapted accordingly.

The method in [19] for computing the $\mathscr{H}_\infty$ norm of (3) relies on a two-directional search in a modified singular value plot, induced by a spectral discretization of the operator $\mathscr{L}_\xi$, followed by a local correction of the peak value up to the desired accuracy. The latter is based on the nonlinear equation (12), supplemented with a local optimality condition. The main steps are sketched in Figure 1.

A related problem is the computation of the pseudospectral abscissa, for which we refer to [11] and the references therein.

2.3 Computation of $\mathscr{H}_2$ Norms

We assume that (1) is internally stable and that $D = 0$. Then the $\mathscr{H}_2$ norm of G is finite and it satisfies

$$\|G(j\omega)\|_{\mathscr{H}_2} = \sqrt{\frac{1}{2\pi} \int_{-\infty}^{\infty} \mathrm{tr}(G(j\omega)^* G(j\omega)) \, d\omega}.$$

We present two different approaches for its computation, which once again stem from the two descriptions of the time-delay system, by the functional differential equation (1) and by the abstract linear equation (5), respectively.

The first approach makes use of so-called delay-Lyapunov equations, which were introduced in the context of constructing complete-type Lyapunov-Krasovskii functionals (see, e.g. [16]). The following result is a special case of [15, Theorem 1].

Theorem 1. *Assume that $D = 0$. The $\mathscr{H}_2$ norm of (3) satisfies*

$$\|G(j\omega)\|_{\mathscr{H}_2} = \mathrm{tr}(B^T U(0)B), \tag{13}$$
$$= \mathrm{tr}(CV(0)C^T), \tag{14}$$

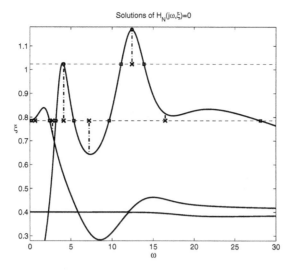

Fig. 1 Principles of the method in [19]. The function H_N is an approximation of H, induced by a spectral discretization of $\mathcal{L}_\xi$ on a grid consisting of $2N + 1$ points over the interval $[-\tau_m,\ \tau_m]$. In the first step the peak value is found for a fixed value of N by the iterative algorithm of [4]: for a given level ξ all intersections with the approximate singular value curves are computed (squares). In the geometric midpoints of these intersections (crosses) a vertical search for intersections is performed. The maximum value of ξ over all the intersections gives rise to the new value of the level. In the second step, the effect of the discretization is removed by a local corrector based on the nonlinear eigenvalue problem (12) (circles).

if U,V are unique solutions of the delay Lyapunov equation

$$\begin{cases} U'(t) = U(t)A_0 + \sum_{k=1}^{m} U(t - \tau_k)A_k, & t \in [0,\ \tau_{\max}], \\ U(-t) = U^T(t), \\ -C^T C = U(0)A_0 + A_0^T U(0) + \sum_{k=1}^{m} \left(U^T(\tau_k)A_k + A_k^T U(\tau_k) \right), \end{cases} \tag{15}$$

and the dual delay Lyapunov equation

$$\begin{cases} V'(t) = V(t)A_0^T + \sum_{k=1}^{m} V(t - \tau_k)A_k^T, & t \in [0,\ \tau_{\max}], \\ V(-t) = V^T(t), \\ -BB^T = V(0)A_0^T + A_0 V(0) + \sum_{k=1}^{m} \left(V^T(\tau_k)A_k^T + A_k V(\tau_k) \right). \end{cases} \tag{16}$$

The underlying idea of the proof is that the solution of (15)-(16), as well as the $\mathcal{H}_2$ norm, can be expressed in terms of the *fundamental solution* of the delay equation.

Theorem 1 opens the possibility to compute $\mathcal{H}_2$ norms by solving delay Lyapunov equations numerically. An approach based on spectral collocation on a Chebyshev grid is presented in [15]. This approach is generally applicable, but the convergence rate of the approximation (as a function of the number of grid points) depends on the smoothness properties of the solutions, which are on their turn determined by the interdependence of the delays (see Section 4 of this reference for a complete characterization). It is also shown that, in the case of commensurate delays, an analytic solution of (15)-(16) can be obtained, which leads to an explicit expression for the $\mathscr{H}_2$ norm.

The second approach is based on discretizing (5). A spectral discretization on the grid (8) leads to the linear system (assuming once again that $D = 0$)

$$\begin{cases} \dot{\mathbf{x}}_N(t) = \mathscr{A}_N \mathbf{x}_N(t) + \mathscr{B}_N \zeta(t), \\ \eta(t) = \mathscr{C}_N \mathbf{x}_N(t), \end{cases} \tag{17}$$

where $\mathscr{A}_N$ is given by (9) and

$$\mathscr{B}_N = [0 \cdots 0\, I]^T \otimes B, \ \mathscr{C}_N = [0 \cdots 0\, I] \otimes C.$$

From the fact that (17) is a standard LTI system we can approximately compute

$$\|G(j\omega)\|_{\mathscr{H}_2} \approx \|G_N(j\omega)\|_{\mathscr{H}_2} = \operatorname{tr}(\mathscr{B}^T Q_N \mathscr{B}_N), \tag{18}$$
$$= \operatorname{tr}(\mathscr{C}_N P_N \mathscr{C}^T), \tag{19}$$

where G_N is the transfer function of (17) and the pair (P_N, Q_N) satisfies (see, e.g. [32]),

$$\begin{aligned} \mathscr{A}_N P_N + P_N \mathscr{A}_N^T &= -\mathscr{B}_N \mathscr{B}_N^T, \\ \mathscr{A}_N^T Q_N + Q_N \mathscr{A}_N &= -\mathscr{C}_N^T \mathscr{C}_N. \end{aligned} \tag{20}$$

In [26] it has been shown that the approximation error satisfies

$$\|G(j\omega)\|_{\mathscr{H}_2} - \|G_N(j\omega)\|_{\mathscr{H}_2} = \mathscr{O}(N^{-3}), \ N \to \infty,$$

and arguments are provided why fairly accurate results are expected for a moderate value of N.

Although the second approach is essentially a "discretize first" approach it is to be preferred for most control problems because, unlike the approach of [15], it does not involve an explicit vectorization of matrix equations (which squares the dimensions of the problem), and because expressions of derivatives of the $\mathscr{H}_2$ norm with respect to the elements of the system matrices in (3) can easily be obtained as a by-product of solving (20), see [26].

3 Solving Synthesis Problems

In what follows we assume that the system matrices in (1) smoothly depend on a finite number of parameters $p = (p_1, \ldots, p_{n_p}) \in \mathbb{R}^{n_p}$. Making the dependence explicit in the notation leads to

$$\begin{cases} \dot{x}(t) = A_0(p)x(t) + \sum_{i=1}^m A_i(p)x(t - \tau_i) + B(p)\zeta(t), \\ \eta(t) = C(p)x(t) + D(p)\zeta(t). \end{cases} \tag{21}$$

In many control design problems the closed-loop system can be brought into the form (21), where the parameters p have an interpretation in terms of a parametrization of the controller and ζ and η appear as external inputs and outputs, used in the description of performance and robustness specifications. Note that both static and dynamic controllers can be addressed in this framework. It is also possible to impose additional structure on the controller, e.g., a proportional-integral-derivative (PID) like structure.

Because time-delay systems are inherently infinite-dimensional, illustrated by the representation (5) and by the presence of infinitely many characteristic roots, *any* control design problem involving the determination of a finite number of parameters can be interpreted as a reduced-order controller synthesis problem. This partially explains the difficulties and limitations in controlling time-delay systems [22, 25]. The proposed control synthesis methods are based on a direct optimization of appropriately defined cost functions as a function of the parameters p. They are inspired by recent work on the design of reduced-order controller for LTI systems within an eigenvalue optimization framework, as, e.g., implemented in the software HIFOO [12].

3.1 Stabilization

To impose exponential stability of the null solution of (21), with $\zeta \equiv 0$, it is necessary to find values of p for which the spectral abscissa

$$\alpha(p) := \sup_{\lambda \in \mathbb{C}} \left\{ \Re(\lambda) : \det\left(\lambda I - A_0(p) - \sum_{i=1}^{m} A_i(p) e^{-\lambda \tau_i} \right) = 0 \right\}$$

is strictly negative. The approach of [28] is based on minimizing the function

$$p \to \alpha(p). \tag{22}$$

The function (22) is in general non convex. It may be not everywhere differentiable, even not everywhere Lipschitz continuous. A lack of differentiability may occur when there is more than one *active* characteristic root, i.e., a characteristic root whose real part equals the spectral abscissa. A lack of Lipschitz continuity may occur when an active characteristic roots is multiple and non-semisimple. On the contrary, the spectral abscissa function is differentiable at points where there is only one active characteristic root with multiplicity one. Since this is the case with probability one when randomly sampling parameter values, the spectral abscissa is smooth almost everywhere [28].

The properties of the function (22) preclude the use of standard optimization methods, developed for smooth problems. Instead we propose a combination of BFGS with weak Wolfe line search, whose favorable properties for nonsmooth problems have been reported in [17], and gradient sampling [6], as implemented in the MATLAB code HANSO [23]. The overall algorithm only requires the evaluation of the objective function, as well as its derivatives with respect to parameters, *whenever* it is differentiable.

The value of the spectral abscissa can obtained by computing the rightmost characteristic roots, using the methods described in §2.1. If there is only one active characteristic root λ_a with multiplicity one, the spectral abscissa is differentiable and we can express

$$\frac{\partial \alpha}{\partial p_k}(p) = \Re \left(\frac{w^* \left(\frac{\partial A_0}{\partial p_k}(p) + \sum_{i=1}^{m} \frac{\partial A_i}{\partial p_k}(p) e^{-\lambda_a \tau_i} \right) \upsilon}{w^* \left(I + \sum_{i=1}^{m} A_i(p) \tau_i e^{-\lambda_a \tau_i} \right) \upsilon} \right), \quad k = 1, \ldots, n_p, \quad (23)$$

where w and υ are the left and right eigenvector corresponding to λ_a. An alternative to formula (23) consists of a finite difference approximation.

3.2 $\mathscr{H}_\infty$ and $\mathscr{H}_2$ Optimization Problems

The properties of the function

$$p \mapsto \|G(\lambda; p)\|_{\mathscr{H}_\infty}, \quad (24)$$

where $G(\lambda; p)$ is the transfer function of (21), are very similar to the spectral abscissa function. In particular (24) is in general not convex, not everywhere differentiable, but it is smooth almost everywhere. Consequently, the methods described in §3.1 can also be applied to (24). For almost all p derivatives exist and they can be computed from the sensitivity of an individual singular value of $G(j\omega; p)$ with respect to p, for a fixed value of ω, see [10]. Unlike (24), the function

$$p \mapsto \|G(\lambda; p)\|_{\mathscr{H}_2}, \quad (25)$$

is smooth whenever it is finite, which allows an embedding in a derivative based optimization framework. As mentioned before, derivatives of (25) can be obtained from the solution of Lyapunov equations [26].

Both the minimization problem of (24) and (25) contain an implicit constraint, $\alpha(p) < 0$, because the norms are only finite if the system is internally stable. This leads to a two-stage approach: if the initial values of the parameters are not stabilizing, then the overall procedure contains a preliminary stabilization phase, using the methods of §3.1. For the next phase, the actual minimization of (24)-(25), the line-search mechanism in BFGS and the gradient sampling algorithm are adapted in order to discard trial steps or samples outside the feasible set, defined by the implicit constraint[10].

Instead of directly optimizing the spectral abscissa as in §3.1, which requires methods for nonsmooth problems, it is also possible to optimize the smooth relaxation of the spectral abscissa function proposed in [29], which is defined in terms of a relaxed $\mathscr{H}_2$ criterion. In this way stability optimization can also be performed within a derivative based framework. Moreover, an adaptation of the approach makes it possible to solve $\mathcal{H}_2$ optimization problems without the explicit need for a preliminary stabilization phase [27].

4 Software and Applications

Software for solving the analysis and synthesis problems discussed in Sections 2-3, as well as a many benchmark problems, can be downloaded from

```
http://twr.cs.kuleuven.be/research/software/
delay-control/.
```

In what follows we illustrate a flexibility of the presented control design approach, which allows to incorporate control design techniques for LTI systems based on pole placement. The flexibility consists of assigning a finite number of characteristic roots, smaller or equal to the number of controller parameters, and using the remaining degrees of freedom to optimize the criteria presented in the previous sections.

The characteristic matrix of (21) is given by

$$M(\lambda;\ p) := \lambda I - A_0(p) - \sum_{i=1}^{m} A_i(p)e^{-\lambda \tau_i}. \tag{26}$$

Assigning a real characteristic root to the location c yields the constraint on the parameter values,

$$\det(M(c;\ p)) = 0. \tag{27}$$

Similarly, assigning a complex conjugate pair of characteristic roots, $c \pm di$, results in

$$\Re(\det(M(c \pm dj;\ p))) = 0; \quad \Im(\det(M(c \pm dj;\ p))) = 0. \tag{28}$$

It it easy to see that if M affinely depends on p and if the condition

$$\text{rank}\left(\left[\frac{\partial M}{\partial p_1}(\lambda;\ p) \cdots \frac{\partial M}{\partial p_{n_p}}(\lambda;\ p)\right]\right) = 1, \ \forall \lambda \in \mathbb{C}, \tag{29}$$

is satisfied, then the constraints (27)-(28) are linear in p. Hence, assigning k characteristic roots, with $k \leq n_p$, can be expressed by constraints of the form

$$Sp = R, \tag{30}$$

where $S \in \mathbb{R}^{k \times n_p}$ and $R \in \mathbb{R}^{k \times 1}$. It is important to note that rank condition (29) is satisfied in many problems where the closed-loop characteristic matrix results from control through a single input.

In [21] it is shown how the constraints (30) on the parameters can be eliminated. Subsequently, the optimization problem

$$\min_{p \in \mathbb{R}^{n_p},\ Sp=R} \bar{\alpha}(p) \tag{31}$$

is addressed, where

$$\bar{\alpha}(p) := \sup_{\lambda \in \mathbb{C}} \left\{ \Re(\lambda) : \frac{\det\left(\lambda I - A_0(p) - \sum_{i=1}^{m} A_i(p)e^{-\lambda \tau_i}\right)}{\Pi_{i=1}^{k}(\lambda - \lambda_i)} = 0 \right\} \quad (32)$$

and $\lambda_1, \ldots, \lambda_k$ are the assigned characteristic roots. Problem (31) is a modification of the spectral abscissa minimization problem discussed in §3.1. It corresponds to minimizing the real part of the rightmost non-assigned characteristic root. As the assigned roots are specified by the designer, the value of (32), where p satisfies (30), can be obtained by computing the rightmost characteristic roots and removing the assigned ones.

To conclude the section, let us solve problem (31) for the model of an experimental heat transfer set-up at the Czech Technical University in Prague, comprehensively described in [31]. The model consists of 10 delay differential equations. The inclusion of an integrator, to achieve a zero steady state error of one of the state variables w.r.t. a set-point, eventually leads to equations of the form

$$\dot{x}(t) = A_0 x(t) + \sum_{i=1}^{5} A_i x(t - \tau_i) + Bu(t - \tau_6), \quad x(t) \in \mathbb{R}^{11 \times 11}, \ u(t) \in \mathbb{R}. \quad (33)$$

See [31] for the corresponding matrices and delay values. In Figure 2, the rightmost characteristic roots of the open-loop system are shown. For the control law

$$u(t) = \sum_{i=1}^{11} p_i x_i(t),$$

the solutions of the optimization problem (31) are presented in Table 1. The setting SN1 corresponds to the (unconstrained) minimization of the spectral abscissa (22).

Table 1 Controller parameters corresponding to the solution of (31).

SN	1	2	3	4
λ_i	- - - -	-0.01	$-0.02 \pm 0.02i$	-0.02, -0.03
				$-0.03 \pm 0.03i$
$\min \bar{\alpha}$	-0.0565	-0.0629	-0.0659	-0.0736
p_1	-5.4349	-0.0732	-4.1420	-0.3521
p_2	3.5879	8.1865	5.9345	8.6190
p_3	-1.4411	-1.2503	-2.3820	-4.8822
p_4	-3.7043	-7.1472	-7.9449	-17.2747
p_5	24.616	32.8003	27.8585	35.1494
p_6	-2.1778	4.4977	0.4490	-1.3188
p_7	9.6924	10.3140	8.4887	6.0338
p_8	-4.5121	-2.6572	-0.2605	5.4190
p_9	-14.631	-21.6711	-20.5152	-24.6596
p_{10}	11.351	4.1244	5.4531	2.3754
p_{11}	-0.7562	-0.2749	-0.3635	-0.1360

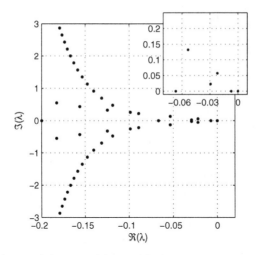

Fig. 2 Rightmost characteristic roots of the open-loop system.

The other settings correspond to assigning one real characteristic root (SN2), one pair of complex conjugate characteristic roots (SN3) and, finally, two real roots and two complex conjugate roots (SN4). The assigned characteristic roots were chosen to the right of the minimum of the spectral abscissa function, because these root were intended to become the rightmost roots after solving (31), and their positions were optimized to achieve a properly damped set-point response and disturbance rejection [21]. The optimized characteristic root locations are shown in Figure 3 for settings SN1 and SN4.

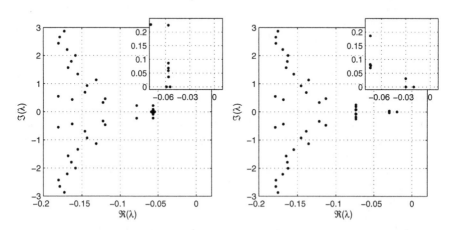

Fig. 3 Rightmost characteristic roots corresponding to the solution of (31), for settings SN1 and SN4 in Table 1.

5 Concluding Remarks

An eigenvalue based solution to the robust control of linear time-delay systems has been presented. Because any controller characterized by a finite number of parameters can be seen as a reduced-order controller, a direct optimization approach has been taken.

The advantages of the approach are two-fold. First, the methods are generally applicable and can, e.g., handle systems with multiple delays in states, inputs and outputs, as well as a broad class of controllers, where the computational complexity does not depend on the number of delays. Moreover, they are not restricted to a description of the closed-loop system by equations of the form (21). For instance, an extension to a system description by delay-differential algebraic equations has recently been proposed (see, e.g., [10]), which allows to incorporate neutral dynamics, delays in the inputs and outputs describing the performance specifications, and controllers characterized by a nontrivial feedthrough. As a second advantage the approach is not conservative in the sense that a stabilizing or optimal fixed-structure $\mathscr{H}_2 - \mathscr{H}_\infty$ controller can be computed whenever it exists, in contrast to approaches inferred from sufficient (but not necessary) conditions for a stabilizing or a suboptimal solution.

As a price to pay for the beneficial properties, the optimization problems encountered in Section 3 are in general non-convex, hence, there is no guarantee that the optima found by the local methods are global. A comparison with existing LMI based techniques once again illustrates the following dilemma inherent to reduced-order control design: a possibly local solution of the original nonconvex problem versus a global solution of a convex over-relaxation of the problem.

Acknowledgements. This work has been supported by the Programme of Interuniversity Attraction Poles of the Belgian Federal Science Policy Office (IAP P6- DYSCO), by OPTEC, the Optimization in Engineering Center of the K.U.Leuven, by the project STRT1- 09/33 of the K.U.Leuven Research Council and the project G.0712.11N of the Research Foundation - Flanders (FWO).

References

1. Boyd, S., Balakrishnan, V.: A regularity result for the singular values of a transfer matrix and a quadratically convergent algorithm for computing its L_∞-norm. Systems & Control Letters 15, 1–7 (1990)
2. Breda, D., Maset, S., Vermiglio, R.: Pseudospectral differencing methods for characteristic roots of delay differential equations. SIAM Journal on Scientific Computing 27(2), 482–495 (2005)
3. Breda, D., Maset, S., Vermiglio, R.: TRACE-DDE: a Tool for Robust Analysis and Characteristic Equations for Delay Differential Equations. In: Loiseau, J.J., Michiels, W., Niculescu, S.-I., Sipahi, R. (eds.) Topics in Time Delay Systems. LNCIS, vol. 388, pp. 145–155. Springer, Heidelberg (2009)
4. Bruinsma, N.A., Steinbuch, M.: A fast algorithm to compute the $\mathscr{H}_\infty$-norm of a transfer function matrix. Systems and Control Letters 14, 287–293 (1990)

5. Burke, J.V., Henrion, D., Lewis, A.S., Overton, M.L.: HIFOO - a MATLAB package for fixed-order controller design and H-infinity optimization. In: Proceedings of the 5th IFAC Symposium on Robust Control Design, Toulouse, France (2006)

6. Burke, J.V., Lewis, A.S., Overton, M.L.: A robust gradient sampling algorithm for nonsmooth, nonconvex optimization. SIAM Journal on Optimization 15(3), 751–779 (2005)

7. Curtain, R.F., Zwart, H.: An introduction to infinite-dimensional linear systems theory. Texts in Applied Mathematics, vol. 21. Springer, Heidelberg (1995)

8. Engelborghs, K., Luzyanina, T., Roose, D.: Numerical bifurcation analysis of delay differential equations using DDE-BIFTOOL. ACM Trans. Math. Softw. 28(1), 1–24 (2002)

9. Gu, K., Kharitonov, V.L., Chen, J.: Stability of time-delay systems. Birkhauser (2003)

10. Gumussoy, S., Michiels, W.: Fixed-order strong H-infinity control of interconnected systems with time-delays. TW Report 579, Department of Computer Science, Katholieke Universiteit Leuven, Belgium (October 2010)

11. Gumussoy, S., Michiels, W.: A predictor corrector type algorithm for the pseudospectral abscissa computation of time-delay systems. Automatica 46(4), 657–664 (2010)

12. Gumussoy, S., Overton, M.L.: Fixed-order H-infinity controller design via HIFOO, a specialized nonsmooth optimization package. In: Proceedings of the American Control Conference, Seattle, USA, pp. 2750–2754 (2008)

13. Jarlebring, E., Meerbergen, K., Michiels, W.: A Krylov method for the delay eigenvalue problem. SIAM Journal on Scientific Computing 32(6), 3278–3300 (2010)

14. Krstic, M.: Delay compensation for nonlinear, adaptive, and PDE systems. Birkhauser (2009)

15. Jarlebring, E., Vanbiervliet, J., Michiels, W.: Characterizing and computing the $\mathcal{H}_2$ norm of time-delay systems by solving the delay Lyapunov equation. IEEE Transactions on Automatic Control, Technical Report 553, K.U.Leuven, Leuven (2009)

16. Kharitonov, V., Plischke, E.: Lyapunov matrices for time-delay systems. Syst. Control Lett. 55(9), 697–706 (2006)

17. Lewis, A., Overton, M.L.: Nonsmooth optimization via BFGS (2009), http://cs.nyu.edu/overton/papers.html

18. Mahmoud, M.S.: Robust control and filtering for time-delay systems. Control Engineering, vol. 5. Dekker, New York (2000)

19. Michiels, W., Gumussoy, S.: Characterization and computation of h-infinity norms of time-delay systems. SIAM Journal on Matrix Analysis and Applications 31(4), 2093–2115 (2010)

20. Michiels, W., Niculescu, S.-I.: Stability and stabilization of time-delay systems. An eigenvalue based approach. An Eigenvalue Based Approach. SIAM (2007)

21. Michiels, W., Vyhlídal, T., Zítek, P.: Control design for time-delay systems based on quasi-direct pole placement. Journal of Process Control 20(3), 337–343 (2010)

22. Niculescu, S.-I.: Delay effects on stability. A robust control approach. LNCIS, vol. 269. Springer, Heidelberg (2001)

23. Overton, M.: HANSO: a hybrid algorithm for nonsmooth optimization (2009), http://cs.nyu.edu/overton/software/hanso/

24. Saad, Y.: Numerical methods for large eigenvalue problems. Manchester University Press (1992)

25. Sipahi, R., Niculescu, S.I., Abdallah, C.T., Michiels, W., Gu, K.: Stability and stabilization of systems with time delay. Control Systems Magazine 31(38-65), 3278–3300 (2011)

26. Vanbiervliet, J., Michiels, W., Jarlebring, E.: Using spectral discretisation for the optimal $\mathcal{H}_2$ design of time-delay systems. International Journal of Control (in press, 2011)

27. Vanbiervliet, J., Michiels, W., Vandewalle, S.: Smooth stabilization and optimal h2 design. In: Proceedings of the IFAC Workshop on Control Applications of Optimization, Jyväskylä, Finland (2009)

28. Vanbiervliet, J., Vandereycken, B., Michiels, W., Vandewalle, S.: A nonsmooth optimization approach for the stabilization of time-delay systems. ESAIM Control, Optimisation and Calcalus of Variations 14(3), 478–493 (2008)

29. Vanbiervliet, J., Vandereycken, B., Michiels, W., Vandewalle, S., Diehl, M.: The smoothed spectral abscissa for robust stability optimization. SIAM Journal on Optimization 20(1), 156–171 (2009)

30. Vyhlídal, T., Zítek, P.: Mapping based algorithm for large-scale computation of quasi-polynomial zeros. IEEE Transactions on Automatic Control 54(1), 171–177 (2009)

31. Vyhlídal, T., Zítek, P., Paulů, K.: Design, modelling and control of the experimental heat transfer set-up. In: Loiseau, J.J., et al. (eds.) Topics in Time Delay Systems. Analysis, Algorithms, and Control. LNCIS, vol. 308, pp. 303–314. Springer, Heidelberg (2009)

32. Zhou, K., Doyle, J.C., Glover, K.: Robust and optimal control. Prentice-Hall (1995)

Discretization of Solution Operators for Linear Time Invariant - Time Delay Systems in Hilbert Spaces

Dimitri Breda, Stefano Maset and Rossana Vermiglio

Abstract. In this paper a numerical scheme to discretize the solution operators of linear time invariant - time delay systems is proposed and analyzed. Following previous work of the authors on the classic state space of continuous functions, here the focus is on working in product Hilbert state spaces. The method is based on a combination of collocation and Fourier projection. Full discretization details for constructing the approximation matrices are given for the sake of implementation. Moreover, convergence results are proved and discussed, with particular attention to their pros and cons with regards to fundamental targets such as time-integration and detection of asymptotic stability of equilibria.

1 Introduction

In this work we consider the Linear Time Invariant - Time Delay System (LTI-TDS)

$$x'(t) = Ax(t) + \sum_{\ell=1}^{k} \left(B_\ell x(t - \tau_\ell) + \int_{-\tau_\ell}^{-\tau_{\ell-1}} C_\ell(\theta)x(t + \theta)d\theta \right) \tag{1}$$

Dimitri Breda
Dipartimento di Matematica e Informatica
Università degli Studi di Udine
via delle Scienze 206 I-33100 Udine, Italy
e-mail: dimitri.breda@uniud.it

Stefano Maset
Dipartimento di Matematica e Informatica
Università degli Studi di Trieste
via Valerio 12 I-34127 Trieste, Italy
e-mail: maset@units.it

Rossana Vermiglio
Dipartimento di Matematica e Informatica
Università degli Studi di Udine
via delle Scienze 206 I-33100 Udine, Italy
e-mail: rossana.vermiglio@uniud.it

R. Sipahi et al. (Eds.): Time Delay Sys.: Methods, Appli. and New Trends, LNCIS 423, pp. 217–228.
springerlink.com   © Springer-Verlag Berlin Heidelberg 2012

where $0 := \tau_0 < \tau_1 < \cdots < \tau_{k-1} < \tau_k =: \tau$, $A \in \mathbb{R}^{d \times d}$ and, for $\ell = 1, \ldots, k$, $B_\ell \in \mathbb{R}^{d \times d}$ and $C_\ell \in L^2\left(-\tau_\ell, -\tau_{\ell-1}; \mathbb{R}^{d \times d}\right)$.

In the recent decades, LTI-TDS (and their Linear Time Varying (LTV) counterpart) have attracted the attention of diverse scientific communities, automatic control and mathematics above all. A central question from a dynamical point of view is that of asymptotic stability for the zero solution of (1). Despite the great effort, analytical results are rather lacking in generality and, at best, suitable for restricted subclasses (e.g. single delay or second order systems). As a natural consequence [19, p.109], a number of approximation techniques have been proposed, mostly based on computing the characteristic values (read roots, multipliers, Lyapunov exponents) associated to the system, see e.g. [6, 9, 15, 16, 17, 20, 24, 26].

When investigating on stability (but not only), the *state space* description of (1) is advantageous, and the classic literature resorts to the Banach space of continuous functions $\mathscr{C} := C(-\tau, 0; \mathbb{R}^d)$, [2, 13, 19]. This choice seems to be motivated by the fact that, for rather general selections of the space of initial data, the "smoothing effect" makes the (forward) solution be continuous anyway: "...if some other space than continuous functions is used for initial data, then the solution lies in C...Therefore, for the fundamental theory, the space of initial data does not play a role which is too significant." [19, p.33]. Anyway, Hale continues his comment by adding "However, in the applications, it is sometimes convenient to take initial data with fewer or more restrictions." In this sense, an alternative which has been quite studied is represented by the Hilbert product space $X := \mathbb{R}^d \times L^2(-\tau, 0; \mathbb{R}^d)$, [3, 4, 12, 18, 23]. This second choice is often justified in the context of quadratic feedback control and linear filtering for retarded systems [11, 18, 25], for approximation reasons [21, 22], or when orthogonality is necessary [5].

Once the proper state space is chosen, the long-time behavior of the evolution can be determined through the knowledge of the spectrum of suitable infinite dimensional linear operators such as the solution operators, their generator, the monodromy operator for periodic problems and so forth. Part of this manuscript is then devoted to resume the basic features of the semigroup approach for model (1) in X. The case with state space $\mathscr{C}$ has been treated by the authors in [8]. We refer the interested readers to [14] for a comprehensive treatment of the theory of general one-parameter linear semigroups.

The reduction of such operators to finite dimension allows to consider standard eigenvalue problems which can be easily solved, hopefully providing accurate estimates for the stability indicators (e.g. the rightmost root or the dominant multiplier). Following [8], we present an approach for discretizing the solution operators in X, discussing computational issues and convergence as well as investigating to what extent this latter can be obtained. The numerical scheme is a combination of polynomial collocation and Fourier projection.

The paper is structured as follows. Section 2 resumes the semigroup theory for (1) and introduces the necessary notation. Section 3 presents the steps of the discretization, together with the resulting matrices necessary for code implementations. Section 3.1 is about the convergence analysis. Concluding remarks are given in 4 and an Appendix collects some preparatory results.

2 Solution Operators and Notation

Given $(u, \varphi) \in X$ and fixed $r > 0$, a well-posed Initial Value Problem (IVP) for (1) reads

$$
\begin{cases}
x'(t) = (Gx)(t), & \text{a.e. in } [0,r] \\
x(0) = u \\
x(\theta) = \varphi(\theta), & \text{a.e. in } [-\tau,0),
\end{cases}
\tag{2}
$$

where $G : L^2(-\tau, r; \mathbb{R}^d) \to L^2(0, r; \mathbb{R}^d)$ corresponds to the right-hand side of (1). The problem (2) admits a unique solution x (e.g. [1, 5]). This allows to introduce the *solution operator* as the one-parameter linear and bounded operator $T(r) : X \to X$ given by

$$
T(r)(u, \varphi) = (x(r), x_r)
\tag{3}
$$

where, according to the standard Hale-Krasovskii notation, $x_r(\theta) := x(r + \theta)$ for $\theta \in [-\tau, 0]$.

The *infinitesimal generator* associated to the semigroup of the solution operators $\{T(r)\}_{r \geq 0}$ reads

$$
\begin{cases}
\mathscr{D}(\mathscr{A}) = \{(v, \psi) \in X^1 : \psi(0) = v\} \\
\mathscr{A}(v, \psi) = \left(Av + \sum_{\ell=1}^{k} \left(B_\ell \psi(-\tau_\ell) + \int_{-\tau_\ell}^{-\tau_{\ell-1}} C_\ell(\theta) \psi(\theta) d\theta \right), \psi' \right)
\end{cases}
\tag{4}
$$

where $X^1 := \mathbb{R}^d \times H^1(-\tau, 0; \mathbb{R}^d)$ with H^1 the Sobolev space of L^2 functions with first derivative in L^2.

Remark 1. Let us underline that as far as time varying coefficients are considered, $T(r)$ in (3) can be extended to $T(r, s)$ where s is the initial time of the relevant IVP, which inevitably matters in the LTV case. Also (4) can be trivially extended since the domain of the generator is not altered by the time-dependency. This latter reflects only in the vector component of the action of $\mathscr{A}$. As observed in [7], this is not possible when the state space is $\mathscr{C}$, since $\mathscr{D}(\mathscr{A})$ would be time-dependent. Current research of the authors is focused on the extension of the numerical scheme here proposed to the LTV case in X (as for $\mathscr{C}$ in [8]), as well as on the study of a full spectral method not resorting to collocation.

We introduce now the necessary notation. Set $L := L^2(-\tau, 0; \mathbb{R}^d)$, $L^+ := L^2(0, r; \mathbb{R}^d)$ and $L^\pm := L^2(-\tau, r; \mathbb{R}^d)$. The relevant functions are denoted respectively as f, f^+ and $f^\pm$, these latter being tacitly intended as divided into $f := f^\pm_{|[-\tau,0]}$ and $f^+ := f^\pm_{|[0,r]}$. The same notation is adopted for spaces and functions other than L^2, e.g. H^1, X, X^1 and Π_N, the set of polynomials of degree at most N. For $(Y, \|\cdot\|_Y)$ and $(Z, \|\cdot\|_Z)$ normed linear spaces, $\mathscr{B}(Y, Z)$ is the set of linear and bounded operators from Y to Z and $\|A\|_{Z \leftarrow Y}$ is the operator norm of $A \in \mathscr{B}(Y, Z)$ (simply $\mathscr{B}(Y)$ and $\|A\|_Y$ when $Z = Y$).

For a given positive integer M, let $\{\phi_i\}_{i=0}^{\infty}$ be the system of Legendre polynomials of L with $-\tau < \theta_M < \cdots < \theta_1 < 0$ the M zeros of $\phi_M \in \Pi_M$. Set $L_M := (\mathbb{R}^d)^{M+1}$ be the discrete counterpart of L, i.e. a function $f \in L$ with Fourier coefficients $\{f_i\}_{i=0}^{\infty}$ is discretized by the vector $\boldsymbol{f}_M = \mathscr{R}_M f = (f_0, \ldots, f_M)^T \in L_M$, $\mathscr{R}_M : L \to L_M$ being the restriction operator. Correspondingly, let $\mathscr{P}_M \boldsymbol{f}_M \in \Pi_M$ be the polynomial of degree at most M for f, i.e.

$$\mathscr{P}_M \boldsymbol{f}_M := \sum_{j=0}^{M} f_j \phi_j,$$

$\mathscr{P}_M : L_M \to \Pi_M \subset L$ being the prolongation operator. Observe that $\mathscr{R}_M \mathscr{P}_M = I_{L_M} : L_M \to L_M$, the identity in L_M, while $\mathscr{P}_M \mathscr{R}_M = \mathscr{F}_M : L \to \Pi_M \subset L$, the Fourier projection operator relevant to $\{\phi_i\}_{i=0}^{\infty}$.

Exactly the same notation is adopted for L^+ with a positive integer N, except for the opposite numbering of the zeros $0 < \theta_1^+ < \cdots < \theta_N^+ < r$ of the Legendre polynomial $\phi_N^+ \in \Pi_N^+$. Such zeros will be used as collocation nodes.

In a similar fashion, the same notation is extended to the state space X by setting $X_M := \mathbb{R}^d \times L_M = (\mathbb{R}^d)^{M+2}$, $\hat{\mathscr{R}}_M := (\text{diag}(I_{\mathbb{R}^d}, \mathscr{R}_M), 0) : X \to X_M$ and $\hat{\mathscr{P}}_M := (\text{diag}(I_{\mathbb{R}^d}, \mathscr{P}_M), 0)^T : X_M \to X$, together with $\hat{\mathscr{R}}_M \hat{\mathscr{P}}_M = I_{X_M} : X_M \to X_M$ and $\hat{\mathscr{P}}_M \hat{\mathscr{R}}_M = \hat{\mathscr{F}}_M : X \to \mathbb{R}^d \times \Pi_M \subset X$. The indexing of a block-vector in X_M will be $-1, 0, 1 \ldots, M$ to take into account for the presence of the vector element of $\mathbb{R}^d$ (index -1) and the discrete functional element (indices $0, \ldots, M$). This convention will be used in Section 3.2.

3 Projection and Collocation

Let M and N be positive integers. The numerical method we propose is based on first projecting the initial data (u, φ) by $\hat{\mathscr{F}}_M$, second collocating the corresponding IVP on the Legendre-Gauss zeros θ_i^+, $i = 1, \ldots, N$, and third projecting again the result by $\hat{\mathscr{F}}_M$. We thus define the approximation $T_{M,N}(r) : X \to X$ of $T(r) : X \to X$ in (3) as

$$T_{M,N}(r) = \hat{\mathscr{F}}_M T_N(r) \hat{\mathscr{F}}_M, \tag{5}$$

with $T_N(r) : X \to X$ given by

$$T_N(r)(u, \varphi) = (p_N^{\pm}(r), (p_N^{\pm})_r), \tag{6}$$

where $p_N^{\pm}$ is split into φ in $[-\tau, 0)$ and $p_N^+ \in \Pi_N^+$ in $[0, r]$ determined by the collocation of (2):

$$\begin{cases} (p_N^+)'(\theta_i^+) = (G p_N^{\pm})(\theta_i^+), & \text{if } i = 1, \ldots, N, \\ p_N^+(0) = u. \end{cases} \tag{7}$$

Notice that the definition of $T_N(r)$ in (6) makes sense only for an initial function φ everywhere defined, condition ensured by its use in (5). In general, instead, and in view of analyzing the convergence of $T_{M,N}(r)$ to $T(r)$ in Section 3.1, we will assume $(u, \varphi) \in X^1$.

3.1 Convergence Analysis

We aim at studying the error $T(r) - T_{M,N}(r)$. From (5) we have

$$T(r) - T_{M,N}(r) = E_1 + E_2 + E_3$$

where

$$E_1 = T(r) - T_N(r)$$

is the error due to the collocation (7) while

$$E_2 = (I_X - \hat{\mathscr{F}}_M)T_N(r) \tag{8}$$

and

$$E_3 = \hat{\mathscr{F}}_M T_N(r)(I_X - \hat{\mathscr{F}}_M). \tag{9}$$

are the errors due to the projection in (5) of the collocated state at time r and of the initial state, respectively.

As for E_1 the following bound holds, where $\mathscr{L}_N^+$ is the interpolation operator based on the Legendre-Gauss zeros in L^+.

Theorem 1. *Let x be the solution of (2) for a given $(u, \varphi) \in X^1$. Then, for sufficiently large N, $T_N(r)$ in (6) is well-defined and there exists a constant c independent of N such that*

$$\|[T(r) - T_N(r)](u, \varphi)\|_X \le c \left\|(I_{L^+} - \mathscr{L}_N^+)Gx^{\pm}\right\|_{L^+}. \tag{10}$$

Proof. From (3) and (6) we have

$$[T(r) - T_N(r)](u, \varphi) = \left(e_N^{\pm}(r), (e_N^{\pm})_r\right), \tag{11}$$

where $e_N^{\pm} = x^{\pm} - p_N^{\pm} \in L^{\pm}$.

The solution x exiting from $(u, \varphi) \in X^1$ satisfies the functional equation in $L^{\pm}$

$$x^{\pm} = u_{\varphi} + VGx^{\pm} \tag{12}$$

as soon as we consider $u_{\varphi} \in L^{\pm}$ as the function

$$u_{\varphi}(t) := \begin{cases} u, & t \in [0, r] \\ \varphi(t), & \text{a.e. in } [-\tau, 0) \end{cases}$$

and $V : L^+ \to L^{\pm}$ as the integral operator

$$(Vx)(t) := \begin{cases} \displaystyle\int_0^t x(s)ds, & t \in [0, r] \\ 0, & t \in [-\tau, 0]. \end{cases} \tag{13}$$

Similarly, we obtain from (7)

$$p_N^{\pm} = u_{\varphi} + V\mathscr{L}_N^+ Gp_N^{\pm}. \tag{14}$$

This follows from the fact that $\mathscr{L}_N^+ (p_N^+)' = (p_N^+)'$ being $p_N^+ \in \Pi_N^+$.

From (12) and (14) we see that

$$e_N^\pm = V \varepsilon_N^+ \tag{15}$$

where ε_N^+ is the unique solution of the functional equation in L^+

$$\varepsilon_N^+ = \mathscr{L}_N^+ GV \varepsilon_N^+ + (I_{L^+} - \mathscr{L}_N^+)Gx^\pm.$$

Lemma 3 in the Appendix provides uniqueness for N sufficiently large and, together with Lemma 1, also the bounds

$$\left\| \varepsilon_N^\pm \right\|_{L^+} \leq 2 \left\| (I_{L^+} - GV)^{-1} \right\|_{L^+} \left\| (I_{L^+} - \mathscr{L}_N^+)Gx^\pm \right\|_{L^+}$$

and

$$\left\| e_N^\pm \right\|_{L^\pm} \leq \|V\|_{L^\pm \leftarrow L^+} \left\| \varepsilon_N^+ \right\|_{L^+}.$$

The thesis now follows from (11) by observing first that

$$x^\pm(r) - p_N^\pm(r) = \int_0^r [G(x^\pm - p_N^\pm)](s)ds$$

is well-defined for $(u, \varphi) \in X^1$ and implies

$$|x^\pm(r) - p_N^\pm(r)| \leq \sqrt{r} \|G\|_{L^+ \leftarrow L^\pm} \left\| e_N^\pm \right\|_{L^\pm}$$

by Hölder's inequality, and second that

$$\left\| (x^\pm - p_N^\pm)_r \right\|_L \leq \left\| e_N^\pm \right\|_{L^\pm}. \tag{16}$$

Theorem 1 implies the two following fundamental corollaries on the convergence of the collocation scheme, respectively pointwise and in norm.

Corollary 1. $T_N(r)$ converges pointwise to $T(r)$ in X^1 w.r.t. $\|\cdot\|_X$, precisely

$$\|[T(r) - T_N(r)](u, \varphi)\|_X = O\left(N^{-\frac{1}{2}}\right)$$

for all $(u, \varphi) \in X^1$.

Proof. If $(u, \varphi) \in X^1$ then the solution x to (2) is in H^2 and $Gx^\pm = (x^+)' \in H^{1,+}$, see [5, Theorem 3], except at most at τ if $r \geq \tau$. The thesis follows from (24) in Theorem 4 of the Appendix as applied to (10), in a piecewise fashion if the mentioned exception occurs.

Corollary 2. $T_N(r)$ converges in norm to $T(r)$ in X^1 w.r.t. $\|\cdot\|_{X^1}$, precisely

$$\|T(r) - T_N(r)\|_{X^1} = O\left(N^{-\frac{1}{2}}\right). \tag{17}$$

Proof. Since

$$\|[T(r) - T_N(r)](u, \varphi)\|_{X^1}^2 = \|x^{\pm}(r) - p_N^{\pm}(r)\|_{\mathbb{R}^d}^2$$
$$+ \|(x^{\pm} - p_N^{\pm})_r\|_L^2 + \|[(x^{\pm})' - (p_N^{\pm})']_r\|_L^2,$$

it is enough to observe that, similarly to (16), we have

$$\|[(x^{\pm})' - (p_N^{\pm})']_r\|_L \le \|\varepsilon_N^{+}\|_{L^+}$$

by virtue of (15). The thesis eventually follows by proceeding as in the proof of Theorem 1 and Corollary 1.

Now we analyze the projection errors (8) and (9), either pointwise and in norm.

Theorem 2. *Let $M \ge N$. Then*

$$\left\|(I_X - \hat{\mathscr{F}}_M)T_N(r)(u, \varphi)\right\|_X = O\left(M^{-1}\right) \tag{18}$$

for all $(u, \varphi) \in X^1$ and $r \ge 0$ while, for $r \ge \tau$,

$$(I_X - \hat{\mathscr{F}}_M)T_N(r) = 0. \tag{19}$$

Proof. As long as $(u, \varphi) \in X^1$ we have

$$(I_X - \hat{\mathscr{F}}_M)T_N(r)(u, \varphi) = (0, (I_L - \mathscr{F}_M)(p_N^{\pm})_r),$$

so if $r \ge \tau$ the thesis is trivial since $p_N^{\pm} = p_N^{+} \in \Pi_N$ and $M \ge N$. If $r < \tau$, instead, we have

$$\left\|(I_X - \hat{\mathscr{F}}_M)T_N(r)(u, \varphi)\right\|_X \le \|(I_L - \mathscr{F}_M)\varphi\|_L + \|(I_{L^+} - \mathscr{F}_M^+)p_N^+\|_{L^+}$$
$$= \|(I_L - \mathscr{F}_M)\varphi\|_L$$

and (18) follows from (23) in Theorem 4 of the Appendix. As for (19), note that it trivially follows from the fact that $T_N(r)$ maps into Π_N^+ and $M \ge N$ holds.

Theorem 3. *Let $M \ge N$. Then*

$$\left\|\hat{\mathscr{F}}_M T_N(r)(I_X - \hat{\mathscr{F}}_M)(u, \varphi)\right\|_X = O\left(M^{-1}\right)$$

for all $(u, \varphi) \in X^1$ and $r \ge 0$ while, for $r \ge \tau$,

$$\left\|\hat{\mathscr{F}}_M T_N(r)(I_X - \hat{\mathscr{F}}_M)\right\|_{X^1} = O\left(M^{-1}\right). \tag{20}$$

Proof. From Bessel's inequality $\|\hat{\mathscr{F}}_M\|_{X^1} = O(1)$ follows and for the second assertion, after replacing $T_N(r)$ with $T(r)$, we can use (23) since $r \ge \tau$.

3.2 The Matrix Form

We aim at finding now a matrix representation $\boldsymbol{T}_{M,N}(r)$ of the approximation $T_{M,N}(r)$. This is useful to implement codes based on the presented algorithm if one is interested in the approximation of the spectrum of $T(r)$ as commented in the Conclusions.

Let us observe that, from (5),

$$T_{M,N}(r) = \mathscr{P}_M U_{M,N}(r) \hat{\mathscr{R}}_M$$

where $U_{M,N}(r) : X_M \to X_M$ given by

$$U_{M,N}(r) = \hat{\mathscr{R}}_M T_N(r) \mathscr{P}_M$$

is the finite dimensional operator corresponding to $T_{M,N}(r)$. For a positive integer K, we identify $(\mathbb{R}^d)^K$ with $\mathbb{R}^{dK}$. Then $\boldsymbol{T}_{M,N}(r)$, and the other matrices denoted in boldface, are nothing else than the canonical representation according to the Legendre polynomials in L and L^+ as described in the following and according to Section 2.

We first construct matrices $\boldsymbol{V}_{M,N} : X_M \to X_N^+$ and $\boldsymbol{V}_N^+ : X_N^+ \to X_N^+$ such that

$$\boldsymbol{V}_N^+ (p_{M,N}^+(r), \mathscr{R}_N^+ p_{M,N}^+) = \boldsymbol{V}_{M,N}(u, \mathscr{R}_M \varphi) \qquad (21)$$

where $p_{M,N}^+ \in \Pi_N^+$ is the part of $p_{M,N}^\pm$ determined by the collocation equations

$$\begin{cases} (p_{M,N}^+)'(\theta_i^+) = (G p_{M,N}^\pm)(\theta_i^+), & \text{if } i = 1,\dots,N, \\ p_{M,N}^+(0) = u \end{cases}$$

where $p_{M,N}^+$ is $\mathscr{F}_M \varphi$ in $[-\tau, 0)$. It is not difficult, yet technical, to check that the above matrices have entries, respectively,

$$[\boldsymbol{V}_N^+]_{ij} := \begin{cases} -\phi_j^+(r), & \text{if } i = -1 \\ \phi_j^+(0), & \text{if } i = 0 \\ \begin{aligned} &((\phi_j^+)' - A\phi_j^+)(\theta_i^+) - \sum_{\ell=1}^{k_i}\left(B_\ell \phi_j^+(\theta_i^+ - \tau_\ell)\right. \\ &\left. \quad + \int_{-\tau_\ell}^{-\tau_{\ell-1}} C_\ell(\theta)\phi_j^+(\theta_i^+ + \theta)d\theta \right) \\ &\quad + \int_0^{-\tau_{k_i}} C_{k_i+1}(\theta)\phi_j^+(\theta_i^+ + \theta)d\theta, \quad \text{if } i = 1,\dots,N \end{aligned} \end{cases}$$

for all $j = 0,\dots,N$, plus the first column ($j = -1$) as $(1,0,\dots,0)^T \in \mathbb{R}^{d(N+2)}$, and

$$[\boldsymbol{V}_{M,N}]_{ij} := B_{k_i+1}\phi_j(\theta_i^+ - \tau_{k_i+1}) + \int_{-\tau_{k_i+1}}^{0} C_{k_i+1}(\theta)\phi_j(\theta_i^+ + \theta)d\theta$$

$$+ \sum_{\ell=k_i+2}^{k} \left(B_\ell\phi_j(\theta_i^+ - \tau_\ell) + \int_{-\tau_\ell}^{-\tau_{\ell-1}} C_\ell(\theta)\phi_j(\theta_i^+ + \theta)d\theta \right)$$

for all $i = 0, \ldots, N$ and $j = 0, \ldots, N$, plus $[\boldsymbol{V}_{M,N}]_{0,-1} = 1$ and zero elsewhere. Above we set

$$k_i := \max_{\ell=1,\ldots,k} \{\theta_i^+ - \tau_\ell \geq 0\}, \; i = 0, \ldots, N.$$

Second, and independently of the model coefficients in (1), we construct matrices $\boldsymbol{W}_M, \boldsymbol{W}_M^- : X_M \to X_M$ and $\boldsymbol{W}_{M,N}^+ : X_N^+ \to X_M$ such that

$$\boldsymbol{W}_M(p_{M,N}^\pm(r), \mathscr{R}_M(p_{M,N}^\pm)r) = \boldsymbol{W}_{M,N}^+(p_{M,N}^+(r), \mathscr{R}_N^+ p_{M,N}^+) + \boldsymbol{W}_M^-(u, \mathscr{R}_M\varphi) \quad (22)$$

by restriction of $p_{M,N}^\pm$ to $[r - \tau, r]$ when $r \geq \tau$, and possibly prolongation by $\mathscr{P}_M\varphi_M$ when $r < \tau$. In particular, it is sufficient to define the above matrices with entries, respectively,

$$[\boldsymbol{W}_M]_{ij} := \begin{cases} \phi_j(0), & \text{if } i = 0 \\ \phi_j(\theta_i), & \text{if } i = 1, \ldots, M \end{cases}$$

for all $j = 0, \ldots, N$, plus $[\boldsymbol{W}_M]_{-1,-1} = 1$ and 0 elsewhere,

$$[\boldsymbol{W}_{M,N}^+]_{ij} := \begin{cases} \phi_j^+(r), & \text{if } i = 0 \\ \phi_j^+(r + \theta_i), & \text{if } i = 1, \ldots, M_r \\ 0, & \text{if } i = M_r + 1, \ldots, M \end{cases}$$

for all $j = 0, \ldots, N$, plus $[\boldsymbol{W}_{M,N}^+]_{-1,-1} = 1$ and 0 elsewhere, and

$$[\boldsymbol{W}_M^-]_{ij} := \begin{cases} 0, & \text{if } i = 0, \ldots, M_r \\ \phi_j(r + \theta_i), & \text{if } i = M_r + 1, \ldots, M \end{cases}$$

for all $j = 0, \ldots, N$ and 0 elsewhere. Above we set

$$M_r := \max_{i=0,\ldots,M} \{r + \theta_i \geq 0\},$$

with the convention that $\boldsymbol{W}_{M,N}^+$ is full and $\boldsymbol{W}_M^-$ is empty when $M_r = M$, i.e. for $r \geq \tau$. Eventually, it follows from (21) and (22) that

$$\boldsymbol{T}_{M,N}(r)(u, \mathscr{R}_M\varphi) = (p_{M,N}^\pm(r), \mathscr{R}_M(p_{M,N}^\pm)r)$$

is the sought matrix approximation of (3) with

$$\boldsymbol{T}_{M,N}(r) = (\boldsymbol{W}_M)^{-1}[\boldsymbol{W}_{M,N}^+(\boldsymbol{V}_N^+)^{-1}\boldsymbol{V}_{M,N} + \boldsymbol{W}_M^-].$$

Standard approximation arguments ensure that W_M and V_N^+ are invertible for sufficiently large M and N, respectively.

4 Conclusions

In this paper we have studied the discretization of the solution operators of LTI-TDS in L^2. We have obtained both pointwise and norm convergence results, which hold similarly for other families of orthogonal polynomials and relevant zeros w.r.t. their weighted norms. Let us observe first that the pointwise convergence order can be raised as soon as more regular initial data are considered, i.e. $(u, \phi) \in \mathbb{R}^d \times H^m$. Second, by (17), (19) and (20) we have the norm convergence of $T_{M,N}(r)$ to $T(r)$ for $r \geq \tau$. This fact is fundamental when one decides to approximate the spectrum of $T(r)$ with that of $T_{M,N}(r)$ for stability analysis of equilibria of nonlinear TDS. For this reason we have given also the matrix representation of $T_{M,N}(r)$.

Appendix

Besides the notation introduced in Section 2, we recall that G refers to (2), V is given in (13) and $\mathscr{L}_N^+$ before Theorem 1. It is easy to see that $G \in \mathscr{B}(L^\pm, L^+)$ and $V \in \mathscr{B}(L^+, L^\pm)$ is a contraction. We first recall basic approximation results from [10], namely (9.4.6) and (9.4.24), and then prove some technical lemmas.

Theorem 4. *Let $\varphi \in H^m$, $m \geq 1$. There exists a constant $c_{\mathscr{F}}$ independent of M such that*

$$\|\varphi - \mathscr{F}_M \varphi\|_L \leq c_{\mathscr{F}} M^{-m} \|\varphi\|_{H^m} \tag{23}$$

where $\mathscr{F}_M$ is the Fourier projection operator w.r.t. the Legendre polynomials and a constant $c_{\mathscr{L}}$ independent of N such that

$$\|\varphi - \mathscr{L}_N \varphi\|_L \leq c_{\mathscr{L}} N^{\frac{1}{2}} N^{-m} \|\varphi\|_{H^m} \tag{24}$$

where $\mathscr{L}_N$ is the interpolation operator w.r.t. the Legendre-Gauss zeros.

Lemma 1. $(I_{L^\pm} - VG)^{-1} \in \mathscr{B}(L^\pm)$ *and* $(I_{L^+} - GV)^{-1} \in \mathscr{B}(L^+)$.

Proof. The first assertion corresponds to proving that for any given $f \in L^\pm$ there exists a unique $y \in L^\pm$ solution of $(I_{L^\pm} - VG)y = f$. This follows from standard results on Volterra integral equations. The second assertion follows similarly once a change of integration order is applied w.r.t. V and the distributed terms in (1).

Lemma 2. $\left\|(I_{L^+} - \mathscr{L}_N^+)GV\right\|_{L^+} \to 0$ *as $N \to \infty$.*

Proof. For any $y \in L^+$, (13) implies $Vy \in H^{1,+}$ and, therefore, Vy and GVy are (absolutely) continuous. Now apply the Theorem of Erdös-Turàn.

Lemma 3. *For sufficiently large N, $(I_{L^+} - \mathscr{L}_N^+ GV)^{-1} \in \mathscr{B}(L^+)$ and*

$$\left\|(I_{L^+} - \mathscr{L}_N^+ GV)^{-1}\right\|_{L^+} \leq 2\left\|(I_{L^+} - GV)^{-1}\right\|_{L^+}.$$

Proof. The thesis follows by applying the Banach's Perturbation Lemma, Lemma 1 and Lemma 2 since $I_{L^+} - \mathscr{L}_N^+ GV = (I_{L^+} - GV) + (I_{L^+} - \mathscr{L}_N^+)GV$.

References

1. Batkai, A., Piazzera, S.: Semigroups for delay equations. Number 10 in Research Notes in Mathematics. A K Peters Ltd., Canada (2005)
2. Bellen, A., Zennaro, M.: Numerical methods for delay differential equations. Numerical Mathemathics and Scientifing Computing Series. Oxford University Press (2003)
3. Bensoussan, A., Da Prato, G., Delfour, M.C., Mitter, S.K.: Representation and control of infinite dimensional systems, vol. I, II. Birkhäuser (1992, 1993)
4. Borisovič, J.G., Turbabin, A.S.: On the Cauchy problem for linear nonhomogeneous differential equations with retarded arguments. Dokl. Akad. Nauk SSSR 185(4), 741–744 (1969); English Transl. Soviet Math. Dokl. 10(2), 401-405 (1969)
5. Breda, D.: Nonautonomous delay differential equations in Hilbert spaces and Lyapunov exponents. Diff. Int. Equations 23(9-10), 935–956 (2010)
6. Breda, D., Maset, S., Vermiglio, R.: Pseudospectral differencing methods for characteristic roots of delay differential equations. SIAM J. Sci. Comput. 27(2), 482–495 (2005)
7. Breda, D., Maset, S., Vermiglio, R.: On discretizing the semigroup of solution operators for linear time invariant - time delay systems. In: IFAC 2010 Proceedings of Time Delay Systems (2010) (in Press)
8. Breda, D., Maset, S., Vermiglio, R.: Approximation of eigenvalues of evolution operators for linear retarded functional differential equations (submitted, 2011)
9. Butcher, E.A., Ma, H.T., Bueler, E., Averina, V., Szabo, Z.: Stability of linear time-periodic delay-differential equations via Chebyshev polynomials. Int. J. Numer. Meth. Engng. 59, 895–922 (2004)
10. Canuto, C., Hussaini, M.Y., Quarteroni, A., Zang, T.: Spectral Methods. Evolution to Complex Geometries and Applications to Fluid Dynamics. Scientific Computation Series. Springer, Berlin (2007)
11. Delfour, M.C.: State theory of linear hereditary differential systems. J. Differ. Equations 60, 8–35 (1977)
12. Delfour, M.C., Mitter, S.K.: Hereditary differential systems with constant delays. I. General case. J. Differ. Equations 12, 213–235 (1972)
13. Diekmann, O., van Gils, S.A., Verduyn Lunel, S.M., Walther, H.O.: Delay Equations - Functional, Complex and Nonlinear Analysis. Number 110 in AMS series. Springer, New York (1995)
14. Engel, K., Nagel, R.: One-Parameter Semigroups for Linear Evolution Equations. Number 194 in Graduate texts in mathematics. Springer, New York (1999)
15. Engelborghs, K., Luzyanina, T., Roose, D.: Numerical bifurcation analysis of delay differential equations using DDE-BIFTOOL. ACM T. Math. Software 28(1), 1–21 (2002)
16. Engelborghs, K., Roose, D.: On stability of LMS methods and characteristic roots of delay differential equations. SIAM J. Numer. Anal. 40(2), 629–650 (2002)
17. Farmer, D.: Chaotic attractors of an infinite-dimensional dynamical system. Physica D 4, 605–617 (1982)
18. Hadd, S., Rhandi, A., Schnaubelt, R.: Feedback theory for time-varying regular linear systems with input and state delays. IMA J. Math. Control Inform. 25(1), 85–110 (2008)
19. Hale, J.K.: Introduction to functional differential equations, 1st edn. Number 99 in AMS series. Springer, New York (1977)

20. Insperger, T., Stépán, G.: Semi-discretization method for delayed systems. Int. J. Numer. Meth. Engng 55, 503–518 (2002)
21. Ito, K., Kappel, F.: A uniformly differentiable approximation scheme for delay systems using splines. Appl. Math. Opt. 23, 217–262 (1991)
22. Kappel, F.: Semigroups and delay equations. Number 152 (Trieste, 1984) in Pitman Res. Notes Math. Ser. Longman Sci. Tech., Harlow (1986)
23. Peichl, G.H.: A kind of "history space" for retarded functional differential equations and representation of solutions. Funkc. Ekvacioj-SER I 25, 245–256 (1982)
24. Verheyden, K., Luzyanina, T., Roose, D.: Efficient computation of characteristic roots of delay differential equations using LMS methods. J. Comput. Appl. Math. 214(1), 209–226 (2008)
25. Vinter, R.B.: On the evolution of the state of linear differential delay equations in M^2: properties of the generator. J. Inst. Maths. Applics. 21, 13–23 (1978)
26. Vyhlídal, T., Zítek, P.: Mapping based algorithm for large-scale computation of quasi-polynomial zeros. IEEE T. Automat. Cont. 54(1), 171–177 (2009)

The Infinite Arnoldi Method and an Application to Time-Delay Systems with Distributed Delays

Elias Jarlebring, Wim Michiels, and Karl Meerbergen

Abstract. The Arnoldi method, which is a well-established numerical method for standard and generalized eigenvalue problems, can conceptually be applied to standard but infinite-dimensional eigenvalue problems associated with an operator. In this work, we show how such a construction can be used to compute the eigenvalues of a time-delay system with distributed delays, here given by $\dot{x}(t) = A_0 x(t) + A_1 x(t - \tau) + \int_{-\tau}^{0} F(s)x(t + s)\,ds$, where $A_0, A_1, F(s) \in \mathbb{C}^{n \times n}$. The adaption is based on formulating a more general problem as an eigenvalue problem associated with an operator and showing that the action of this operator has a finite-dimensional representation when applied to polynomials. This allows us to implement the infinite-dimensional algorithm using only (finite-dimensional) operations with matrices and vectors of size n. We show, in particular, that for the case of distributed delays, the action can be computed from the Fourier cosine transform of a function associated with F, which in many cases can be formed explicitly or computed efficiently.

1 Introduction

Consider a linear time-invariant time-delay system with a distributed delay term,

$$\dot{x}(t) = A_0 x(t) + A_1 x(t - \tau) + \int_{-\tau}^{0} F(s)x(t + s)\,ds. \tag{1}$$

Time-delay systems without distributed terms ($F(s) = 0$) have been extensively studied in the literature. See, e.g., the survey papers [7, 19]. There are however numerous applications where the delay is not localized to one value, e.g., when the derivative of the current state depends on an expectation value of a continuum of

Elias Jarlebring · Wim Michiels · Karl Meerbergen
KU Leuven, Celestijnenlaan 200 A, 3001 Heverlee, Belgium
e-mail: {elias.jarlebring,wim.michiels}@cs.kuleuven.be,
        karl.meerbergen@cs.kuleuven.be

R. Sipahi et al. (Eds.): Time Delay Sys.: Methods, Appli. and New Trends, LNCIS 423, pp. 229–239.
springerlink.com   © Springer-Verlag Berlin Heidelberg 2012

previous states. In such situations, it is natural to consider the generalized form (1). See [13] and [4] and references therein for literature on time-delay systems with distributed delays.

The eigenvalues of (1) are given by the solutions to the nonlinear eigenvalue problem,

$$M(\lambda)v = 0, \tag{2}$$

where $\lambda \in \mathbb{C}$ and $v \in \mathbb{C} \setminus \{0\}$ and

$$M(\lambda) := -\lambda I + A_0 + A_1 e^{-\lambda \tau} + \int_{-\tau}^{0} F(s) e^{\lambda s} ds. \tag{3}$$

Our approach is based on first (in Section 2) reformulating the problem (2) as a linear infinite-dimensional eigenvalue problem. The associated operator is an integration operator and we show that for this operator we can carry out the method, which for standard eigenvalue problems is known as the *Arnoldi method*.

We show that the action of the operator (and the corresponding Arnoldi algorithm) can be carried out with standard linear algebra operations, including operations with the matrices

$$R_j := \int_{-\tau}^{0} F(s) \hat{T}_j(s), j = 0, \ldots, N,$$

where $\hat{T}_j$ are scaled and shifted Chebyshev polynomials. The coefficients R_j can be computed before the iteration starts and can often be computed accurately and cheaply.

The attractive properties of the method which are a consequence of the equivalence with the Arnoldi method are illustrated with examples in Section 5.

Although there are numerous methods for the eigenvalues of (1) with $F(s) = 0$, e.g., [6, 22], only a few approachs have been generalized to the distributed case, e.g., [4, 13]. There exist also many numerical methods for the nonlininear eigenvalue problem (2), presented in different generality settings. See reviews in [15, 20] and problem collection [2]. There exists methods based on subspace projection [1, 14, 21], and iterative locally convergent methods [12, 17].

2   Operator Formulation

Suppose $\lambda = 0$ is not a solution and consider the transformation of the nonlinear eigenvalue problem (2) by introducing

$$B(\lambda) := \frac{1}{\lambda} M(0)^{-1} (M(0) - M(\lambda)). \tag{4}$$

This implies that

$$\lambda B(\lambda) x = x. \tag{5}$$

We will assume that B is an entire function, where in the case of (4) we define B such that it is analytic also in $\lambda = 0$, by also assuming that $M(0)$ is non-singular. With this definition of B we can introduce an operator $\mathscr{B}$, which reciprocal eigenvalues turn out to be the solutions to the nonlinear eigenvalue problem (5) or equivalently (2).

Definition 1 (The operator $\mathscr{B}$). Let $\mathscr{B}$ denote the operator defined by the domain $\mathscr{D}(\mathscr{B}) := \{ \varphi \in C_\infty(\mathbb{C}, \mathbb{C}^n) : \sum_{i=0}^{\infty} B_i \varphi^{(i)}(0) \text{ is finite} \}$ and the action

$$(\mathscr{B}\varphi)(\theta) = C(\varphi) + \int_0^\theta \varphi(\theta)\, d\theta, \qquad (6)$$

where

$$C(\varphi) := \sum_{i=0}^{\infty} B_i \varphi^{(i)}(0) = \left(B(\frac{d}{d\theta})\varphi \right)_{\theta=0}. \qquad (7)$$

Theorem 1 (Operator equivalence). *Let $x \in \mathbb{C}^n \backslash \{0\}$, $\lambda \in \mathbb{C}$ and denote $\varphi(\theta) := x e^{\lambda \theta}$. Then the following statements are equivalent.*

i) The pair (λ, x) is a solution to the nonlinear eigenvalue problem (5).
ii) The pair (λ, φ) is a solution to the infinite dimensional eigenvalue problem

$$\lambda \mathscr{B} \varphi = \varphi. \qquad (8)$$

Moreover, all eigenfunctions of $\mathscr{B}$ depend exponentially on θ, i.e., if $\lambda \mathscr{B} \psi = \psi$ then $\psi(\theta) = x e^{\lambda \theta}$.

Proof. See [11, Theorem 1].

3 The Infinite Arnoldi Method

In the previous section we saw that the solutions of the nonlinear eigenvalue problem (2) could be characterized as solutions to a linear, but infinite-dimensional eigenvalue problem corresponding to the operator $\mathscr{B}$. We will now carry out the Arnoldi method for the operator $\mathscr{B}$, which by the equivalence above (Theorem 1) will yield solutions of (2).

The Arnoldi method applied to $\mathscr{B}$, can be interpreted as first constructing an orthogonal basis of the space

$$K_k(\mathscr{B}, \varphi) := \operatorname{span}(\varphi, \mathscr{B}\varphi, \dots, \mathscr{B}^{k-1}\varphi),$$

which is known as a *Krylov subspace*. The orthogonal basis is used to project the operator $\mathscr{B}$ to this space, from which the eigenvalue approximations can be computed.

If φ is a polynomial, then $\mathscr{B}\varphi$ is also a polynomial. In the following theorem we see that the action of $\mathscr{B}$ on a polynomial can be carried out by operations with matrices and vectors. In this work we will focus on representing the polynomials as linear combinations of scaled and shifted Chebyshev polynomials for the interval $[-\tau, 0]$,

$$\hat{T}_j(\theta) := \cos\left(j \arccos\left(\frac{2}{\tau}\theta + 1\right)\right).$$

Theorem 2 (Chebyshev coefficient mapping). *Let φ denote an arbitrary vector of polynomials of degree N and $(x_0, \ldots, x_{N-1}) =: X$ the corresponding coefficients in a Chebyshev basis, for Chebyshev polynomials scaled to the interval $I = [a, b]$, i.e.,*

$$\varphi(\theta) =: \sum_{i=0}^{N-1} \hat{T}_i(\theta) x_i.$$

Moreover, let $y_0, \ldots, y_N$ denote the coefficients of $\mathscr{B}\varphi$, i.e.,

$$(\mathscr{B}\varphi)(\theta) =: \sum_{i=0}^{N} \hat{T}_i(\theta) y_i.$$

Then,

$$(y_1, \ldots, y_N) = X L_{N,N}, \tag{9}$$

where

$$L_{N,N}^T = \frac{b-a}{4} \begin{pmatrix} 2 & 0 & -1 & & & & \\ \frac{1}{2} & 0 & -\frac{1}{2} & & & \\ & \frac{1}{3} & 0 & \ddots & & \\ & & \frac{1}{4} & \ddots & -\frac{1}{N-2} & \\ & & & \ddots & 0 & \\ & & & & \frac{1}{N} \end{pmatrix} \tag{10}$$

and

$$y_0 = \left(\sum_{i=0}^{N-1} B\left(\frac{d}{d\theta}\right)\hat{T}_i(\theta) x_i \right)_{\theta=0} - \sum_{i=1}^{N} \hat{T}_i(0) y_i. \tag{11}$$

Proof. See [11, Theorem 2]. $\quad\square$

Note that all components of the action can be carried out in finite arithmetic, if a finite-dimensional expression for y_0 in (11) can be found. This can indeed be done in many cases. An explicit formula for the case of distributed delays will be derived in Section 4. See [11, Remark 3] for other approaches to compute y_0.

We now use this action to construct an Arnoldi algorithm for $\mathscr{B}$. In this construction, we have (as for the standard Arnoldi method) that the projected eigenvalue problem can be obtained as a by-product of the orthogonalization. We arrive at the algorithm summarized in Algorithm 1, which we will refer to as the *infinite Arnoldi method*.

We use the following notation. The upper block of the rectangular Hessenberg matrix $\underline{H}_k \in \mathbb{C}^{(k+1) \times k}$ is denoted $H_k \in \mathbb{C}^{k \times k}$ and the (i, j) element of $\underline{H}_k$ is denoted $h_{i,j}$.

Algorithm 1: The infinite Arnoldi method

1. Let $V_1 = x_0/\|x_0\|_2, k = 1, \underline{H}_0 =$ empty matrix
2. For $k = 1, 2, \ldots$ until converged
3.   Let $\text{vec}(X) = v_k$
4.   Compute $y_1, \ldots, y_{k+1}$ according to (9) with sparse L_k
5.   Compute y_0 according to (11)
6.   Expand V_k with one block row (zeros)
7.   Let $w_k := \text{vec}(y_0, \ldots, y_{k+1})$, compute $h_k = V_k^* w_k$ and then $\hat{w}_k = w_k - V_k h_k$
8.   Compute $\beta_k = \|\hat{w}_k\|_2$ and let $v_{k+1} = \hat{w}_k/\beta_k$
9.   Let $\underline{H}_k = \begin{bmatrix} H_{k-1} & h_k \\ 0 & \beta_k \end{bmatrix} \in \mathbb{C}^{(k+1) \times k}$
10. Expand V_k into $V_{k+1} = [V_k, v_{k+1}]$
11. Compute the eigenvalues $\{\mu_i\}_{i=1}^k$ of the Hessenberg matrix H_k
12. Return approximations $\{1/\mu_i\}_{i=1}^k$

Note that the scalar product, used for the orthogonalization of functions, corresponds to the scalar product of the coefficient vectors in the Chebyshev basis. In [11, Section 6] this choice is motivated, based on a connection with a spectral discretization.

4 Adaption for Time-Delay Systems with Distributed Delays

The infinite Arnoldi method is a general method and in order to apply it, for each given problem, we have to establish

- the interval on to which the Chebyshev polynomials are scaled, and
- a formula for y_0.

We now wish to adapt the infinite Arnoldi method to (1) and need to consider the interval and y_0. In general, for (linear) functional differential equations the natural interval is $I = [-\tau, 0]$. This follows from the connection of the algorithm as a discretized problem [11, Section 6.2]. Formulas for y_0 are derived next.

4.1 Computing y_0 for Distributed Delays

In order to implement Algorithm 1 on a computer, we need to compute y_0, defined by (11), for given values of $x_0, \ldots, x_k$ and possibly $y_1, \ldots, y_{k+1}$.

For notational convenience, we will use the function q defined by

$$q(\lambda, s) := \frac{1 - e^{\lambda s}}{\lambda}. \qquad (12)$$

The transformed nonlinear eigenvalue problem (5) can now be expressed as

$$B(\lambda) = \frac{1}{\lambda}M(0)^{-1}(M(0) - M(\lambda)) =$$

$$(A_0 + A_1 + R_0)^{-1}(I + A_1 q(\lambda, -\tau) + R(\lambda)), \quad (13)$$

where

$$R(\lambda) := \int_{-\tau}^{0} F(s)q(\lambda, s)\,ds \text{ and } R_0 = \int_{-\tau}^{0} F(s)\,ds.$$

Note that $A_0 + A_1 + R_0$ is invertible if and only if $\lambda = 0$ is not a solution to the nonlinear eigenvalue problem (2). This condition can easily be checked by hand before starting the algorithm.

The last term in (11) is already given in a form which is numerically tractable and we now need to study the first term,

$$\left(\sum_{i=0}^{N-1} B(\frac{d}{d\theta})\hat{T}_i(\theta)x_i\right)_{\theta=0}. \quad (14)$$

The derivation is based on the fact that the function q corresponds to integration in the sense that

$$\left(q(\frac{d}{d\theta}, s)\varphi\right)_{\theta=0} = \int_{s}^{0} \varphi(\theta)\,d\theta. \quad (15)$$

This follows from inserting the Taylor expansion of φ on the right-hand side and the Taylor expansion of q on the left-hand side. The Taylor expansion of q follows from insertion of the Taylor expansion of $e^{\lambda s}$ in (12). The derivation is also available in [11, Equation (11)].

Now note that in Theorem 2, the function $\psi(\theta) = \sum_{i=1}^{N} \hat{T}_i y_i$ is a primitive function of $\varphi = \sum_{i=0}^{N-1} \hat{T}_i x_i$. By combining this observation with (15), the term involving R in (14) can be reduced to

$$\sum_{i=0}^{N-1} (R(\frac{d}{d\theta})\hat{T}_i x_i)_{\theta=0} = \int_{-\tau}^{0} R(s) \sum_{i=1}^{N} (\hat{T}_i(0) - \hat{T}_i(s))y_i\,ds =$$

$$\sum_{i=1}^{N} \left(R_0 \hat{T}_i(0) - \int_{-\tau}^{0} F(s)\hat{T}_i(s)\,ds\right) y_i. \quad (16)$$

We will now for notational convenience define,

$$R_i := \int_{-\tau}^{0} F(s)\hat{T}_i(s)\,ds \text{ for } i = 0.\dots \quad (17)$$

Later, (in Section 4.2) we will see how R_i can be computed in practice. From (16) it follows that the formula for y_0 is equivalently,

$$y_0 = (A_0 + A_1 + R_0)^{-1} \left(\sum_{i=0}^{N-1} x_i + A_1 \sum_{i=1}^{N} (1 - \hat{T}_i(-\tau)) y_i + \sum_{i=1}^{N} (R_0 - R_i) y_i \right.$$

$$\left. - \sum_{i=1}^{N} y_i, \quad (18) \right.$$

where we used that $\hat{T}_i(0) = 1$. By factorizing the last term into the expression involving $(A_0 + A_1 + R_0)^{-1}$, we have established a compact formula for y_0. We summarize it as follows.

Theorem 3. *Consider the nonlinear eigenvalue problem* (2) *corresponding to the time-delay system with distributed delays* (1). *Let B be defined by* (4). *Let x_i and y_i be as in Theorem 2. Then, $y_0 \in \mathbb{C}^n$, defined in* (11), *is given by*

$$y_0 = (A_0 + A_1 + R_0)^{-1} \left(\sum_{i=0}^{N-1} x_i - A_0 \sum_{i=1}^{N} y_i - A_1 \sum_{i=1}^{N} \hat{T}_i(-\tau) y_i - \sum_{i=1}^{N} R_i y_i \right), \quad (19)$$

where R_i is defined by (17).

4.2 Connection with the Fourier Cosine Transform

The formula (19) involves R_i, which is given as the integral (17). In many problems, R has such a structure that it can be written as a sum of a low number of constant matrices, times scalar nonlinearities. In those situations the computation reduces to numerically computing integrals of scalar functions. This can usually be done cheaply and accurately, in comparison to the computation time of Algorithm 1, if n is sufficiently large.

In some situations the integral can be computed analytically. For instance, if $F(s) = C$ is constant, then

$$R_k = C \frac{\tau}{2} \frac{(-1)^k - 1}{k}$$

More generally, if $F(s) = Cf(s)$, where

$$f(s) = \begin{cases} 1 & s \in [a, b] \\ 0 & \text{otherwise,} \end{cases}$$

then from the properties of Chebyshev polynomials we have,

$$R_0 = C(b - a) \tag{20a}$$

$$R_1 = C \left(\frac{1}{\tau}(b^2 - a^2) + b - a \right) \tag{20b}$$

$$R_k = \frac{\tau}{4} C \left(\frac{\hat{T}_{k+1}(b) - \hat{T}_{k+1}(a)}{k+1} - \frac{\hat{T}_{k-1}(b) - \hat{T}_{k-1}(a)}{k-1} \right), k > 1. \tag{20c}$$

We now provide an alternative way to compute R_i, by deriving a relation with the Fourier cosine transform. Let $\cos(\theta) = \frac{2}{\tau}s + 1$. Then

$$R_i = \int_{-\tau}^{0} F(s) \cos\left(i \arccos\left(\frac{2}{\tau}s + 1\right)\right) ds =$$
$$\frac{\tau}{2} \int_{0}^{\pi} \left[\sin(\theta)F(\frac{\tau}{2}(\cos(\theta) - 1))\right]\cos(i\theta)\,d\theta. \quad (21)$$

Note that (17) is essentially the Fourier cosine transform of $\sin(\theta)F(\frac{\tau}{2}(\cos(\theta) - 1))$. For the computation of the right-hand side of (21), numerical integration methods can be used. For small values of i, this integration is fairly cheap with standard integration software provided F is not oscillating heavily in the integration interval. For large values of i, the integrand is definitely highly oscillating and dedicated integration techniques are highly recommended in order to keep the computational cost low. For further literature on oscillatary integrals see [8, 9, 18] and the references therein.

5 Examples

5.1 Example 1: A Rectangular Function

We consider the example in [3, 5]

$$\dot{x}(t) = \begin{pmatrix} -3 & 1 \\ -24.646 & -35.430 \end{pmatrix} x(t) + \begin{pmatrix} 1 & 0 \\ 2.35553 & 2.00365 \end{pmatrix} x(t-1) +$$
$$\begin{pmatrix} 2 & 2.5 \\ 0 & -0.5 \end{pmatrix} \int_{-0.3}^{-0.1} x(t+s)\,ds - \int_{-1}^{-0.5} x(t+s)\,ds. \quad (22)$$

In order to carry out the algorithm, we compute the coefficients R_i analytically. The coefficients are given by

$$R_0 = 0.2 \begin{pmatrix} 2 & 2.5 \\ 0 & -0.5 \end{pmatrix} - 0.5I$$

and

$$R_i = \beta_{0,i} \begin{pmatrix} 2 & 2.5 \\ 0 & -0.5 \end{pmatrix} - \beta_{1,i}I, \quad i = 1, \ldots, 100$$

where we use (20) to compute $\beta_{0,i}, \beta_{1,i}, i = 0, \ldots, 100$,

$$(\beta_{0,1}, \beta_{0,2}, \ldots, \beta_{0,100}) \approx (0.1200, -0.0507, -0.1680, -0.1434, \ldots),$$
$$(\beta_{1,1}, \beta_{1,2}, \ldots, \beta_{1,100}) \approx (-0.2500, -0.1667, 0.2500, -0.0333\ldots).$$

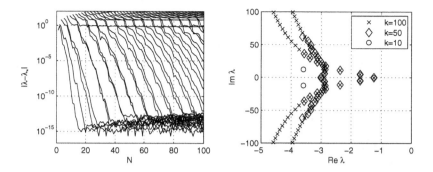

Fig. 1 Convergence and computed solution of the example given in Section 5.1

In Figure 1, we observe convergence similar to the standard Arnoldi method, i.e., simultaneous convergence to several eigenvalues. The eigenvalues closest to the origin are found first. After $k = 100$ iterations, 42 eigenvalues are found to an accuracy of 10^{-10}.

5.2 Example 2: A Gaussian Distribution

Now consider the following system where the distributed term is a Gaussian distribution,

$$\dot{x}(t) = \frac{1}{10} \begin{pmatrix} 25 & 28 & -5 \\ 18 & 3 & 3 \\ -23 & -14 & 35 \end{pmatrix} x(t) + \frac{1}{10} \begin{pmatrix} 17 & 7 & -3 \\ -24 & -21 & -2 \\ 20 & 7 & 4 \end{pmatrix} x(t-\tau) +$$

$$\int_{-\tau}^{0} \begin{pmatrix} 14 & -13 & 4 \\ 14 & 7 & 10 \\ 6 & 16 & 17 \end{pmatrix} \frac{e^{(s+\frac{1}{2})^2} - e^{\frac{1}{4}}}{10} x(t+s)\, ds. \quad (23)$$

with $\tau = 1$. Unlike Example 1, there is no simple analytic expression for R_i. The coefficients are

$$R_i = \frac{1}{10} \begin{pmatrix} 14 & -13 & 4 \\ 14 & 7 & 10 \\ 6 & 16 & 17 \end{pmatrix} \beta_i,$$

where we compute β_i, $i = 0, \dots, 100$ by numerical integration,

$$(\beta_0, \beta_1, \dots, \beta_{100}) \approx (0, 0.1142, 0, -0.0138, 0, -0.0024, \dots)$$

Similar to the previous example, we observe (in Figure 2) the convergence characteristic typical for the Arnoldi method. After 100 iterations, we have found 44 eigenvalues to an accuracy 10^{-10}. This success justifies that numerical integration can be used to compute the coefficients R_i.

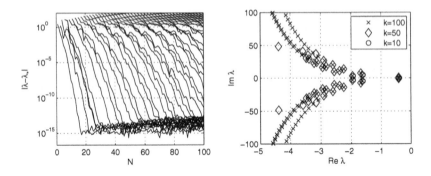

Fig. 2 Convergence and computed solution of the example given in Section 5.2

6   Conclusions

The infinite Arnoldi method is a general method for nonlinear eigenvalue problems which needs to be adapted to the problem at hand. In this work, we considered (in particular) systems with distributed delays (1). The adaption consisted of deriving a computational formula for y_0 in Theorem 3. When comparing to other problems in [10, 11], it is clear that the distributed terms appear as additional terms in the formula for y_0. From this observation it is straightforward to generalize the formula by combining results for, e.g., neutral systems with distributed delays.

We also wish to point out that we have here, for clarity, focused on the classical variant of the Arnoldi method. When considering large-scale problems, it is important to consider implementation and other technical aspects of the Arnoldi method. Issues such as restarting and deflation are currently being investigated by the authors. See also the discussion in the conclusions in [10, 11]. The presented algorithm can also be used for model order reduction, similar to the time-delays without distributed delays [16].

Acknowledgements. This work represents results of the Belgian Programme on Interuniversity Poles of Attraction, initiated by the Belgian State, Prime Minister's Office for Science, Technology and Culture, the Optimization in Engineering Centre OPTEC of the K.U.Leuven, the projects STRT1-09/33, OT/10/038 of the K.U.Leuven Research Foundation and G0712.11 of the Research Council Flanders (FWO). We thank Prof. Daan Huybrechs and Andreas Asheim of the department of computer science, KU Leuven, for clarifying aspects related to oscillatory integrals.

References

1. Bai, Z., Su, Y.: SOAR: A second-order Arnoldi method for the solution of the quadratic eigenvalue problem. SIAM J. Matrix Anal. Appl. 26(3), 640–659 (2005)
2. Betcke, T., Higham, N.J., Mehrmann, V., Schröder, C., Tisseur, F.: NLEVP: A collection of nonlinear eigenvalue problems. Technical report, Manchester Institute for Mathematical Sciences (2008)

3. Breda, D.: Solution operator approximations for characteristic roots of delay differential equations. Appl. Numer. Math. 56, 305–317 (2006)
4. Breda, D., Maset, S., Vermiglio, R.: Pseudospectral differencing methods for characteristic roots of delay differential equations. SIAM J. Sci. Comput. 27(2), 482–495 (2005)
5. Breda, D., Maset, S., Vermiglio, R.: Pseudospectral approximation of eigenvalues of derivative operators with non-local boundary conditions. Appl. Numer. Math. 56, 318–331 (2006)
6. Engelborghs, K., Luzyanina, T., Samaey, G.: DDE-BIFTOOL v. 2.00: a Matlab package for bifurcation analysis of delay differential equations. Technical report, K.U.Leuven, Leuven, Belgium (2001)
7. Gu, K., Niculescu, S.-I.: Survey on recent results in the stability and control of time-delay systems. J. Dyn. Syst. T.- ASME 125, 158–165 (2003)
8. Huybrechs, D.: On the evaluation of highly oscillatory integrals by analytic continuation. SIAM Journal on Numerical Analysis 44(3), 1026–1048 (2006)
9. Iserles, A., Nørsett, P.: Efficient quadrature of highly oscillatory integrals using derivatives. Proceedings of the Royal Society A 461, 1383–1399 (2005)
10. Jarlebring, E., Meerbergen, K., Michiels, W.: A Krylov method for the delay eigenvalue problem. SIAM J. Sci. Comput. 32(6), 3278–3300 (2010)
11. Jarlebring, E., Michiels, W., Meerbergen, K.: A linear eigenvalue algorithm for the nonlinear eigenvalue problem. Technical report, Dept. Comp. Sci., KU Leuven (2010) (submitted)
12. Kressner, D.: A block Newton method for nonlinear eigenvalue problems. Numer. Math. 114(2), 355–372 (2009)
13. Luzyanina, T., Roose, D., Engelborghs, K.: Numerical stability analysis of steady state solutions of integral equations with distributed delays. Appl. Numer. Math. 50(1), 75–92 (2004)
14. Meerbergen, K.: The quadratic Arnoldi method for the solution of the quadratic eigenvalue problem. SIAM J. Matrix Anal. Appl. 30(4), 1463–1482 (2008)
15. Mehrmann, V., Voss, H.: Nonlinear eigenvalue problems: A challenge for modern eigenvalue methods. Gamm Mitteilungen 27, 121–152 (2004)
16. Michiels, W., Jarlebring, E., Meerbergen, K.: Krylov based model order reduction of time-delay systems. Technical report, Dept. Comp. Sci., KU Leuven (2010) (submitted)
17. Neumaier, A.: Residual inverse iteration for the nonlinear eigenvalue problem. SIAM J. Numer. Anal. 22, 914–923 (1985)
18. Olver, S.: Moment-free numerical integration of highly oscillatory functions. IMA Journal of Numerical Analysis 26, 213–227 (2006)
19. Richard, J.-P.: Time-delay systems: an overview of some recent advances and open problems. Automatica 39(10), 1667–1694 (2003)
20. Ruhe, A.: Algorithms for the nonlinear eigenvalue problem. SIAM J. Numer. Anal. 10, 674–689 (1973)
21. Voss, H.: An Arnoldi method for nonlinear eigenvalue problems. BIT 44, 387–401 (2004)
22. Vyhlídal, T., Zítek, P.: Mapping based algorithm for large-scale computation of quasipolynomial zeros. IEEE Trans. Autom. Control 54(1), 171–177 (2009)

A New Method for Delay-Independent Stability of Time-Delayed Systems

Ali Fuat Ergenc

Abstract. A new method is presented for determining delay-independent stability zones of the general LTI dynamics with multiple delays against parametric uncertainties. This method utilizes extended kronecker summation and unique properties of self-inversive polynomials. Self-inversive polynomials are special polynomials which exert useful tools for examination of the distribution of its zeros. A sufficient condition for delay-independent stability is presented. The main foci in this paper is a novel approach to the robustness of the time-delayed systems. A new sufficient condition for delay-independent stability is introduced. These new concepts are also demonstrated via some example case studies.

1 Introduction and the Problem Statement

We consider linear time invariant, retarded multiple time delayed systems (LTI-MTDS), general form of which is given as,

$$\dot{\mathbf{x}}(t) = \mathbf{A}(\mathbf{q})\mathbf{x}(t) + \sum_{j=1}^{p} \mathbf{B}_j(\mathbf{q})\mathbf{x}(t - \tau_j) \qquad (1)$$

where $\mathbf{x} \in \mathbb{R}^n$, $\mathbf{A}(\mathbf{q})$, $\mathbf{B}_j(\mathbf{q})$, $j = 1 \ldots p$ are matrices in $\mathbb{R}^{n \times n}$, $\mathbf{q} \in \mathbb{R}^r$ and the vector of time delays $\tau = (\tau_1, \tau_2, \ldots, \tau_j, \ldots, \tau_p) \in \mathbb{R}^{p+}$ the elements of which are rationally independent from each other. As a note of formalism we use boldface capital notation for vector and matrix quantities in the text. In the text, open unit disc, unit circle and outside of unit circle are referred as $\mathbb{D}, \mathbb{T}, \mathbb{S}$, respectively. Therefore, $\mathbb{D} \cup \mathbb{T} \cup \mathbb{S} = \mathbb{C}$ represents the entire complex plane. In addition to these, complex domain can be separated into to open left $\mathbb{C}_-$ and open right $\mathbb{C}_+$ half planes and $\mathbb{C}_0$ imaginary axis.

Ali Fuat Ergenc
Istanbul Technical University, Control Engineering Department, 34469, Istanbul, Turkey
e-mail: ali.ergenc@itu.edu.tr
http://web.itu.edu.tr/ergenca

R. Sipahi et al. (Eds.): Time Delay Sys.: Methods, Appli. and New Trends, LNCIS 423, pp. 241–252.
springerlink.com © Springer-Verlag Berlin Heidelberg 2012

The stability of this general class of systems has been widely studied over the years [1, 2, 3, 11, 21]. The determination of the robustness of such systems against uncertainties in delay and other parameters (i.e., uncertain τ, $\mathbf{A}$ and $\mathbf{B}_j$) are also investigated by many researchers [5, 6, 7, 8, 9, 24, 27]. The stability of the time-delayed systems can be investigated in two classes: delay-independent and delay-dependent [1, 4, 20]. The asymptotic stability of linear delay-differential systems independent of delay is declared as N-P hard (nondeterministic-polynomial time hard) problem [10]. Many studies appeared on this particular subject, on some simplified forms of the delay-independent stability problem given here [4, 20, 22, 23, 25, 26, 28, 29, 30, 31, 32, 33]. In the mainstream, there are usually two approaches to the problem. First one is Lyapunov based approaches, such as Lyapunov -Krasovskii functionals which provide some sufficient delay-independent stability conditions using specific matrix inequalities. Another one is Lyapunov-Razumikhin which can be derived from the first one with a matrix operation. These methods are based on finding arbitrary positive definite matrices and it may be hard to find an appropriate one. On the other hand, they provide sufficient conditions for the stability with a conservative manner. Second approach is basically stating that the delay free system has all its characteristic roots on the left half plane and for all the delays there are no root crossings of the imaginary axis. There are several methods for the determination of the delay-independent stability criteria such as frequency (ω) sweeping methods or matrix pencil approaches. In a recent study, a new method for delay independent stability test is also presented as a natural result of determination of the crossing frequency set [28].

In this study, the problem of dictating delay-independent stability criteria is transferred to assigning a certain number of the zeros in $\mathbb{D}$ of a polynomial derived from the system equations. The key novelty introduced by the method is that it doesn't have any restrictions on the number of the delays (p) and the number of the parametric uncertainties (r). It is based on unique properties of a self-inversive polynomial which represents the infinite dimensional delayed system in terms of a finite dimensional polynomial with interspersed zeros on the unit-circle.

The text is structured as follows: In section 2, preliminaries of the study are given. Section 3 states the delay-independent stability criteria for LTI system with multiple delays under parametric uncertainties. Section 4 contains illustrative example case studies.

2   Preliminaries

The characteristic equation of the system in (1) is

$$
\begin{aligned}
CE(s, \mathbf{q}, \tau_1, \ldots, \tau_p) &= \det\left[s\mathbf{I} - \mathbf{A}(\mathbf{q}) - \sum_{j=1}^{p} \mathbf{B}_j(\mathbf{q})e^{-\tau_j s}\right] \\
&= \mathbf{A}_0(s, \mathbf{q}) + \mathbf{A}_{p+1}(s, \mathbf{q}, \tau_1, \ldots, \tau_p) + \\
&\quad \sum_{j=1}^{p} e^{-n_j \tau_j s} \mathbf{A}_j\left(s, \mathbf{q}, \tau_1, \ldots, \tau_{j-1}, \tau_{j+1}, \ldots, \tau_p\right) = 0
\end{aligned}
\tag{2}
$$

where $A_0(s, \mathbf{q})$ is an nth degree polynomial in s , A_j's $(j = 1 \ldots p)$ are quasi-polynomials in s with the parametric uncertainties $\mathbf{q}$'s and all the delays except τ_j. n_j is the highest order of commensuracy of delay τ_j in the dynamics $(n_j \leq n)$. A_j contains s terms with the highest degree of $n - 1$ and they are the factors multiplying the representative exponential of the highest commensuracy of τ_j, i.e., $e^{-n_j \tau_j s}$. Since the system is 'retarded', s^n term appears only in $A_0(s)$ which is free of delays. A_{p+1} is another quasi-polynomial which contains all the remaining terms with lower commensuracy levels (in τ_j) than n_j, $j = 1 \ldots p$.

Definition 1. The stability posture of the system in (1) is determined by the number of characteristic roots of (2) in $\mathbb{C}_+$. This number is naturally a function of the delays and the parametric uncertainties, which are the parameters in (1). Wherever the number is equal to 0, the system is labeled as "stable". For any stability switching a characteristic root $\pm \omega i$ must exist on $\mathbb{C}_0$. This is a direct result of the "root continuity" argument in the parametric space [1, 2].

As per the earlier discussions for the delay-independent stability of the system we need to determine <u>all</u> the control parameter set which provides robustness against delay and parametric uncertainties. For this mission we convert the problem of examining imaginary axis crossing of infinitely many characteristic roots of the system into determination of the root locations of the auxiliary equation of which representing the system. Before further explanation of the concept, it is necessary to reveal some properties of Kronecker Summation.

Kronecker Summation of Two Matrices and Its Properties:
Kronecker sum of two square matrices $\mathbf{M}_1$ $(n_1 \times n_1)$ and $\mathbf{M}_2$ $(n_2 \times n_2)$ is defined as [12, 13]

$$\mathbf{M}_1 \oplus \mathbf{M}_2 = \mathbf{M}_1 \otimes \mathbf{I}_{n_2} + \mathbf{I}_{n_1} \otimes \mathbf{M}_2$$

where $\mathbf{M}_1 \in \mathbb{R}^{n_1 \times n_1}, \mathbf{M}_2 \in \mathbb{R}^{n_2 \times n_2}$. Here $\oplus$ denotes the *Kronecker summation* and $\otimes$ the *Kronecker product* operations. The important property of the Kronecker summation of $\mathbf{M}_1$ and $\mathbf{M}_2$ is that this new square matrix

$$\mathbf{M}_1 \oplus \mathbf{M}_2 \in \mathbb{R}^{(n_1 \cdot n_2) \times (n_1 \cdot n_2)}$$

has $n_1 \cdot n_2$ eigenvalues which are indeed pair-wise combinatoric summations of the n_1 eigenvalues of $\mathbf{M}_1$ and n_2 eigenvalues of $\mathbf{M}_2$. That is, the Kronecker sum operation, in fact, induces the "eigenvalue addition" character to the matrices. We take advantage of this feature as discussed next in a definition and the highlight theorem:

Definition 2. Auxiliary Characteristic Equation (*ACE*) of the system in (1), with $z_j = e^{-\tau_j s}$ is defined in [14] as follows:

$$ACE(\mathbf{z}, \mathbf{q}) = \det \left[\begin{array}{c} \mathbf{A}(\mathbf{q}) \otimes \mathbf{I} + \mathbf{I} \otimes \mathbf{A}^{\mathrm{T}}(\mathbf{q}) + \\ \sum_{j=1}^{p} (\mathbf{B}_j(\mathbf{q}) \otimes \mathbf{I} z_j + \mathbf{I} \otimes \mathbf{B}^{\mathrm{T}}{}_j(\mathbf{q}) z_j^{-1}) \end{array} \right] = 0 \qquad (3)$$

Theorem 1. *For the system given in (1) the following findings are equivalent:*

1. *A vector of p-dimensional unitary complex numbers* $\mathbf{z} = \{z_j\} \in \mathbb{T}^p$, $|z_j| = 1, \forall j = 1 \ldots p$ *satisfies ACE.*
2. *There exists at least one pair of imaginary characteristic roots,* $\pm \omega i$, *of (2).*
3. *There exists a corresponding delay vector* $\tau = \{\tau_j\} \in \mathbb{R}^{p+}$ *and a parameter vector* $\mathbf{q} \in \mathbb{R}^r$, *where* $\langle \tau, \mathbf{q}, \omega \rangle$ *holds.*

Proof. Proof of the theorem can be found in details in [14] considering that $\mathbf{q}$ is a constant parameter vector, thus $\mathbf{A}$ and $\mathbf{B}$ are constant matrices. Shortly, *ACE* is the determinant of the Kronecker sum of state space equation of the system and its complex conjugate. The positive imaginary eigenvalues of the system eliminate negative ones, thus the auxiliary characteristic equations is free of s terms.

Equation (3) is both necessary and sufficient condition for a point $\tau = (\tau_1, \ldots, \tau_p) \in \mathbb{R}^{p+}$ generating an $\omega i \in \mathbb{C}_0$ for a certain $\mathbf{q} \in Q$. Since this equation is completely free from the delays, and a function of $\mathbf{z}$ and $\mathbf{q}$, the procedure is now considerably simplified to find $\mathbf{z} \in \mathbb{T}^P$ solutions of (3) with respect to certain $\mathbf{q}$. To determine the imaginary characteristic roots of (2) one simply plugs such a $\mathbf{z}$ and $\mathbf{q}$ into (2) and solves for s. These roots reveal the crossing frequencies we are interested in and form the set, i.e.,

$$\Omega = \{\omega | CE(s = \omega i, \mathbf{z}, \mathbf{q}) = 0, \ \mathbf{z} \in Z, \ \mathbf{q} \in Q\} \tag{4}$$

One then uses the individual components of s to determine the respective delays which are

$$\tau_{jk} = \frac{\arg(z_j) \mp 2k\pi}{\omega} \ j = 1 \ldots p, \ k = 0, 1, 2, \ldots \tag{5}$$

where τ_{jk} implies the j the delay value for various k values.

After construction of some of the base points, we propose a new approach for the determination of $\mathbf{q}$ of parameter space Q where the system (1) is delay-independent stable.

3   Main Results

A linear time delayed system is delay-independent stable when all the characteristic roots of (2) lie on the $\mathbb{C}^-$ for all $\tau \in \mathbb{R}^{p+}$. As noted above, to determine the characteristic roots for all τ is almost impossible and even harder when the system has parametric variations. In this study, we purpose a relatively simple theorem which presents the necessary and sufficient conditions for delay-independent stability of (1). A similar theorem for single delay case exists in an earlier study [20].

Theorem 2. *(Delay-Independent Stability) A system given in (1) is delay independent stable if*

1. $Re\left(\sigma\left[\mathbf{A}(\mathbf{q}) + \Sigma_{j=1}^{p}(\mathbf{B}_j(\mathbf{q}))\right]\right) < 0$
2. *ACE has no roots* $\mathbf{z} = \{z_j\} \in \mathbb{T}^P$.

Proof. Proof of the theorem can be derived using root crossing concept. In the first item we establish the foundation for the stability of the system for non-delay case. In the second item, stating that *ACE* has no roots on $\mathbb{T}^p$, we guarantee that characteristic equation (2) has no $i\omega$ roots on the imaginary axis.

This is a very general theorem for delay-independent stability. In following part of the paper, we would like to stress on some unique properties of *ACE* which will open new fronts for the robustness analysis of the time delayed systems against parametric variations. *ACE* is a special multinomial of z_j $(j = 1\ldots p)$. We can rewrite *ACE* in another structure where we embed z_j $\quad j = 1\ldots k-1, k+1\ldots p$ in $b_j(\mathbf{q}, z_1, z_2, \ldots, z_{k-1}, z_{k+1}, \ldots, z_p)$. Here b_j's are coefficients of a polynomial in terms of z_k. Formally, it is as the following:

$$ACE(\mathbf{z}, \mathbf{q}) = \sum_{j=0}^{m} b_j(\mathbf{q}, z_1, z_2, \ldots, z_{k-1}, z_{k+1}, \ldots, z_p) z_k^j \tag{6}$$

Before we move along further analysis of this polynomial it is useful to give some definitions and lemmas.

Definition 3. A polynomial P is called self-inversive if

$$P^*(z) = z^n \overline{P(1/z)} = P(z) \; for \; all \; z \neq 0. \tag{7}$$

Equivalently, writing $P(z) = \sum_j^p b_j z^j$ we have for a self-inversive polynomial

$$\sum_{j=0}^{p} \overline{b_{p-j}} z^j = \sum_{j=0}^{p} b_j z^j \; for \; all \; z \neq 0. \tag{8}$$

and therefore $\overline{b_{p-j}} = b_j$ for $0 \leq j \leq p$.

Inherently, (6) is a self-inversive polynomial in terms of z_k^j's. The zeros of this class of polynomials lie either on the unit circle $\mathbb{T}$ or occur in pairs conjugate to $\mathbb{T}$ (symmetric pair of roots wrt unit circle). This is a natural property of *ACE* that we use when we determine crossing points of (2). Conversely, for delay-independent stability we desire that none of the roots of (6) lie on the unit circle. Our goal is to determine control parameter space which satisfies this condition. Examination of $\mathbf{q}$ space for which (6) have no unitary zeros is a very hard problem. At this point, it is very useful to use some of the unique properties of self-inversive polynomials. Critical points of the polynomial P (i.e. zeros of P', where P' denotes derivative of P wrt z) and the zeros of P have a remarkable relationship. It is stated as in the theorem below [15].

Theorem 3. *Let P is a self-inversive polynomial of degree p. Suppose that P has exactly β zeros on the unit circle $\mathbb{T}$ (multiplicity included) and exactly μ critical points in the closed unit disc $\mathbb{U}$ (counted according to multiplicity). Then*

$$\beta = 2(\mu + 1) - p. \tag{9}$$

Proof. The proof of this theorem can be done using argument principle. It can be found in [15] in details.

This theorem is a key tool for establishing some criterion for delay-independent stability. It is stated before that a system as described in (1) is delay-independent stable if its *ACE* has no zeros on $\mathbb{T}$ and it is cumbersome to check if zeros are unitary. Notice that theorem describes the number of the unitary roots of *ACE* in terms of the number of critical points in $\mathbb{U}$. In other words, let (6) be a self-inversive polynomial degree of 8 and our condition for delay-independent stability is $\beta = 0$. Then, by the theorem it clear to find the number of the critical points (μ) of which should lie in $\mathbb{U}$ is equal to 3. It is a crucial result, because there are many methods analyze the number of the zeros of a polynomial in the unit-circle. A theorem taken from [16] and [17] is given below for such analysis.

Theorem 4. *Let P(z) a polynomial equation,*

$$P(z) = \sum_{j=0}^{p} b_j z^j \tag{10}$$

where $b_j \in \mathbb{C}$ and $b_p \neq 0$
If

$$|b_k| > \sum_{\substack{j \neq k}}^{p} |b_j| \tag{11}$$

then P(z) has exactly k zeros in the unit circle and noting that P(z) , under the above condition has no unitary zeros (i.e. $z \in \mathbb{T}$).

Proof. Proof of this theorem is easily achieved by substituting $r = R = 1$ in Pellet's Theorem in [18]. The proof is also given in [16] and [17].

The theorem states that if the condition is satisfied the polynomial has no unitary roots. It is an essential condition for our aim to find μ number of critical roots of *ACE* in the unit circle. Notice that if a critical root of *ACE* is on the unit-circle, *ACE* has that root with a multiplicity at least equal to 2.

Following the theorems, we shall summarize delay independent stability conditions. For a system given in (1) to be delay-independent stable, its *ACE* has to have a certain number of critical roots (i.e. roots of the derivative of *ACE* $(D(\alpha, z))$ on the open unit disc ($\mathbb{D}$)). According to (4), the condition of delay-independent stability is the following corollary.

Corollary 1. *A linear time invariant system with multiple time delays described in (1) is delay-independent stable if*

1. $Re\left(\sigma\left[\mathbf{A}(\mathbf{q}) + \sum_{j=1}^{p}(\mathbf{B}_j(\mathbf{q}))\right]\right) < 0$
2. Critical equation of ACE satisfies

$$| b_\mu (\mathbf{q}, z_1, z_2, \ldots, z_{k-1}, z_{k+1}, \ldots, z_p)| >$$
$$\sum_{j \neq \mu}^{p} |b_j (\mathbf{q}, z_1, z_2, \ldots, z_{k-1}, z_{k+1}, \ldots, z_p)| \qquad (12)$$

where $\mu \leq (p/2) - 1$ and μ is an integer number.

The parameter set $\mathbf{q} \in \mathbf{Q}$ that satisfies the inequality above is the required for delay-independent stability of the system (1). In other words, if the inequality is satisfied, *ACE* has roots $\mathbf{z} \notin \mathbb{T}^P$ and τ is an empty set. Thus, the system (1) is stable for all $\tau \in \mathbb{R}^{p+}$.

In the following section we give some example cases to explain our claim clearly.

4   Example Case Studies

Example 1: A case is borrowed from [20], with $n = 1$ and $p = 1$.

$$\dot{x}(t) = -ax(t) - bx(t - \tau) \qquad (13)$$

The system is stable for $\tau = 0$ if $a + b > 0$ is satisfied. In order to find delay independent stability we use the second statement of the corollary. The corresponding auxiliary characteristic equation and its derivative are

$$ACE(z, a, b) = bz^2 + 2az + b = 0 \quad D(z, a, b) = bz + 2a \qquad (14)$$

Now, since the degree of *ACE* p=2 the number (μ) of the critical roots of *ACE* must be equal to zero so that there would be no unitary roots of *ACE*. Thus, using corollary (1) we reach the following inequality,

$$|b| < |a| \qquad (15)$$

Notice that the result coincides with [20] .

Example 2: We take a higher degree system with single delay. ($n = 3$)

$$\dot{\mathbf{x}}(t) = \mathbf{A}\mathbf{x}(t) + \mathbf{G}(\alpha)\mathbf{x}(t - \tau) \qquad (16)$$

where,

$$\mathbf{A} = \begin{bmatrix} 0 & 1 & 0 \\ 0 & 0 & 1 \\ -2.5 & -1.8 & -4 \end{bmatrix} \quad \mathbf{G} = \begin{bmatrix} 0 & 0 & 0 \\ 0 & 0 & 0 \\ -2\alpha & 0 & -1 \end{bmatrix} \qquad (17)$$

The corresponding auxiliary characteristic equation and its derivative are

$ACE(z, \alpha) = (-33.520\alpha + 6.250 + 49.60\alpha^2 + 8.0\alpha^3)z^6 + (456.80\alpha^2 + 66.20 - 288.320\alpha)z^5 + (24.0\alpha^3 + 162.950 - 107.20\alpha^2 + 48.160\alpha)z^4 + (-1134.40\alpha^2 + 703.36\alpha + 374.20)z^3 + (24.0\alpha^3 + 162.950 - 107.20\alpha^2 + 48.160\alpha)z^2 + (456.8000000\alpha^2 + 66.20 - 288.320\alpha)z - 33.520\alpha + 6.250 + 49.60\alpha^2 + 8.0\alpha^3$

$$D(z,\alpha) = (37.50 + 47.9999999\alpha^3 - 201.120\alpha + 297.60\alpha^2)z^5 + (2284.0\alpha^2 + 331.0 - 1441.60\alpha)z^4 + (192.640\alpha - 428.8\alpha^2 + 96.0\alpha^3 + 651.7999997)z^3 + (-3403.20\alpha^2 + 2110.080\alpha + 1122.60)z^2 + (-214.40\alpha^2 + 325.9 + 47.99999999\alpha^3 + 96.31999997\alpha)z + 456.7999999\alpha^2 + 66.19999997 - 288.320\alpha$$

Keeping in mind that *ACE* is sixth order , the number of roots of $D(z,\alpha)$ that should lie in the unit circle is equal to 2 (using (9)). The inequality (1) is constructed for k=2;

$$|-3403.20\alpha^2 + 2110.080\alpha + 1122.60| > |456.7999999\alpha^2 + 66.19999997 - 288.320\alpha| - |-214.40\alpha^2 + 325.8999999 + 47.9999999\alpha^3 + 96.31999997\alpha| - |192.640\alpha - 428.8\alpha^2 + 96.0\alpha^3 + 651.7999997| - |2284.0\alpha^2 + 330.999999 - 1441.60\alpha| - |37.50 + 47.9999999\alpha^3 - 201.120\alpha + 297.60\alpha^2|0.00004288\alpha^2|$$

The system given here is delay-independent stable for the α values those satisfy this inequality. In Fig.1, we present the number of the roots of *ACE* on the unit-circle by sweeping α (thick-line). In the same figure, inequality is also depicted (thin-dots). Notice, the region, where inequality is satisfied, coincides with zero number unitary roots of *ACE*. Thus, the system is delay-independent stable for that region. We would like to point out that (1) is a sufficient condition, so inequality provides a subset of the α's for delay-independent stability.

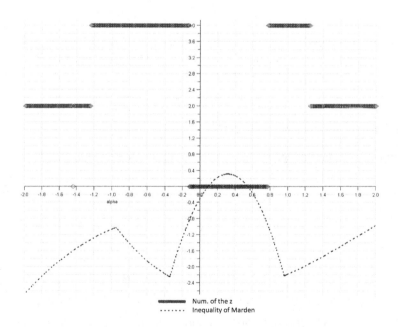

Fig. 1 Number of the roots of *ACE* that $z \in \mathbb{T}$ and inequality given in corollary (1)

Example 3: A motor driven pendulum with multiple delays in feedback is considered [19]. The state space representation of the system is the following.

$$\dot{x}(t) = A(\delta)x(t) + G_1(b)x(t - \tau_1) + G_2x(t - \tau_2) \tag{18}$$

where,

$$A = \begin{bmatrix} 0 & 1 \\ -58.95680415 & -1.45842014 2\delta \end{bmatrix}, \tag{19}$$

$$G_1 = \begin{bmatrix} 0 & 0 \\ 13.56025912b & 0 \end{bmatrix}, \quad G_2 = \begin{bmatrix} 0 & 0 \\ 0 & 1.233983580 \end{bmatrix} \tag{20}$$

In the system, δ represents the variation of the viscous friction coefficient of the pendulum shaft and b is the control parameter in the feedback matrix G_1 . We can assign a value to b that can make the system delay-independent stable under parametric uncertainties. Using (6) auxiliary characteristic equation is derived.

$ACE(\delta,b,z_1,z_2) \quad = \quad 183.8806274b^2z_2{}^2z_1{}^4 \quad + \quad (73.21183253b\delta z_2 +$
$24.40394419b\delta z_2{}^3$
$+ \quad 20.64841642bz_2{}^2 \quad + \quad 57.68505239b\delta^2z_2{}^2 \quad + \quad 20.64841642b)z_1{}^3 \quad +$
$(424.4103433\delta z_2{}^3 + 424.4103433\delta z_2 - 367.7612548b^2z_2{}^2 + 89.77443809 +$
$501.6019688\delta^2z_2{}^2$
$+ \quad 179.5488762z_2{}^2 \quad + \quad 89.77443809z_2{}^4)z_1{}^2 \quad + \quad (73.21183253b\delta z_2{}^3 \quad +$
$20.64841642bz_2{}^4$
$+ \quad 24.40394419b\delta z_2 \quad + \quad 20.64841642bz_2{}^2 \quad + \quad 57.68505239b\delta^2z_2{}^2)z_1 \quad +$
$183.8806274b^2z_2{}^2$

In order to find sufficient condition for delay independent control scheme for the pendulum following inequality must be satisfied.

$| - 179.5488761 - 848.8206862\delta z_2{}^3 - 848.8206862\delta z_2 + 735.5225094b^2z_2{}^2$
$- \quad 1003.203938\delta^2z_2{}^2 \quad - \quad 359.0977523z_2{}^2 \quad - \quad 179.5488761z_2{}^4| \quad >$
$|73.21183250b\delta z_2{}^3 + 20.64841641bz_2{}^4 + 24.40394418b\delta z_2 + 20.64841641bz_2{}^2 +$
$57.68505237b\delta^2z_2{}^2|$
$- \quad |61.94524924b \quad + \quad 173.0551572b\delta^2z_2{}^2 \quad + \quad 219.6354975b\delta z_2 \quad +$
$73.21183255b\delta z_2{}^3 + 61.94524924bz_2{}^2| - |735.5225094b^2z_2{}^2|$

Here, z_2 is considered to be on the unit-circle(i.e., $|z_2| = 1$). It is known that if the system is delay independent stable z_2 may not be on the unit circle. We shall state the proposition clearly; we investigate the delay independent stability conditions over z_1 with the assumption that z_2 is on the unit-circle and there are no unitary values for z_1. The inverse proposition is also valid. If there were z_1 on the unit-circle there should not be unitary z_2. If they are both not on the unit-circle, delay-independent stability is self-evident. In that case ACE is not self-inversive and it doesn't represent the system. In the Fig.2 the delay independent stability switching with respect to b and δ is depicted. The contours represent values of the inequality above. System is stable for all delays when inequality is satisfied.

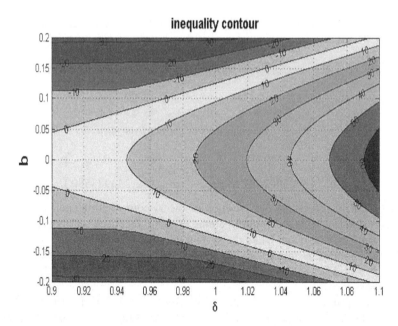

Fig. 2 The contour plot of the inequality in corollary (1) wrt b and δ

5 Conclusions

This paper is on the delay-independent stability robustness of LTI systems with multiple delays against parametric uncertainties. In the core of the main idea, unique properties of self-inversive polynomials are utilized. The problem of delay-independent stability of LTI systems with time delays are associated with self-inversive polynomials and their root distribution relative to the unit-circle. It is shown that the relationship of the self-inversive polynomial and its derivative is a very useful tool to analyze root locations of the auxiliary characteristic equation of the system. A simple inequality condition is presented which, in turn, resolves the delay-independent stability robustness of LTI systems with multiple delays against parametric uncertainties. The application of the concept to various systems is explained in example case studies.

References

1. Hale, J.K.: Theory of Functional Differential Equations. Springer, New York (1977)
2. Hale, J.K., Verduyn Lunel, S.M.: An Introduction to Functional Differential Equations. Springer, New York (1993)
3. Cooke, K.L., van den Driessche, P.: On Zeros of Some Transcendental Equations. Funkcialaj Ekvacioj 29, 77–90 (1986)

4. Kamen, E.W.: Linear Systems with Commensurate Time Delays: Stability and Stabilization Independent of Delay. IEEE Transactions on Automatic Control 25, 367–375 (1982)

5. Fu, M., Olbrot, A.W., Polis, M.P.: Robust stability for time-delayed systems: The edge theorem and graphical tests. IEEE Transactions on Automatic Control 34, 813–819 (1989)

6. Kharitonov, V.L., Zhabko, A.P.: Robust Stability of Time-Delay Systems. IEEE Transactions on Automatic Control 39, 2388–2397 (1994)

7. Kharitonov, V.L.: Robust Stability Analysis of Time Delay Systems: A Survey. Annual Reviews in Control 23, 185–196 (1999)

8. Richard, J.P.: Time-Delay systems: an overview of some recent advances and open problems. Automatica 39, 1667–1694 (2003)

9. Silva, G.J., Datta, A., Bhattacharyya, S.P.: PI Stabilization of first-order systems with time delay. Automatica 37, 2025–2031 (2001)

10. Toker, O., Ozbay, H.: Complexity Issues in Robust Stability of Linear Delay-Differential Systems. Math. Control Signals Systems 9, 386–400 (1996)

11. Stepan, G.: Retarded Dynamical Systems: Stability and Characteristic Function. Longman Scientific & Technical, co-publisher John Wiley & Sons Inc., New York, US (1989)

12. Bernstein, D.S.: Matrix Mathematics. Princeton University Press (2005)

13. Brewer, J.W.: Kronecker products and matrix calculus in system theory. IEEE Transactions on Circuits and Systems CAS-25, 772–781 (1978)

14. Ergenc, A.F., Olgac, N., Fazelinia, H.: Extended Kronecker Summation for Cluster Treatment of LTI Systems with Multiple Delays. SIAM Journal of Control and Optimization 46, 143–155 (2007)

15. Sheil-Small, T.: Complex Polynomials. Cambridge University Press (2002)

16. Rajan, P.K., Reddy, H.C.: Comments on Note on the Absolute Value of the Roots of a Polynomial. IEEE Transactions of Automatic Control AC-30(1), 80 (1985)

17. Mori, T.: Note on the Absolute Value of the Roots of a Polynomial. IEEE Transactions of Automatic Control AC-29(1), 54–55 (1984)

18. Marden, M.: The Geometry of the Zeros of a Polynomial in a Complex Variable. American Mathematical Society (1949)

19. Olgac, N., Ergenc, A.F., Sipahi, R.: Delay Scheduling: A New Concept for Stabilization in Multiple Delay Systems. Journal of Vibration and Control 11(9), 1159–1172 (2005)

20. Niculescu, S.I.: Stability and hyperbolicity of linear systems with delayed state: a matrix-pencil approach. IMA Journal of Mathematical Control & Information (15), 331–347 (1998)

21. Niculescu, S.I.: Delay Effects on Stability: A Robust Control Approach. LNCIS, vol. 269. Springer, Berlin (2001)

22. Niculescu, S.I., Ionescu, V.: On delay-independent stability criteria: a matrix-pencil approach. IMA Journal of Mathematical Control and Information 14, 299–306 (1997)

23. Chen, J., Xu, D., Shafai, B.: On Sufficient Conditions for Stability Independent of Delay. IEEE Transactions of Automatic Control 40(9) (1995)

24. Chapellat, H., Bhattacharyya, S.P.: A Generalization of Kharitonov's Theorem: Robust Stability of Interval Plants. IEEE Transactions on Automatic Control 34, 306–311 (1989)

25. Chen, J., Latchman, H.A.: Frequency Sweeping Tests for Stability Independent of Delay. IEEE Transactions of Automatic Control 40(9) (1995)

26. Tuzcu, I., Ahmadian, M.: Delay-independent Stability of Uncertain Control Systems. Journal of Vibration and Acoustics 124, 277–283 (2002)

27. Gu, K., Kharitonov, V.L., Chen, J.: Stability of Time-Delay Systems. Birkhuser, Boston (2003)

28. Delice, I.I., Sipahi, R.: Exact Upper and Lower Bounds of Crossing Frequency Set and Delay Independent Stability Test for Multiple Time Delayed Systems. In: IFAC TDS Workshop, Romania (2009)

29. Hertz, D., Jury, E.I., Zeheb, E.: Stability independent and dependent of delay for delay differential systems. Journal of The Franklin Institute 150, 143–150 (1984)

30. Wang, Z.H., Hu, H.Y.: Delay-independent stability of retarded dynamic systems of multiple degrees of freedom. Journal of Sound and Vibration 226(1), 57–81 (1999)

31. Wu, S., Ren, G.: Delay-independent stability criteria for a class of retarded dynamical systems with two delays. Journal of Sound and Vibration 270(4-5), 625–638 (2004)

32. Thowsen, A.: Delay-independent asymptotic stability of linear systems. IEEE Proceedings Control Theory and Applications 129(3), 73–75 (1982)

33. Michiels, W., Niculescu, S.I.: Stability and Stabilization of Time-Delay Systems: An Eigenvalue-Based Approach. Advances in Design and Control, vol. 12. SIAM, Philadelphia (2007)

A Hybrid Method for the Analysis
of Non-uniformly Sampled Systems

Laurentiu Hetel, Alexandre Kruszewski, and Jean-Pierre Richard

Abstract. In this chapter we propose a method for the analysis of sampled-data systems with sampling jitter. We consider that the sampling interval is unknown and time-varying and we provide a method for estimating the Lyapunov exponent. The proposed method is hybrid, in the sense that it combines continuous-time models (based on time delay systems) with polytopic embedding methods, specific to discrete-time approaches. The approach exploits the fact that the command is a piecewise constant signal and leads to less conservative stability conditions with respect to the existing literature. Using geometrical arguments, a lower bound of the Lyapunov exponent can be expressed as a generalized eigenvalue problem. Numerical examples are given to illustrate the approach.

1 Introduction

This chapter is dedicated to the analysis of sampled data systems with non-uniform sampling periods . This problem represents an abstraction of more complex sample-and-hold phenomena that can be encountered in networked control / embedded systems [22]. In the literature, several authors addressed the control under time-varying

Laurentiu Hetel
LAGIS, FRE CNRS 3303, Ecole Centrale de Lille, BP48,
59651 Villeneuve d'Ascq Cedex, France
e-mail: laurentiu.hetel@ec-lille.fr

Alexandre Kruszewski
LAGIS, FRE CNRS 3303, Ecole Centrale de Lille, BP48,
59651 Villeneuve d'Ascq Cedex, France
e-mail: alexandre.kruszewski@ec-lille.fr

Jean-Pierre Richard
INRIA Non-A; LAGIS, FRE CNRS 3303, Ecole Centrale de Lille, BP48,
59651 Villeneuve d'Ascq Cedex, France
e-mail: jean-pierre.richard@ec-lille.fr

R. Sipahi et al. (Eds.): Time Delay Sys.: Methods, Appli. and New Trends, LNCIS 423, pp. 253–264.
springerlink.com   © Springer-Verlag Berlin Heidelberg 2012

sampling period. This control problem is not easy even for linear time invariant (LTI) systems. In fact, it has been shown that, under variations of the sampling interval, the trajectory of a system may be unstable although for each value of the sampling period the equivalent discrete-time model is Schur stable [11].

The stability problem for systems with sampling jitter has been addressed both from the discrete- and continuous-time points of view. In fact, most of the existing stability analysis approaches address the problem using discrete-time tools. The first methods that are related to this problem can be found in the work of [21] and [20]. In [18, 19], the authors used the gridding of the set of possible sampling intervals to derive LMI-based stability conditions and performance criteria. This simplified modeling approach has been applied by [5], in order to obtain sufficient algebraic conditions for existence of quadratic Lyapunov functions. In this case, the Lyapunov functions can be analytically constructed using Lie algebra arguments. However, the gridding based-models are not so accurate as, in practice, one encounters an infinite number of possible sampling periods. More accurate discrete-time models have been obtained by combining the gridding approach with a discrete time equivalent of the bounded real lemma [1, 7] or by using a convex embedding modeling approach [10, 11]. The main drawback of the discrete-time analysis is the fact that it ignores the inter-sample behavior of the system, in the sense that it does not take into account the existence of side-effects of sampled-data control, such as ripples. Moreover, these methods are numerically inaccurate when the sampling interval goes to zero (see [7, 17]).

This stability analysis of systems with sampling jitter has also been addressed in continuous time using a time-delay system modeling of the sample-and-hold function. The main advantage of such methods is the fact that they take into account the inter-sampling behavior of the system. A notable approach is the descriptor system modeling applied by [6] for which stability conditions can be obtained using the Lyapunov Krasovskii functional method. It has been shown by [14] that this approach is related to the one of [12] that is based on a norm bounded uncertainty modeling of the sample-and-hold operator. However, such an approach may induce conservatism, due to the symmetry of ellipsoidal norms used for norm bounded uncertainties. A different method, using impulsive delay differential equations, has been proposed in [15, 16]. In general, for the existing continuous-time approaches, the analysis does not take into account the particular variation of the sampling induced delay. This is why, faced to numerical benchmarks, they seem to be more conservative than the discrete-time approaches (see [7, 11]).

From the literature survey, we can conclude that both continuous-time and discrete-time approaches have drawbacks and advantages. It is desirable to provide one method that is able to treat the problem in continuous-time (for inter-sampling performance issues) and to take into account the particular variation of the sampling induced delay (in order to provide less conservative stability conditions, such as in the case of discrete-time approaches). Here we intend to propose such a method. In order to explicitly analyze the inter-sampling behavior, the proposed method is based on the evaluation of the Lyapunov exponent (also called decay-rate, see [2]). It represents a measure that can be used both for stability and performance analysis. For linear time invariant system it can be directly deduced from the state-matrix eigenvalues. However, in the case of sampled-data systems with non-uniform sampling its computation is not a trivial task. The contribution is to show how to take into account the piecewise constant nature of the control signal in order to determine an efficient estimation of the Lyapunov exponent.

This chapter is structured as follows:

In section 2 we mathematically formalize the problem under study. Section 3 presents a method for estimating the Lyapunov exponent in the case of sampling variation. Next, we introduce a LMI method for the computation of the Lyapunov exponent. In Section 4, we present numerical examples illustrating our method. Finally, conclusions are given in Section 5.

Notations: For a matrix M we denote by $\|M\|$ the induced matrix norm. By $M \succ 0$ or $M \prec 0$ we mean that the symmetric matrix M is positive or negative definite respectively. We denote the transpose of a matrix M by M^T. By $\mathbf{I}$ (or $\mathbf{0}$) we denote the identity (or the null) matrix with the appropriate dimension. By $\lambda_{max}(M)$ we denote the maximum eigenvalue of a square matrix. For a given set $\mathscr{S}$, $co(\mathscr{S})$ denotes the convex hull of $\mathscr{S}$. By $diag(v_1, v_2, \ldots, v_n)$ we denote a diagonal matrix with $v_1, v_2, \ldots, v_n$ on the main diagonal and zeros elsewhere.

2 Problem Description

In this section we provide a mathematical description of the problem under study. Consider n, m, two positive integers and the matrices $A \in \mathbb{R}^{n \times n}$, $B \in \mathbb{R}^{n \times m}$. We are interested in the class of systems described by the equation:

$$\dot{x}(t) = Ax(t) + Bu(t), \ \forall t \in \mathbb{R}^+ \tag{1}$$

$$x(0) = x_0 \in \mathbb{R}^n. \tag{2}$$

Here $x(\cdot) : \mathbb{R}^+ \to \mathbb{R}^n$ represents the system state and $u(\cdot) : \mathbb{R}^+ \to \mathbb{R}^m$ represents the control. In order to take the sampling effect into account, we consider that the command is a piecewise constant state feedback, i.e.

$$u(t) = Kx(t_k), \ \forall t \in [t_k, t_{k+1}) \tag{3}$$

where $\{t_k\}_{k \in \mathbb{N}}$ represents the set of sampling instants such that

$$\lim_{k \to \infty} t_k = \infty, \ 0 = t_0 < t_1 < \ldots < t_k < \ldots, \tag{4}$$

that is, no accumulation is allowed. We consider that the sampling interval $t_{k+1} - t_k$ is bounded parameter, i.e.

$$0 \leq h_{min} \leq t_{k+1} - t_k \leq h_{max} < \infty. \tag{5}$$

We assume that the gain $K \in \mathbb{R}^{m \times n}$ in equation (3) represents a known constant matrix, computed such as the $A + BK$ is Hurwitz.

The Lyapunov exponent [2] is defined to be the largest α such that

$$\lim_{t \longrightarrow \infty} e^{\alpha t} \|x(t)\| = 0.$$

It represents a basic measure of system performance. It shows how fast the norm of the state vector converges to zero, since

$$\|x(t)\| \leq e^{-\alpha t} c \|x(0)\|, \forall t > 0 \tag{6}$$

for some positive constant $c \in \mathbb{R}$ (for more details, see [2], chapter 5). It can also represent a parameter for analysis the system stability : a system is stable if and only if its Lyapunov exponent is strictly positive. The goal is provide a method for computing the Lyapunov exponent. This problem is formalized as follows :

Problem: Consider the system (1) with the control law (3) and all the possible sequences of sampling under the assumptions (4) and (5). Obtain the maximum computable α that satisfies (6).

3   Lower Estimate of the Lyapunov Exponent

In this section, we present a method for computing a lower estimate of the Lyapunov exponent for the system (1) with the control law (3). First we present theoretical results for the computation of the Lyapunov exponent using a quadratic Lyapunov function; next we show how these results can be expresses a convex optimization problem the may be addressed using classical numerical optimization tools.

3.1   Time-Delay Model of the System

Following the work [6], system (1) with the control law (3) can be represented as a time-delay system with piecewise-continuous time varying delay:

$$\frac{dx(t)}{dt} = Ax(t) + BKx(t - \tau(t)) \tag{7}$$

with

$$\tau(t) := t - t_k, \ \forall t \in [t_k, t_{k+1}). \tag{8}$$

The delay is bounded in a interval specified by the maximum sampling period, i.e. $\tau(t) \in [0, h_{max}]$. It represents a piecewise continuous and piecewise derivable function. Notice that the derivative of the delay satisfies the relation $\dot{\tau}(t) = 1, \forall t \in [t_k, t_{k+1})$, and that $\tau(t_k) = 0$, i.e the delay signal has a sawtooth form.

3.2 Theoretical Evaluation

As follows we show how to use quadratic Lyapunov functions in order to estimate a lower bound of the largest Lyapunov exponent of the system (7) with the particular definition of the delay given in (8).

Proposition 1. *Consider the following matrix*

$$\Lambda(\tau) := I + \int_0^\tau e^{sA} ds \, (A + BK). \tag{9}$$

and let $\mathcal{M}_{[0,h_{max}]} \subset \mathbb{R}^{n\times n}$ *denote the convex hull of* $\Lambda(\tau)$ *for all* $\tau \in [0, h_{max}]$, *i.e.*

$$\mathcal{M}_{[a,b]} := co\{\Lambda(\tau), \forall \tau \in [a,b]\}, \tag{10}$$

with $a, b \in \mathbb{R}$. *Moreover, consider the following optimization problem:*

$$\max \ \alpha \ subject \ to \ \exists P \succ 0, X_1, X_2, \ and$$
$$\begin{pmatrix} A^T P + PA + 2\alpha P + X_1 + X_1^T & PBK - X_1 M + X_2^T \\ K^T B^T P - M^T X_1^T + X_2 & -X_2 M - M^T X_2^T \end{pmatrix} \prec 0 \tag{11}$$

for all $M \in \mathcal{M}_{[0,h_{max}]}$. *Then the largest Lyapunov exponent for the system (7) is at least* α.

Proof. Consider the following candidate Lyapunov function :

$$V(x) = x^T P x, \text{ where } P = P^T \succ 0. \tag{12}$$

If the derivative of the Lyapunov function satisfies the relation $\dfrac{dV(x)}{dt} < -2\alpha V(x)$ for all system's trajectories except $x \neq 0$, then $V(x(t)) < e^{-2\alpha t} V(0)$ and $\|x(t)\| < e^{-\alpha t} \sqrt{\dfrac{\lambda_{max}(P)}{\lambda_{min}(P)}} \|x(0)\|$, i.e. the Lyapunov exponent is at least α. The derivative of the Lyapunov function (12) along the solution of system (1) with the control law (3) is given by :

$$\frac{dV(x)}{dt} = \begin{pmatrix} x(t) \\ x(t-\tau(t)) \end{pmatrix}^T \begin{pmatrix} A^T P + PA & PBK \\ K^T B^T P & 0 \end{pmatrix} \begin{pmatrix} x(t) \\ x(t-\tau(t)) \end{pmatrix}. \tag{13}$$

Notice that for all $t \in [t_k, t_{k+1})$ the solutions of system (1) with the control law (3) satisfy the relation:

$$x(t) = e^{(t-t_k)A} x(t_k) + \int_0^{(t-t_k)} e^{sA} ds BK x(t_k) \tag{14}$$

Using the definition of the delay given in equation (8) this is the same as

$$x(t) - \left(I + \int_0^{\tau(t)} e^{sA} ds \, (A+BK)\right) x(t-\tau(t)) = 0 \tag{15}$$

From (13), (15), (9) and the Finsler lemma we obtain that the relation $\dfrac{dV(x)}{dt} <$ $-2\alpha V(x)$ is satisfied if there exists matrices X_1 X_2 such that:

$$\begin{pmatrix} A^T P + PA + 2\alpha P + X_1 + X_1^T & PBK - X_1\Lambda(\tau) + X_2^T \\ K^T B^T P - \Lambda^T(\tau)X_1^T + X_2 & -X_2\Lambda(\tau) - \Lambda^T(\tau)X_2^T \end{pmatrix} \prec 0 \qquad (16)$$

$\forall \tau \in [0, h_{max}]$ with $\Lambda(\tau)$ defined in (9). Then one can show that the previous condition holds if there exists a solution for the inequality (11) for all $M \in \mathcal{M}_{[0,h_{max}]}$ which ends the proof. $\qquad\qquad\qquad\qquad\qquad\qquad\qquad\qquad\qquad\qquad\qquad\qquad\qquad\square$

3.3 Numerical Evaluation

In this subsection, we show how to solve numerically the optimization problem presented in the previous subsection (the evaluation of the Lyapunov exponent). The difficulty of solving the problem is the fact that it leads to solving an infinite number of linear matrix inequalities, one for each point M in the set $\mathcal{M}_{[0,h_{max}]}$ defined in equation (10). To derive a finite number of LMI conditions, one has to deal with the nonlinear representation of the *exponential uncertainties* [10]

$$\Omega(\tau) = \int_0^\tau e^{As} ds \qquad (17)$$

that appear in the definition of the set

$$\mathcal{M}_{[0,h_{max}]} = co\{\Lambda(\tau), \forall \tau \in [0, h_{max}]\},$$

with $\Lambda(\tau) := I + \Omega(\tau)(A + BK)$ defined in (9). Analytical methods exist in the literature for dealing with such uncertainties [3, 4, 8, 9, 10]. The basic idea is to embed this set in a polytopic set of matrices $\mathcal{S}$, i.e. to find a set of N matrices M_l such that

$$\mathcal{M}_{[0,h_{max}]} \subset \mathcal{S} = co\{M_1, M_2, \ldots, M_N\}. \qquad (18)$$

In order to provide some insight about how such a convex embedding can be computed, consider the method proposed by [3], based on the Jordan normal form of the state matrix A. We chose it here because, in comparison with the other approaches, it can be explained in quite an easy manner. For the sake of simplicity we present the case when the matrix A has n real distinct eigenvalues $\lambda_1, \ldots, \lambda_n$. In this case there exists an invertible matrix T such that

$$A = T^{-1} diag(\lambda_1, \ldots, \lambda_n) T.$$

The method leads to re-expressing the exponential uncertainty (17) as

$$\Omega(\tau) = T^{-1} diag\left(\int_0^\tau e^{\lambda_1 s} ds, \ldots, \int_0^\tau e^{\lambda_n s} ds\right) T.$$

Computing the different minimum and the maximum bounds on the scalar terms $\delta_i(\tau) = \int_0^\tau e^{\lambda_i s} ds$, $i = 1,\ldots,n$ for $\tau \in [0, h_{max}]$ and replacing for all the possible combinations of extreme values leads to a polytopic representation, a hypercube with $N = 2^n$ vertices:

$$\Omega(\tau) \in co\{\Omega_i, i = 1,\ldots,N = 2^n\},$$

with

$$\Omega_i = T^{-1} diag(\omega_1, \omega_2, \ldots, \omega_n) T \qquad (19)$$

and

$$\omega_j \in \left\{ \min_{\tau \in [0,h_{max}]} \int_0^\tau e^{\lambda_i s} ds, \max_{\tau \in [0,h_{max}]} \int_0^\tau e^{\lambda_i s} ds \right\},$$

$\forall j = 1,\ldots,n.$

Example 1. Consider that the state matrix:

$$A = \begin{pmatrix} \lambda_1 & 0 \\ 0 & \lambda_2 \end{pmatrix}$$

with $\lambda_1 = 2$, $\lambda_2 = -1.5$ and $\tau \in [0,1]$. In this case the exponential uncertainty $\Omega(\tau)$ is also a diagonal matrix of the form

$$\Omega(\tau) = \begin{pmatrix} \delta_1(\tau) & 0 \\ 0 & \delta_2(\tau) \end{pmatrix} \qquad (20)$$

$$= \begin{pmatrix} \int_0^\tau e^{\lambda_1 s} ds & 0 \\ 0 & \int_0^\tau e^{\lambda_2 s} ds \end{pmatrix} \qquad (21)$$

$$= \begin{pmatrix} \frac{1}{\lambda_1}(e^{\lambda_1 \tau} - 1) & 0 \\ 0 & \frac{1}{\lambda_2}(e^{\lambda_2 \tau} - 1) \end{pmatrix} \qquad (22)$$

For the given numerical values, the unknown scalar parameters $\delta_1(\tau), \delta_2(\tau)$ are bounded in the intervals $[0,0.51]$ and $[0,3.2]$, respectively. An illustration of the polytopic description for this example (based on the Jordan development) is given in Figure 1. The convex description (19) can be obtained using the 4 vertices of the rectangle in the figure, i.e. $\Omega_1 = diag(0,0)$, $\Omega_2 = diag(0,0.51)$, $\Omega_3 = diag(3.2,0)$ and $\Omega_4 = diag(3.2,0.51)$.

In the case of the Jordan normal form, the polytopic set (18) can be obtained using equations (19), (9) with

$$M_i = I + \Omega_i(A + BK).$$

Using such a polytopic set, with a finite number of N vertices M_i, a finite number of LMI conditions can be obtained and the computation of the Lyapunov exponent can be expressed as the following generalized eigenvalue minimization problem [2] in P, X_1, X_2 and α

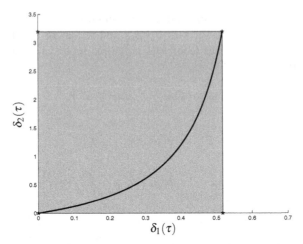

Fig. 1 Illustration of a polytopic embedding for the exponential uncertainty $\Omega(\tau)$ from Example 1. The exponential uncertainty is represented with the bold blue curve and the vertex of the polytopic descriptions are marked by stars.

$$\max \ \alpha \text{ subject to } \exists P \succ 0, X_1, X_2, \text{ s.t.}$$

$$\begin{pmatrix} A^T P + PA + 2\alpha P + X_1 + X_1^T & PBK - X_1 M_i + X_2^T \\ K^T B^T P - M_i^T X_1^T + X_2 & -X_2 M_i - M_i^T X_2^T \end{pmatrix} \prec 0. \tag{23}$$

$$\forall i = 1, \dots, N.$$

3.4 Numerical Improvements

For any of the existing methods [3, 8, 9], a tighter over-approximation of the exponential uncertainty can be obtained by combining the convex embedding method with the gridding method proposed by [7, 18]. The idea is to define a grid $0 = h_1 < h_2 < \dots < h_P = h_{max}$ over the interval $[0, h_{max}]$ and to use the existing analytical methods for constructing locally a convex embedding $\mathscr{S}_j$,

$$\mathscr{S}_j = co\left\{M_{j1}, M_{j2}, \dots, M_{jN^*}\right\} \supset \mathscr{M}_{[h_j, h_{j+1}]},$$

for each sub-interval $[h_j, h_{j+1}]$ with $j = 1, \dots, P$. The global convex embedding $\mathscr{S} \supset \mathscr{M}_{[0, h_{max}]}$ is obtained with

$$\mathscr{S} = co\left\{\mathscr{S}_j, j = 1, \dots, P\right\}.$$

In this case, some of the local vertex can be removed using numerical methods for convex hull and vertex computation. An illustration of this method for the case presented in Example 1 is given in Figure 2 for the case of the Jordan normal form and in Figure 3 for the case of the Taylor development method [9]. The method is applied by replacing the M_i matrices in (23) with the vertex of the polytope $\mathscr{S}$.

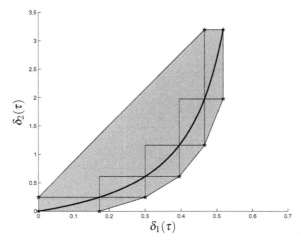

Fig. 2 Illustration of a polytopic embedding for the exponential uncertainty $\Omega(\tau)$ from Example 1 using the Jordan normal form with 5 sub-interval. The global polytope is represented in gray, its vertex are marked by stars. The local polytopes are represented using thin lines.

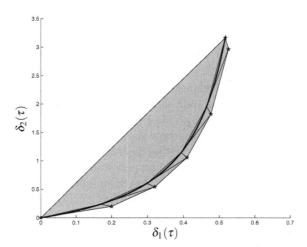

Fig. 3 Illustration of a polytopic embedding for the exponential uncertainty $\Omega(\tau)$ from Example 1 using the Taylor expansion method with 5 sub-interval. The global polytope is represented in gray, its vertex are marked by stars. The local polytopes are represented using thin lines.

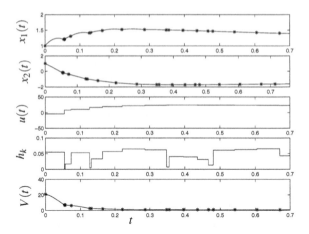

Fig. 4 Simulation with $h_{min} = 0$ and $h_{max} = 0.08$ for the Example 2. The sampling instances are marked by stars.

4 Numerical Example

Example 2. Consider a continuous-time system described by:

$$A = \begin{pmatrix} 1 & 15 \\ -15 & 1 \end{pmatrix} \text{ and } B = \begin{pmatrix} 1 \\ 1 \end{pmatrix}, \ K = (5.33 \ -9.33).$$

We use our method in order to characterize an allowable sampling interval h_{max} for which the stability is ensured despite sampling variations. This comes to finding the maximum value of h_{max} for which the Lyapunov exponent is positive. For this example, the methods of [14] and [16] show that the system is stable for the intervals $[0, 0.014]$ and $[0, 0.033]$, respectively. Using the method proposed here, the stability can be ensured for a time-varying sampling period in the interval $[0, 0.08]$ which shows a significant conservatism reduction. An example of system evolution is given in Figure 4. We use a polytopic embedding based on the Taylor method [10] (10th order development) with 100 local polytopes obtained from an equidistant gridding. The corresponding optimization problems is solved using SEDUMI and YAMLIP [13], a PC with Intel Centrino 2 and 4G RAM, in less than 1 minute.

5 Conclusion

This chapter proposed a method for computing the Lyapunov exponent of sampled-data systems with sampling jitter. The problem was addressed from the continuous time point of view. The basic idea was to take into account the evolution of the delay τ by using an integration operator $\Lambda(\tau)$ that can be treated by means of the exponential uncertainty approach. Numerical examples are given to present the approach and illustrate the improvement in comparison with other classical approaches. This preliminary research has been exploited more generally in [11].

Acknowledgements. The research leading to these results has received funding from the European Community's Seventh Framework Program (FP7/2007-2013) under grant agreement n. 257462: HYCON2 Network of Excellence "Highly-Complex and Networked Control Systems".

References

1. Balluchi, A., Murrieri, P., Sangiovanni-Vincentelli, A.L.: Controller Synthesis on Non-uniform and Uncertain Discrete–Time Domains. In: Morari, M., Thiele, L. (eds.) HSCC 2005. LNCS, vol. 3414, pp. 118–133. Springer, Heidelberg (2005)
2. Boyd, S., El Ghaoui, L., Feron, E., Balakrishnan, V.: Linear Matrix Inequalities in System and Control Theory. SIAM, Philadelphia (1994)
3. Cloosterman, M.B.G., Hetel, L., van de Wouw, N., Heemels, W.P.M.H., Daafouz, J., Nijmeijer, H.: Controller synthesis for networked control systems. Automatica 46(10), 1584–1594 (2010)
4. Donkers, M.C.F., Hetel, L., Heemels, W.P.M.H., van de Wouw, N., Steinbuch, M.: Stability Analysis of Networked Control Systems Using a Switched Linear Systems Approach. In: Majumdar, R., Tabuada, P. (eds.) HSCC 2009. LNCS, vol. 5469, pp. 150–164. Springer, Heidelberg (2009)
5. Felicioni, F.E., Junco, S.J.: A Lie algebraic approach to design of stable feedback control systems with varying sampling rate. In: Proceedings of the 17th IFAC World Congress, Seoul, Korea, pp. 4881–4886 (July 2008)
6. Fridman, E., Seuret, A., Richard, J.-P.: Robust sampled-data stabilization of linear systems: an input delay approach. Automatica 40, 1441–1446 (2004)
7. Fujioka, H.: Stability analysis for a class of networked / embedded control systems: Output feedback case. In: Proceedings of the 17th IFAC World Congress, Seoul, Korea, pp. 4210–4215 (July 2008)
8. Gielen, R., Olaru, S., Lazar, M.: On polytopic embeddings as a modeling framework for networked control systems. In: Proc. 3rd Int. Workshop on Assessment and Future Directions of Nonlinear Model Predictive Control, Pavia, Italy (2008)
9. Hetel, L., Daafouz, J., Iung, C.: Stabilization of arbitrary switched linear systems with unknown time varying delays. IEEE Transactions on Automatic Control (2006)
10. Hetel, L., Daafouz, J., Iung, C.: LMI control design for a class of exponential uncertain systems with application to network controlled switched systems. In: Proceedings of 2007 IEEE American Control Conference (2007)
11. Hetel, L., Kruszewski, A., Perruquetti, W., Richard, J.P.: Discrete and intersample analysis of systems with aperiodic sampling. IEEE Transactions on Automatic Control (2011)
12. Kao, C.-Y., Lincoln, B.: Simple stability criteria for systems with time-varying delays. Automatica 40(8), 1429–1434 (2004)
13. Lofberg, J.: Yalmip: A toolbox for modeling and optimization in MATLAB. In: Proceedings of the CACSD Conference, Taipei, Taiwan (2004), http://control.ee.ethz.ch/~joloef/yalmip.php
14. Mirkin, L.: Some remarks on the use of time-varying delay to model sample-and-hold circuits. IEEE Transactions on Automatic Control 52(6), 1109–1112 (2007) ISSN 0018-9286

15. Naghshtabrizi, P., Hespanha, J.P.: Stability of network control systems with variable sampling and delays. In: Singer, A., Hadjicostis, C. (eds.) Proc. of the Forty-Fourth Annual Allerton Conference on Communication, Control, and Computing (September 2006)
16. Naghshtabrizi, P., Teel, A., Hespanha, J.P.: Exponential stability of impulsive systems with application to uncertain sampled-data systems. Systems and Control Letters 57(5), 378–385 (2008)
17. Oishi, Y., Fujioka, H.: Stability and stabilization of aperiodic sampled-data control systems using robust linear matrix inequalities. Technical Report of the Nanzan Academic Society Information Sciences and Engineering (2009)
18. Sala, A.: Computer control under time-varying sampling period: An LMI gridding approach. Automatica 41(12), 2077–2082 (2005)
19. Skaf, J., Boyd, S.: Analysis and synthesis of state-feedback controllers with timing jitter. IEEE Transactions on Automatic Control 54(3), 652–657 (2009) ISSN 0018-9286
20. Walsh, G.C., Ye, H., Bushnell, L.: Stability analysis of networked control systems. IEEE Transactions on Control Systems Technology, 2876–2880 (1999)
21. Wittenmark, B., Nilsson, J., Torngren, M.: Timing problems in real-time control systems. In: Proceedings of the 1995 American Control Conference, Seattle, WA, USA (1995)
22. Zhang, W., Branicky, M.S., Phillips, S.M.: Stability of networked control systems. IEEE Control Systems Magazine, 84–99 (2001)

A Numerical Method for the Construction of Lyapunov Matrices for Linear Periodic Systems with Time Delay

Olga N. Letyagina and Alexey P. Zhabko

Abstract. A numerical procedure for the construction of Lyapunov matrices is proposed. It is shown that the matrices satisfy an auxiliary two-point boundary value problem for a special delay free system of matrix equations. Some applications of the functionals are also given. In this paper we study the problem of computation of Lyapunov functionals with a prescribed time derivative for the case of delay systems with periodic coefficients. Similar to the case of time invariant systems the functionals are defined by special Lyapunov matrices.

1 Introduction

In numerical analysis, the Runge-Kutta methods form are an important class of implicit and explicit iterative methods for approximation of solutions of ordinary differential equations.

The problems of construction of Lyapunov quadratic functional and analysis of it's properties for linear periodic systems with time delay are treated in this paper. It is shown that these functionals can be used for stability and robust stability analysis of time delay systems.

One of the basic methods for the stability analysis of systems with time delay is the method of Lyapunov functionals [6]. For the case of linear systems a problem of finding quadratic functionals with a preassigned derivative was given in work [2]. One of the basic difficulties in applications of the functionals is a lack of computational algorithms for the corresponding Lyapunov matrices. It has been shown in [3]

Olga N. Letyagina
Faculty of Applied Mathematics and Control Processes,
Saint-Petersburg State University
e-mail: olga.spbu@gmail.com

Alexey P. Zhabko
Faculty of Applied Mathematics and Control Processes,
Saint-Petersburg State University
e-mail: zhabko@apmath.spbu.ru

R. Sipahi et al. (Eds.): Time Delay Sys.: Methods, Appli. and New Trends, LNCIS 423, pp. 265–275.
springerlink.com   © Springer-Verlag Berlin Heidelberg 2012

that the matrices satisfy an auxiliary delay free system of matrix equations with a certain set of two-point boundary conditions. There has been shown that Lyapunov quadratic functionals admit only cubic estimations. In [5] a certain modification has been proposed in order to compute functionals that admit global quadratic lower bounds. The modified functionals have been named as the complete type ones. The complete type Lyapunov quadratic functionals are shown to be useful in solving such problems as estimation of robustness bounds, and calculation of exponential estimates for time delay systems [4].

In this paper we present an extension of the complete type functional theory to the case of systems with periodic coefficients. We show how one can use such functionals for the robust stability analysis of time-delay systems. It is also shown in Section 5 that these functionals admit a lower quadratic bound.

In Section 6 the main attention is paid to the computation of the Lyapunov matrices.

An illustrative example is given in Section 7. Some concluding remarks end the contribution.

2 Preliminaries

In this paper we consider the following time delay system

$$\dot{x}(t) = \sum_{i=0}^{m} A_i(t)x(t - h_i), \tag{1}$$

here $A_j(t)$ are $n \times n$ matrices with periodic, continuous coefficients, i.e., $A_j(t+T) = A_j(t)$, $j = \overline{0,m}$, $0 = h_0 < h_1 < h_2 < ... < h_m$ are positive delays.

We will denote by $x(t,t_0,\varphi)$ the solution of system (1) with a initial conditions $t_0, \varphi \in C([-h_m,0],R^n)$, and by $x_t(t_0,\varphi)$ the following segment $\{x(t+\theta,t_0,\varphi)|\theta \in [-h_m,0]\}$ of the solution. If the initial function is not important or clear of the context we will drop the argument φ in these notations.

The space of initial functions $C([-h_m;0],R^n)$ is provided with the supremum norm
$$||\varphi||_{h_m} = \max_{\theta \in [t_0-h_m,t_0]} ||\varphi(\theta)||.$$

We introduce some basic definitions.

Definition: Let $K(t,s)$ be $n \times n$ matrix function which is defined for $t \geq s$ and satisfy the matrix equation

$$-\frac{\partial K(t,s)}{\partial s} = \sum_{i=0}^{m} K(t,s+h_i)A_i(s+h_i), s < t \tag{2}$$

with the following initial condition: $K(t,s) = 0, s < t, K(t,t) = I$. $K(t,s)$ is known as the fundamental matrix of the system (1).

Definition: System (1) is said to be exponentially stable if there exist $\sigma > 0$ and $\gamma \geq 1$ such that every solution $x(t, \varphi)$ of the system satisfies the following exponential estimate

$$||x(t, \varphi)|| \leq \gamma e^{-\sigma(t-t_0)}||\varphi||_{h_m}, \quad t \geq t_0. \tag{3}$$

Claim. The equation $K(t+T, s+T) \equiv K(t, s)$ is true.

3 Complete Type Lyapunov Functionals

Given symmetric definite positive $n \times n$ matrices $W_j, j = \overline{0,m}, R_j, j = \overline{1,m}$. Let be an initial function. Consider the functional

$$
\omega(x_t) = \sum_{i=0}^{m} x^T (t - h_i) W_i x(t - h_i) +
$$
$$
+ \sum_{i=1}^{m} \int_{-h_i}^{0} x^T (t + \tau) R_i x(t + \tau) d\tau. \tag{4}
$$

We are looking for the functional such that

$$\frac{dv(t, x_t(t_0, \varphi))}{dt} = -\omega(x(t, t_0, \varphi)), \quad t \geq t_0.$$

Using Cauchy formula, see [1], we obtain

$$
v(t, x_t) = x^T (t) U(t, t) x(t) + 2x^T (t) \times
$$
$$
\times \sum_{i=1}^{m} \int_{-h_i}^{0} U(t, \tau + h_i + t) A_i(t + h_i + \tau) x(t + \tau) d\tau +
$$
$$
+ \sum_{i,k=1}^{m} \int_{-h_i}^{0} \int_{-h_k}^{0} \left(x^T (t + \tau_1) A_i^T (t + h_i + \tau_1) \times \right.
$$
$$
\times U(\tau_1 + h_i + t, \tau_2 + h_k + t) A_k(t + h_k + \tau_2) \times
$$
$$
\left. \times x(t + \tau_2) d\tau_1 d\tau_2 \right) + \sum_{i=1}^{m} \int_{-h_i}^{0} x^T (t + \tau) \times
$$
$$
\times [W_i + (h_i + \tau) R_i] x(t + \tau) d\tau. \tag{5}
$$

Here

$$
U(\tau_1, \tau_2) = \int_{\max\{\tau_1, \tau_2\}}^{\infty} K^T(t, \tau_1) W K(t, \tau_2) dt, \tag{6}
$$

and $W = \left[W_0 + \sum_{i=1}^{m} (W_i + h_i R_i) \right].$

The Lyapunov-Krasovskii functional (5) is of the complete type. For the case of exponentially stable system functional (5) is such that there exists $\mu > 0$ for which the following inequality holds

$$\omega(t,\varphi) \leq 2\mu\upsilon(t,\varphi), \quad t \geq 0, \varphi \in C([-h_m,0],R^n).$$

In addition, the existence of Lyapunov-Krasovskii functionals of the complete type is necessary and sufficient for the exponential stability of the differential delay equation (1). Since matrix $U(\tau_1,\tau_2)$ plays in (5) the same role as the classical Lyapunov matrix does for the case of delay free system we shall call it Lyapunov matrix associated with W. In the following we present its properties:
- the dynamic property

Assume that $\tau_2 > \tau_1$, then $\max\{\tau_1,\tau_2\} = \tau_2$.

Partial derivative of (6) with respect to τ_1 and τ_2 are the following:

$$\frac{\partial U}{\partial \tau_1}(\tau_1,\tau_2) = -A_0^T(\tau_1)U(\tau_1,\tau_2) - \sum_{i=1}^{m} A_i^T(\tau_1 + h_i)U(\tau_1 + h_i,\tau_2). \tag{7}$$

$$\frac{\partial U}{\partial \tau_2}(\tau_1,\tau_2) = -U(\tau_1,\tau_2)A_0(\tau_2) - \sum_{i=1}^{m} U(\tau_1,\tau_2 + h_i)A_i(\tau_2 + h_i) - K(\tau_2,\tau_1)W. \tag{8}$$

- the symmetry property

$$U^T(\tau_1,\tau_2) = U(\tau_2,\tau_1); \tag{9}$$

- the algebraic property

If $\tau_1 = \tau_2 = \tau$, then from the dynamic property we have

$$\frac{dU}{d\tau}(\tau,\tau) = -A_0^T(\tau)U(\tau,\tau) - \sum_{i=1}^{m} A_i^T(\tau + h_i)U(\tau + h_i,\tau) - U(\tau,\tau)A_0(\tau) - \sum_{i=1}^{m} U(\tau,\tau + h_i)A_i(\tau + h_i) - W. \tag{10}$$

- the periodic property

$$U(\tau_1 + T,\tau_2 + T) = U(\tau_1,\tau_2). \tag{11}$$

This property follows from periodic property of the fundamental matrix: $K(t + T,\tau + T) = K(t,\tau)$.

Claim. Periodic property of $A_j(t), j = \overline{0,m}$ has not been used for construction of the functionals, Lyapunov matrix and derivation of the matrices properties $(7) - (10)$.

Now we provide lower and upper bounds for the functional. Let $P_j, j = \overline{0, m}$ be definite positive $n \times n$ matrices. We consider a functional

$$u(x_t) = x^T(t)P_0x(t) + \sum_{i=1}^{m}\int_{-h_i}^{0} x^T(t + \theta)P_ix(t + \theta)d\theta. \tag{12}$$

Lemma 1. *Let* $W_j > 0, j = \overline{0, m}, R_j > 0, j = \overline{1, m}$. *If the system (1) is exponentially stable, then for the functional of (5) the following condition is satisfied along the solutions of system (1)*

$$\beta_1 u(x_t) \leq v(x_t), \quad t \geq t_0, \tag{13}$$

where β_1 *is positive constant.*

Lemma 2. *The following condition is satisfied for the functional (5)*

$$v(x_t) \leq \beta_2 u(x_t), \quad t \geq t_0,$$

where β_2 *is a positive constant.*

Theorem 1. *If the system (1) is exponentially stable,* $W_j > 0, j = \overline{0, m}, R_j > 0, j = \overline{1, m}$, *then the following condition is satisfied along the solutions of the system (1)*

$$\alpha_1 ||x(t)||^2 \leq v(x_t) \leq \alpha_2 ||x_t||_{h_m}^2, \quad t \geq t_0.$$

Here α_1, α_2 *are positive constants.*

4   Existence Issue

Lemma 3. *Let* $\hat{U}(\tau_1, \tau_2)$ *be a solution of the equation (7) with initial conditions (9) – (11). Then functional (5) where U replaced by* $\hat{U}$ *satisfies the equality*

$$\frac{dv(t, x_t)}{dt}\bigg|_{(1)} = -\sum_{i=0}^{m} x^T(t - h_i)W_ix(t - h_i) -$$

$$- \sum_{i=1}^{m}\int_{-h_i}^{0} x^T(t + \tau)R_ix(t + \tau)d\tau.$$

Theorem 2. *Let system (1) be exponentially stable, then matrix (6) is the unique solution of matrix equations (7), (9) – (11).*

Proof. The fact that matrix (6) satisfies these equation is trivial. Assume by contradiction that matrix equations (7), (9) – (11) admit at least two solutions $U^{(1)}(\tau_1, \tau_2)$ and $U^{(2)}(\tau_1, \tau_2)$. Each one of the solutions defines the corresponding functional $v^{(j)}(t, x_t), j = 1, 2$, of the form

$$v(t,x_t) = x^T(t)U(t,t)x(t)+$$

$$+2x^T(t)\sum_{i=1}^{m}\int_{-h_i}^{0}\Big[U(t,\tau+h_i+t)A_i(t+h_i+\tau)\times$$

$$\times x(t+\tau)\Big]d\tau + \sum_{i,k=1}^{m}\int_{-h_i}^{0}\int_{-h_k}^{0}\Big(x^T(t+\tau_1)\times \qquad (14)$$

$$\times A_i^T(t+h_i+\tau_1)U(\tau_1+h_i+t,\tau_2+h_k+t)\times$$

$$\times A_k(t+h_k+\tau_2)x(t+\tau_2)d\tau_1\,d\tau_2\Big).$$

Exponential stability of system (1) implies that

$$\Delta v(t_0,\varphi) = v^{(2)}(t_0,\varphi) - v^{(1)}(t_0,\varphi) = 0,$$

for any initial function φ. If we define the difference

$$\Delta U(\tau_1,\tau_2) = U^{(2)}(\tau_1,\tau_2) - U^{(1)}(\tau_1,\tau_2),$$

then

$$0 = \Delta v(t_0,\varphi) = \varphi^T(t_0)U(t_0,t_0)\varphi(t_0)+$$

$$+2\varphi^T(t_0)\sum_{i=1}^{m}\int_{-h_i}^{0}\Big(U(t_0,\tau+h_i+t_0)\times$$
$$\times \Lambda_i(t_0 \mid h_l+\tau)\varphi(t_0+\tau)\Big)d\tau+$$

$$+ \sum_{i,k=1}^{m}\int_{-h_i}^{0}\int_{-h_k}^{0}\Big(\varphi^T(t_0+\tau_1)A_i^T(t_0+h_i+\tau_1)\times \qquad (15)$$

$$\times U(\tau_1+h_i+t_0,\tau_2+h_k+t_0)A_k(t_0+h_k+\tau_2)\times$$

$$\times \varphi(t_0+\tau_2)d\tau_1\,d\tau_2\Big).$$

Given a vector γ, define the piecewise continuous initial function

$$\varphi(t_0+\theta) = \begin{cases} \gamma, & \text{for } \theta = 0, \\ 0, & \text{for } \theta \in [-h_m,0). \end{cases}$$

For this function equality (15) takes the form

$$\gamma^T \Delta U(t_0,t_0)\gamma = 0.$$

As the last equality holds for any vector γ and matrix $\Delta U(t_0, t_0)$ is symmetric one may easily conclude that

$$\Delta U(t_0, t_0) = 0. \tag{16}$$

Now, given two vectors γ and μ, let us select $\tau \in (h_{j-1}, h_j]$ and sufficiently small $\varepsilon > 0$ such that $\tau + \varepsilon < h_{j-1}$. Then we can define the following piece-wise continuous initial function

$$\varphi(t_0 + \theta) = \begin{cases} \gamma, & \text{for } \theta = 0, \\ \mu, & \text{for } \theta \in [-\tau, -\tau + \varepsilon], \\ 0, & \text{for all other points of segment } [-h_m, 0]. \end{cases}$$

If $\varepsilon > 0$ is small then the first integral in (14) is proportional to ε while the double integral is proportional ε^2 so that (14) for this initial function can be written as

$$0 = 2\varepsilon\gamma^T \sum_{i=j}^{m} \left[\Delta U(h_j, \tau + h_i) A_i(\tau + h_i) \right] \mu + o(\varepsilon),$$

where

$$\lim_{\varepsilon \to +0} \frac{o(\varepsilon)}{\varepsilon} = 0.$$

The fact that γ and μ are arbitrary vectors and that ε can be made arbitrarily small implies that

$$\sum_{i=j}^{m} \Delta U(h_j, \tau + h_i) A_i(\tau + h_i) = 0.$$

The last equality holds for all $\tau \in (h_{j-1}, h_j]$, and using continuity of the arguments we arrive at the following condition

$$\sum_{i=j}^{m} A_i^T(\tau + h_i) \Delta U(\tau + h_i, h_j) = 0, \tau \in (h_{j-1}, h_j]. \tag{17}$$

Now, (17) holds for all $j = \overline{1, m}$. For $j = 1$, we therefore obtain the differential equation

$$\frac{\partial \Delta U(\tau_1, \tau_2)}{\partial \tau_1} = -\sum_{i=0}^{m} A_i^T(\tau_1 + h_i) \Delta U(\tau_1 + h_i, \tau_2),$$

if $\tau_2 > \tau_1$, or if $\tau_2 = h_1$ concludes that

$$\frac{d\Delta U(\tau_1, h_1)}{d\tau_1} = -A_0^T(\tau_1) \Delta U(\tau_1, h_1), \tau_1 \in [0, h_1].$$

We are looking for a solution of this equation which satisfies condition (16). The solution of this equation is trivial, that is, $\Delta U(\tau_1, \tau_2) = 0, \tau_1 \in [0, h_1]$, or $U^{(2)}(\tau_1, \tau_2) = U^{(1)}(\tau_1, \tau_2), \tau_1 \in [0, h_1]$.

Considering eqs.(7) and (18) on the interval $(h_1, h_2]$ and for $j = 2$ we obtain the following delay equation $\dfrac{d\Delta U(\tau_1, h_2)}{d\tau_1} = -A_0^T(\tau_1)\Delta U(\tau_1, h_2) - A_1^T(\tau_1)\Delta U(\tau_1 + h_1, h_2), \tau_1 \in [h_1, h_2]$. But on the interval $[0, h_1]$, $\Delta U(\tau_1, \tau_2)$ is constantly 0, therefore $\Delta U(\tau_1, \tau_2) = 0$ for $\tau_1 \in (h_1, h_2]$. Continuing this process, we conclude that $\Delta U(\tau_1, \tau_2) = 0$, $\tau_1, \in [0, h_m]$, i.e. $U_1(\tau_1, \tau_2) = U_2(\tau_1, \tau_2)$ for all $\tau_1 \in [-h_m, h_m]$. The same way we can proof it for variable τ_2. Hence, whenever (1) is exponentially stable the unique solution of (7), (9) – (11) is given by the integral equation (6).

Theorem 3. *Let $x^{(1)}(t) = e^{st}\varphi(t)$ and $x^{(2)}(t) = e^{-st}\psi(t)$ be the solutions of the system (1), where $\varphi(t), \psi(t)$ are periodic vectors and $\varphi(t_0) \neq 0$, $\psi(t_0) \neq 0$. Then there exists W that satisfies (9) – (11) such that system (7) has no solution.*

5 Exponential Estimation and Robust Stability

5.1 Exponential Estimation

We are going to demonstrate that for the case of exponentially stable system (1) functional (5) can be used for the computation of the constants τ and γ. Actually, let in (12) $P_0 = W_0$ and $P_j = R_j$, $j = \overline{1, m}$. From Lemma 2 we get that functional (5) satisfies the estimation $\upsilon(x_t) \leq \beta_2 u(x_t)$. Then one can easily verify, that (12) implies the inequality

$$\frac{d\upsilon(x_t)}{dt} + u(x_t) \leq 0, \quad t \geq t_0,$$

therefore

$$\frac{d\upsilon(x_t)}{dt} + \frac{1}{\beta_2}\upsilon(x_t) \leq 0, \quad t \geq t_0,$$

and finally

$$\upsilon(x_t(\varphi)) \leq \upsilon(\varphi)e^{-\frac{1}{\beta_2}(t - t_0)}.$$

From Theorem 1 we get

$$\alpha_1||x(t)||^2 \leq \upsilon(x_t), \quad \upsilon(\varphi) \leq \alpha_2||\varphi||_{h_m}^2.$$

And we arrive at the following constants

$$\gamma = \sqrt{\frac{\alpha_2}{\alpha_1}}, \quad \sigma = \frac{1}{2\beta_2}.$$

5.2 Robust Stability

Let the nominal system (1) be exponentially stable. In this section we consider the perturbed system

$$\dot{y}(t) = \sum_{j=0}^{m} (A_j(t) + \Delta_j(t)) y(t - h_j). \tag{18}$$

Matrices $\Delta_j(t), j = \overline{0,m}$ are assumed to be piecewise continuous and satisfy the following inequalities

$$\sup_t ||\Delta_j(t)|| \le \rho_j, \quad j = \overline{0,m}, \tag{19}$$

where ρ_j are positive numbers.

The principal goal of the section is to find lower bounds for $\rho_j, j = \overline{0,m}$, such that the perturbed equation remains stable for all $\Delta_j(t), j = \overline{0,m}$, within restrictions (19).

To derive such lower bounds we will use functional (5) constructed for the nominal system (1). The first time derivative of functional (5) along the solutions of Eq. (18) is

$$\frac{d\upsilon(y_t)}{dt} = -\omega(y_t) + 2 \left[\sum_{i=0}^{m} \Delta_i(t) y(t - h_i) \right]^T l(t, y_t), t \ge t_0.$$

Here

$$l(t, y_t) = U(t,t)y(t) +$$

$$+ \sum_{i=1}^{m} \int_{-h_i}^{0} U(t, t + h_i + \theta) A_j(t + h_i + \theta) y(t + \theta) d\theta.$$

Define the constants

$$\upsilon = \max_{\tau_1, \tau_2 \in [0, h_m]} ||U(\tau_1, \tau_2)||; \quad a_j = \sup_t ||A_j(t)||, j = \overline{0,m};$$

$$\lambda_{\min} = \min\{ \min_{0 \le j \le m} [\lambda_{\min}(W_j)], \min_{1 \le j \le m} [\lambda_{\min}(R_j)] \}.$$

The following estimations hold along the solutions of (18)

$$\left\| \sum_{i=0}^{m} \Delta_i y(t - h_i) \right\| \le \left(\sum_{i=0}^{m} \rho_i^2 \right)^{1/2} \left[\sum_{k=0}^{m} ||y(t - h_k)||^2 \right]^{1/2} \times$$

$$\times \left(\sum_{i=0}^{m} \rho_i^2 \right)^{1/2} \sqrt{\frac{\omega(y_t)}{\lambda_{\min}}},$$

and

$$||l(t, y_t)|| \le \upsilon \left(1 + \sum_{i=1}^{m} a_i^2 h_i^2 \right)^{1/2} \times$$

$$\times \left[||y(t)||^2 + \sum_{i=1}^{m} \int_{-h_i}^{0} ||y(t + \theta)||^2 d\theta \right]^{1/2} \le$$

$$\le \upsilon \left(1 + \sum_{i=1}^{m} a_i^2 h_i^2 \right)^{1/2} \sqrt{\frac{\omega(y_t)}{\lambda_{\min}}}.$$

In result we have

$$\frac{dv(y_t)}{dt} \leq -\omega(y_t)\left[1 - \frac{2v}{\lambda_{min}}\left(\sum_{i=1}^m \rho_i^2\right)^{1/2} \times \right.$$

$$\left. \times \left(1 + \sum_{i=1}^m a_i^2 h_i^2\right)^{1/2}\right].$$

So, system (18) remains stable for all perturbations satisfying (19) if values $\rho_j, j = \overline{0,m}$ are such that

$$\sum_{i=0}^m \rho_i^2 \leq \frac{\lambda_{min}^2}{4v^2}\left(1 + \sum_{i=1}^m a_i^2 h_i^2\right).$$

6 Computation Issue

Here we will assume that $T = h$ and system (1) has only one delay term

$$\dot{x}(t) = A_0(t)x(t) + A_1(t)x(t - h), t \geq 0. \tag{20}$$

Let us [7] define matrices
$D(\tau) = U(\tau, \tau)$,
$U_1(\tau_1, \tau_2) = U(\tau_1, \tau_2)$,
$U_2(\tau_1, \tau_2) = A_1^T(\tau_1 + h)U^T(\tau_2, \tau_1 + h)$.
It is evident matrix $D(\tau)$ is symmetrical, $D^T(\tau) = D(\tau)$.
Property (7),

$$\frac{\partial U_1}{\partial \tau_1}(\tau_1, \tau_2) = -A_0^T(\tau_1)U(\tau_1, \tau_2) - A_1^T(\tau_1 + h)U(\tau_1 + h, \tau_2)$$

and

$$\frac{\partial U_2}{\partial \tau_2}(\tau_1, \tau_2) = A_1^T(\tau_1 + h)\frac{\partial U}{\partial \tau_2}(\tau_2, \tau_1 + h) =$$
$$= A_1^T(\tau_1 + h)\left[-A_0^T(\tau_2)U(\tau_2, \tau_1 + h) - \right.$$
$$\left. -A_1^T(\tau_2 + h)U(\tau_1 + h, \tau_2 + h)\right]^T,$$

how takes the form

$$\begin{cases} \dfrac{\partial U_1}{\partial \tau_1}(\tau_1, \tau_2) = -A_0^T(\tau_1)U_1(\tau_1, \tau_2) - U_2(\tau_1, \tau_2), \\[4mm] \dfrac{\partial U_2}{\partial \tau_2}(\tau_1, \tau_2) = -U_2(\tau_1, \tau_2)A_0(\tau_2) - \\[2mm] \qquad\qquad\qquad -A_1^T(\tau_1)U_1(\tau_1, \tau_2)A_1(\tau_2). \end{cases} \tag{21}$$

$$
\begin{cases}
U_1(\tau, \tau) = D(\tau), \\[2mm]
U_2(\tau, \tau) = \dfrac{1}{2}(W + D'(\tau)) - P(\tau) - A_0^T(\tau)D(\tau), \\[2mm]
U_2(0, \tau) = A_1^T(h)U_1^T(\tau, h).
\end{cases}
\tag{22}
$$

Here $P^T(\tau) = -P(\tau)$.

We apply the modified second-order Runge-Kutta method for solving (21). It is assumed here that matrices $D(\tau)$, $P(\tau)$ define the initial conditions for U_1 and U_2.

Let's consider a system of the form

$$
\begin{cases}
\dfrac{\partial U_1}{\partial \tau_1} = f_1(\tau_1, \tau_2, U_1, U_2), \\[3mm]
\dfrac{\partial U_2}{\partial \tau_2} = f_2(\tau_1, \tau_2, U_1, U_2).
\end{cases}
\tag{23}
$$

Let vector functions $f_j, j = 1, 2$ be k times continuously differentiable by τ_j and $(k+1)$ times continuously differentiable by the other variable.

Here matrices $\Phi_1(\tau)$ and $\Phi_2(\tau)$ define the initial condition

$$
U_1(\tau, \tau) = \Phi_1(\tau), U_2(\tau, \tau) = \Phi_2(\tau).
\tag{24}
$$

Theorem 4. *If $\Phi_j \in C^{k+1}(R^1), j = 1, 2$. Then there exist a unique solution of the system (23) with initial conditions (24) which is $(k+1)$ times continuously differentiable.*

References

1. Bellman, R., Cooke, K.: Differential-Difference Equations. Mathematics in Science and Engineering, vol. 6, pp. 480–494. Academic Press, New York (1963)
2. Infante, E., Castelan, W.: A Lyapunov functional for a matrix difference-differential equation. Journal Differential Equations 29, 439–451 (1978)
3. Kharitonov, V.: Lyapunov matrices for a class of time delay systems. Systems and Control Letters 55, 697–706 (2006)
4. Kharitonov, V.: Lyapunov functionals which derivative is set. Herald of the SPbGU 10, 110–117 (2005)
5. Kharitonov, V., Zhabko, A.: Lyapunov- Krasovskii approach to the robust stability analysis of time delay systems. Automatica 39, 15–20 (2003)
6. Krasovskii, N.: On the application of the second method of Lyapunov for equations with time delays. Prikladnaya Matematika i Mekhanika 20, 315–327 (1956)
7. Zhabko, A., Letyagina, O.: Robust stability analysis of linear periodic systems with time delay. Modern Physics 24, 893–907 (2009)

Polytopic Discrete-Time Models for Systems with Time-Varying Delays

Warody Lombardi, Sorin Olaru, and Silviu-Iulian Niculescu

Abstract. This paper focuses on the modeling of linear systems with time-varying input delays. During the discretization process, three cases can be differentiated with respect to the structure of the continuous-state transition matrix: non-defective with real eigenvalues, non-defective with complex-conjugated eigenvalues and defective matrices with real/complex-conjugated eigenvalues. The goal is to model the variable input delays as a polytopic uncertainty in order to preserve a linear difference inclusion framework. It is shown that the *embedding problem* can be translated into a *polytopic containment problem* and an iterative approach to the construction of tight, low complexity embedding (in terms of a simplex body) is proposed.

1 Introduction

It is well-known that the presence of a delay in a system model leads often to complex behaviors (oscillations, instabilities, sensitivity of th closed-loop performances) that do not occur for the systems free of delays (see e.g. [13, 14] and the references therein). In this context, the discrete control of continuous systems affected by variable delays is even more difficult due to the uncertainties introduced by the sampling in the corresponding discrete-time models.

Warody Lombardi
SUPELEC Systems Sciences (E3S) - Automatic Control Department, INRIA - DISCO
and Laboratoire de Signaux et Systemes, CNRS-SUPELEC, Gif-sur-Yvette, France
e-mail: warody.lombardi@supelec.fr

Sorin Olaru
SUPELEC Systems Sciences (E3S) - Automatic Control Department and INRIA - DISCO,
Gif-sur-Yvette, France
e-mail: sorin.olaru@supelec.fr

Silviu-Iulian Niculescu
Laboratoire de Signaux et Systemes, CNRS-SUPELEC and INRIA - DISCO,
Gif-sur-Yvette, France
e-mail: silviu.niculescu@lss.supelec.fr

R. Sipahi et al. (Eds.): Time Delay Sys.: Methods, Appli. and New Trends, LNCIS 423, pp. 277–288.
springerlink.com   © Springer-Verlag Berlin Heidelberg 2012

Some approaches to handle such kind of systems have been proposed recently. In [2, 3] and [15], basic models were discussed and analyzed for the case of non-defective state transition matrices with real eigenvalues. Global approximation-based methods based on Taylor expansion [9], Cayley-Hamilton decomposition [4] can be applied (after an appropriate adjustment of the approximations) and prove that deeper analysis of the nonlinear implications in the model structure can bring valuable understanding on the embedding procedure and further to the stabilizability properties [8].

The aim of the present chapter is to review the linear polytopic models for systems with time-varying input delays derived using the Jordan decomposition and analyze each particular case: non-defective system matrices with real eigenvalues [15], non-defective system matrices with complex-conjugated eigenvalues [11] and defective system matrices with real eigenvalues[12]. As a complement to these basic constructions, we show how it can be optimized the shape of a fix complexity polytopic embedding in order to refine the degree of approximation. The novelty of the approach is to consider the embedding problem from a different angle: computational geometry of the convex bodies [17] and related containment problems [6] using appropriate volume measures [7]. To the best of the authors' knowledge, such an approach is new and it offers the advantage of exploiting the convexity properties in appropriate normed vector spaces.

2   Problem Formulation

Consider a linear continuous-time system with input delay:

$$\dot{x}(t) = A_c x(t) + B_c u(t - h), \tag{1}$$

with $A_c \in \mathbb{R}^{n \times n}$, $B_c \in \mathbb{R}^{n \times m}$ and $h > 0$. Consider a sampling period T_e, the time-instants $t_k = kT_e$ and the state at sampling instants as $x_k = x(t_k)$. As discussed in [18], a certain degree of uncertainty is imposed when dealing with delays, $h = dT_e - \varepsilon$. It is worth mentioning that the variation ε can be time-varying, but it will be supposed in the sequel that the choice of "d" is such that it ensures the boundedness $0 < \varepsilon \le \bar{\varepsilon}$. The discretized version of the model is (1) is given by:

$$x_{k+1} = Ax_k + Bu_{k-d} - \Delta(\varepsilon)(u_{k-d} - u_{k-d+1}), \tag{2}$$

where $\Delta(\varepsilon)$ accounts for the model mismatch imposed by the uncertainty as a function of ε. The matrices $A, B, \Delta(\varepsilon)$ are given by:

$$A = e^{A_c T_e} \; ; \; B = \int_0^{T_e} e^{A_c(T_e - \theta)} B_c d\theta \; ; \; \Delta(\varepsilon) = \int_{-|\varepsilon|}^0 e^{-A_c \tau} B_c d\tau, \tag{3}$$

obtained by assuming piecewise constant control action $u(t) = u_k, \forall t \in [t_k, t_{k+1})$. The objective in the rest of the chapter is to model the uncertainty $\Delta(\varepsilon)$ in terms of a polytopic linear differential inclusion , which covers all the realizations of the delay variation ε.

3   Polytopic Embeddings

The dynamics (2) can be rewritten in an extended form:

$$\xi_{k+1} = A_\Delta \xi_k + B_\Delta u_k, \tag{4}$$

with:

$$\xi_k^T = \begin{bmatrix} x_k \\ u_{k-d} \\ \vdots \\ u_{k-2} \\ u_{k-1} \end{bmatrix} ; A_\Delta = \begin{bmatrix} A & B-\Delta(\varepsilon) & \Delta(\varepsilon) & \cdots & 0 \\ 0 & 0 & I_m & \cdots & 0 \\ \cdots & & \cdots & & \cdots \\ 0 & 0 & 0 & \cdots & I_m \\ 0 & 0 & 0 & \cdots & 0 \end{bmatrix} ; B_\Delta = \begin{bmatrix} 0 \\ 0 \\ \vdots \\ 0 \\ I_m \end{bmatrix}. \tag{5}$$

The confinement of the function $\Delta(\varepsilon)$ expressed by (3) in a polytope in order to cover all the possible realizations in terms of the variable delay ε will use as a basic element the Jordan canonical form decomposition. We consider here the the Jordan decomposition of the transition matrix $A_c = V\Sigma V^{-1}$, with block diagonal matrix $\Sigma \in \mathbb{R}^{n \times n}$. As developed by [12], one has:

$$\Sigma = \begin{bmatrix} \Sigma_{1,m_1} & \cdots & 0 \\ \vdots & \ddots & \vdots \\ 0 & \cdots & \Sigma_{p,m_p} \end{bmatrix}, \tag{6}$$

for all $i \in [1,...,p]$ which leads $\Sigma_{i,m_i} \in \mathbb{R}^{m_i \times m_i}$ with:

$$\Sigma_{i,m_i} = \begin{bmatrix} \sigma_i & 1 & \cdots & 0 & 0 \\ \vdots & \vdots & \ddots & \vdots & \vdots \\ 0 & 0 & \cdots & \sigma_i & 1 \\ 0 & 0 & \cdots & 0 & \sigma_i \end{bmatrix} = \sigma_i \underbrace{\begin{bmatrix} 1 & & \\ & \ddots & \\ & & 1 \end{bmatrix}}_{\Lambda_i} + \underbrace{\begin{bmatrix} 0 & 1 & 0 & \cdots & 0 \\ \vdots & \vdots & \ddots & \ddots & \vdots \\ 0 & 0 & 0 & \cdots & 1 \\ 0 & 0 & 0 & \cdots & 0 \end{bmatrix}}_{\Gamma_i}.$$

One can further decompose (6) explicitly as follows:

$$\Sigma = \sum_{i=1}^{p} \sigma_i \underbrace{\begin{bmatrix} 0 & & & \\ & \ddots & & \\ & & \Lambda_i & \\ & & & \ddots \\ & & & & 0 \end{bmatrix}}_{L_i} + \sum_{i=1}^{p} \underbrace{\begin{bmatrix} 0 & & & \\ & \ddots & & \\ & & \Sigma_i & \\ & & & \ddots \\ & & & & 0 \end{bmatrix}}_{G_i},$$

leading to a compact form:

$$\Sigma = \sum_{i=1}^{p} \sigma_i L_i + G_i,$$

where L_i are diagonal matrices of the size of Σ with "1"s on the diagonal of the cell corresponding to the Jordan block Σ_{i,m_i}. Similarly, G_i are diagonal matrices of the size of Σ with "1"s on the upper-diagonal of the cell corresponding to the Jordan block Σ_{i,m_i}.

With these notations the function $\Delta(\varepsilon)$ in equation (3) can be rewritten as:

$$\Delta(\varepsilon) = \int_0^\varepsilon e^{A_c \tau} B_c d\tau = V \int_0^\varepsilon e^{\Sigma \tau} d\tau \, V^{-1} B_c.$$

By exploiting the structure of the Jordan blocks one gets:

$$e^{\Sigma_{i,m_i} \tau} = exp(\sigma_i \Lambda_i \tau) exp(\Gamma_i \tau), \tag{7}$$

where $exp(X)$ is the matrix $Y = exp(X) \in \mathbb{R}^{n \times m}$ which satisfies:

$$Y_{i,j} = \begin{cases} 0 & \text{if } X_{i,j} = 0, i \in \mathbb{Z}_{[1,n]}, j \in \mathbb{Z}_{[1,m]} \\ e^{X_{i,j}} & \text{if } X_{i,j} \neq 0, i \in \mathbb{Z}_{[1,n]}, j \in \mathbb{Z}_{[1,m]}. \end{cases} \tag{8}$$

Three cases (developed in the next subsections) can be distinguished:

1. Jordan blocks Σ_{i,m_i} with $m_i = 1$ corresponding to real eigenvalues with algebraic multiplicity equal to the geometric multiplicity. In this case G_i is identically zero and Λ_i are diagonal.
2. Jordan blocks Σ_{i,m_i} corresponds to a repeated eigenvalue with geometrical multiplicity m_i strictly inferior to the algebraic multiplicity.
3. Jordan blocks complex eigenvalues for which a direct simplex embedding will be replaced by a zonotopic embedding and further refined.

3.1 Non-defective System Matrix with Real Eigenvalues

The term $\Delta(\varepsilon)$ of Eq. (3) can be written as:

$$\Delta(\varepsilon) = \int_0^\varepsilon e^{A_c \tau} B_c d\tau = A_c^{-1}(e^{A_c \varepsilon} - I_n) B_c,$$

then for the limit values of ε, the extreme realizations can be obtained by the Jordan canonical form of A_c [15]:

$$\Delta_0 = \mathbf{0}_{n \times m}; \ \Delta_i = V \int_{-\bar\varepsilon}^0 exp(-L_i \tau) d\tau V^{-1} B_c, \forall i = 1, \dots, n.$$

Theorem 1. For any $0 \leq \varepsilon \leq \bar\varepsilon$ the state matrix A_Δ satisfies:

$$A_\Delta \in Co\{A_{\Delta_0}, A_{\Delta_1}, \dots, A_{\Delta_n}\}, \tag{9}$$

where Co{.} denotes the convex hull and vertices A_i are given by:

$$A_{\Delta_i} = \begin{bmatrix} A & B - \Delta_i & \Delta_i & \cdots & 0 \\ 0 & 0 & I_m & \cdots & 0 \\ \cdots & \cdots & \cdots & \ddots & \cdots \\ 0 & 0 & 0 & \cdots & I_m \\ 0 & 0 & 0 & \cdots & 0 \end{bmatrix}, \quad i = 0,\ldots,n. \tag{10}$$

Proof. For any $0 \le \varepsilon \le \bar{\varepsilon}$ and $i = 1,\ldots,n$ there exists $0 \le \beta_i \le 1$ such that:

$$\Delta(\varepsilon) = \int_{-\varepsilon}^{0} e^{-A_c\tau} B_c d\tau = V \int_{-\varepsilon}^{0} e^{-\Sigma\tau} d\tau V^{-1} B_c =$$

$$= (n - \sum_{i=1}^{n} \beta_i) \underbrace{0_{n \times m}}_{\Delta_0} + \sum_{i=1}^{n} \beta_i V \underbrace{\int_{-\bar{\varepsilon}}^{0} \exp(-L_i\tau) d\tau V^{-1} B_c}_{\Delta_i}$$

$$= \underbrace{\frac{(n - \sum_{i=1}^{n} \beta_i)}{n}}_{\alpha_0} n\Delta_0 + \sum_{i=1}^{n} \underbrace{\frac{\beta_i}{n}}_{\alpha_i} n\Delta_i.$$

The matrix $\Delta(\varepsilon)$ appears in a linear manner in the structure of A_Δ as it can be seen in (5) and using the scalars $\alpha_i \ge 0, i = 0,\ldots,n$ found before, one can write $A_\Delta = \sum_{i=0}^{n} \alpha_i A_i$. By noting that $\sum_{i=0}^{n} \alpha_i = 1$ the proof is completed. ∎

3.2 Defective System Matrix with Real Eigenvalues

When the Jordan blocks Σ_{i,m_i} correspond to a repeated eigenvalue with geometrical multiplicity strictly superior to the geometrical multiplicity, by using the Taylor expansion of $e^{\Sigma_{i,m_i}\tau}$ up to the $(m_i - 1)^{th}$ term, Eq. (6) becomes:

$$e^{\Sigma_{i,m_i}\tau} = \exp(\sigma_i\Lambda_i\tau)\left(I + \frac{1}{1!}\Gamma_i\tau + \ldots + \frac{1}{(m_i-1)!}\Gamma_j^{m_i-1}\tau^{m_i-1}\right).$$

The development is restricted to the $(m_i - 1)^{th}$ term by the fact Σ_i is nilpotent with degree m_i. In this case, the embedding problem can be decoupled on the embedding of the blocks:

$$\Delta_i(\varepsilon) = V L_i \int_0^\varepsilon \exp(\sigma_i\tau) d\tau V^{-1} B_c;$$

$$\Delta_{i,1}(\varepsilon) = V \tfrac{1}{1!} G_i \int_0^\varepsilon \tau\exp(\sigma_i\tau) d\tau V^{-1} B_c; \tag{11}$$

$$\vdots$$

$$\Delta_{i,m_i-1}(\varepsilon) = V \tfrac{1}{(m_i-1)!} G_i^{m_i-1} \int_0^\varepsilon \tau^{m_i-1}\exp(\sigma_i\tau) d\tau V^{-1} B_c.$$

The functions $\tau^k e^{\sigma_i \tau}$ with $i \in [1,...,p]$, $k \in [0,...,m_i - 1]$ are monotones and positive functions on $\tau \in [0,\bar{\varepsilon}]$. For a given $\bar{\varepsilon}$ we have m_i monotone and positive functions on $\tau \in [0,\bar{\varepsilon}]$. Noting these functions as $g_1(\tau),...,g_{m_i}(\tau) : [0,\bar{\varepsilon}] \to \mathbb{R}_+$, there exists an ordered sequence $i_1,...,i_{m_i}$ with $i_k \in [0,...,m_i - 1]$ such that:

$$g_{i_1}(\bar{\varepsilon}) \geq g_{i_2}(\bar{\varepsilon}) \geq ... \geq g_{i_{m_i}}(\bar{\varepsilon}).$$

For all $g_i(\varepsilon)$ there exists a $\Delta(\varepsilon)$ and by its correspondences there exists $\Delta_{g_{i_k}}(\varepsilon)$ for all $g_{i_k}(\varepsilon)$.

Theorem 2. *For any $0 \leq \varepsilon \leq \bar{\varepsilon}$ there exists Δ_i satisfying:*

$$\Delta(\varepsilon) \in Co\{\Delta_0, \Delta_{g_1}(\bar{\varepsilon}),...,\Delta_{g_n}(\bar{\varepsilon})\}.$$

Proof. For all $\varepsilon \in [0,\bar{\varepsilon}]$, there exists $0 \leq \beta_i \leq 1$ such that:

$$\Delta_{g_i}(\varepsilon) = \beta_i \Delta_{g_i}(\bar{\varepsilon}),$$

which can be proved by the monotonicity of $\Delta_{g_i}(\varepsilon)$:

$$\Delta(\varepsilon) = \sum_{i=1}^{n} \{\beta_i \Delta_{g_i} + (1 - \beta_i)\Delta_0\}.$$

It can be easily checked that $\alpha_i \leq 1, \forall i = 1,...,n$ and $\sum_{i=0}^{n} \alpha_i = 1$.

Then for a given $\varepsilon \in [0, \bar{\varepsilon}]$ we have an ordering of the convex combination coefficients β_i:

$$\beta_1 \leq \beta_2 \leq \cdots \leq \beta_n,$$

which implies $0 \leq \alpha_i$ and concludes the proof. ∎

The previous approach provides a (simple) automatic construction of the embedding but it may be proven to be conservative. A better understanding of the structure of $\Delta(\varepsilon)$ can be reached by exploiting the monotonicity characteristics of the exponential functions, and further conduct to a less conservative embedding. A candidate for a less conservative approach of (11) can be obtained using the following combinations:

$$\Delta_j(\varepsilon) = V \left(I + \tfrac{1}{1!}\Gamma + ... + \tfrac{1}{(p-1)!}\Gamma^{p-1}\right) \int_0^{\varepsilon} \tau^{p-1} exp(\sigma_j \tau) d\tau \, V^{-1} B_c;$$

$$\Delta_{j,1}(\varepsilon) = V \left(I + ... + \tfrac{1}{(p-2)!}\Gamma^{p-2}\right) \int_0^{\varepsilon} \left(-\tfrac{1}{(p-1)!}\tau^{p-1} + \tau^{p-2}\right) exp(\sigma_j \tau) d\tau \, V^{-1} B_c;$$

$$\vdots$$

$$\Delta_{j,m_j-1} = V \int_0^{\varepsilon} (-\tau + 1) exp(\sigma_j \tau) d\tau \, V^{-1} B_c.$$

The remaining Δ_{j,m_k}, $k \in [1,...,n-p]$ elements can be derived by an appropriate permutation between the above elements:

$$\Delta_{j,m_j} = \Delta_j(\varepsilon) + \Delta_{j,1}(\varepsilon),$$
$$\Delta_{j,m_j+1} = \Delta_j(\varepsilon) + \Delta_{j,2}(\varepsilon),$$
$$\vdots$$
$$\Delta_{j,m_k} = \Delta_j(\varepsilon) + \Delta_{j,1}(\varepsilon) + \ldots + \Delta_{j,k-1}(\varepsilon).$$

3.3 Non-defective System Matrix with Complex-Conjugated Eigenvalues

For this class of systems, the function $\Delta(\varepsilon)$ is non-monotone (this can be assimilated with is of spiral-type curve). The containment by a simplex is a difficult task, and direct methods based on hypercubes (elementwise bounds) will provide a basic (sometimes conservative) embeddings [11]. By exploiting the Jordan decomposition of $A_c = V \Sigma V^{-1}$:

$$\Delta(\varepsilon) = A_c^{-1} \left(e^{A_c \bar{\varepsilon}} - 1 \right) B_c = \underbrace{V \Sigma^{-1} e^{\Sigma \bar{\varepsilon}} V^{-1} B_c}_{\theta} - V \Sigma^{-1} V^{-1} B_c,$$

two independent cases need to be treated:

(i) A_c stable - In this case, the bounds corresponding to each Jordan block can be obtained by considering the forward dynamic for the autonomous system $\dot{\theta} = \Sigma \theta$, and $A_c^{-1} B_c$ as a settling point. A polyhedral containment can be obtained in terms of:

$$|\theta| \leq \left| V \Sigma^{-1} \right| \left| V^{-1} B_c \right|, \tag{12}$$

where the notation $|\theta|$ represents the elementwise absolute value of the vector θ.

(ii) A_c instable - In this case, ε varies in the interval $[0, \bar{\varepsilon}]$. The non-monotone behavior can be represented by the trajectory of a linear system driven in the reverse time by a transfer matrix $-A_c$ which is stable. Similar to the case i), a polytopic embedding can be obtained

$$|\theta| \leq \left| V \Sigma^{-1} \right| e^{\Sigma \bar{\varepsilon}} \left| V^{-1} B_c \right|. \tag{13}$$

These basic embeddings can be used in a recursive procedure in order to construct a *tight* polytopic model using the methodology described in the sequel (see, for instance, Section 4).

4 Towards a Less Conservative Simplex Embedding

Less conservative polytopic embeddings with $n + 1$ generators in a n-dimensional space can be obtained by defining an appropriate optimization problem, as developed in [11] and [12]. For low complexity embeddings, the Jordan normal form represents the main ingredient, and we will consider here the case $D(\varepsilon) = \int_0^\varepsilon e^{\Sigma \tau} d\tau$ (the case of repeated eigenvalues can be treated in a similar manner and for the sake of brevity is omitted). A cost function taking into account the topology of the embedding has to measure the *"size"* of an n-dimensional object created as a convex hull of generators. By the diagonal structure of D, one can map the diagonal elements in a n-dimensional vector and cast the embedding problem in the form of volume minimization for a simplex body in a n-dimensional space (represented by $n+1$ vertices).

Consider now a polytopic embedding of $D(\varepsilon)$ with $n+1$ generators:

$$\mathscr{E} = Co\{D_0, D_1, \dots, D_n\},$$

where $D_i = diag(d_{i,1}, \dots, d_{i,n})$ will be represented by the following vectors:

$$d_i^\top = \begin{bmatrix} d_{i,1} & d_{i,2} & \dots & d_{i,n} \end{bmatrix}^\top ; \forall i \in \mathbb{Z}_{[0,n]}.$$

The less conservative simplex embedding is obtained as solution of the following *optimization problem*:

$$\min_{d_0,\dots,d_n} \quad Vol(Co\{d_0,\dots,d_n\}) = \frac{1}{n!} \left| det \begin{bmatrix} d_0 & \dots & d_n \\ 1 & \dots & 1 \end{bmatrix} \right|.$$
$$\text{subject to } D(\varepsilon) \in Co\{d_0,\dots,d_n\}$$

This problem is similar to the approximation of a geometrical body by a combinatorially simpler shape. The nonlinear structure (especially for the constraints) will influence the convergence of the optimization, making the global approach not very interesting at least for large n. Local embeddings can be constructed upon the splitting of $\Delta(\bar{\varepsilon})$ like $[0, \bar{\varepsilon}]$ into ℓ subintervals $[0, \varepsilon_1] \cup [\varepsilon_1, \varepsilon_2] \cup \dots \cup [\varepsilon_{\ell-1}, \bar{\varepsilon}]$.

Then problem of finding the minimum volume simplex enclosing a given polyhedral set is to be solved for the set of extreme points generated by the local embeddings $D(\varepsilon) \in \bigcup_{i=0}^{\ell-1} \mathscr{E}_{\varepsilon_i, \varepsilon_{i+1}}$, instead of the conservative nonlinear expression of $D(\varepsilon)$:

$$\min_{d_0,\dots,d_n} \quad \frac{1}{n!} \left| det \begin{bmatrix} d_0 & \dots & d_n \\ 1 & \dots & 1 \end{bmatrix} \right|$$
$$\text{subject to } d_i^{[\varepsilon_k, \varepsilon_{k+1}]} \in Co\{d_0, \dots, d_n\}, k = 0, \dots, \ell$$

For this problem, efficient optimization routines can be obtained by reformulating it as an instance of polynomial programming with linear constraints. A straightforward way of expressing the enclosing of a set of points is to use the half-space representation which expresses a simplex in $\mathbb{R}^n$ by the $n+1$ bounding hyperplanes:

$$h_{i0} + \sum_{i=0}^{n} h_{ij} x_j = 0 \qquad (i = 0, 1, \dots, n).$$

By noting $H \in \mathbb{R}^{(n+1) \times (n+1)}$ the matrix with elements h_{ij}, $0 \leq i, j \leq n$, the simplicial constraints are translated in:

$$Ad_i^{[\varepsilon_k, \varepsilon_{k+1}]} \leq 0, k = 0, \dots, \ell$$

The cost function can be adapted using the next result:

Theorem 3. *The volume of a simplex S described by its bounding hyperplanes is given by:*

$$Vol(S) = \frac{|det(H)|^n}{n! \Pi_{i=0}^n H_{i0}},$$

where H_{i0} is the cofactor of h_{i0} in H.

Proof. See [6] for details.   ∎

With these elements, a tractable optimization problem can be formulated for the reduced complexity embedding of $\Delta(\varepsilon)$:

$$
\begin{aligned}
\min_{H} \quad & \frac{|det(H)|^n}{n! \Pi_{i=0}^n H_{i0}} \\
\text{subject to} \quad & H d_i^{[\varepsilon_k, \varepsilon_{k+1}]} \leq 0, k = 0, \ldots, \ell; \ i = 1, \ldots, n.
\end{aligned}
\tag{14}
$$

Algorithm 1 describes the recursive procedure which provides a δ-approximation of the minimal-volume embedding with $n + 1$ generators for the function $\Delta(\varepsilon)$.

Algorithm 1. Computes $\mathscr{E}^a(\delta)$.

Require: $A_c, B_c, \bar{\varepsilon}, \delta$
Ensure: $\{\Delta_0, \Delta_1, \ldots, \Delta_n\}$
1: Compute an initial embedding based on Sections 3.1, 3.3 and 3.2
2: Sample the intervals $[0, \bar{\varepsilon}/2]$ and $[\bar{\varepsilon}/2, \bar{\varepsilon}]$
3: For each pair of consecutive points in $[0, \bar{\varepsilon}]$, construct a local embedding. Collect all the resulting extreme points by eliminating the redundancies.
4: Solve (14) with the constraints given by the points obtained at step 3 using as initial conditions the existing embedding or the one obtained in step 1
5: **if** $\delta < precision$ **then**
6:     **return** Polytope Δ_i
7: **else**
8:     Identify the local embedding (i^*) with maximum volume and extreme point placed on the boundary of the global simplex embedding $\mathscr{E}^a$
9:     Divide the i^*-th interval $[\varepsilon_{i^*}, \varepsilon_{i^*+1}]$
10:    Return to step 2
11: **end if**

The generators of the embedding will be obtained from a classical transformation from half-space representation to the vertex representation (inexpensive in the case of a simplex). The last points to be clarified is the convergence of the algorithm and the degree of approximation of the solution found by solving the problems proposed in Sections 3.1, 3.3 and 3.2, with respect to the minimum volume ellipsoid.

Proposition 1. *Let $\mathscr{E}_\ell^a$ and $\mathscr{E}_{\ell+1}^a$ be the approximations obtained at two consecutive iterations of the Algorithm 1.*

$$Vol(\mathscr{E}_\ell^a) \geq Vol(\mathscr{E}_{\ell+1}^a).$$

Proof. These two simplex approximations are built upon the decomposition of the interval $[0,\bar{\varepsilon}]$ in: $[0,\varepsilon_1^\ell] \cup [\varepsilon_1^\ell, \varepsilon_2^\ell] \cup \cdots \cup \cdots [\varepsilon_{i-1}^\ell, \varepsilon_i^\ell] \cup [\varepsilon_{\ell-1}^\ell, \bar{\varepsilon}]$ and $[0, \varepsilon_1^{\ell+1}] \cup [\varepsilon_1^{\ell+1}, \varepsilon_2^{\ell+1}] \cup \cdots \cup \cdots \cup [\varepsilon_{i-1}^{\ell+1}, \varepsilon_i^{\ell+1}] \cup [\varepsilon_i^{\ell+1}, \varepsilon_{i+1}^{\ell+1}] \cup [\varepsilon_\ell^{\ell+1}, \bar{\varepsilon}]$. The subintervals are identical excepting the fact that $[\varepsilon_{i-1}^\ell, \varepsilon_i^\ell]$ is decomposed in $[\varepsilon_{i-1}^{\ell+1}, \varepsilon_i^{\ell+1}] \cup [\varepsilon_i^{\ell+1}, \varepsilon_{i+1}^{\ell+1}]$ at the second iteration. Next the local embedding for $\Delta(\varepsilon) = \int_{\varepsilon_i}^\varepsilon e^{A_c\tau} B_c d\tau$, with $\forall \varepsilon \in [\varepsilon_{i-1}, \varepsilon_i]$ is denoted $\mathscr{E}_\ell^i$ and the ones corresponding to $[\varepsilon_{i-1}^{\ell+1}, \varepsilon_i^{\ell+1}] \cup [\varepsilon_i^{\ell+1}, \varepsilon_{i+1}^{\ell+1}]$ by $\mathscr{E}_{\ell+1}^i \cup \mathscr{E}_{\ell+1}^{i+1}$. But $\mathscr{E}_\ell^i \supset \{\mathscr{E}_{\ell+1}^i \cup \mathscr{E}_{\ell+1}^{i+1}\}$ and recalling $\Delta(\varepsilon)$ which can be rewritten compactly as the volume minimization of $Vol(\mathscr{E}^a)$ subject to $\mathscr{E}^a \supset Co\{\mathscr{E}_\ell^0, \ldots, \mathscr{E}_\ell^\ell\}$ it becomes obvious that the set of constraints for $\mathscr{E}_{\ell+1}^a$ is redundant with respect to the one for $\mathscr{E}_\ell^a$. By consequence the corresponding optimum satisfies $Vol(\mathscr{E}_\ell^a) \geq Vol(\mathscr{E}_{\ell+1}^a)$. $\blacksquare$

Definition 1. Consider the extreme point d_i of a local embedding $\mathscr{E}_\ell^i$ that are placed on the border of $\mathscr{E}_\ell^a$. By noting d_i^c the most distanced extreme point from d_i inside $\mathscr{E}_\ell^i$, and $d_i^\mathscr{E} \in d(\varepsilon)$ the most distanced point from d_i inside $d(\varepsilon)$ for $\forall \varepsilon \in [0, \bar{\varepsilon}]$. One can define c_i as the ratio:

$$c_i = \frac{dist(d_i, d_i^c)}{dist(d_i, d_i^\mathscr{E})} \tag{15}$$

where $dist(.,.)$ is understood as the Euclidean distance between 2 points in $\mathbb{R}^n$.

Theorem 4. *Let the minimum volume simplex embedding for $\int_0^\varepsilon e^{\Sigma\tau} d\tau$, $\varepsilon \in [0, \bar{\varepsilon}]$ be $\mathscr{E}$ and the simplex resulting from the solution of Δ_i with respect to the local embedding on $[0, \varepsilon_1] \cup [\varepsilon_1, \varepsilon_2] \cup \cdots \cup [\varepsilon_{\ell-1}, \bar{\varepsilon}]$ be $\mathscr{E}^a$. Then, by considering the constant $c^* = \max c_j \in \mathbb{R}_+$ over all the scalars defined by (15), the following relationship is verified:*

$$Vol(c^* \mathscr{E}^a) \geq \left[(Vol(\mathscr{E}^a)^{\frac{1}{n}}) - (Vol(\mathscr{E})^{\frac{1}{n}})\right]^n$$

Proof. Consider $\mathscr{C} = Co\{d(\varepsilon), \forall \varepsilon \in [0, \bar{\varepsilon}]\}$. Then there exist a homotetic transformation to cover all the local embeddings $\mathscr{C} + c^* \mathscr{C} \supseteq Co\{\mathscr{E}_i, \forall i = 1, \ldots, l\}$ where the set addition is considered in the Minkowski sense. The ideal simplex embedding $\mathscr{E} \supset \mathscr{C}$ implies that $\{\mathscr{E} + c^* \mathscr{E}\} \supset \{\mathscr{C} + c^* \mathscr{C}\}$ and using the Brunn-Minkowski theorem [17] with equality instead of inequality due to the homotetic transformation and the fact that $\mathscr{E}^a$ is minimum-volume for the approximated containment problem:

$$Vol\{\mathscr{E} + c^* \mathscr{E}\} \geq Vol\{\mathscr{E}^a\},$$
$$Vol\{\mathscr{E} + c^* \mathscr{E}\}^{\frac{1}{n}} \geq Vol\{\mathscr{E}^a\}^{\frac{1}{n}},$$
$$Vol\{\mathscr{E}\}^{\frac{1}{n}} + Vol\{c^* \mathscr{E}\}^{\frac{1}{n}} \geq Vol\{\mathscr{E}^a\}^{\frac{1}{n}},$$
$$Vol\{c^* \mathscr{E}\} \geq \left[Vol\{\mathscr{E}^a\}^{\frac{1}{n}} - Vol\{\mathscr{E}\}^{\frac{1}{n}}\right]^n.$$

But $Vol\{c^* \mathscr{E}^a\} \geq Vol\{c^* \mathscr{E}\}$ and the proof is complete. $\blacksquare$

The previous result provides an effective way of measuring the degree of approximation for the step 5 of Algorithm 1. It should be however noticed that c^* can increase between two iterations of the algorithm but this phenomenon will not affect the convergence of the algorithm as long as the Proposition 1 guarantees the monotone decreasing of the volume of the approximation.

5 Conclusion

A generic approach to obtain polytopic models for systems with input affected by time-delays was proposed. At first, the uncertainty containment by polyhedra (simplex or hyperbox type) were based on the Jordan decomposition using very simple calculations. In a second stage, the conservativeness is reduced upon an iterative procedure which refine the local embeddings by splitting the delay variation in subintervals. The technique guarantees a fixed complexity (simplex-type) global embedding. Subsequently, this can be used to characterize the discrete-time dynamics by a polytopic difference inclusion.

References

1. Blanchini, F.: Set invariance in control. Automatica 35, 1747–1767 (1999)
2. Cloosterman, M.B.G., Van de Wouw, N., Heemels, W.M.P.H., Nijmeijer, H.: Robust Stability of Networked Control Systems with Time-varying Network-induced Delays. In: Proceedings of the 45th IEEE Conference on Decision and Control (2006)
3. Cloosterman, M.B.G., Van de Wouw, N., Heemels, W.M.P.H., Nijmeijer, H.: Stability of Networked Control Systems with Large Delays. In: Proceedings of the 46th IEEE Conference on Decision and Control (2007)
4. Gielen, R.H., Olaru, S., Lazar, M., Heemels, W.P.M.H., van de Wouw, N., Niculescu, S.-I.: On polytopic inclusions as a modeling framework for systems with time-varying delays. Automatica 46(3), 615–619 (2010)
5. Gilbert, E.G., Tan, K.T.: Linear systems with state and control constraints, The theory and application of maximal output admissible sets. IEEE Transactions on Automatic Control 36, 1008–1020 (1991)
6. Gritzmann, P., Klee, V.: On the complexity of some basic problems in computational convexity: I. Containment problems. Discrete Mathematics 136(1-3), 129–174 (1994)
7. Gritzmann, P., Klee, V.: On the complexity of some basic problems in computational convexity: II. Volume and mixed volumes. NATO ASI Series C Mathematical and Physical Sciences-Advanced Study Institute 440, 373–466 (1994)
8. Heemels, W.P.M.H., van de Wouw, N., Gielen, R.H., Donkers, M.C.F., Hetel, L., Olaru, S., Lazar, M., Daafouz, J., Niculescu, S.: Comparison of overapproximation methods for stability analysis of networked control systems. In: Proceedings of the 13th ACM International Conference on Hybrid Systems: Computation and Control, pp. 181–190. ACM (2010)
9. Hetel, L., Daafouz, J., Iung, C.: Stabilization of arbitrary switched linear systems with unknown time varying delays. IEEE Transactions on Automatic Control 169(10), 1668–1674 (2006)
10. Kothare, M.V., Balakrishnan, V., Morari, M.: Robust Constrained Model Predictive Control using Linear Matrix Inequalities. Automatica 32(10), 1361–1379 (1996)

11. Lombardi, W., Olaru, S., Niculescu, S.-I.: Invariant sets for a class of linear systems with variable time-delay. In: Proceedings of the European Control Conference (2009)
12. Lombardi, W., Olaru, S., Niculescu, S.-I.: Robust invariance for a class of time-delay systems with repeated eigenvalues. In: Proceedings of the IFAC TDS (2009)
13. Michiels, W., Niculescu, S.-I.: Stability and stabilization of time-delay systems. SIAM (2007)
14. Niculescu, S.-I.: Delay Effects on Stability: A Robust Control Approach. LNCIS. Springer, Heidelberg (2001)
15. Olaru, S., Niculescu, S.-I.: Predictive control for systems with delay and constraints. In: Proceedings of the IFAC World Congress (2008)
16. Olaru, S., Niculescu, S.-I.: Predictive control for systems with variable time-delay. Robust positive invariant set approximations. In: Proceedings of the 5th International Conference on Informatics in Control (2008)
17. Scheneider, R.: Convex bodies: The Brunn-Minkowski Theory. Cambridge University Press (1993)
18. Lozano, R., Gil, P.G., Castillo, P., Dzul, A.: Robust Prediction-Based Control for Unstable Delay Systems. In: Niculescu, S.I., Gu, K. (eds.) Advances in Time- Delay Systems, pp. 311–326. Springer, Heidelberg (2004)

Part IV
Predictor-Based Control and Compensation

Predictor Feedback: Time-Varying, Adaptive, and Nonlinear

Miroslav Krstic

Abstract. We present a tutorial introduction to methods for stabilization of systems with long input delays—the so-called "predictor feedback" techniques. The methods are based on techniques originally developed for boundary control of partial differential equations using the "backstepping" approach. We start with a consideration of linear systems, first with a known delay and then subject to a small uncertainty in the delay. Then we study linear systems with constant delays that are completely unknown, which requires an adaptive control approach. For linear systems, we also present a method for compensating arbitrarily large but known time-varying delays. Next, we consider nonlinear control problems in the presence of arbitrarily long input delays. Finally, we close with a design for general nonlinear systems with delays that have a general dependency on the system state.

1 Introduction

An enormous wealth of knowledge and research results exists for control of systems with state delays and input delays [1]. Problems with long input delays, for unstable plants, represent a particular challenge. In fact, they were the first challenge to be tackled, in Otto J. M. Smith's article [2], where the compensator known as the Smith predictor was introduced five decades ago. The Smith predictor's value is in its ability to compensate for a long input or output delay in set point regulation or constant disturbance rejection problems. However, its major limitation is that, when the plant is unstable, it fails to recover the stabilizing property of a nominal controller when delay is introduced.

A substantial modification to the Smith predictor, which removes its limitation to stable plants was developed three decades ago in the form of finite spectrum

Miroslav Krstic
Department of Mechanical and Aerospace Engineering, University of California,
San Diego, La Jolla, CA 92093-0411, USA
e-mail: krstic@ucsd.edu

R. Sipahi et al. (Eds.): Time Delay Sys.: Methods, Appli. and New Trends, LNCIS 423, pp. 291–306.
springerlink.com   © Springer-Verlag Berlin Heidelberg 2012

assignment (FSA) controllers [3, 4, 5]. More recent treatment of this subject can also be found in the books [6, 7]. In the FSA approach, the system

$$\dot{X}(t) = AX(t) + BU(t - D),$$ (1)

where X is the state vector, U is the control input (scalar in our consideration here), D is an arbitrarily long delay, and (A, B) is a controllable pair, is stabilized with the infinite-dimensional predictor feedback

$$U(t) = K \left[e^{AD} X(t) + \int_{t-D}^{t} e^{A(t-\theta)} BU(\theta) d\theta \right],$$ (2)

where the gain K is chosen so that the matrix $A + BK$ is Hurwitz. The word 'predictor' comes from the fact that the bracketed quantity is actually the future state $X(t + D)$, expressed using the current state $X(t)$ as the initial condition and using the controls $U(\theta)$ from the past time window $[t - D, t]$. Concerns are raised in [8] regarding the robustness of the feedback law (2) to digital implementation of the distributed delay (integral) term but are resolved with appropriate discretization schemes [9, 10].

One can view the feedback law (2) as being given implicitly, since U appears both on the left and on the right, however, one should observe that the input memory $U(\theta), \theta \in [t - D, t]$ is actually a part of the state of the overall infinite-dimensional system, so the control law is actually given by an explicit full-state feedback formula. The predictor feedback (2) actually represents a particular form of boundary control , commonly encountered in the context of control of partial differential equations.

Motivated by our recent efforts in solving boundary control problems for various classes of partial differential equations (PDEs) using the continuum version of the backstepping method [11, 12], we review in this article various extensions to the predictor feedback design that we have recently developed. Some of these extensions are the subject of our new book [13], our articles [14, 15, 16, 17, 18, 19, 20, 21, 22, 23, 24, 25, 26, 27, 28], and some of the extensions are even more recent. They include the extension of predictor feedback to nonlinear systems and PDEs with input delays, a delay-adaptive design, an extension to time-varying delays, and predictor feedback for state-dependent delays.

2 Lyapunov Functional and Its Immediate Benefits

The key to various extensions to the predictor feedback that we present here is the observation that the invertible backstepping transformation

$$w(x,t) = u(x,t) - \int_0^x K e^{A(x-y)} Bu(y,t) dy - K e^{Ax} X(t),$$ (3)

$$u(x,t) = w(x,t) + \int_0^x K e^{(A+BK)(x-y)} Bw(y,t) dy + K e^{(A+BK)x} X(t),$$ (4)

where

$$u(x,t) = U(t+x-D),\tag{5}$$

can transform the system (1), (2) into the *target system*

$$\dot{X}(t) = (A+BK)X(t)+Bw(0,t),\tag{6}$$
$$w_t(x,t) = w_x(x,t),\tag{7}$$
$$w(D,t) = 0,\tag{8}$$

which is a cascade of an undriven transport PDE w-subsystem and the exponentially stable X-system.

Since the undriven transport PDE (7), (8) is exponentially stable, the overall cascade is exponentially stable. This fact is established with a Lyapunov functional

$$V(t) = X(t)^T P X(t) + 2\frac{|PB|^2}{\lambda_{\min}(Q)}\int_0^D (1+x)w(x,t)^2\,dx,\tag{9}$$

where P is the solution of the Lyapunov equation

$$P(A+BK)+(A+BK)^T P = -Q.\tag{10}$$

Taking the derivative of the Lyapunov function, we obtain

$$\frac{dV}{dt} = -\begin{bmatrix} X(t) & w(0,t) \end{bmatrix}\begin{bmatrix} Q & -PB \\ -B^T P^T & \dfrac{2|PB|^2}{\lambda_{\min}(Q)} \end{bmatrix}\begin{bmatrix} X(t) \\ w(0,t) \end{bmatrix}$$
$$-\frac{2|PB|^2}{\lambda_{\min}(Q)}\int_0^D w^2(x,t)dx.\tag{11}$$

Since the matrix

$$\begin{bmatrix} Q & -PB \\ -B^T P^T & \dfrac{2|PB|^2}{\lambda_{\min}(Q)} \end{bmatrix}\tag{12}$$

is positive definite, it follows that the equilibrium $X = 0, w(x) \equiv 0$ is exponentially stable in the sense of the euclidean norm on X and the L_2 norm of w. With further analysis, exponential stability of the equilibrium $X = 0, u(x) \equiv 0$ is also established, obtaining the following theorem.

Theorem 1. *There exist positive constants G and g such that the solutions of the closed-loop system (1), (2) satisfy $\Gamma(t) \leq Ge^{-gt}\Gamma(0)$ for all $t \geq 0$, where*

$$\Gamma(t) = |X(t)|^2 + \int_0^D u(x,t)^2 dx.\tag{13}$$

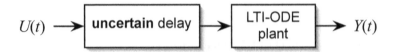

Fig. 1 An ODE with input delay which is known up to a small mismatch error ΔD, which can be either positive or negative. Stability is preserved under predictor feedback (15) for sufficiently small $|\Delta D|$ but arbitrarily large D, as stated in Theorem 2.

3 Delay-Robustness, Delay-Adaptivity, and Time-Varying Delays

In control systems with input delay, the length of the delay is the most significant possible uncertainty, both in the sense of robustness to a small mismatch in the delay D when designing constant predictor feedback and in the sense of designing delay-adaptive predictor feedback for a large uncertainty in the delay D.

3.1 Robustness to Delay Mismatch

We first discuss the problem of robustness to delay mismatch, as depicted in Figure 1, and consider the feedback system

$$\dot{X}(t) = AX(t) + BU(t - D_0 - \Delta D), \tag{14}$$

$$U(t) = K\left[e^{AD_0}X(t) + \int_{t-D_0}^{t} e^{A(t-\theta)}BU(\theta)d\theta\right]. \tag{15}$$

The actuator delay mismatch ΔD can be either positive or negative relative to the assumed actuator delay $D_0 > 0$. However, the actual delay must be nonnegative, $D_0 + \Delta D \geq 0$. For the study of robustness to a small ΔD, we use two different Lyapunov functionals, one for $\Delta D > 0$, which is the easier of the two cases, and another for $\Delta D < 0$, in which case we employ

$$V(t) = X(t)^T P X(t) + \frac{a}{2}\int_0^{D_0+\Delta D}(1+x)w(x,t)^2\,dx + \frac{1}{2}\int_{\Delta D}^0(D_0+x)w(x,t)^2dx \tag{16}$$

with a sufficiently large a.

Theorem 2. *There exists a positive constant δ such that for all $\Delta D \in (-\delta, \delta)$ there exist positive constants G and g such that the solutions of the closed-loop system (14), (15) satisfy $\Gamma(t) \leq Ge^{-gt}\Gamma(0)$ for all $t \geq 0$, where*

$$\Gamma(t) = |X(t)|^2 + \int_{t-\bar{D}}^{t} U(\theta)^2 d\theta \tag{17}$$

and where

$$\bar{D} = D_0 + \max\{0, \Delta D\}. \tag{18}$$

The significance of this robustness result can be assessed based on the intuition drawn from existing results. For example, finite-dimensional feedback laws for finite-dimensional plants are robust to small delays [29], however, this result does not apply to our infinite-dimensional problem. The delay perturbation to predictor feedback incorporates the possibility of two different classes of perturbations, depending on whether ΔD is positive or negative, so off-the-shelf results cannot be used.

The result of Theorem 2 may be surprising in light of Datko's negative result on delay-robustness for certain examples of hyperbolic PDEs with boundary control [30]. Even though the input-delay problem also involves a hyperbolic PDE, such a negative result does not hold for predictor feedback because of a significant difference between first-order and second-order hyperbolic PDEs. The second-order hyperbolic PDEs in Datko's work have infinitely many eigenvalues on the imaginary axis, whereas is not the case with an ODE with input delay, even when the ODE is unstable, only a finite number of open-loop eigenvalues may be in the closed right-half plane.

3.2 Delay-Adaptive Predictor

Now we turn our attention from robustness to small delay mismatch to adaptivity for large delay uncertainty. Several results exist on adaptive control of systems with input delays, including [31, 32]. However, existing results deal with parametric uncertainties in the ODE plant, whereas the key challenge is uncertainty in the delay.

Let us consider the plant (1) but with a transport PDE representation of the input delay given as

$$\dot{X}(t) = AX(t) + Bu(0,t), \tag{19}$$
$$Du_t(x,t) = u_x(x,t), \tag{20}$$
$$u(1,t) = U(t). \tag{21}$$

Here, the actuator state is defined as

$$u(x,t) = U(t + D(x-1)), \qquad x \in [0,1] \tag{22}$$

instead of the definition (5) with $x \in [0,D]$. We take the predictor feedback in the certainty equivalence form

$$U(t) = K \left[e^{A\hat{D}(t)} X(t) + \hat{D}(t) \int_0^1 e^{A\hat{D}(t)(1-y)} Bu(y,t)dy \right], \tag{23}$$

where the update law for the estimate $\hat{D}(t)$ is designed as

$$\dot{\hat{D}}(t) = \gamma \mathrm{Proj}_{[0,\bar{D}]}\{\tau(t)\}, \tag{24}$$

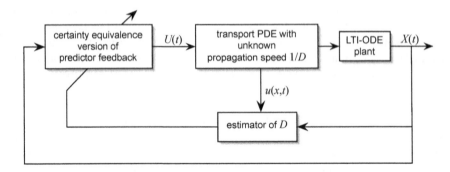

Fig. 2 Delay-adaptive predictor feedback for a true delay D varying in a broad range from 0 to a possibly large value $\bar{D}$. The certainty-equivalence controller (23) is combined with the update law (24)–(26). Global stability and regulation of the state and control are achieved, as specified in Theorem 3.

$$\tau(t) = -\frac{\int_0^1 (1+x)w(x,t)Ke^{A\hat{D}(t)x}dx\,(AX(t)+Bu(0,t))}{1+X(t)^TPX(t)+b\int_0^1(1+x)w(x,t)^2dx}, \tag{25}$$

$$w(x,t) = u(x,t) - \hat{D}(t)\int_0^x Ke^{A\hat{D}(t)(x-y)}Bu(y,t)dy - Ke^{A\hat{D}(t)x}X(t), \tag{26}$$

where $b \geq \frac{4\bar{D}|PB|^2}{\lambda_{\min}(Q)}$ and where $\bar{D}$ is an a priori known upper bound on D. The standard projection operator projects $\hat{D}(t)$ into the interval $[0,\bar{D}]$. The structure of the adaptive control system is shown in Figure 2. The choice of the update law (24)–(26) is motivated by a rather subtle Lyapunov analysis, resulting in a normalization of the update law, without the use of any filters or overparametrization.

Theorem 3. *Consider the closed-loop adaptive system (19)–(26). There exists $\gamma^* > 0$ such that for all $\gamma \in (0,\gamma^*)$ there exist positive constants R and ρ (independent of the initial conditions) such that for all initial conditions satisfying $(X_0,u_0,\hat{D}_0) \in \mathbb{R}^n \times L_2[0,1] \times [0,\bar{D}]$, the norm of the solutions obeys an exponential bound relative to the norm of initial conditions, namely*

$$\Upsilon(t) \leq R\left(e^{\rho\Upsilon(0)} - 1\right), \quad \text{for all } t \geq 0, \tag{27}$$

where

$$\Upsilon(t) = |X(t)|^2 + \int_0^1 u(x,t)^2dx + \left(D - \hat{D}(t)\right)^2. \tag{28}$$

Furthermore

$$\lim_{t\to\infty} X(t) = 0, \quad \lim_{t\to\infty} U(t) = 0. \tag{29}$$

Example 1. We illustrate the delay-adaptive design for the example plant

$$X(s) = \frac{e^{-s}}{(s-0.75)} U(s) \qquad (30)$$

with the simulation results given in Figure 3. The period up to 1 sec is the dead time, the parameter estimation is active until about 3 sec, the control evolution is exponential (corresponding to a predominantly LTI system) after 3 sec, and the state evolves exponentially after 4 sec. The adaptive controller is successful both with $\hat{D}(0) = 0$ and with $\hat{D}(0) = 2D$ (100% parameter error in both cases). ∎

The controller (23)–(26) uses full state measurement of the transport PDE state. In the absence of such measurement, a slightly different design guarantees local stability, which is the strongest result achievable in that case due to a nonlinear parametrization of the operator e^{-Ds}.

3.3 Time-Varying Input Delay

Before we close this section on uncertain delays, let us briefly turn our attention to the problem of *time-varying* known input delays, which is depicted in Figure 4. We consider the system

$$\dot{X}(t) = AX(t) + BU(\phi(t)). \qquad (31)$$

A predictor feedback for this system is

$$U(t) = K\left[e^{A\left(\phi^{-1}(t)-t\right)}X(t) + \int_{\phi(t)}^{t} e^{A\left(\phi^{-1}(t)-\phi^{-1}(\theta)\right)}B\frac{U(\theta)}{\phi'\left(\phi^{-1}(\theta)\right)}d\theta \right], \qquad \text{for all } t \geq 0. \qquad (32)$$

With rather extensive effort, going through a transport PDE representation with $u(x,t) = U\left(\phi\left(t+x\left(\phi^{-1}(t)-t\right)\right)\right)$ and the time-varying backstepping transformation

$$w(x,t) = u(x,t) - Ke^{Ax\left(\phi^{-1}(t)-t\right)}X(t) - K\int_{0}^{x} e^{A(x-y)\left(\phi^{-1}(t)-t\right)}Bu(y,t)\left(\phi^{-1}(t)-t\right)dy \qquad (33)$$

into the target system

$$\dot{X}(t) = (A+BK)X(t) + Bw(0,t), \qquad (34)$$
$$w_t(x,t) = \pi(x,t)w_x(x,t), \qquad (35)$$
$$w(1,t) = 0, \qquad (36)$$

where the variable speed of propagation of the transport equation w is given by

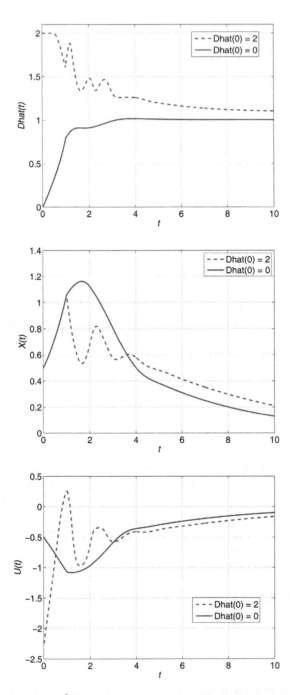

Fig. 3 Time responses of $\hat{D}(t)$, $X(t)$, and $U(t)$ under delay-adaptive predictor feedback for an unstable first-order plant. Stabilization is achieved both with $\hat{D}(0) = 0$ and with a $\hat{D}(0)$ that heavily overestimates the true D.

Fig. 4 Linear system $\dot{X}(t) = AX(t) + BU(\phi(t))$ with time-varying actuator delay $\delta(t) = t - \phi(t)$. The predictor feedback (32) with compensation of the time-varying delay achieves exponential stabilization in the sense of Theorem 4.

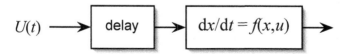

Fig. 5 Nonlinear control in the presence of arbitrarily long input delay. Global stabilization is achieved with the predictor feedback (40)–(42) if the plant is forward complete and globally asymptotically stabilizable in the absence of delay, as stated in Theorem 5.

$$\pi(x,t) = \frac{1 + x\left(\dfrac{d\left(\phi^{-1}(t) \right)}{dt} - 1 \right)}{\phi^{-1}(t) - t}, \tag{37}$$

we obtain the following stabilization result.

Theorem 4. *Consider the closed-loop system (31), (32). Let the delay function $\delta(t) = t - \phi(t)$ be strictly positive and uniformly bounded from above. Let the delay rate function $\delta'(t)$ be strictly smaller than 1 and uniformly bounded from below. There exist positive constants G and g (the latter one being independent of ϕ) such that*

$$|X(t)|^2 + \int_{\phi(t)}^{t} U^2(\theta)d\theta \le Ge^{-gt}\left(|X_0|^2 + \int_{\phi(0)}^{0} U^2(\theta)d\theta \right), \quad \text{for all } t \ge 0. \tag{38}$$

4 Predictor Feedback for Nonlinear Systems

In the area of robust nonlinear control various types of uncertainties are considered—unmeasurable disturbances, static nonlinear functional perturbations, dynamic perturbations on the state, and dynamic perturbations on the input. The unmodeled *input* dynamics are represent the greatest challenge in robust nonlinear control. It is for this reason no surprise that long delays at the input of nonlinear systems, as depicted in Figure 5, have remained an unsolved challenge in nonlinear control. Considerable success has been achieved in recent years with control of nonlinear systems with state delay [33, 34, 35, 36], however, only one result exists where input delay of arbitrary length is being addressed [37]. Systematic *compensation* of input delays of arbitrary length is non-existent.

A conceptually easy and natural way to compensate input delays in nonlinear control is through an extension of predictor feedback to nonlinear systems, which we present next. Consider the general class of nonlinear systems

$$\dot{X}(t) = f(X(t), U(t-D)), \quad f(0,0) = 0, \tag{39}$$

and assume that a feedback law $U = \kappa(X)$ with $\kappa(0) = 0$ is known which globally asymptotically stabilizes the system at the origin when $D = 0$. Denote the initial conditions as $X_0 = X(0)$ and $U_0(\theta) = U(\theta), \theta \in [-D,0]$. A predictor feedback is given by

$$U(t) = \kappa(P(t)), \tag{40}$$

where the predictor is defined as

$$P(t) = \int_{t-D}^{t} f(P(\theta), U(\theta)) d\theta + X(t), \quad t \geq 0, \tag{41}$$

$$P(\theta) = \int_{-D}^{\theta} f(P(\sigma), U_0(\sigma)) d\sigma + X_0, \quad \theta \in [-D,0]. \tag{42}$$

A key feature to note about the predictor $P(t)$ is that it is defined implicitly, through a nonlinear integral equation, rather than explicitly, through matrix exponentials and the variation of constants formula, as is the case when the plant is linear. The lack of an explicit formula for $P(t)$ is not necessarily an obstacle numerically, since $P(t)$ is defined in terms of its past values.

The more serious question is conceptual, does a solution for $P(t)$ always exist? Fortunately, the answer to this question is rather simple. Since control has no effect for D seconds after it has been applied, the system can indeed exhibit finite escape over that period, resulting in a finite escape for $P(t)$ since the predictor is governed by the same model as the plant. Hence, a natural way to ensure global existence of the predictor state is to assume that the plant is *forward complete*.

A system is said to be forward complete if, for all initial conditions and all locally bounded input signals, its solutions exist for all time. This definition does not require the solutions to be *uniformly* bounded. They can be growing to infinity as time goes to infinity. For example, all LTI systems, stable or unstable, driven by inputs of exponential growth, are forward complete. The same is true of nonlinear systems with globally Lipschitz right-hand sides, but also of many systems that are neither globally Lipschitz nor stable but contain super-linear nonlinearities that induce limit cycles, rather than finite escape.

The nonlinear predictor design is developed for two classes of systems. For the broad class of *forward complete* systems, that is, systems that do not exhibit a finite escape time for any initial condition and any input signals that remain finite over finite time intervals, which includes many mechanical and other systems, predictor feedback is developed which achieves global asymptotic stability, as long as the system without delay is globally asymptotically stabilizable. However, the predictor requires the solution of a nonlinear integral equation, or a nonlinear DDE, in real time.

Theorem 5. *Let $\dot{X} = f(X,U)$ be forward complete and $\dot{X} = f(X, \kappa(X))$ be globally asymptotically stable at $X = 0$. Consider the closed-loop system (39)–(42). There exists a function $\hat{\beta} \in \mathcal{KL}$ such that*

$$|X(t)| + \|U\|_{L_\infty[t-D,t]} \leq \hat{\beta}\left(|X(0)| + \|U_0\|_{L_\infty[-D,0]}, t\right) \tag{43}$$

for all $(X_0, U_0) \in \mathbb{R}^n \times L_\infty[-D,0]$ and for all $t \geq 0$.

As we have mentioned above, the only weakness of predictor feedback laws is that $P(t)$ may not be explicitly computable. Fortunately, a significant class of nonlinear system exists which are not only forward complete and globally stabilizable, but where $P(t)$ is also explicitly computable. This is the class of *strict-feedforward* systems.

Example 2. We illustrate the explicit computability of the predictor, and thus of the feedback law, for an example of a strict-feedforward system. Consider the third-order system

$$\dot{X}_1(t) = X_2(t) + X_3^2(t), \tag{44}$$
$$\dot{X}_2(t) = X_3(t) + X_3(t)U(t-D), \tag{45}$$
$$\dot{X}_3(t) = U(t-D), \tag{46}$$

which is not feedback linearizable and is in the strict-feedforward class. The globally asymptotically stabilizing predictor feedback for this system is given by

$$\begin{aligned}
U(t) = {}& -P_1(t) - 3P_2(t) - 3P_3(t) - \frac{3}{8}P_2^2(t) \\
& + \frac{3}{4}P_3(t)\left(-P_1(t) - 2P_2(t) + \frac{1}{2}P_3(t) + \frac{P_2(t)P_3(t)}{2}\right. \\
& \left. + \frac{5}{8}P_3^2(t) - \frac{1}{4}P_3^3(t) - \frac{3}{8}\left(P_2(t) - \frac{P_3^2(t)}{2}\right)^2\right), \tag{47}
\end{aligned}$$

where the D-second-ahead predictor of $(X_1(t), X_2(t), X_3(t))$ is given explicitly by

$$\begin{aligned}
P_1(t) = {}& X_1(t) + DX_2(t) + \frac{1}{2}D^2X_3(t) + DX_3^2(t) + 3X_3(t)\int_{t-D}^t (t-\theta)U(\theta)d\theta \\
& + \frac{1}{2}\int_{t-D}^t (t-\theta)^2 U(\theta)d\theta + \frac{3}{2}\int_{t-D}^t \left(\int_{t-D}^\theta U(\sigma)d\sigma\right)^2 d\theta, \tag{48} \\
P_2(t) = {}& X_2(t) + DX_3(t) + X_3(t)\int_{t-D}^t U(\theta)d\theta + \int_{t-D}^t (t-\theta)U(\theta)d\theta \\
& + \frac{1}{2}\left(\int_{t-D}^t U(\theta)d\theta\right)^2, \tag{49}
\end{aligned}$$

$$P_3(t) = X_3(t) + \int_{t-D}^{t} U(\theta)d\theta. \tag{50}$$

Note that the nonlinear infinite-dimensional feedback operator employs a finite Volterra series in $U(\theta)$. ∎

5 Non-holonomic Unicycle Controlled over a Network

In this section we show that predictor feedback can be used also for sampled-data nonlinear systems and also for systems that involve measurement delay, in addition to actuation delay.

We consider the non-holonomic unicycle

$$\dot{x}(t) = v(t-D)\cos(\theta(t)) \tag{51}$$
$$\dot{y}(t) = v(t-D)\sin(\theta(t)) \tag{52}$$
$$\dot{\theta}(t) = \omega(t-D), \tag{53}$$

where

$$(x,y) \quad \text{position of robot}$$
$$\theta \quad \text{heading of robot}$$
$$v \quad \text{speed}$$
$$\omega \quad \text{turning rate}.$$

We can compensate any input and output delay, using any sampling time, provided that the input delay is an integer multiple of the sampling time. For simplicity, in this article we assume that

$$D = \text{transmission delay in both directions} = \text{sampling time}.$$

We construct the controller

$$v(t) = \frac{1}{D}(k_1(P,Q,\Theta) + Qk_2(P,Q,\Theta)), \quad \text{for } t \in [iD,(i+1)D) \tag{54}$$
$$\omega(t) = \frac{1}{D}k_2(P,Q,\Theta), \quad \text{for } t \in [iD,(i+1)D) \tag{55}$$

with the transformation transformation

$$P = X\cos(\Theta) + Y\sin(\Theta) \tag{56}$$
$$Q = X\sin(\Theta) - Y\cos(\Theta) \tag{57}$$

and with the exact predictor of $(x,y,\theta)((i+1)D)$

$$X = x((i-1)D) + \int_{(i-2)D}^{iD} v(s)\cos\left(\theta((i-1)D) + \int_{(i-2)D}^{s} \omega(z)dz\right)ds \tag{58}$$
$$Y = y((i-1)D) + \int_{(i-2)D}^{iD} v(s)\sin\left(\theta((i-1)D) + \int_{(i-2)D}^{s} \omega(z)dz\right)ds \tag{59}$$

$$\Theta = \theta((i-1)D) + \int_{(i-2)D}^{iD} \omega(s)ds \tag{60}$$

where the feedback functions k_1 and k_2 are defined as

$$k_1(P,Q,\Theta) = - \begin{cases} |Q|^{1/2}, & Q(2Q-P\Theta) \neq 0 \\ \dfrac{P^2\Theta}{P^2+\Theta^2}, & Q=0, P\Theta \neq 0 \\ \Theta, & 2Q=P\Theta \end{cases} \tag{61}$$

$$k_2(P,Q,\Theta) = - \begin{cases} 2\left(P+\text{sgn}(Q)|Q|^{1/2}\right), & Q(2Q-P\Theta) \neq 0 \\ \dfrac{P\Theta^2}{P^2+\Theta^2}, & Q=0, P\Theta \neq 0 \\ P, & 2Q=P\Theta . \end{cases} \tag{62}$$

The feedback functions k_1, k_2 were constructed by Iasson Karafyllis. These functions are discontinuous and they ensure sampled-data dead-beat stabilization in the delay-free case.

Theorem 6. *For any $D>0$, the closed-loop system is globally asymptotically stable at the origin. Moreover, $x(t) = y(t) = \theta(t) = 0$ for $t \geq 5D$.*

Predictor (58)–(60) enables stabilization of the unicycle using only state values from two arbitrarily long sample periods ago.

6 State-Dependent Delay

In this section we bring together the techniques for linear systems with time varying delays and techniques for nonlinear systems. We consider systems with state-dependent delays modeled as

$$\dot{X}(t) = f\left(X(t), U\left(t - D\left(X(t)\right)\right)\right), \tag{63}$$

where the system $\dot{X} = f(X,\omega)$ is assumed forward complete and D is a positive-valued continuously differentiable function.

State-dependent delays require special care because the prediction horizon is not known a priori. The prediction time depends on the future state (the predictor state), whereas the predictor state depends on the prediction horizon itself. Fortunately, this circuitous dependency can be broke with the predictor given by

$$P(t) = X(t) + \int_{t-D(X(t))}^{t} \frac{f(P(s), U(s))}{1 - \nabla D(P(s)) f(P(s), U(s))} ds \tag{64}$$

$$\sigma(t) = t + D(P(t)), \tag{65}$$

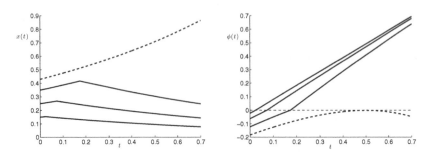

Fig. 6 Stabilization by nonlinear predictor feedback for a linear system with a delay that grows with the square of the state value. For larger initial conditions, the state-dependent delay prevents the control from ever kicking in. Thus, only local stabilization is achieved. The variable $\phi(t)$ represents $t - D(X(t))$.

where $\sigma(t)$ represents the prediction time, namely, $X(\sigma(t)) = P(t) = X(t + D(P(t)))$, and where the denominator $1 - \nabla D(P(s)) f(P(s), U(s))$ plays the key role in predicting X.

The predictor feedback is given by

$$U(t) = \kappa(\sigma(t), P(t)), \tag{66}$$

with an assumption that $\kappa(t, X)$ is globally stabilizing in the absence of delay, namely, that $dX(t)/dt = f(X(t), \kappa(t, X(t)))$ is globally asymptotically stable at the origin.

Theorem 7. *There exist a constant $\psi_{RoA} > 0$ and a class $\mathcal{KL}$ function β such that for all initial conditions of the plant that satisfy*

$$|X(0)| + \sup_{-D(X(0)) \leq \theta \leq 0} |U(\theta)| < \psi_{RoA} \tag{67}$$

the following holds

$$|X(t)| + \sup_{t - D(X(t)) \leq \theta \leq t} |U(\theta)| \leq \beta\left(|X(0)| + \sup_{-D(X(0)) \leq \theta \leq 0} |U(\theta)|, t\right), \qquad \forall t \geq 0. \tag{68}$$

Example 3. In the presence of a state-dependent delay, even linear systems act as nonlinear systems. We consider the unstable plant

$$\dot{X}(t) = X(t) + U\left(t - X(t)^2\right). \tag{69}$$

We choose the initial conditions for the input as $U(\theta) = 0, -X(0)^2 \leq \theta \leq 0$. For $X(0) \geq X^* = \frac{1}{\sqrt{2e}} = 0.43$, the controller never "kicks in," as displayed in Figure 6. However, for smaller initial conditions the predictor feedback achieves stabilization.

7   Conclusions

The PDE backstepping approach is a potentially powerful tool in advancing the design techniques for systems with input and output delays. Two ideas presented in this article may be of interest to researchers in delay systems. The first idea is the construction of backstepping transformations that allow one to deal with delays and PDE dynamics at the input, as well as in the main line of applying control action, such as in the chain of integrators for systems in triangular forms. The second idea is the construction of Lyapunov functionals and explicit stability estimates, with the help of direct and inverse backstepping transformations.

Acknowledgements. I thank Iasson Karafyllis for the result in Section 5 and Nikolaos Bekiaris-Liberis for contributing much of Section 6.

References

1. Sipahi, R., Niculescu, S.-I., Abdallah, C.T., Michiels, W., Gu, K.: Stability and stabilization of systems with time delay. Control Systems Magazine 31, 38–65 (2011)
2. Smith, O.J.M.: A controller to overcome dead time. ISA 6, 28–33 (1959)
3. Manitius, A.Z., Olbrot, A.W.: Finite spectrum assignment for systems with delays. IEEE Trans. on Automatic Control 24, 541–553 (1979)
4. Kwon, W.H., Pearson, A.E.: Feedback stabilization of linear systems with delayed control. IEEE Trans. on Automatic Control 25, 266–269 (1980)
5. Artstein, Z.: Linear systems with delayed controls: a reduction. IEEE Trans. on Automatic Control 27, 869–879 (1982)
6. Zhong, Q.-C.: Robust Control of Time-delay Systems. Springer, Heidelberg (2006)
7. Michiels, W., Niculescu, S.-I.: Stability and Stabilization of Time-Delay Systems: An Eigenvalue-Based Approach. SIAM (2007)
8. Mondie, S., Michiels, W.: Finite spectrum assignment of unstable time-delay systems with a safe implementation. IEEE Trans. on Automatic Control 48, 2207–2212 (2003)
9. Zhong, Q.-C.: On distributed delay in linear control laws—Part I: Discrete-delay implementation. IEEE Transactions on Automatic Control 49, 2074–2080 (2006)
10. Zhong, Q.-C., Mirkin, L.: Control of integral processes with dead time—Part 2: Quantitative analysis. IEE Proc. Control Theory & Appl. 149, 291–296 (2002)
11. Krstic, M., Smyshlyaev, A.: Boundary Control of PDEs: A Course on Backstepping Designs. SIAM (2008)
12. Vazquez, R., Krstic, M.: Control of Turbulent and Magnetohydrodynamic Channel Flows. Birkhäuser (2007)
13. Krstic, M.: Delay Compensation for Nonlinear, Adaptive, and PDE Systems. Birkhäuser (2009)
14. Krstic, M.: On compensating long actuator delays in nonlinear control. IEEE Transactions on Automatic Control 53, 1684–1688 (2008)
15. Krstic, M.: Lyapunov tools for predictor feedbacks for delay systems: Inverse optimality and robustness to delay mismatch. Automatica 44, 2930–2935 (2008)
16. Krstic, M.: Compensating actuator and sensor dynamics governed by diffusion PDEs. Systems and Control Letters 58, 372–377 (2009)
17. Krstic, M.: Compensating a string PDE in the actuation or sensing path of an unstable ODE. IEEE Transactions on Automatic Control 54, 1362–1368 (2009)

18. Krstic, M.: Input delay compensation for forward complete and feedforward nonlinear systems. IEEE Transactions on Automatic Control 55, 287–303 (2010)

19. Krstic, M.: Lyapunov stability of linear predictor feedback for time-varying input delay. IEEE Transactions on Automatic Control 55, 554–559 (2010)

20. Krstic, M.: Control of an unstable reaction-diffusion PDE with long input delay. Systems and Control Letters 58, 773–782 (2009)

21. Krstic, M.: Compensation of infinite-dimensional actuator and sensor dynamics: Nonlinear and delay-adaptive systems. IEEE Control Systems Magazine 30, 22–41 (2010)

22. Bresch-Pietri, D., Krstic, M.: Adaptive trajectory tracking despite unknown input delay and plant parameters. Automatica 45, 2075–2081 (2009)

23. Bresch-Pietri, D., Krstic, M.: Delay-adaptive predictor feedback for systems with unknown long actuator delay. IEEE Transactions on Automatic Control 55, 2106–2112 (2010)

24. Bekiaris-Liberis, N., Krstic, M.: Delay-adaptive feedback for linear feedforward systems. Systems and Control Letters 59, 277–283 (2010)

25. Bekiaris-Liberis, N., Krstic, M.: Compensating the distributed effect of a wave PDE in the actuation or sensing path of MIMO LTI Systems. Systems & Control Letters 59, 713–719 (2010)

26. Bekiaris-Liberis, N., Krstic, M.: Stabilization of linear strict-feedback systems with delayed integrators. Automatica 46, 1902–1910 (2010)

27. Bekiaris-Liberis, N., Krstic, M.: Lyapunov stability of linear predictor feedback for distributed input delay. IEEE Transactions on Automatic Control 56, 655–660 (2011)

28. Bekiaris-Liberis, N., Krstic, M.: Compensating distributed effect of diffusion and counter-convection in multi-input and multi-output LTI systems. IEEE Transactions on Automatic Control 56, 637–642 (2011)

29. Teel, A.R.: Connections between Razumikhin-type theorems and the ISS nonlinear small gain theorem. IEEE Transactions on Automatic Control 43, 960–964 (1998)

30. Datko, R.: Not all feedback stabilized hyperbolic systems are robust with respect to small time delays in their feedbacks. SIAM Journal on Control and Optimization 26, 697–713 (1988)

31. Ortega, R., Lozano, R.: Globally stable adaptive controller for systems with delay. Internat. J. Control 47, 17–23 (1988)

32. Niculescu, S.-I., Annaswamy, A.M.: An Adaptive Smith-Controller for Time-delay Systems with Relative Degree $n* \geq 2$. Systems and Control Letters 49, 347–358 (2003)

33. Jankovic, M.: Control Lyapunov-Razumikhin functions and robust stabilization of time delay systems. IEEE Trans. on Automatic Control 46, 1048–1060 (2001)

34. Germani, A., Manes, C., Pepe, P.: Input-output linearization with delay cancellation for nonlinear delay systems: the problem of the internal stability. International Journal of Robust and Nonlinear Control 13, 909–937 (2003)

35. Karafyllis, I.: Finite-time global stabilization by means of time-varying distributed delay feedback. SIAM Journal of Control and Optimization 45, 320–342 (2006)

36. Mazenc, F., Bliman, P.-A.: Backstepping design for time-delay nonlinear systems. IEEE Transactions on Automatic Control 51, 149–154 (2006)

37. Mazenc, F., Mondie, S., Francisco, R.: Global asymptotic stabilization of feedforward systems with delay at the input. IEEE Trans. Automatic Control 49, 844–850 (2004)

38. Smyshlyaev, A., Krstic, M.: Closed form boundary state feedbacks for a class of 1D partial integro-differential equations. IEEE Trans. on Automatic Control 49(12), 2185–2202 (2004)

Smoothing Techniques-Based Distributed Model Predictive Control Algorithms for Networks

Ion Necoara, Ioan Dumitrache, and Johan A.K. Suykens

Abstract. In this chapter we propose dual distributed methods based on smoothing techniques, called here the proximal center method and the interior-point Lagrangian method, to solve distributively separable convex problems. We also present an extension of these methods to the case of separable non-convex problems but with a convex objective function using a sequential convex programming framework. We prove that some relevant centralized model predictive control (MPC) problems for a network of coupling linear (non-linear) dynamical systems can be recast as separable convex (non-convex) problems for which our distributed methods can be applied. We also show that the solution of our distributed algorithms converges to the solution of the centralized problem and we provide estimates for the rate of convergence, which improve the estimates of some existing methods.

1 Introduction

Model predictive control (MPC) is one of the most successful advanced control technology implemented in industry due to its ability to handle complex systems with hard constraints [5]. The essence of MPC is to determine a control profile that optimizes a cost criterion over a prediction window and then to apply this control profile until new measurements become available. Then the whole procedure is repeated. Feedback is incorporated by using the measurements to update the optimization problem for the next step.

In many application fields, the notion of networks has emerged as a central, unifying concept for solving different problems in systems and control theory such as control and estimation. These large-scale networks are often composed of multiple

Ion Necoara · Ioan Dumitrache
Automation and Systems Engineering Department, University Politehnica Bucharest,
060042 Bucharest, Romania
e-mail: `ion.necoara@upb.ro, ioan.dumitrache@upb.ro`

Johan A.K. Suykens
Electrical Engineering Department (ESAT-SCD), Katholieke Universiteit Leuven,
B–3001 Leuven, Belgium
e-mail: `johan.suykens@esat.kuleuven.be`

R. Sipahi et al. (Eds.): Time Delay Sys.: Methods, Appli. and New Trends, LNCIS 423, pp. 307–318.
springerlink.com   © Springer-Verlag Berlin Heidelberg 2012

subsystems characterized by complex dynamics and mutual interactions such that local decisions have long-range effects throughout the entire network. A centralized decision unit for such network systems is usually considered impractical and inflexible due to information requirements and computational aspects. Many problems associated to network systems, such as optimal control (in particular MPC), can be posed as *separable optimization problems* (see e.g. [2, 3, 9, 16], etc). When the control problem is formulated as a separable optimization problem, a valid alternative to centralized control can be achieved using decomposition techniques.

Approaches to distributed MPC design differ from each other in the problem setup or in the corresponding distributed algorithm used for solving the problem. Several standard distributed control methods can be found in the textbooks [4, 7]. More recently, in [11, 15] the actual status of research in the field of coordinated and distributed optimization-based control is presented. When the dynamics of the subsystems are coupled, a common approach to distribute the computation is based on Jacobi or dual subgradient type algorithms [1, 2, 3, 6, 16].

However, it is known that most of the methods based on Jacobi or dual subgradient algorithms have slow convergence (see e.g. [1, 13]). In [9, 10] it has been shown that for separable convex problems (e.g arising from linear optimal control) we can use smoothing techniques to devise distributed optimization algorithms with a great improvement of the convergence rate compared to the previous methods. The algorithms, called in Section 2 the proximal center method (PCM) [9] and the interior-point Lagrangian method (IPLM) [10], involve for every subsystem optimizing an objective function that is the sum of his own objective and a smoothing term while the coordination between the subsystems is performed via the Lagrange multipliers. We also generalize these results to separable non-convex optimization problems by combining sequential convex programming and the PCM or IPLM algorithms, and thus we obtain a distributed algorithm also for non-linear control problems. One of the key contribution of this paper is given in Section 3 where we prove that some optimal control problems for network systems leads to separable optimization problems with particular structure and to show how this specific control problem structure can be exploited in our distributed algorithms. Moreover, we prove that the solution of these distributed algorithms converges to the solution of the original problem and we also provide estimates for the rate of convergence, which improve the estimates of some existing methods with at least one order of magnitude.

2 Application of Smoothing Techniques to Separable Convex Problems

In this section we consider a network of agents (e.g. subsystems) that cooperatively has to minimize an additive cost but with coupling constraints that express the interactions between agents. In particular, the agents want to cooperatively solve the following *separable convex* optimization problem:

$$f^* = \min_{x^i \in X^i, G^i x^i = a^i} \sum_{i=1}^{M} f^i(x^i), \quad \text{s.t.} \quad \sum_{i=1}^{M} F^i x^i = a, \tag{1}$$

where $f^i : \mathbb{R}^n \to \mathbb{R}$ are convex functions, X^i are closed convex sets, $G^i \in \mathbb{R}^{m \times n}$, $F^i \in \mathbb{R}^{p \times n}$, $a^i \in \mathbb{R}^m$ and $a \in \mathbb{R}^p$. For simplicity we define: $x = [x^{1T} \cdots x^{MT}]^T$, $f(x) = \sum_{i=1}^{M} f^i(x^i)$, $X = \prod_{i=1}^{M} X^i$, $F = [F^1 \cdots F^M]$. We assume that the communication network among agents (subsystems) is described by a graph $G = (V, E)$, where the nodes in $V = \{1, \cdots, M\}$ represent the agents and the edge $(i, j) \in E \subseteq V \times V$ models that agent j sends information to agent i. Then, the main challenge is to provide distributed algorithms for solving (1) such that the computations can be distributed among the agents and the amount of information that the agents must exchange is limited, according to the communication graph G. The following assumptions hold for the optimization problem (1):

Assumption 1
1. Functions f^i are linear/convex quadratic[1] and X^i are compact convex sets.
2. Matrix $[\text{diag}(G^1 \cdots G^M); F]$ has full row rank and the feasible set $\{x \in \text{int}(X) : G^i x^i = a^i \ \forall i, Fx = a\} \neq \emptyset$.

Note that we do not assume strict/strong convexity of any function f^i. Let also $\langle \cdot, \cdot \rangle / \| \cdot \|$ denote the Euclidian inner product/norm on $\mathbb{R}^n$.

A well-known method for solving the optimization problem (1) distributively is based on the dual decomposition . The potential advantages of solving the corresponding dual problem of (1) are the following: (i) the dual problem is usually unconstrained or has simple constraints, and (ii) the dual problem decomposes into smaller subproblems. In the dual decomposition method we first move the coupling constraints in the cost and form the Lagrangian corresponding to these constraints: $L_0(x, \lambda) = f(x) + \langle \lambda, Fx - a \rangle$. Then, we define the standard dual function $d_0(\lambda) = \min_{x^i \in X^i, G^i x^i = a^i} L_0(x, \lambda)$ and we need to find an optimal multiplier λ^* of the dual problem $\max_\lambda d_0(\lambda)$. The constraint qualification condition from Assumption 1 guarantees that there exists a primal-dual optimal solution (x^*, λ^*). Note that in general (e.g. when f is not strictly convex) d_0 is not differentiable. Therefore, for maximizing d_0 we have to use involved nonsmooth optimization techniques. For example, in the dual subgradient method (DSM), whose convergence rate is of order $\mathcal{O}(\frac{1}{\varepsilon^2})$, where $\varepsilon > 0$ is the required accuracy of the approximation of the optimal value f^* (see e.g. [13]), we need to perform the following iterations:

Algorithm DSM

1. given λ_k compute in *parallel* $x^i_{k+1} = \arg\min_{x^i \in X^i, G^i x^i = a^i} f^i(x^i) + \langle \lambda_k, F^i x^i \rangle$
2. update the Lagrange multipliers $\lambda_{k+1} = \lambda_k + \alpha_k (\sum_{i=1}^{M} F^i x^i_{k+1} - a)$

where α_k is an appropriate step-size [1, 13]. Although in the subgradient methods the assumptions for convergence are weaker, they usually convergence slow, the step-size selection is difficult and it does not handle nontrivial dual feasibility constraints. Decomposition methods obtained by smoothing techniques can address

---

[1] For a general treatment of convex functions see [9, 10].

these issues. In order to obtain a smooth dual function we need to use smoothing techniques applied to the ordinary Lagrangian L_0 [14]. In [9, 10] we proposed two dual decomposition methods for solving (1), in which we add to the Lagrangian L_0 a smoothing term $\mu \sum_{i=1}^{M} \phi_{X^i}$, where each function ϕ_{X^i} associated to the set X^i (called *prox function*) must have certain properties explained below. The two algorithms differ in the choice of the prox functions ϕ_{X^i}. In this case we define the *augmented Lagrangian*:

$$L_\mu(x, \lambda) = \sum_{i=1}^{M} [f^i(x^i) + \mu \phi_{X^i}(x^i)] + \langle \lambda, Fx - a \rangle$$

and the *augmented dual function* $d_\mu(\lambda) = \min_{x^i \in X^i, G^i x^i = a^i} L_\mu(x, \lambda)$. Denote $x^i(\mu, \lambda)$ the optimal solution of the minimization problem in x^i:

$$x^i(\mu, \lambda) = \arg \min_{x^i \in X^i, G^i x^i = a^i} [f^i(x^i) + \mu \phi_{X^i}(x^i) + \langle \lambda, F^i x^i \rangle].$$

2.1 Proximal Center Decomposition Method

In the sequel we describe the *proximal center decomposition* method whose efficiency estimates improve with one order of magnitude the bounds on the number of iterations of the classical dual subgradient method (see [9] for more details). In this method, the functions ϕ_{X^i} are chosen to be continuous, nonnegative and strongly convex on X^i with strong parameter σ_i. Since X^i are compact, we can choose finite and positive constants D_{X^i} such that $D_{X^i} \geq \max_{x^i \in X^i} \phi_{X^i}(x^i)$ for all i. In the following lemma we present the main smoothness properties of the dual function $d_\mu(\cdot)$:

Lemma 1. *[9] If Assumption 1 holds and ϕ_{X^i} are continuous, nonnegative and strongly convex on X^i, then the family of dual functions $\{d_\mu(\cdot)\}_{\mu>0}$ is concave, differentiable and the gradient $\nabla d_\mu(\lambda) = \sum_{i=1}^{M} F^i x^i(\mu, \lambda) - a$ is Lipschitz continuous with Lipschitz constant $D_\mu = \sum_{i=1}^{M} \frac{\|F^i\|^2}{\mu \sigma_i}$. The following inequalities also hold: $d_\mu(\lambda) \geq d_0(\lambda) \geq d_\mu(\lambda) - \mu \sum_{i=1}^{M} D_{X^i}$, for all λ.*

We now describe a distributed optimization method for (1), called the *proximal center decomposition method*, that consists in the following iterations:

Algorithm PCM

1. given λ_k compute in *parallel*
   $x_{k+1}^i = \arg \min_{x^i \in X^i, G^i x^i = a^i} f^i(x^i) + \mu \phi_{X^i}(x^i) + \langle \lambda_k, F^i x^i \rangle$
2. compute $\bar{\lambda}_k = \lambda_k + \frac{1}{D_\mu} (\sum_{i=1}^{M} F^i x_{k+1}^i - a)$ and
   $\tilde{\lambda}_k = \frac{1}{D_\mu} \sum_{l=0}^{k} \frac{l+1}{2} \sum_{i=1}^{M} (F^i x_{l+1}^i - a)$
3. update the Lagrange multiplier $\lambda_{k+1} = \frac{k+1}{k+3} \bar{\lambda}_k + \frac{2}{k+3} \tilde{\lambda}_k$.

The main computational effort is done in Step 1. However, in some applications, e.g. distributed MPC, Step 1 can be performed also very efficiently (see Section 3), making the MPC algorithm suitable for online implementation. After k iterations of

Algorithm PCM we define:

$$\hat{x}^i = \sum_{l=0}^{k} \frac{2(l+1)}{(k+1)(k+2)} x_{l+1}^i \text{ and } \hat{\lambda} = \lambda_k.$$

Theorem 1. *[9] Under the hypothesis of Lemma 1 and taking $\mu = \frac{\varepsilon}{\Sigma_i D_{X^i}}$ and $k = 2\sqrt{(\Sigma_i \frac{\|F^i\|^2}{\sigma_i})(\Sigma_i D_{X^i})} \frac{1}{\varepsilon}$, then after k iterations $-\|\lambda^*\| \| \Sigma_i F^i \hat{x}^i - a\| \leq f(\hat{x}) - f^* \leq \varepsilon$ and the coupling constraints satisfy $\|\Sigma_i F^i \hat{x}^i - a\| \leq \varepsilon(\|\lambda^*\| + \sqrt{\|\lambda^*\|^2 + 2})$, where λ^* is the minimum norm optimal multiplier.*

Therefore, the efficiency estimates of the Algorithm PCM is of order $\mathscr{O}(\frac{1}{\varepsilon})$ and thus improves with one order of magnitude the complexity of the dual subgradient algorithm, whose efficiency estimates is $\mathscr{O}(\frac{1}{\varepsilon^2})$ [13]. Note that instead of a static selection of the smoothing parameter μ in the Algorithm PCM, we can select it dynamically, which usually gives a significant speed-up. Similarly, instead of using a fixed Lipschitz constant D_μ, we can adaptively estimate this constant during the iterations of the Algorithm PCM, which in general leads to a faster convergence than the fixed one.

2.2 Interior-Point Lagrangian Decomposition Method

In this section we describe the *interior-point Lagrangian decomposition* algorithm. In this method we add to the standard Lagrangian a smoothing term $\mu \sum_{i=1}^{M} \phi_{X^i}$, where each function ϕ_{X^i} is a N^i-self-concordant barrier associated to the convex set X^i (see [12] for precise definitions).

Lemma 2. *If Assumption 1 holds and ϕ_{X^i}'s are N^i-self-concordant barriers associated to X^i, then the family of dual functions $\{-d_\mu(\cdot)\}_{\mu>0}$ is self-concordant and with positive definite Hessian.*

Proof. Since f^i's are linear or convex quadratic functions, then they are also self-concordant functions. Moreover, since ϕ_{X^i}'s are N^i-self-concordant barriers, it follows that $f^i + \mu \phi_{X^i}$ are also self-concordant and with positive definite Hessians and thus for each λ the augmented Lagrangian $L_\mu(\cdot, \lambda)$ is self-concordant as a function of x. Since $d_\mu(\cdot)$ is basically the Legendre transformation of $\Sigma_i f^i + \mu \phi_{X^i}$, in view of well-known properties of the Legendre transformation [12], it follows that the augmented dual function is self-concordant and its Hessian is also positive definite. The expression for $\nabla^2 d_\mu(\lambda)$ can be found in Theorem 3.1 in [10]. $\qquad\square$

From Lemma 2 we can conclude that the family of augmented Lagrangians $\{L_\mu(\cdot, \lambda)\}_{\mu>0}$ and the family of dual functions $\{-d_\mu(\cdot)\}_{\mu>0}$ are self-concordant. Therefore, in both optimization problems $\min_x L_\mu(x, \lambda)$ and $\max_\lambda d_\mu(\lambda)$, respectively, the objective functions are self-concordant. This opens the possibility of deriving a dual interior-point based method for (1) using Newton directions for updating the multipliers to speed up the convergence rate. We define the following algorithm, where $0 < \tau < 1$:

Algorithm IPLM

1. update $\mu_k = \tau\mu_{k-1}$
2. find $\lambda_{k+1} = \arg\max_\lambda d_{\mu_k}(\lambda)$ and define $x_{k+1} = x(\mu_k, \lambda_{k+1})$.

The main computations are done in Step 2 of the algorithm where we need to solve with the Newton method an optimization problem with self-concordant objective function: $\max_\lambda d_{\mu_k}(\lambda)$. At each outer iteration k we have an inner iteration in p that consists of solving in parallel M small optimization problems in the variables x^i's and with self-concordant cost functions:

$$x^i_{p+1|k+1} = \arg\min_{x^i \in X^i, G^i x^i = a^i} f^i(x^i) + \mu\phi_{X^i}(x^i) + \langle\lambda_{p|k+1}, F^i x^i\rangle.$$

It is straightforward to see that $\lambda_{p|k+1}$ converges to λ_{k+1} and $x_{p+1|k+1}$ converges to x_{k+1}. Note that the cost of computing extremely accurate optimizers for the problems that need to be solved at the outer and inner iterations as compared to the cost of computing a good maximizer of them is only marginally more, i.e. a few Newton steps at most (due to quadratic convergence of the Newton method close to the solution). Therefore, it is not unreasonable to assume exact computations in the proposed algorithm. For an algorithm based on approximate iterations see [10] for more details.

Theorem 2. [10] *Under the hypothesis of Lemma 2 the following convergence rate holds for the Algorithm IPLM:* $0 \leq f(x_k) - f^* \leq N_\phi\mu_k$, *where* $N_\phi = \sum_i N^i$. *Moreover,* x_k *is feasible for the original problem* (1) *at each k, so that the coupling constraints are satisfied over all outer iterations.*

Therefore, the complexity of Algorithm IPLM is of order $\mathcal{O}(\ln(\frac{\mu_0}{\varepsilon}))$. Note however that at each iteration we have to invert matrices of dimension equal to the dimension of the local variables x^i and thus such an algorithm is not suitable when the dimensions of the variables x^i are very large. As we will show in the next section, this algorithm is still suitable for distributed MPC.

2.3   Application of Smoothing Techniques to Separable Non-convex Problems

In this section we show how the two distributed algorithms, based on smoothing techniques, presented above can be used for solving the *separable non-convex* optimization problem:

$$\min_{x^i \in X^i, G^i x^i = a^i} \sum_{i=1}^M f^i(x^i), \quad \text{s.t. } h(x^1, \cdots, x^M) = 0, \tag{2}$$

where the same assumptions hold as in Assumption 1. Note that the non-convexity in problems (2) is concentrated in the equality constraints $h(x) = 0$.

Since the objective function in (2) is convex we can use the framework of sequential convex programming (SCP) for solving this separable non-convex problem. The underlying idea of SCP is that we can solve the non-convex optimization problem

(2) by iteratively solving a convex approximation of the original problem. Starting from an initial guess $x_0 = [x_0^{1T} \cdots x_0^{MT}]^T$ satisfying $x_0^i \in X^i, G^i x_0^i = a^i$ for all i, the SCP algorithm calculates a sequence $\{x_k\}_{k \geq 0}$ by solving the convex approximation of (2):

$$\min_{x^i \in X^i, G^i x^i = a^i} \sum_{i=1}^{M} f^i(x^i), \text{ s.t. } h(x_k) + \nabla h(x_k)^T (x - x_k) = 0, \tag{3}$$

where $\nabla h(x)$ denotes the Jacobian of h at x. Note that we implicitly assumed that the function h is differentiable. The optimal solution of the separable convex problem (3) is denoted with x_{k+1}. The separable optimization problem (3) that we have to solve in each SCP iteration is convex and has a decomposable structure as in (1), and therefore is suitable for decomposition using the smoothing techniques from Section 2. Sections 2.1 and 2.2 illustrate how we can distributively compute x_{k+1} using the proximal center or interior-point Lagrangian decomposition algorithms for problem (3). Local linear convergence of the SCP method can be established under some mild regularity assumptions (see [8] for more details).

3 Distributed Model Predictive Control

In the rest of the paper we explore the potential of the previous distributed algorithms based on smoothing, the proximal center and interior-point Lagrangian method, in MPC problems for networks. We show that many relevant centralized MPC schemes for non-linear/linear network systems can be recast as separable optimization problems for which our distributed algorithms can be applied but exploiting the specific control problem structure.

3.1 Distributed MPC for Coupling Non-linear Dynamics

The application that we will discuss in this section is distributed control (MPC) of large-scale network systems with interacting subsystem dynamics, which can be found in a broad spectrum of applications ranging from robotics to electrical distribution and transportation networks. In such applications there are not only the difficulties caused by complex interacting dynamics, but also the limitation of information structure due to organizational aspects.

We consider non-linear discrete-time systems which can be decomposed into M subsystems described by difference equations of the form:

$$x_{t+1}^i = \phi^i(x_t^j, u_t^j; j \in \mathcal{N}^i) \; \forall i = 1 \cdots M. \tag{4}$$

The vectors $x_t^i \in \mathbb{R}^{n_{x^i}}$, $u_t^i \in \mathbb{R}^{n_{u^i}}$ represent the state and the input of the subsystem[2] i at time t. The index set $\mathcal{N}^i$ denotes the set of neighbors of subsystem i in the graph

---

[2] Throughout the paper we will use the convention that every superscript indicates a subsystem index.

G, graph that describes the interaction between subsystems. In the problem formulation we assume that the control and state vectors must satisfy local constraints:

$$x_t^i \in \mathbb{X}^i, \ u_t^i \in \mathbb{U}^i, \ x_N^i \in \mathbb{X}_f^i \qquad \forall t = 0 \cdots N-1 \quad \forall i = 1 \cdots M, \tag{5}$$

where all these sets are assumed to be convex, compact and N is the prediction horizon. The system performance is expressed via a stage and a final cost, which are composed of individual linear or quadratic convex costs for each subsystem i and have the form:

$$\sum_{t=0}^{N-1} \ell^i(x_t^i, u_t^i) + \ell_f^i(x_N^i). \tag{6}$$

In the centralized MPC we must solve at each step k, given $x^i(k) = \bar{x}^i$, a finite horizon non-linear optimal control problem over a prediction horizon of length N (since we deal with time invariant systems we can consider $k = 0$):

$$\min_{x_t^i, u_t^i} \sum_{i=1}^{M} \sum_{t=0}^{N-1} \ell^i(x_t^i, u_t^i) + \sum_{i=1}^{M} \ell_f^i(x_N^i) \tag{7}$$

$$\text{s.t. } x_0^i = \bar{x}^i, \ x_{t+1}^i = \phi^i(x_t^j, u_t^j; j \in \mathcal{N}^i) \tag{7.1}$$

$$x_t^i \in \mathbb{X}^i, \ u_t^i \in \mathbb{U}^i, \ x_N^i \in \mathbb{X}_f^i \ \forall t = 1 \cdots N-1, \ \forall i = 1 \cdots M. \tag{7.3}$$

The centralized non-linear optimal control problem (7) becomes interesting if the computations can be distributed among the subsystems and the amount of information that the agents must exchange is limited. In comparison with the centralized approach, a distributed strategy offers a series of advantages: first, the numerical effort is considerably smaller since we solve low dimension problems in parallel and secondly such a design is modular, i.e. adding or removing subsystems does not require any controller redesign. Now, we show that the non-linear optimal control problem (7) can be recast as a separable non-convex problem of the form (2) but with a particular structure. To this purpose, define the variables

$$\mathbf{x}^i = \begin{bmatrix} x_1^{iT} \cdots x_{N-1}^{iT} \ x_N^{iT} \ u_0^{iT} \cdots u_{N-1}^{iT} \end{bmatrix}^T, \quad \mathbf{x} = \begin{bmatrix} \mathbf{x}^{1T} \cdots \mathbf{x}^{MT} \end{bmatrix}^T,$$

the sets $X^i = \Pi_{i=1}^{N-1} \mathbb{X}^i \times \mathbb{X}_f^i \times \Pi_{i=1}^{N} \mathbb{U}^i$ and the convex functions $f^i(\mathbf{x}^i) = \sum_{t=0}^{N-1} \ell^i(x_t^i, u_t^i) + \ell_f^i(x_N^i) \ \forall i$. With this notation, the non-linear optimal control problem (7) now reads:

$$\min_{\mathbf{x}^i \in X^i} \sum_{i=1}^{M} f^i(\mathbf{x}^i), \ \text{s.t. } h^i(\mathbf{x}^j; \ j \in \mathcal{N}^i) = 0 \ \forall i = 1 \cdots M, \tag{8}$$

where $h^i(\mathbf{x}^j; \ j \in \mathcal{N}^i) = 0$ are obtained by stacking the constraints (7.1) for an i over the prediction horizon N. From the previous discussion we obtain:

Theorem 3. *The centralized non-linear MPC problem (7) can be written as a particular case of a separable non-convex problem (2), i.e. in the form (8).*

Note that in these settings the local equality constraints of the form $G^i x^i = a^i$ are not present since the dynamics are non-linear and coupled.

3.2 Distributed MPC for Coupling Linear Dynamics

Many network systems, e.g. interconnected chemical processes [16] or urban traffic systems [2], can be decomposed into M appropriate linear subsystems:

$$x^i_{t+1} = A_i x^i_t + B_i u^i_t + \sum_{j \in \mathcal{N}^{-i}} A_{ij} x^j_t + B_{ij} u^j_t, \quad \forall i = 1 \cdots M, \tag{9}$$

where the index set $\mathcal{N}^{-i} = \mathcal{N}^i - \{i\}$, i.e. it contains all the indices of the subsystems which interact with the ith subsystem. If we introduce an auxiliary variable $w^i_t \in \mathbb{R}^{n_{w_i}}$ to represent the influence of the neighboring subsystems on the ith subsystem (in applications we usually have $n_{w_i} \ll n_i$), we can rewrite the dynamics (9) as $x^i_{t+1} = A_i x^i_t + B_i u^i_t + E_i w^i_t$ for all i, where the matrices E_i are of appropriate dimensions and

$$w^i_t = \sum_{j \in \mathcal{N}^{-i}} A^-_{ij} x^j_t + B^-_{ij} u^j_t,$$

with the matrices A^-_{ij}, B^-_{ij} being obtained from the matrices A_{ij}, B_{ij} by removing the rows with all entries equal to zero. The system performance is expressed through a linear or quadratic stage cost and a final cost for each subsystem i as in (6). We also assume that the sets $\mathbb{X}^i, \mathbb{X}^i_f$ and $\mathbb{U}^i$ that define the local state and input constraints (5) are compact convex sets. The centralized linear optimal control problem over the prediction horizon N for this application can be formulated as follows:

$$\min_{x^i_t, u^i_t, w^i_t} \sum_{i=1}^{M} \sum_{t=0}^{N-1} \ell^i(x^i_t, u^i_t) + \sum_{i=1}^{M} \ell^i_f(x^i_N) \tag{10}$$

$$\text{s.t.} : x^i_0 = \bar{x}^i, \ x^i_{t+1} = A_i x^i_t + B_i u^i_t + E_i w^i_t, \ w^i_t = \sum_{j \in \mathcal{N}^{-i}} A^-_{ij} x^j_t + B^-_{ij} u^j_t \tag{10.1}$$

$$x^i_t \in \mathbb{X}^i, \ u^i_t \in \mathbb{U}^i, \ x^i_N \in \mathbb{X}^i_f \ \forall t = 1 \cdots N-1, \ \forall i = 1 \cdots M. \tag{10.2}$$

We can eliminate the state variables in the control problem (10) using the dynamics (10.1). In this case we can define: $\mathbf{x}^i = [w^{iT}_0 \cdots w^{iT}_{N-1} \ u^{iT}_0 \cdots u^{iT}_{N-1}]^T$. Then, the linear control problem (10) can be recast as a separable convex program[3] with decoupling cost and coupling constraints in the form (1):

$$\min_{x^i \in X^i} \sum_{i=1}^{M} f^i(\mathbf{x}^i), \quad \text{s.t.} \ \sum_{i=1}^{M} F^i \mathbf{x}^i = a. \tag{11}$$

where the coupling constraints $\sum_{i=1}^{M} F^i \mathbf{x}^i = a$ are obtained by stacking the coupling constraints (10.1) for all i, t. It is easy to see that the number of rows of the matrices

---

[3] Note that we can also use the approach from Section 3.1 to obtain a separable convex problem in the variables $\mathbf{x}^i = [x^{iT}_1 \cdots x^{iT}_{N-1} \ x^{iT}_N \ u^{iT}_0 \cdots u^{iT}_{N-1}]^T$ (see [9] for more details).

F^i are equal to $N \sum_{i=1}^{M} n_{w_i}$. Note also that the functions f^i are convex quadratic when we consider quadratic performance indexes in (6) and the local constraint sets X^i are convex. Based on the previous discussion, we obtain the following result:

Theorem 4. *The centralized linear MPC optimization problem* (10) *can be written as a particular case of the separable convex problem* (1), *more specifically in the form* (11).

3.3 Practical Implementation

In this section we describe the practical implementation of the Algorithm PCM and IPLM to the non-linear MPC problem (7) and linear MPC problem (10) in terms of distributed computations and feasibility. Our algorithms can be an alternative to the classical methods (e.g. Jacobi algorithm, incremental subgradient algorithm, dual subgradient algorithm, etc), leading to new methods of solution, but with a faster convergence rate.

We first prove that the update rules in the algorithms DSM and PCM are completely distributed, according to the communication graph between subsystems. Indeed, we recall that the non-linear coupling constraints $h^i(\mathbf{x}^j; \ j \in \mathcal{N}^i) = 0$ in (8) are obtained by stacking the non-linear dynamics (7.1) for a given i over the prediction horizon N. Similarly, the linear coupling constraints $\sum_{i=1}^{M} F^i \mathbf{x}^i = a$ in (11) are obtained by stacking the coupling between subsystems (10.1) for all i, t. Therefore, after linearizing $h^i(\mathbf{x}^j; \ j \in \mathcal{N}^i) = 0$ at some x_k, the coupling constraints in the separable convex problems (3) and also in (11) can be rearranged in the form: $F_i[\mathbf{x}^j]_{j \in \mathcal{N}^i} = a_i$ for all i, i.e. we have $[F^1 \cdots F^M] = [F_1^T \cdots F_M^T]^T$. Let λ^i be the Lagrange multipliers for the constraints $F_i[\mathbf{x}^j]_{j \in \mathcal{N}^i} = a_i$, and thus $\lambda = [\lambda^{1T} \cdots \lambda^{MT}]^T$. Then, the main update rules in Algorithms DSM and PCM are distributed, each subsystem i using information only from its neighbors according to the communication graph G, e.g. Step 2 of Algorithm PCM becomes:

$$\lambda_{k+1}^i = \lambda_k^i + \frac{1}{D_\mu} \left(F_i[\mathbf{x}_{k+1}^j]_{j \in \mathcal{N}^i} - a_i \right).$$

However, for the Algorithm IPLM, the update of the Lagrange multipliers has to be done by a central agent, i.e. in this case we have a star-shaped topology for the communication graph G among subsystems.

We now show that due to the special structure of the MPC problems (7) and (10), our two algorithms leads to decomposition in both "space" and "time", i.e. not only over M but also over N. For simplicity of the exposition we assume that the sets are normalized Euclidian balls: $\mathbb{X}^i = \{x^i \in \mathbb{R}^{n_{x^i}} : \|x^i\| \leq 1\}$, $\mathbb{U}^i = \{u^i \in \mathbb{R}^{n_{u^i}} : \|u^i\| \leq 1\}$. In the Algorithm PCM we need to properly choose the function ϕ_{X^i} according to the structure of X^i. Whenever the set X^i has the structure of a Cartesian product, e.g as in the MPC formulations (7), we choose: $\phi_{X^i}(\mathbf{x}^i) = \sum_{l=1}^{N} \|x_l^i\|^2 + \sum_{l=0}^{N-1} \|u_l^i\|^2$. Similarly, in Algorithm IPLM the prox function ϕ_{X^i} must be chosen as self-concordant barrier functions for the set X^i. Let $b(x)$ be the self-concordant barrier function for the Euclidian set $\{x : \|x\| \leq 1\}$. When the set X^i has the structure of a Cartesian product, we choose: $\phi_{X^i}(\mathbf{x}^i) = \sum_{l=1}^{N} b(x_l^i) + \sum_{l=0}^{N-1} b(u_l^i)$. If we further assume that the stage costs are given by quadratic expressions, then the centralized MPC problems take the form: $\min_{\mathbf{x}^i \in X^i} \{ \sum_{i=1}^{M} \mathbf{x}^{iT} \mathbf{Q}_i \mathbf{x}_i : \sum_{i=1}^{M} F^i \mathbf{x}^i = a \}$, where $\mathbf{Q}_i$ has a block diagonal structure.

In both Algorithms PCM and IPLM the most expensive computations consists in solving for each i, the minimization problem $\min_{\mathbf{x}^i \in X^i} \mathbf{x}^{iT} \mathbf{Q}_i \mathbf{x}^i + \langle \lambda_k, F^i \mathbf{x}^i \rangle + \mu \phi_{X^i}(\mathbf{x}^i)$, with λ_k computed at previous iteration. Since $\mathbf{Q}_i$ has a diagonal structure, we can further decompose each minimization problem into $2N$ quadratic cost quadratic constraints problems with a particular structure:

$$\min_{\|x\| \le 1} x^T Q x + \langle q, x \rangle + b(x),$$

where Q is a positive semidefinite matrix, x represents the state variable x_l^i or control variable u_l^i at step l and $b(\cdot)$ is quadratic or barrier function. In summary, the special structure of the centralized MPC problems shows that our two algorithms lead to decomposition in both "space" and "time", i.e. the MPC problem can be decomposed into small subproblems corresponding to spatial structure of the system (M) but also to prediction horizon (N).

We should note that in the algorithm IPLM we maintain feasibility of the coupling constraints (that represent the dynamics of the network system over N) at each outer iteration (see Theorem 2), while for first-order algorithms DSM and PCM feasibility holds only at convergence of these algorithms and not at the intermediate iterations. Note that this issue is solved in the context of MPC for networks where the neighboring subsystems are interacting with subsystem i only throug states, i.e. where we have dynamics of the form:

$$x_{t+1}^i = \phi^i(x_t^j, u_t^i; j \in \mathcal{N}^i) \quad \text{or} \quad x_{t+1}^i = A_i x_t^i + B_i u_t^i + \sum_{j \in \mathcal{N}^{-i}} A_{ij} x_t^j.$$

In this case the coupling constraint $w_0^i = \sum_{j \in \mathcal{N}^{-i}} A_{ij}^- x_0^j$ holds in the MPC framework since $x_0^j = \bar{x}^j$ (the measured state).

4 Conclusions

The proximal center and the interior-point Lagrangian decomposition algorithms were applied for solving distributed MPC problems for non-linear or linear networks. We showed that the corresponding centralized optimal control problems can be recast as separable optimization problems for which our two algorithms can be applied. We proved that the solution generated by our distributed algorithms converges to the solution of the centralized control problem and we provided also estimates for the rate of convergence. We also proved that the main steps of the two algorithms can be computed efficiently for MPC problems by using the specific structure of the control problem.

Acknowledgements. The research leading to these results has received funding from: the European Union, Seventh Framework Programme (FP7/2007–2013) under grant agreement no 248940; CNCSIS-UEFISCSU (project TE, no. 19/11.08.2010); ANCS (project PN II, no. 80EU/2010); Sectoral Operational Programme Human Resources Development 2007-2013 POSDRU/89/1.5/S/62557; GOA-MaNet; CoE EF/05/006; FWO G.0226.06.

References

1. Bertsekas, D.P., Tsitsiklis, J.N.: Parallel and distributed computation: Numerical methods. Prentice-Hall (1989)
2. Camponogara, E., De Oliveira, L.: Distributed optimization for model predictive control of linear-dynamic networks. IEEE Transactions on Systems, Man, and Cybernetics, Part A: Systems and Humans 39(6), 1331–1338 (2009)
3. Dunbar, W.B.: Distributed receding horizon control of dynamically coupling non-linear systems. IEEE Transactions on Automatic Control 52(7), 1249–1263 (2007)
4. Lasdon, L.S.: Optimization Theory for Large Systems. Macmillan Series for Operations Research (1970)
5. Maciejowski, J.M.: Predictive Control with Constraints. Prentice Hall, England (2002)
6. Keviczky, T., Johansson, K.H.: A study on Distributed Model Predictive Consensus. In: 17th IFAC World Congress, Seoul, South Korea (2008)
7. Mesarovic, M.D., Macko, D., Takahara, Y.: Theory of Hierarchical, Multilevel, Systems. Academic Press (1970)
8. Necoara, I., Savorgnan, C., Tran-Dinh, Q., Suykens, J., Diehl, M.: Distributed Non-linear Optimal Control using Sequential Convex Programming and Smoothing Techniques. In: Proceedings of the 48th IEEE Conference on Decision and Control (2009)
9. Necoara, I., Suykens, J.: Application of smoothing technique to decomposition in convex optimization. IEEE Transactions on Automatic Control 53(11), 2674–2679 (2008)
10. Necoara, I., Suykens, J.A.K.: An interior-point Lagrangian decomposition method for separable convex optimization. Journal of Optimization Theory and Applications 143(3), 567–588 (2009)
11. Necoara, I., Nedelcu, V., Dumitrache, I.: Parallel and distributed optimization methods for estimation and control in networks. Journal of Process Control (2011)
12. Nesterov, Y., Nemirovskii, A.: Interior Point Polynomial Algorithms in Convex Programming. SIAM Studies in Applied Mathematics, Philadelphia (1994)
13. Nesterov, Y.: Introductory Lectures on Convex Optimization. Kluwer (2004)
14. Nesterov, Y.: Smooth minimization of non-smooth functions. Mathematical Programming A 103(1), 127–152 (2005)
15. Rawlings, J.B., Stewart, B.T.: Coordinating multiple optimization-based controllers: New opportunities and challenges. Journal of Process Control 18, 839–845 (2008)
16. Venkat, A.N., Rawlings, J.B., Wright, S.J.: Distributed MPC of large-scale systems. Assessment and Future Directions of Non-linear Model Predictive Control (2005)

Model Predictive Control with Delay Compensation for Air-to-Fuel Ratio Control

Sergio Trimboli, Stefano Di Cairano, Alberto Bemporad, and Ilya V. Kolmanovsky

Abstract. To meet increasingly stringent emission regulations modern internal combustion engines require highly accurate control of the air-to-fuel ratio. The performance of the conventional air-to-fuel ratio feedback loop is limited by the combustion delay between fuel injection and engine exhaust, and by the transport delay for the exhaust gas to propagate to the air-to-fuel ratio sensor location. The combined delay is variable, since it depends on engine speed and airflow. Drivability, fuel economy and emission requirements result in constraints on the deviations of the air-to-fuel ratio, stored oxygen in the three-way catalyst, and fuel injection. This paper proposes an approach for air-to-fuel ratio control based on Model Predictive Control (MPC). The approach systematically handles both variable time delays and pointwise-in-time constraints. A delay-free model is considered first, which takes into account the dynamic relations between the injected fuel and the air-to-fuel ratio and the dynamics of the oxygen stored in the catalyst. For the delay-free model, the explicit MPC law is computed. Delay compensation is obtained by estimating the delay online from engine operating conditions, and feeding the MPC law with the state predicted ahead over the time interval of the estimated delay. The predicted state is computed by combining measurement filtering with forward iterations of the nonlinear dynamic equations of the model. The achieved performance in tracking the air-to-fuel ratio and the oxygen storage setpoints while enforcing the constraints is demonstrated in simulation using real data profiles.

Sergio Trimboli · Alberto Bemporad
Department of Mechanical and Structural Engineering, University of Trento, Italy
e-mail: trimboli@ieee.org, bemporad@ing.unitn.it

Stefano Di Cairano
Powertrain Control R&A, Ford Motor Co., Dearborn, MI, US
e-mail: dicairano@ieee.org

Ilya V. Kolmanovsky
Department of Aerospace Engineering, University of Michigan, Ann Arbor, MI, US
e-mail: ilya@umich.edu

R. Sipahi et al. (Eds.): Time Delay Sys.: Methods, Appli. and New Trends, LNCIS 423, pp. 319–330.

1 Introduction

During recent years tightened emission regulations have been imposed to reduce pollution. In order to meet these regulations without degrading fuel efficiency and vehicle performance, and without embedding excessively expensive hardware components, engineers increasingly rely on improvements in software and controls.

High performance Air-to-Fuel Ratio (AFR) control is critical for emission reductions in spark ignited (SI) engines [5]. The AFR has to be maintained within tight bounds not to degrade vehicle driveability and fuel economy, and to ensure proper functioning of the exhaust aftertreatment system.

The combustion that takes place in the cylinders involves a mixture of air and fuel. For the combustion to be complete, the mass of air and fuel in the mixture must be in a precise ratio, referred to as the *stoichiometric AFR*. The mixture with excess fuel is referred to as *rich*, while the mixture with excess air is referred to as *lean*.

The exhaust gas chemical composition depends on the AFR. The pollutants are converted in the exhaust aftertreatment system, which typically consists of one or more three-way-catalysts (TWC). The TWC converts nitrogen-oxides (NOx), hydrocarbons (HC) and carbon-monoxide (CO) emissions to carbon dioxide and water. In order to maintain high catalyst conversion efficiency, the feedgas AFR must be carefully controlled [5].

The AFR is not directly measured in the cylinder, where the combustion takes place. The feedgas (exhaust gas) AFR is measured by the universal exhaust gas oxygen (UEGO) sensor that is placed close to the TWC inlet, see Fig. 1 where the plant and the control architecture introduced below are shown. Hence, the AFR in the cylinder is controlled by regulating the AFR as measured in the exhaust gas, after the combustion took place and the gas has travelled through the manifold up to the UEGO sensor location.

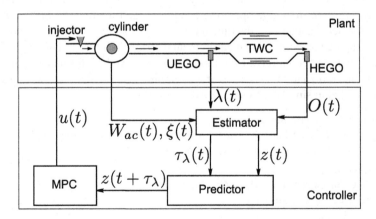

Fig. 1 Schematics of the gas propagation path and AFR-TWC controller architecture.

The dynamics of the air-to-fuel ratio can be represented by a first-order time-delayed system that models the combustion process and the gas transport delay. The transport delay varies up to a factor of 10 depending on the engine operating conditions, as a consequence of the variable exhaust gas pressure and velocity. Since the AFR is affected by disturbances, estimation and actuation errors, and model imprecisions, feedback control is needed. Also, the limits on the injector flow impose constraints on the injected fuel, limiting the actuation range.

Several approaches for controlling AFR have been considered in the literature, see for instance [1, 6, 9] and the references therein. In all of these approaches, AFR has to be regulated to track a time-varying set-point reference. The modulation of the AFR reference may be used to maintain the oxygen storage in the TWC at a desired level and improve catalyst conversion efficiency. A rich mixture causes excess generation of CO and HC in the feedgas, that are converted by the TWC while consuming stored oxygen. A lean mixture may cause an excess of NOx in the feedgas. This NOx is reduced in the TWC and the excess oxygen is stored. If the oxygen storage in the TWC is completely full or completely empty, the TWC cannot efficiently convert the pollutants resulting in emission breakthrough into the tailpipe. Hence, the purpose of controlling the AFR to a given reference is to control the oxygen storage at its desired level, to avoid saturation and depletion. The control of AFR and TWC oxygen is usually solved by an inner-outer loop strategy [6], where the outer-loop controller generates the AFR reference to maintain the oxygen storage at a desired value and the inner-loop controller regulates the AFR to track such a reference, while maintaining it within a range where the vehicle driveability and fuel economy are not degraded. Different design have been proposed for the outer controller, see for instance [4, 7] and the references therein.

In this paper we propose a model predictive control (MPC) approach for controlling the single-input multiple-output (SIMO) system that represents the AFR and oxygen storage dynamics. Even though the system is underactuated, tracking the two setpoints is possible thanks to the integral dynamics of oxygen storage. The MPC can regulate AFR and oxygen storage to track independent references by negotiating the overall system behavior depending on a predefined performance criterion. Furthermore, MPC can explicitly enforce actuation limits and desired operating ranges. Finally, by using explicit model predictive control techniques, the MPC feedback law can be computed in the form of a piecewise affine function [2], which can be evaluated without the need of executing optimization algorithms online, and hence it is suitable for execution on conventional automotive Electronic Control Units (ECU). Due to the time-varying nature of the system and to the presence of time delays, a control architecture is proposed, where a predictor is used to compensate the delays, so that the MPC design can be based on a delay-free model where the state and input bounds are time-varying. As a result of the proposed modelling approach, the explicit MPC law is a function of the current states, references, and time-varying bounds.

The paper is organized as follows. In Sect. 2 the MPC-oriented models of the AFR and of the oxygen storage dynamics are formulated. The delay compensator

and the MPC controller are designed in Sect. 3 and the closed-loop system behavior
is simulated in Sect. 4. Conclusions are summarized in Sect. 5.

2   Model of AFR and Oxygen Storage

In this section we discuss the model of the air-to-fuel ratio and of the oxygen storage
that will be used for prediction in the model predictive control strategy.

2.1   *Air-to-Fuel Ratio Dynamics*

The (normalized) Air-to-Fuel Ratio (AFR) λ is defined by

$$\lambda = \frac{1}{\sigma_0} \frac{W_{ac}}{W_{fc}} , \tag{1}$$

where W_{ac} [kg/s] is the cylinder air mass flow, W_{fc} [kg/s] is the cylinder base fuel
mass flow, and σ_0 is the stoichiometric air-to-fuel ratio, where for the gasoline engines considered in this paper $\sigma_0 = 14.64$. When $\lambda = 1$ the mixture is *stoichiometric*,
hence all the air and fuel present in the cylinder are burned. When $\lambda < 1$ the mixture
is *rich*, when $\lambda > 1$ the mixture is *lean*.

In AFR control, the objective is for λ to track a reference setpoint by command-
ing the fuel injectors. The AFR is not measured in the cylinder but after the combustion took place and the exhaust gas has reached the UEGO sensor location. Due
to UEGO sensor time-constant, exhaust mixing and cylinder-by-cylinder firing, the
AFR dynamics are modelled as a first-order system with delayed input, formulated
in discrete time as

$$\lambda(k+1) = \alpha\lambda(k) + \beta u(k - \tau_\lambda(k)) , \tag{2}$$

where α, β are such that the dc-gain of (2) is 1 (i.e., $\beta = 1 - \alpha$), $\tau_\lambda(k)$ is the input
time delay, and we consider a sampling period $T_s = 25$ms. The time delay $\tau_\lambda(k)$
in (2) varies depending on the engine operating conditions [6], the airflow W_{ac} and
the engine speed ξ [rpm]

$$\tau_\lambda(k) = \alpha_0 + \frac{\alpha_1}{\xi(kT_s)} + \frac{\alpha_2 M_{\max}(\xi(kT_s))}{W_{ac}(kT_s)} , \tag{3}$$

where $M_{\max}(\xi)$ defines the maximum achievable load W_{ac}/ξ as a function of the
engine speed ξ, and α_i, $i = 1,2,3$, are coefficients that depend on the engine and
that can be identified by regression on experimental data.

The control input u is the ratio of in-cylinder air flow W_{ac} and in-cylinder fuel
flow W_{fc},

$$u(k) = \frac{W_{ac}(kT_s)}{W_{fc}(kT_s)} . \tag{4}$$

The bounds on u depend on the transient fuel dynamics

$$\dot{m}_p(t) = -\frac{m_p(t)}{\tau_p} + X W_{fi}(t) , \tag{5a}$$

$$W_{fc}(t) = \frac{m_p(t)}{\tau_p} + (1-X)W_{fi}(t) , \tag{5b}$$

where m_p [kg] is the mass of the fuel in the fuel puddle formed at the ports, $\tau_p > 0$ [s] is the time-constant of puddle fuel evaporation, $X \in (0,1)$ is a fraction of the injected fuel that replenishes the liquid fuel puddle. In (5), W_{fi} [kg/s] is the mass flow of injected fuel,

$$0 \leq W_{fi}(t) \leq W_{fMAX} , \tag{6}$$

where W_{fMAX} [kg/s] is the maximum flow that the injectors can provide. Thus, from (4) and (5b) the input at step k is constrained in

$$\frac{W_{ac}(kT_s)}{\frac{m_p(kT_s)}{\tau_p} + (1-X)W_{fMAX}} \leq u(k) \leq \frac{W_{ac}(kT_s)}{m_p(kT_s)}\tau_p , \tag{7}$$

where $m_p(kT_s)$ is computed from (5a) formulated as

$$\dot{m}_p(t) = -\frac{1}{1-X}\frac{m_p(t)}{\tau_p} + \frac{X}{1-X}\frac{W_{ac}(t)}{u(t)} .$$

2.2 Oxygen Storage Dynamics

For control purposes, the dynamics of the oxygen stored in the catalyst may be modelled by a constrained integrator [5]. The normalized catalyst oxygen storage $0 \leq O(t) \leq 1$ in the three-way-catalyst (TWC) evolves depending on the AFR,

$$\dot{O}(t) = W_{ac}(t)K\left(1 - \frac{1}{\lambda(t)}\right) , \tag{8}$$

where K is a known parameter, which may depend on the catalyst temperature and change with catalyst aging. Let $\Theta_c = O/W_{ac}$ and assume that W_{ac} is measured and varies slowly with respect to the AFR dynamics, so that for prediction purposes can be considered constant. Then, $\dot{\Theta}_c(t) = \dot{O}(t)/W_{ac}(t)$, and from (8)

$$\dot{\Theta}_c(t) = K\left(1 - \frac{1}{\lambda(t)}\right), \quad 0 \leq \Theta_c(t) \leq \frac{1}{W_{ac}(t)} . \tag{9}$$

Since for spark ignited engines the AFR varies in a small interval around 1, we can linearize (9), and convert it to discrete-time

$$\Theta(k+1) = \Theta(k) + T_s K(\lambda(k) - 1) , \tag{10a}$$

$$0 \leq \Theta(k) \leq \frac{1}{W_{ac}(kT_s)} . \tag{10b}$$

The oxygen storage dynamics (10a) is affine in Θ and λ. The airflow W_{ac} is a measured disturbance that can be assumed constant, hence (10b) is a parameter-dependent linear constraint.

3 Model Predictive Control with Delay Compensation

The most critical factor in AFR control is the time-varying delay that affects the command input $\tau_\lambda(k)$. The numerical value of $\tau_\lambda(k)$ can vary up to a factor of 10 over engine operating conditions, hence the MPC approaches for systems with constant delay [3] are not suitable. In order to cope with such a delay, we propose an architecture composed of the cascade of a delay compensator and a model predictive controller, as shown in Fig. 1. The delay compensator is a predictor based on AFR dynamics (2) and on oxygen dynamics (8).

In order to account for unmeasured disturbances and modeling imperfection due to transient fuel dynamics, a disturbance model [8] is added to the system which provides integral action. Since in AFR control the disturbances mostly take place in the fuel mixture in the cylinder, an input additive disturbance model is used. We define the state vector $z = [\lambda \; w \; \Theta \;]$, where w is the state of the disturbance model, so that the augmented system dynamics are

$$z(k+1) = Az(k) + Bu(k - \tau_\lambda(k)) + v + B_d v(k) \tag{11a}$$
$$\psi(k) = Cz(k) + e(k) . \tag{11b}$$

where v is the affine term in (10a), $v(k), e(k)$ are Gaussian variables with zero means and covariances P_v, P_e, respectively,

$$A = \begin{bmatrix} \alpha & \beta & 0 \\ 0 & 1 & 0 \\ KT_s & 0 & 1 \end{bmatrix}, B = \begin{bmatrix} \beta \\ 0 \\ 0 \end{bmatrix}, B_d = \begin{bmatrix} 0 \\ 1 \\ 0 \end{bmatrix}, v = \begin{bmatrix} 0 \\ 0 \\ KT_S \end{bmatrix} .$$

and $\psi(k)$ is the current measurement, where $C = [1 \; 0 \; 0]$. In (11), the state of the stored oxygen is not observable. The estimation of the oxygen in the catalyst is a rather complex operation [5], which is beyond the scope of this paper. We assume a low rate estimate of the oxygen storage is available, and we update the oxygen state in open-loop during the interval between two of such estimates.

3.1 State Predictor

The delay compensator performs two operations: (i) estimate the system state, and in particular the disturbance state w, (ii) predict the state of model (11) $\tau_\lambda(k)$ steps ahead to compensate for the delay. The first operation is performed by means of a Kalman Filter based on (11). The prediction is then computed by performing forward iterations of the system dynamics starting from the updated state estimate. Due to the discrete-time nature of the approach used here, perfect delay compensation is not possible when $\tau_\lambda(k)$ is not a multiple of the sampling period T_s. In the following we assume that $\tau_\lambda(k) = cT_s$ for some nonnegative integer c. In case $\tau_\lambda(k)$ is not a multiple of T_s, the compensation is done with respect to $\tilde{\tau}_\lambda(k) = cT_s$, where c is the largest integer such that $c \leq \tau_\lambda(k)/T_s$, i.e., $c = \lfloor \tau_\lambda(k)/T_s \rfloor$. The prediction algorithm can be formalized as described in Algorithm 2.

Algorithm 2. Delay compensation by state prediction.

At time step k:

1. from current engine operating conditions $(\xi(k), W_{ac}(k))$ estimates, $\tau_\lambda(k)$ using (3);
2. from current measurement $\psi(k)$ and previous state estimate $z(k-1)$ updates the state estimate $z(k) = [\lambda(k) \ w(k) \ \Theta(k)]$;
3. fix $\tau_\lambda(h) = \tau_\lambda(k)$, $W_{ac}(h) = W_{ac}(k)$ for all $h \in [k, k+\tau_\lambda(k)]$. From $z(k)$ and stored input $u_p(k) = (u(k-\tau_\lambda(k-\tau_\lambda(k))), \ldots, u(k-1))$ predict the value $z(k+\tau_\lambda)$;
4. feed $x(k) = z(k+\tau_\lambda(k))$ to the the controller as the initial state.

In the prediction step of Algorithm 2 the prediction of the oxygen storage is obtained by using the nonlinear dynamics (9) instead of the linearized one (10a). By assuming that the airflow W_{ac} varies slowly with respect to the AFR dynamics, we can consider W_{ac} constant along the prediction horizon. Also, the time delay $\tau_\lambda(k)$ is assumed constant in prediction, since there is no preview on the future engine operating conditions. The prediction algorithm adapts to the time-varying delay, $\tau_\lambda(k)$. When the delay is large, e.g., at idle speed, the prediction will go far ahead in the future. When the delay is small, e.g., at high engine speed and airflow, the prediction will go ahead in the future only for a short time period. Algorithm 2 needs that a buffer is maintained to keep track of the past inputs. The length of the buffer must be dimensioned for a worst-case estimate $\bar{\tau}_\lambda \geq \max_k\{\tau_\lambda(k)\}$ of the delay.

By defining the variable $x(k) = z(k+\tau_\lambda(k))$ and using the certainty equivalence principle to remove the unpredictable unmeasured disturbance v, the system dynamics (11) formulated as prediction model is

$$x(k+1) = Ax(k) + Bu(k) + v , \tag{12}$$

where the delay has been removed and the system matrices A, B, v are the same as in (11).

3.2 Model Predictive Controller

The delay-free model (12) can be used in a model predictive control strategy. At time k, Algorithm 2 is used to compute $x(k) = z(k+\tau_d(k))$. The time-varying constraint bounds on the input and on Θ are computed depending on the current value of the fuel puddle estimate $m_p(kT_s)$ and on the airflow $W_{ac}(kT_s)$.

At time k the input bounds $u_{min}(k)$, $u_{max}(k)$ are

$$u_{min}(k) = \frac{W_{ac}(kT_s)}{\frac{m_p(kT_s)}{\tau_p} + (1-X)W_{f_{MAX}}}, \quad u_{max}(k) = \frac{W_{ac}(kT_s)}{m_p(kT_s)}\tau_p, \qquad (13)$$

and the state bounds $x_{min}(k)$, $x_{max}(k)$ are

$$x_{min}(k) = \begin{bmatrix} \lambda_{min} \\ -\infty \\ 0 \end{bmatrix}, \quad x_{max}(k) = \begin{bmatrix} \lambda_{max} \\ +\infty \\ 1 \\ \overline{W_{ac}(kT_s)} \end{bmatrix}, \qquad (14)$$

where λ_{min}, λ_{max} defines the boundary of the AFR value for which the three-way-catalyst can efficiently operates, and which shall not be crossed, unless for a very short time period. In this paper we use $\lambda_{min} = 0.95$, $\lambda_{max} = 1.05$, that are tightened with respect to the real desired range of operation being $\lambda \in [0.9, 1.1]$, in order to account for modeling imperfection and disturbances.

The MPC finite-time optimal control problem is

$$\min_{\mathbf{u}(k),\varepsilon} \quad \rho\varepsilon^2 + \qquad\qquad (15a)$$

$$\sum_{i=0}^{N-1} (x(i|k) - \bar{x}(k))'Q(x(i|k) - \bar{x}(k)) + R\Delta u(i|k)^2$$

$$\text{s.t.} \quad x(i+1|k) = Ax(i|k) + Bu(i|k) + v \qquad (15b)$$

$$x_{min}(k) - \varepsilon\mathbf{1} \le x(i|k) \le x_{min}(k) + \varepsilon\mathbf{1}, \quad i=1,\dots,N_c \qquad (15c)$$

$$u_{min}(k) \le u(i|k) \le u_{min}(k), \quad i=0,\dots,N \qquad (15d)$$

$$u(i|k) = u(N_u - 1|k), \quad i=N_u,\dots,N-1 \qquad (15e)$$

$$\Delta u(i|k) = u(i|k) - u(i-1|k) \qquad (15f)$$

$$x(0|k) = z(k + \tau_d(k)) \qquad (15g)$$

$$u(-1|k) = u(k-1), \qquad (15h)$$

where N is the prediction horizon, $N_u \le N$ is the control horizon, the number of free moves to select, $N_c \le N$ is the constraint horizon, the number of steps along which state constraints are enforced, $\mathbf{u}(k) = (u(0|k),\dots,u(N-1|k))$ is the input sequence to be optimized, and $\mathbf{1}$ indicates a vector of appropriate dimension entirely composed of 1.

In (15g), the optimization problem is initialized with state $z(k + \tau_d(k)) = \left[\lambda(k+\tau_d(k)) \; w(k) \; \frac{O(k+\tau_d(k))}{W_{ac}(k)} \right]$ computed by Algorithm 2, and (12) is used for prediction by (15b).

In (15a), the vector $\bar{x}(k)$ is the desired state setpoint which remains constant along the prediction horizon, as well as the state and input bounds (15c), (15d). The state constraint (15c) is "softened" by the additional optimization variable ε, so that it can be violated, at the price of a largely increased cost. This prevents infeasibility due to external unknown disturbances, and still enforces the desired range for AFR whenever possible.

We choose the weight R on input increments positive, and the weight matrix Q with the following structure $Q = \begin{bmatrix} q_1 & 0 & 0 \\ 0 & 0 & 0 \\ 0 & 0 & q_3 \end{bmatrix}$, where $q_1 \geq 0$ is the weight for AFR tracking and $q_3 \geq 0$ is the weight for oxygen storage tracking. The constraint violation weight $\rho \geq 0$ is chosen much larger than q_1, q_3, R. In (15) an incremental formulation of input cost is used, where the input variation $\Delta u(i|k)$ is weighted, which is more suitable for reference tracking.

The overall model predictive control algorithm with delay compensation is summarized in Algorithm 3.

Algorithm 3. Model predictive control with delay compensation for AFR and oxygen storage control.

At time step k:

1. execute Algorithm 2 from state $z(k-1)$, operating conditions $(\xi(k), W_{ac}(k))$, AFR measurement $\psi(k)$, and input buffer $\mathbf{u}_p(k)$ $\lambda(k)$ to obtain $z(k + \tau_\lambda(k))$;
2. set $x(k) = z(k + \tau_\lambda(k))$, solve optimal control problem (15a), and obtain the optimal input profile $\mathbf{u}^*(k) = (u^*(0|k), \ldots, u^*(N-1|k))$;
3. set the input $u(k) = u^*(0|k)$ and update $\mathbf{u}_p(k)$ by $u(k)$.

---

4   Simulation Results

In this section we discuss simulation results obtained from a detailed nonlinear model of AFR and oxygen dynamics in closed-loop with the MPC architecture of Sect. 3.2.

The controller objective is to track the references on the AFR and on the oxygen storage. In contrast with more classical approaches based on inner-outer loop control strategies, here we keep the reference of the AFR constantly to the stoichiometric ratio ($\lambda = 1$). The reference of the oxygen is slightly modulated around the middle point, since it is recognized that keeping the catalyst active improves the durability of the TWC. By this reference choice we investigate the capability of MPC in negotiating the different control objectives imposed by AFR and oxygen storage, using a single control input.

For the controller used in the simulations, in (15a) we have set $q_1 = 0.5$, $q_3 = 10^4$, $\rho = 10^7$, $N = 10$, $N_u = 2$, $N_c = 10$. A 10% error on W_{ac} is added, and an estimate of the stored oxygen is available at 0.1Hz, while the stored oxygen state is updated open-loop within two of such estimates. The reference for the AFR is $\lambda = 1$ while the oxygen reference is a square wave between 40% and 60% of its maximum value. We present two simulation scenarios. The first scenario is a demonstration of the controller behavior when the engine speed and airflow vary, in an acceleration and a deceleration. In the second scenario we test the controller with real airflow and engine speed profiles, obtained during the execution of an EPA driving cycle on a

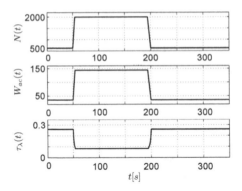

Fig. 2 Scenario 1, acceleration and deceleration. Engine speed, airflow and (estimated) time-delay profiles.

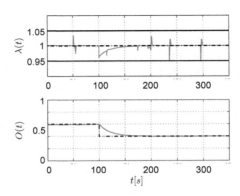

Fig. 3 Scenario 1, acceleration and deceleration. AFR and oxygen storage dynamics.

real vehicle. The first scenario involves acceleration and deceleration. The AFR is affected by an intermittent unmeasured additive input disturbance that models the purge of the fuel canister. When such purge happens an unknown amount of fuel is added to the one commanded by the controller. The engine speed is accelerated from idle speed (625 rpm) to 2000 rpm in 4s, then kept constant, and finally decelerated again to idle speed in 6s. At the same time, the airflow is increased from $W_{ac} = 36$kg/hr to $W_{ac} = 144$kg/hr, then reduced again to $W_{ac} = 36$kg/hr. The profiles of W_{ac}, ξ, and τ_λ along this scenario are shown in Fig. 2. The results for the first scenario are shown in Fig. 3. The controller behaves correctly despite the variability in the time-delay due to different engine speed and airflow. Note that the effect of the fuel purge disturbance at high speed and high airflow is reduced due to the shorter time delay and because the fuel added by the purge is small compared to the commanded injection.

In the second scenario we simulate the controller action with respect to a real experimental airflow and engine speed profile obtained from a vehicle running the EPA drive cycle. In Fig. 4 the operating conditions, airflow and engine speed, and the (estimated) time delay for $t \in [150, 350]$, are shown. In this segment of the profile, which includes acceleration from stationary conditions, cruising at about 60 miles per hour, and deceleration to stationary conditions again, the maximum variation of the time delay is of about a factor 10. As regards the oxygen storage, in this test we assume to have available measurements only when the oxygen level is very low ($O < 0.35$) or very high ($O > 0.75$), simulating the behavior of a heated exhaust gas oxygen sensor (HEGO) placed downstream of the TWC (see Fig. 1).

Despite the large variability in the time delay and the variations in the operating conditions, the controller is able to maintain the AFR in the desired operating range for most of the time, and the situations in which $\lambda \notin [0.9, 1.1]$ are extremely infrequent, as shown in Fig. 5. At the same time, the oxygen stored in the catalyst is maintained close to its desired value. The tracking error, acceptable for this type

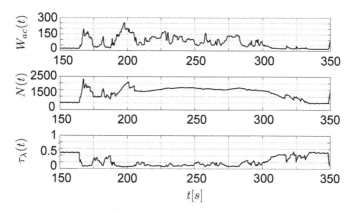

Fig. 4 Scenario 2, EPA test profile from real vehicle. Engine speed, airflow and (estimated) time-delay profiles.

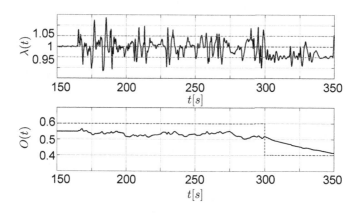

Fig. 5 Scenario 2, EPA test profile from real vehicle. AFR and oxygen storage dynamics.

of test, is mainly due to the fact that during this part of the simulation there is no available feedback from the HEGO for the stored oxygen, since $O \in [0.35, 0.75]$.

5   Conclusions

We have proposed a control architecture for negotiating the control of air-to-fuel ratio and oxygen storage in spark ignited engines. The control architecture is based on a delay-free model predictive controller, that enforces constraints on the actuators and on the operating range of the variables, and a delay compensation strategy based on a state predictor, that counteracts the time-varying delay that affects the system. Simulations in closed-loop with a detailed nonlinear model have been shown.

References

1. Beltrami, C., Chamaillard, Y., Millerioux, G., Higelin, P., Bloch, G.: AFR control in SI engine with neural prediction of cylinder air mass. In: Proc. American Contr. Conf., Denver, CO (2003)
2. Bemporad, A., Morari, M., Dua, V., Pistikopoulos, E.N.: The explicit linear quadratic regulator for constrained systems. Automatica 38(1), 3–20 (2002)
3. Di Cairano, S., Yanakiev, D., Bemporad, A., Kolmanovsky, I.V., Hrovat, D.: An MPC design flow for automotive control and applications to idle speed regulation. In: Proc. 47th IEEE Conf. on Decision and Control, Cancun, Mexico, pp. 5686–5691 (2008)
4. Fiengo, G., Cook, J., Grizzle, J.: Fore-aft oxygen storage control. In: Proc. American Contr. Conf., Anchorage, AK, pp. 1401–1406 (2002)
5. Guzzella, L., Onder, C.H.: Introduction to modeling and control of internal combustion engine systems. Springer, Heidelberg (2004)
6. Muske, K., Peyton-Jones, J., Franceschi, E.: Adaptive Analytical Model-Based Control for SI Engine Air–Fuel Ratio. IEEE Trans. Contr. Systems Technology 16(4), 763–768 (2008)
7. Ohata, A., Ohashi, M., Nasu, M., Inoue, T.: Model based air fuel ratio control for reducing exhaust gas emissions. In: Proc. SAE World Congress, Detroit, MI, page Paper 950075 (1995)
8. Pannocchia, G., Rawlings, J.B.: Disturbance models for offset-free model-predictive control. AIChE Journal 49(2), 426–437 (2003)
9. Powell, J., Fekete, N., Chang, C.: Observer-based air fuel ratio control. IEEE Control Systems Magazine 18(5), 72–83 (1998)

Observer-Based Stabilizing Control for a Class of Nonlinear Retarded Systems

Alfredo Germani, Costanzo Manes, and Pierdomenico Pepe

Abstract. Stabilizing cascade observer-controller schemes for a class of nonlinear retarded systems are presented in this chapter. Conditions for the local and global asymptotic stability of the closed loop system are provided. Such conditions allow the separate design of the observer and of the controller subsystems (separation theorems).

1 Introduction

Many results are available in the literature for the input/output exact linearization problem and for the stabilization problem of retarded nonlinear systems with time-delay in the state (see, for instance, [6, 8, 11, 12, 13, 17, 19, 20, 25]). These papers assume the full knowledge of the system state. On the other hand, few works dealing with the observer problem of retarded nonlinear systems are available in the literature. Contributions can be found in [6, 9, 18]. Observers for nonlinear systems with delayed output have been presented in [7, 14, 16], and an output feedback stabilizing control law is described in [26]. An observer based control law for the human glucose-insulin system, described in [21] by retarded nonlinear equations, is investigated in [22], where the the input/output linearization method of [5, 8] and the nonlinear observer of [9] are used. The local convergence of the glucose evolution to the reference signal is theoretically proved.

In this chapter we investigate an observer based stabilizing control law for the class of retarded systems which admit full relative degree and time-delay matched with the control input (see [9, 23]), where any finite number of discrete time-delays of arbitrary sizes, constant and known, is allowed. Many practical systems belong to this class, like the cited human glucose-insulin system or the two interacting

Alfredo Germani · Costanzo Manes · Pierdomenico Pepe
Dipartimento di Ingegneria Elettrica e dell'Informazione, Universitá degli Studi dell'Aquila, 67040 Poggio di Roio, L'Aquila, Italy
e-mail: {alfredo.germani,costanzo.manes}@univaq.it,
      pierdomenico.pepe@univaq.it

R. Sipahi et al. (Eds.): Time Delay Sys.: Methods, Appli. and New Trends, LNCIS 423, pp. 331–342.
springerlink.com

species system, with control acting on predators [15]. State feedback control laws that achieve the exact input/output linearization are applied to the system using the estimated state instead of the true state. The asymptotic stability of the equilibrium of the closed-loop system is proved under local and global assumptions. In the global case, distributed delay terms are also allowed. An example, concerning the global case, is worked out and numerical simulations are reported.

This chapter results have been presented in a preliminary version in [10].

2 Preliminaries

Notations. Throughout this chapter $\Delta > 0$ denotes the maximum delay involved in the retarded systems under investigation, $\mathbb{R}$ denotes the set of real numbers, and $\mathbb{R}_+$ denotes the set of non negative real numbers. The symbol $|\cdot|$ stands for the Euclidean norm of a real vector, or the induced Euclidean norm of a matrix. For given positive integer n and real Δ, $\mathscr{C}^{(n,\Delta)}$ denotes the space of the continuous functions mapping $[-\Delta,0]$ into $\mathbb{R}^n$. For a function $x : [-\Delta,\bar{t}) \to \mathbb{R}^n$, with $0 < \bar{t} \le \infty$, for any non negative real $t < \bar{t}$, x_t is the function in $\mathscr{C}^{(n,\Delta)}$ defined as $x_t(\tau) = x(t+\tau)$, $\tau \in [-\Delta,0]$. For a given real number $\rho > 0$, $\mathscr{C}_\rho^{(n,\Delta)} \subset \mathscr{C}^{(n,\Delta)}$ denotes the subset of the functions $x \in \mathscr{C}^{(n,\Delta)}$ such that $\|x\|_\infty < \rho$. $\mathscr{B}_\rho(0) \subset \mathbb{R}^n$ denotes the subset of the vectors $x \in \mathbb{R}^n$ such that $|x| < \rho$. For given triple (f,g,h) of smooth functions, where $f,g : \mathbb{R}^n \to \mathbb{R}^n$, $h : \mathbb{R}^n \to \mathbb{R}$, the symbols $L_f^i h(x)$, $L_g L_f^i h(x)$, $x \in \mathbb{R}^n$, $i = 0,1,\ldots$, denote the repeated Lie derivatives, defined as

$$L_f^0 h(x) = h(x), \quad L_f^i h(x) = \frac{\partial L_f^{i-1} h(x)}{\partial x} f(x), \qquad i = 1,2,\ldots$$

$$L_g L_f^j h(x) = \frac{\partial L_f^j h(x)}{\partial x} g(x), \qquad j = 0,1,\ldots \tag{1}$$

A triple (f,g,h) has *full* relative degree in an open set $S \subseteq \mathbb{R}^n$ if, $\forall x \in S$,

$$L_g L_f^i h(x) = 0, \quad i = 0,1,\ldots,n-2, \quad L_g L_f^{n-1} h(x) \ne 0. \tag{2}$$

If $S = \mathbb{R}^n$, then the triple is said to have full *uniform* relative degree. For positive integers n, m, $0_{n \times m}$ denotes a matrix of zeros in $\mathbb{R}^{n \times m}$, I_n denotes the identity matrix in $\mathbb{R}^{n \times n}$. For a positive integer n, A, B, C are a Brunovskii triple of matrices, i.e.

$$A_n = \begin{bmatrix} 0_{n-1 \times 1} & I_{n-1} \\ 0 & 0_{1 \times n-1} \end{bmatrix}, \quad B_n = \begin{bmatrix} 0_{n-1 \times 1} \\ 1 \end{bmatrix}, \quad C_n = \begin{bmatrix} 1 & 0_{1 \times n-1} \end{bmatrix} \tag{3}$$

The class of systems considered in this chapter is the class of nonlinear retarded systems with *matched time-delay and control input* (see [9, 23]), i.e. of the type

$$\begin{aligned} \dot{x}(t) &= f(x(t)) + g(x(t))\,(p_1(x_t)u(t) + p_2(x_t)), \quad t \ge 0 \\ y(t) &= h(x(t)), \qquad\qquad\qquad\qquad\qquad\qquad\quad t \ge -\Delta, \end{aligned} \tag{4}$$

where: $x(t) \in \mathbb{R}^n$, with n positive integer; $u(t), y(t) \in \mathbb{R}$ are the control input and measured output, respectively; $f : \mathbb{R}^n \to \mathbb{R}^n$, $g : \mathbb{R}^n \to \mathbb{R}^n$, $h : \mathbb{R}^n \to \mathbb{R}$ are smooth functions (admit continuous partial derivatives of any order); $p_1 : \mathscr{C}^{(n,\Delta)} \to \mathbb{R}$, $p_2 : \mathscr{C}^{(n,\Delta)} \to \mathbb{R}$ are continuously Fréchet differentiable functionals; $f(0) = 0$, $p_2(0) = 0$.

3 Separation Theorem: Global Results

Let us introduce in this section the following hypotheses for (4):

H_1) the triple f, g, h has full uniform relative degree;

H_2) the nonlinear delay-free system described by the triple (f, g, h) is Globally Uniformly Lipschitz Drift Observable (GULDO, see [3]), i.e., the function $\Phi : \mathbb{R}^n \to \mathbb{R}^n$, defined, for $x \in \mathbb{R}^n$, by

$$\Phi(x) = \left[h(x)\ L_f h(x)\ \cdots\ L_f^{n-1} h(x) \right]^T, \tag{5}$$

is a diffeomorphism in $\mathbb{R}^n$, and there exist positive reals γ_ϕ and $\gamma_{\phi^{-1}}$ such that, $\forall z_1, z_2 \in \mathbb{R}^n$,

$$
\begin{aligned}
|\Phi(z_1) - \Phi(z_2)| &\leq \gamma_\Phi |z_1 - z_2|, \\
|\Phi^{-1}(z_1) - \Phi^{-1}(z_2)| &\leq \gamma_{\phi^{-1}} |z_1 - z_2|;
\end{aligned}
\tag{6}
$$

H_3) there exists a positive real γ_1 such that, $\forall z_1, z_2 \in \mathbb{R}^n$,

$$|L_f^n h(\Phi^{-1}(z_1)) - L_f^n h(\Phi^{-1}(z_2))| \leq \gamma_1 |z_1 - z_2|; \tag{7}$$

H_4) there exists a positive real γ_2 such that, $\forall v_1, v_2 \in \mathscr{C}^{(n,\Delta)}$,

$$
\begin{aligned}
|L_g L_f^{n-1} h(\Phi^{-1}(v_1(0))) p_2(\Psi^{-1}(v_1)) \\
- L_g L_f^{n-1} h(\Phi^{-1}(v_2(0))) p_2(\Psi^{-1}(v_2))| \leq \gamma_2 \|v_1 - v_2\|_\infty,
\end{aligned}
\tag{8}
$$

where $\Psi^{-1}(v_i)(\tau) = \Phi^{-1}(v_i(\tau))$, $\tau \in [-\Delta, 0]$, $i = 1, 2$;

H_5) there exists a function $G : \mathscr{C}^{(n,\Delta)} \to \mathbb{R}$, such that, $\forall v \in \mathscr{C}^{(n,\Delta)}$,

$$L_g L_f^{n-1} h(v(0)) p_1(v) = G(H(v)), \tag{9}$$

where $H : \mathscr{C}^{(n,\Delta)} \to \mathscr{C}^{(n,\Delta)}$ is defined as $H(v)(\tau) = h(v(\tau))$, $\tau \in [-\Delta, 0]$;

H_6) there exists a real $\bar{p} > 0$ such that $p_1(v) \geq \bar{p}$, $\forall v \in \mathscr{C}^{(n,\Delta)}$.

Remark 1. Let $Q : \mathbb{R}^n \mapsto \mathbb{R}^{n \times n}$ denote the Jacobian of the map $\Phi(x)$ defined in (5), i.e.

$$Q(x) = \frac{\partial \Phi(x)}{\partial x} \tag{10}$$

Assumption H_2 implies that $Q(x)$ is nonsingular in all $\mathbb{R}^n$.

Lemma 1. *Consider the following nonlinear retarded system*

$$\dot{v}(t) = L(v_t, v_t + w_t), \tag{11}$$

$$\dot{w}(t) = M(w_t, v_t), \tag{12}$$

where $v_t, w_t \in \mathscr{C}^{(n,\Delta)}$ are the state variables, and the functions $L : \mathscr{C}^{(n,\Delta)} \times \mathscr{C}^{(n,\Delta)} \to \mathbb{R}^n$ and $M : \mathscr{C}^{(n,\Delta)} \times \mathscr{C}^{(n,\Delta)} \to \mathbb{R}^n$ are smooth functions, such that $L(0,0) = M(0,0) = 0$. Moreover, the function L is such that

$$L(\phi, \phi) = F\phi(0), \qquad \forall \phi \in \mathscr{C}^{(n,\Delta)}, \tag{13}$$

where $F \in \mathbb{R}^{n \times n}$ is a Hurwitz matrix, and

$$|L(\phi, \phi + \psi) - L(\phi, \phi)| \leq \gamma \|\psi\|_\infty, \quad \forall \phi, \psi \in \mathscr{C}^{(n,\Delta)}, \tag{14}$$

for some $\gamma \geq 0$. Moreover, assume that the system (11)-(12) is such that there exists a function α of class $\mathscr{K}\mathscr{L}$ such that

$$|w(t)| \leq \alpha(\|w_0\|_\infty, t), \quad t \geq 0. \tag{15}$$

Then, the trivial solution $(0,0) \in \mathscr{C}^{(n,\Delta)} \times \mathscr{C}^{(n,\Delta)}$ of the system (11)-(12) is globally uniformly asymptotically stable .

Proof. The proof follows from the results on ISS of time-delay systems in [24], by the use of the Liapunov-Krasovskii functional $V(\phi) = \phi^T(0)P\phi(0)$, $\phi \in \mathscr{C}^{(n,\Delta)}$, where the matrix P solves the Lyapunov equation $F^T P + PF = -I$.   □

Theorem 1. *Consider the system (4) and let the hypotheses H_1-H_6 be satisfied. Then, there exist a row vector Γ and a column vector K in $\mathbb{R}^n$ such that the trivial solution of the closed loop system*

$$\dot{x}(t) = f(x(t)) + g(x(t))(p_1(x_t)u(t) + p_2(x_t)) \tag{16}$$

$$y(t) = h(x(t)) \tag{17}$$

$$\dot{\hat{x}}(t) = f(\hat{x}(t)) + g(\hat{x}(t))p_2(\hat{x}_t)$$
$$\qquad + \left[Q(\hat{x}(t))\right]^{-1}\left[B_n G(y_t)u(t) + K(y(t) - h(\hat{x}(t)))\right] \tag{18}$$

$$u(t) = \frac{-L_f^n h(\hat{x}(t)) - L_g L_f^{n-1} h(\hat{x}(t))p_2(\hat{x}_t) + \Gamma\Phi(\hat{x}(t))}{G(y_t)}, \tag{19}$$

($G(\cdot)$ is the function defined in H_5, eq. (9)) is globally uniformly asymptotically stable .

Proof. Let us introduce the variables $z(t) = \Phi(x(t))$, $\eta(t) = \Phi(\hat{x}(t)) - \Phi(x(t))$, $t \geq -\Delta$, and define the function $R : \mathscr{C}^{(n,\Delta)} \mapsto \mathbb{R}^n$ as

$$R(v) = L_f^n h(\Phi^{-1}(v(0))) + L_g L_f^{n-1} h(\Phi^{-1}(v(0)))p_2(\Psi^{-1}(v)) \tag{20}$$

Then the couple of variables $z(t), \eta(t)$ satisfy the equations

$$\dot{z}(t) = L(z_t, z_t + \eta_t), \qquad \dot{\eta}(t) = M(\eta_t, z_t), \tag{21}$$

where $L : \mathscr{C}^{(n,\Delta)} \times \mathscr{C}^{(n,\Delta)} \mapsto \mathbb{R}^n$ and $M : \mathscr{C}^{(n,\Delta)} \times \mathscr{C}^{(n,\Delta)} \mapsto \mathbb{R}^n$ are defined as

$$L(v,w) = A_n v(0) + B_n\big(R(v) - R(w) + \Gamma w(0)\big), \tag{22}$$
$$M(v,w) = -L(w, v + w) + (A_n + B_n\Gamma)(v(0) + w(0)) - KC_n v(0) \tag{23}$$

Taking into account the hypotheses H_1-H_6, by the results in [6, 9], it follows that there exist a gain vector K and positive reals a, b such that $|\eta(t)| \le a \cdot e^{-bt}\|\eta_0\|_\infty$. Taking into account that $L(v,v) = (A_n + B_n\Gamma)v(0)$, and that (A_n, B_n) is a controllable pair, we can choose Γ such that $A_n + B_n\Gamma$ is Hurwitz. Thus, the hypotheses of Lemma 1 are satisfied by the equations (21). By applying Lemma 1, the global uniform asymptotic stability of the closed-loop system (16)–(19) is proved. $\qquad\square$

Theorem 2. *Consider the system (4) and let the hypotheses H_1, H_2, H_5, H_6 be satisfied. Let (7) be satisfied $\forall v_1, v_2 \in \mathbb{R}^n$ with $C_n v_1 = C_n v_2$, and let (8) be satisfied $\forall v_1, v_2 \in \mathscr{C}^{(n,\Delta)}$ with $C_n v_1(\tau) = C_n v_2(\tau)$, $\tau \in [-\Delta, 0]$. Let the function Φ^{-1} be known. Then, there exist a row vector Γ and a column vector K in $\mathbb{R}^n$ such that the trivial solution of the closed loop system*

$$\dot{x}(t) = f(x(t)) + g(x(t))(p_1(x_t)u(t) + p_2(x_t)) \tag{24}$$
$$y(t) = h(x(t)) \tag{25}$$
$$\dot{\hat{x}}(t) = f(\hat{x}(t)) + g(\hat{x}(t))p_2(\hat{x}_t) + \big[Q(\hat{x}(t))\big]^{-1} \tag{26}$$
$$\Big[B\big(G(y_t)u(t) + L_f^n h(\tilde{x}(t)) + L_g L_f^{n-1} h(\tilde{x}(t))p_2(\tilde{x}_t)$$
$$-L_f^n h(\hat{x}(t)) - L_g L_f^{n-1} h(\hat{x}(t))p_2(\hat{x}_t)\big) + K(y(t) - h(\hat{x}(t)))\Big] \tag{27}$$
$$u(t) = \frac{-L_f^n h(\tilde{x}(t)) - L_g L_f^{n-1} h(\tilde{x}(t))p_2(\tilde{x}_t) + \Gamma\Phi(\tilde{x}(t))}{G(y_t)}, \tag{28}$$

with $\tilde{x}(t) = \Phi^{-1}\left(\big[y(t)\ L_f h(\hat{x}(t))\ \cdots\ L_f^{n-1}h(\hat{x}(t))\big]^T\right)$, $t \ge -\Delta$, is globally uniformly asymptotically stable.

Proof. The proof is similar to the one given for Theorem 1 and for lack of space is here omitted. $\qquad\square$

Remark 2. Γ and K must be chosen such that $A_n + B_n\Gamma$ and $A_n - KC_n$ are Hurwitz. Hypotheses H_1-H_4 and H_6 concern observability, Lipschitz and full relative degree conditions. Hypothesis H_5 imposes that the term multiplying the input, in the model (4) rewritten in normal form, only depends on the measured output.

4   Separation Theorem: Local Results

Let us introduce in this section the following hypotheses for (4):

J_1) there exists a real number $\rho > 0$ such that the triple (f,g,h) has full relative degree in the ball $\mathscr{B}_\rho(0)$;

J_2) the nonlinear delay-free system described by the triple (f,g,h) is uniformly Lipschitz drift observable (ULDO, see [3]) in $\mathscr{B}_\rho(0)$, i.e., the function $\Phi(x)$ defined in (5) is a diffeomorphism from $\mathscr{B}_\rho(0)$ to $\Phi(\mathscr{B}_\rho(0))$, and there exist positive real numbers γ_ϕ and $\gamma_{\phi^{-1}}$ such that, for all $x_1, x_2 \in \mathscr{B}_\rho(0)$, $z_1, z_2 \in \Phi(\mathscr{B}_\rho(0))$,

$$\begin{aligned}
|\Phi(x_1) - \Phi(x_2)| &\le \gamma_\Phi |x_1 - x_2|, \\
|\Phi^{-1}(z_1) - \Phi^{-1}(z_2)| &\le \gamma_{\Phi^{-1}} |z_1 - z_2|;
\end{aligned} \tag{29}$$

J_3) there exist a positive real γ_1 such that, $\forall z_1, z_2 \in \Phi(\mathscr{B}_\rho(0))$,

$$|L_f^n h(\Phi^{-1}(z_1)) - L_f^n h(\Phi^{-1}(z_2))| \le \gamma_1 |z_1 - z_2|;$$

J_4) there exist a positive real γ_2 such that, $\forall v_1, v_2 \in \mathscr{C}^{(n,\Delta)}$, with $v_i(\tau) \in \Phi(\mathscr{B}_\rho(0)), \tau \in [-\Delta, 0], i = 1,2$,

$$\begin{aligned}
|L_g L_f^{n-1} h(\Phi^{-1}(v_1(0))) p_2(\Psi^{-1}(v_1)) & \\
- L_g L_f^{n-1} h(\Phi^{-1}(v_2(0))) p_2(\Psi^{-1}(v_2))| &\le \gamma_2 \|v_1 - v_2\|_\infty, \tag{30}
\end{aligned}$$

where $\Psi^{-1}(v_i)(\tau) = \Phi^{-1}(v_i(\tau))$, $\tau \in [-\Delta, 0]$, $i = 1,2$;

J_5) there exist a positive integer m and positive reals Δ_k, $k = 1, 2, \ldots, m$, such that, $\forall v \in \mathscr{C}^{(n,\Delta)}$,

$$\begin{aligned}
p_1(v) &= \bar{p}_1(v(0), v(-\Delta_1), \ldots, v(-\Delta_m)), \\
p_2(v) &= \bar{p}_2(v(0), v(-\Delta_1), \ldots, v(-\Delta_m)),
\end{aligned}$$

where $\bar{p}_1, \bar{p}_2 : \mathbb{R}^{n \times m + 1} \to \mathbb{R}$ are smooth functions (admit continuous partial derivatives of any order);

J_6) there exists a real number $\bar{p} > 0$ such that $p_1(\phi) \ge \bar{p}$ for all $\phi \in \mathscr{C}_\rho^{(n,\Delta)}$.

Lemma 2. *Let*

$$S_\rho = \{(z_1, z_2) \in \mathscr{C}^{(n,\Delta)} \times \mathscr{C}^{(n,\Delta)} : z_i(\tau) \in \Phi(\mathscr{B}_\rho(0)), \tau \in [-\Delta, 0]\}$$

Let $L : S_\rho \mapsto \mathbb{R}$ *be the functional defined, for* $(z_1, z_2) \in S_\rho$, *as*

$$\begin{aligned}
L(z_1, z_2) &= L_f^n h(\Phi^{-1}(z_2(0))) + L_g L_f^{n-1} h(\Phi^{-1}(z_2(0))) p_1(\Psi^{-1}(z_2)) \cdot \\
&\quad \frac{1}{L_g L_f^{n-1} h(\Phi^{-1}(z_1(0))) p_1(\Psi^{-1}(z_1))} \cdot \\
&\quad \left[-L_f^n h(\Phi^{-1}(z_1(0))) - L_g L_f^{n-1} h(\Phi^{-1}(z_1(0))) p_2(\Psi^{-1}(z_1)) + W z_1(0) \right]
\end{aligned}$$

$$+L_g L_f^{n-1} h(\Phi^{-1}(z_2(0))) p_2(\Psi^{-1}(z_2)), \tag{31}$$

with W any given row vector in $\mathbb{R}^n$. Then, the functional L admits continuous Fréchet derivative at $(0,0)$, denoted $D_F L$, and, moreover, there exist row vectors $b_k \in \mathbb{R}^n$, $k = 0, 1, \ldots, m$, such that, for all $y_1, y_2 \in S_\rho$, satisfying $y_1 + y_2 \in S_\rho$,

$$D_F L(y_1, y_1 + y_2) - W y_1(0) = b_0 y_2(0) + \sum_{k=1}^{m} b_k y_2(-\Delta_k) \tag{32}$$

Proof. The Fréchet differentiability property follows from the fact that f, g, h, $\bar{p}_1$, $\bar{p}_2$ (see the hypothesis J_5) are smooth, Φ is a diffeomorphism. As far as the equality (32) is concerned, by the hypothesis J_5, there exist row vectors $a_k, b_k \in \mathbb{R}^n$, $k = 0, 1, \ldots, m$, such that

$$D_F L(z_1, z_2) = a_0 z_1(0) + b_0 z_2(0) + \sum_{k=1}^{m} a_k z_1(-\Delta_k) + b_k z_2(-\Delta_k) \tag{33}$$

Since $L(z_1, z_1) = W z_1(0)$, and $(y_1, y_1 + y_2) = (y_1, y_1) + (0, y_2)$, we get

$$D_F L(y_1, y_1 + y_2) - W y_1(0) = D_F L(y_1, y_1 + y_2) - D_F L(y_1, y_1) =$$
$$D_F L(0, y_2) = b_0 y_2(0) + \sum_{k=0}^{m} b_k y_2(-\Delta_k) \tag{34}$$

$\square$

Theorem 3. *Consider the system (4) and let the hypotheses J_1-J_6 be satisfied. Then, there exist a row vector Γ and a column vector K in $\mathbb{R}^n$ such that the trivial solution of the closed loop system*

$$\dot{x}(t) = f(x(t)) + g(x(t))(p_1(x_t)u(t) + p_2(x_t))$$
$$y(t) = h(x(t))$$
$$\dot{\hat{x}}(t) = f(\hat{x}(t)) + g(\hat{x}(t))(p_1(\hat{x}_t)u(t) + p_2(\hat{x}_t)) + [Q(\hat{x}(t))]^{-1} K(y(t) - h(\hat{x}(t)))$$
$$u(t) = \frac{-L_f^n h(\hat{x}(t)) - L_g L_f^{n-1} h(\hat{x}(t)) p_2(\hat{x}_t) + \Gamma \Phi(\hat{x}(t))}{L_g L_f^{n-1} h(\hat{x}(t)) p_1(\hat{x}_t)} \tag{35}$$

is locally asymptotically stable.

Proof. Let us introduce the variables $\hat{z}(t) = \Phi(\hat{x}(t))$, $\eta(t) = \Phi(x(t)) - \Phi(\hat{x}(t))$, $t \geq -\Delta$. Then the couple of variables $\hat{z}(t), \eta(t)$ satisfies the equations

$$\dot{\hat{z}}(t) = (A_n + B_n \Gamma)\hat{z}(t) + K C_n \eta(t),$$
$$\dot{\eta}(t) = (A_n - K C_n)\eta(t) + B_n(L(\hat{z}_t, \hat{z}_t + \eta_t) - \Gamma \hat{z}(t)), \tag{36}$$

where L is the functional defined in (31), with $W = \Gamma$. By Lemma 2, we get that the first order approximation of (36) is given by the following equations

$$\dot{\hat{z}}(t) = (A_n + B_n \Gamma)\hat{z}(t) + KC_n \eta(t),$$

$$\dot{\eta}(t) = (A_n - KC_n)\eta(t) + B_n b_0 \eta(t) + B_n \sum_{k=1}^{m} b_k \eta(t - \Delta_k) \qquad (37)$$

Since we can choose Γ such that $A_n + B_n \Gamma$ has any prescribed eigenvalues in the open complex left semi-plane, the proof of the theorem is completed, provided that the trivial solution of the second equation in (37) is exponentially stable, for some K. We will use the same reasoning used in [22], for the single-delay glucose-insulin system. We get

$$\eta(t) = e^{(A_n - KC_n)t} \eta(0)$$

$$+ \int_0^t e^{(A_n - KC_n)(t - \tau)} B\left(b_0 \eta(\tau) + \sum_{k=1}^{m} b_k \eta(\tau - \Delta_k)\right) d\tau \qquad (38)$$

Let us choose an n-ple of negative real distinct eigenvalues, in decreasing order, $\lambda = (\lambda_1, \lambda_2, \ldots, \lambda_n)$. As in [1], let $V(\lambda)$ be the Vandermonde matrix that diagonalizes $A_n - KC_n$, let $\psi(t) = V(\lambda)\eta(t)$, $t \geq -\Delta$. Then, from (38), we get

$$|\psi(t)| \leq e^{\lambda_1 t}|\psi(0)| + \int_0^t e^{\lambda_1(t - \tau)} \sqrt{n}$$

$$\left(|b_0||V^{-1}(\lambda)||\psi(\tau)| + \sum_{k=1}^{m} |b_k||V^{-1}(\lambda)||\psi(\tau - \Delta_k)|\right) d\tau. \qquad (39)$$

Let $\xi(t) = e^{-\lambda_1 t}|\psi(t)|$. From (39) we obtain, by applying the Bellman-Gronwall inequality,

$$\xi(t) \leq e^{\sqrt{n}|b_0||V^{-1}(\lambda)|t} \xi(0)$$

$$+ \int_0^t e^{\sqrt{n}|b_0||V^{-1}(\lambda)|(t - \tau)} e^{-\lambda_1 \tau} \left(\sum_{k=1}^{m} |b_k||V^{-1}(\lambda)||\psi(\tau - \Delta_k)|\right) d\tau \qquad (40)$$

Turning back to the variable $\psi(t)$, from (40) we get, for $t \geq -\Delta$,

$$|\psi(t)| \leq e^{(\lambda_1 + \sqrt{n}|b_0||V^{-1}(\lambda)|)t}|\psi(0)|$$

$$+ \int_0^t e^{(\lambda_1 + \sqrt{n}|b_0||V^{-1}(\lambda)|)(t - \tau)} \left(\sum_{k=1}^{m} |b_k||V^{-1}(\lambda)||\psi(\tau - \Delta_k)|\right) d\tau \qquad (41)$$

Let $M = \sup_{\tau \in [-\Delta, 0]} |\psi(\tau)|$. Let $p(t) = Me^{\alpha t}$, $t \geq -\Delta$, α a suitable negative real which will be found next. Let us see if there exists a negative α such that $p(t)$ satisfies (41) with the equality sign, i.e.

$$Me^{\alpha t} = e^{(\lambda_1 + \sqrt{n}|b_0||V^{-1}(\lambda)|)t} M$$

$$+ \int_0^t e^{(\lambda_1 + \sqrt{n}|b_0||V^{-1}(\lambda)|)(t - \tau)} \left(\sum_{k=1}^{m} |b_k||V^{-1}(\lambda)|Me^{\alpha(\tau - \Delta_k)} d\tau\right) \qquad (42)$$

If (42) holds, then, from the choice of M and a standard step procedure, it follows that $|\psi(t)| \leq Me^{\alpha t}$, $t \geq 0$. Therefore, if α can be chosen negative, the proof of the exponential stability of the trivial solution of the second equation in (37) is complete. Solving the integral in (42), we get that the equality (42) holds if and only if the following condition can be fulfilled:

$$\sqrt{n} \sum_{k=1}^{m} |b_k| |V^{-1}(\lambda)| e^{-\alpha \Delta_k} = \alpha - \lambda_1 - \sqrt{n} |b_0| |V^{-1}(\lambda)| \tag{43}$$

The equation (43) in the unknown variable α admits a unique negative solution if and only if the value at zero of the right-hand side function of α is greater than the value at zero of the left-hand side function of α, i.e., if and only if

$$\lambda_1 + |V^{-1}(\lambda)| \sqrt{n} \sum_{k=0}^{m} |b_k| < 0 \tag{44}$$

It is proved in [1], that there exists an n-pla of real negative distinct eigenvalues λ such that the inequality (44) holds. Since, for any choice of λ, there exist K such that the set of eigenvalues of $A_n - KC_n$ is equal to λ, the proof of the theorem is complete.   □

5   Numerical Example

Let us consider the unstable system described by the following equations

$$\begin{aligned}
\dot{x}_1(t) &= x_2(t) + \mathrm{sech}(x_1(t)) - 1 \\
\dot{x}_2(t) &= \tanh(x_2(t - \Delta)) + x_1(t) + u(t) \\
y(t) &= x_1(t)
\end{aligned} \tag{45}$$

Straightforward computations give, for $x \in \mathbb{R}^2$ and $w \in \mathscr{C}^{(n,\Delta)}$,

$$\begin{aligned}
\Phi(x) &= \begin{bmatrix} x_1 \\ x_2 + \mathrm{sech}(x_1) - 1 \end{bmatrix}, \\
L_f^2 h(x) &= x_1 - \tanh(x_1)\mathrm{sech}(x_1)(x_2 + \mathrm{sech}(x_1) - 1), \\
L_g L_f h(x) &= 1, \qquad G(H(w))(\tau) = 1, \ \tau \in [-\Delta, 0], \\
\frac{\partial \Phi(x)}{\partial x} &= \begin{bmatrix} 1 & 0 \\ -\tanh(x_1)\mathrm{sech}(x_1) & 1 \end{bmatrix}
\end{aligned} \tag{46}$$

All the hypotheses of Theorem 2 are satisfied by the system (45). The observer based control law for the system (45) is given by (see (28))

$$\begin{bmatrix} \dot{\hat{x}}_1(t) \\ \dot{\hat{x}}_2(t) \end{bmatrix} = \begin{bmatrix} \hat{x}_2(t) + \mathrm{sech}(\hat{x}_1(t)) - 1 \\ \tanh(\hat{x}_2(t - \Delta)) + \hat{x}_1(t) + u(t) \end{bmatrix}$$

$$+ \begin{bmatrix} 1 & 0 \\ \tanh(\hat{x}_1(t))\operatorname{sech}(\hat{x}_1(t)) & 1 \end{bmatrix} K(y(t) - \hat{x}_1(t))$$

$$+B(y(t) - \hat{x}_1(t) + (\hat{x}_2(t) + \operatorname{sech}(\hat{x}_1(t)) - 1) \cdot$$

$$(\tanh(\hat{x}_1(t))\operatorname{sech}(\hat{x}_1(t)) - \tanh(y(t))\operatorname{sech}(y(t)))$$

$$+\tanh(\hat{x}_2(t - \Delta) + \operatorname{sech}(\hat{x}_1(t - \Delta)) - \operatorname{sech}(y(t - \Delta))) - \tanh(\hat{x}_2(t - \Delta)))$$

$$u(t) = -y(t) + \tanh(y(t))\operatorname{sech}(y(t))(\hat{x}_2(t) + \operatorname{sech}(\hat{x}_1(t)) - 1)$$

$$-\tanh(\hat{x}_2(t - \Delta) + \operatorname{sech}(\hat{x}_1(t - \Delta)) - \operatorname{sech}(y(t - \Delta)))$$

$$+\Gamma \begin{bmatrix} y(t) \\ \hat{x}_2(t) + \operatorname{sech}(\hat{x}_1(t)) - 1 \end{bmatrix} \tag{47}$$

By Theorem 2, there exist a row vector Γ and a column vector K (chosen such that $A_n + B_n\Gamma$ and $A_n - KC_n$ are Hurwitz) such that the trivial solution of the closed loop system (45), (47) is globally uniformly asymptotically stable . The performed simulations validate this result. In fig. 1 the evolutions of the true and estimated state are plotted. The initial state variables are set constant and equal to $\begin{bmatrix} 1 \times 10^4 \\ 5 \times 10^3 \end{bmatrix}$, the initial estimated state variables are set constant and equal to $\begin{bmatrix} 0 \\ 0 \end{bmatrix}$. The time-delay Δ is set equal to 50. The gain vector Γ and the gain vector K are chosen such to have eigenvalues -0.1, -0.2, and -1.1, -1.2, for $A_n + B_n\Gamma$ and $A_n - KC_n$, respectively. Both the true and estimated state converge to 0.

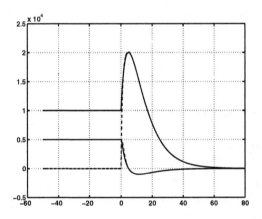

Fig. 1 True (solid line) and Estimated (dashed line) State

6   Conclusions

In this chapter the separation theorem for a class of retarded nonlinear systems is investigated. Global and local convergence results are provided in the case of global Lipschitz property and local Lipschitz property of the functionals describing the equations, respectively. An interesting topic for future work concerns the separation theorem for retarded nonlinear systems with un-matched time-delay and control input. One such a case has been studied in [4]. The separation theorem for nonlinear systems with delayed output is also an interesting topic for future investigation.

References

1. Ciccarella, G., Dalla Mora, M., Germani, A.: A Luenberger-like observer for nonlinear systems. Int. J. of Control 57(3), 537–556 (1993)
2. Ciccarella, G., Dalla Mora, M., Germani, A.: Asymptotic linearization and stabilization for a class of nonlinear systems. J. of Optim.Theory and Appl. 84(3) (1995)
3. Dalla Mora, M., Germani, A., Manes, C.: Design of state observers from a drift-observability property. IEEE Trans. on Aut. Control 45(8), 1536–1540 (2000)
4. Di Ciccio, M.P., Bottini, M., Pepe, P., Foscolo, P.U.: Observed-Based Control of a Continuous Stirred Tank Reactor with Recycle Time-Delay. In: 14th IFAC Conf. on Methods and Models in Autom. and Robotics, Miedzydroje, Poland (August 2009)
5. Germani, A., Manes, C., Pepe, P.: Local Asymptotic Stability for Nonlinear State Feedback Delay Systems. Kybernetika 36(1), 31–42 (2000)
6. Germani, A., Manes, C., Pepe, P.: An Asymptotic State Observer for a Class of Nonlinear Delay Systems. Kybernetika 37(4), 459–478 (2001)
7. Germani, A., Manes, C., Pepe, P.: A New Approach to State Observation of Nonlinear Systems with Delayed Output. IEEE Trans. on Aut. Control 47(1), 96–101 (2002)
8. Germani, A., Manes, C., Pepe, P.: Input-Output Linearization with Delay Cancellation for Nonlinear Delay Systems: the Problem of the Internal Stability. Int. J. of Robust and Nonlinear Control 13(9), 909–937 (2003)
9. Germani, A., Pepe, P.: A State Observer for a Class of Nonlinear Systems with Multiple Discrete and Distributed Time Delays. European J. of Control 11(3), 196–205 (2005)
10. Germani, A., Manes, C., Pepe, P.: Separation Theorems for a Class of Retarded Nonlinear Systems. In: IFAC-Papers OnLine, Workshop on Time-Delay Systems, Praha (2010)
11. Hua, C., Guan, X., Shi, P.: Robust stabilization of a class of nonlinear time-delay systems. Appl. Math. and Comp. 155, 737–752 (2004)
12. Jankovic, M.: Control Lyapunov-Razumikhin Functions and Robust Stabilization of Time Delay Systems. IEEE Trans. on Aut. Control 46(7), 1048–1060 (2001)
13. Karafyllis, I., Jiang, Z.-P.: Necessary and Sufficient Lyapunov-Like Conditions for Robust Nonlinear Stabilization. ESAIM: Control, Optimisation and Calculus of Variations 16, 887–928 (2010)
14. Kazantzis, N., Wright, R.: Nonlinear observer design in the presence of delayed output measurements. Systems & Control Letters 54(9), 877–886 (2005)
15. Kolmanovskii, V., Myshkis, A.: Introduction to the Theory and Applications of Functional Differential Equations. Kluwer Academic Pub., Dordrecht (1999)
16. Koshkouei, A., Burnham, K.: Discontinuous observers for non-linear time-delay systems. Int. J. of Systems Science 40(4), 383–392 (2009)
17. Lien, C.-H.: Global Exponential Stabilization for Several Classes of Uncertain Nonlinear Systems with Time-Varying Delay. Nonlin. Dynamics and Systems Theory 4(1), 15–30 (2004)
18. Márquez-Martinez, L.A., Moog, C.H., Velasco-Villa, M.: Observability and observers for nonlinear systems with time delays. Kybernetika 38(4), 445–456 (2002)
19. Márquez-Martinez, L.A., Moog, C.H.: Input-output feedback linearization of time-delay systems. IEEE Trans. on Aut. Control 49(5), 781–785 (2004)

20. Oguchi, T., Watanabe, A., Nakamizo, T.: Input-Output Linearization of Retarded Nonlinear Systems by Using an Extension of Lie Derivative. Int. J. of Control 75(8), 582–590 (2002)
21. Palumbo, P., Panunzi, S., De Gaetano, A.: Qualitative behavior of a family of delay differential models of the glucose insulin system. Discrete and Continuous Dynamical Systems-Series B 7, 399–424 (2007)
22. Palumbo, P., Pepe, P., Panunzi, S., De Gaetano, A.: Observer-based closed-loop control of plasma glycemia. In: 48th IEEE Conf. on Decision and Control (CDC 2009), Shanghai, China (December 2009)
23. Pepe, P.: Preservation of the Full Relative Degree for a Class of Delay Systems under Sampling. ASME J. of Dynamic Systems, Meas. and Control 125(2), 267–270 (2003)
24. Pepe, P., Jiang, Z.-P.: A Lyapunov-Krasovskii Methodology for ISS and iISS of Time-Delay Systems. Systems and Control Letters 55(12), 1006–1014 (2006)
25. Zhang, X., Cheng, Z.: Global stabilization of a class of time-delay nonlinear systems. Int. J. of Systems Science 36(8), 461–468 (2005)
26. Zhang, X., Zhang, C., Cheng, Z.: Asymptotic stabilization via output feedback for nonlinear systems with delayed output. Int. J. of Systems Science 37(9), 599–607 (2006)

Cascade Control for Time Delay Plants

Pavel Zítek, Vladimír Kučera, and Tomáš Vyhlídal

Abstract. The cascade control architecture is a standard solution in control engineering practice for industrial plants with considerable time delays. In this paper, an affine parameterization based design of cascade controllers for time delay plants is presented. The design rests on the use of the so-called quasi-integrating meromorphic function used to prescribe the desired open-loop behaviour. Due to the parameterization approach both the slave and master controllers are obtained as time delay systems. Unlike most of relevant papers on the subject, the primary controlled output is not considered to be directly dependent on the secondary one. The only property required from the secondary output is its markedly faster response to disturbances to be compensated for.

1 Introduction

The cascade control scheme is a common approach in control engineering practice of overcoming the crucial problem in time delay system control that due to the delays the impact of disturbances on the plant output is detected too late. Finding an auxiliary, the so-called secondary plant output with as prompt as possible response to the disturbances is the key point in closing the secondary feedback loop which provides a pre-compensation of the plant for the primary control circuit. The effect

Pavel Zítek · Tomáš Vyhlídal
Centre for Applied Cybernetics and Dept. of Instrumentation and Control Eng.,
Faculty of Mechanical Eng., Czech Technical University in Prague, Technická 4,
166 07 Praha 6, Czech Republic
e-mail: pavel.zitek@fs.cvut.cz, tomas.vyhlidal@fs.cvut.cz

Vladimír Kučera
Centre for Applied Cybernetics, Faculty of Electrical Eng., Czech Technical University in Prague, Technická 2, 166 27 Praha 6, and Institute of Information Theory and Automation, v.v.i., Academy of Sciences of the Czech Republic, Pod vodárenskou věží 4, 182 08 Praha 8, Czech Republic
e-mail: kucera@fel.cvut.cz

R. Sipahi et al. (Eds.): Time Delay Sys.: Methods, Appli. and New Trends, LNCIS 423, pp. 343–354.
springerlink.com
© Springer-Verlag Berlin Heidelberg 2012

of the well selected secondary output can yield not only a better disturbance rejection but also an enhanced stability and robustness of control [9]. A specific suitability of cascade control scheme for the time-delay systems was anticipated already by Morari and Zafiriou [6] where it is shown that the non-minimum phase character of the plant is an essential condition of efficient application of the cascade control scheme which therefore belongs to specific tools in time-delay system control.

Basically the primary and secondary plant outputs are considered as independent of each other, although in contrary, a dependence of them is assumed by many authors. The secondary plant output is not a really controlled variable and therefore no reference input is supposed for its value. The easy implementation of cascade control by a simple serial linkage of two standard controllers is the advantage which has lead to the widespread application of this principle in industrial control. However, the tuning and auto-tuning of cascade control parameters is not a simple task, investigated e.g. in [3]. An application of Internal Model Control principle to cascade control scheme was presented by Kaya and Atherton in [2]. This chapter on cascade control is based on the author's article [13], where the preliminary results were presented.

Consider the cascade control in the configuration given by the block diagram in Fig. 1, where $G(s)$, $G_S(s)$ and $G_D(s)$, $G_Z(s)$ are the plant transfer functions for the primary and secondary controlled outputs y and z with respect to the control variable u and the disturbance d, respectively. This system can be viewed as a special case of two-variable control system with the input $\mathbf{v} = [u,d]^T$ and the output $\mathbf{y} = [y,z]^T$ where, actually, only the primary output variable y is to be controlled. The four transfer functions are then considered to make up the transfer function matrix of the plant

$$\mathbf{G}(s) = \begin{bmatrix} G(s), & G_D(s) \\ G_S(s), & G_Z(s) \end{bmatrix}. \tag{1}$$

The primary transfer functions are supposed to be the products of the time delay exponentials and rational stable and proper functions which are coprime. The orders of denominator and numerator polynomials of these functions are n_S, m_S for $G_S(s), G_Z(s)$ and n_M, m_M for $G(s), G_D(s)$, respectively. Even though the secondary output z is utilized, the entire control system has the same degree of freedom as a single-loop control. For specific investigations the plant transfer functions are supposed to be stable and aperiodic, in the generic form

$$G(s) = \frac{K\exp(-\tau s)}{A(s)} = \frac{K\exp(-\tau s)}{(T_1 s + 1)(T_2 s + 1).....} \tag{2}$$

i.e. with denominator polynomials having only real and negative zeros. Let the so-called *master* and *slave* controllers be represented by their transfer functions $R_M(s)$ and $R_S(s)$, respectively, in Fig. 1. Then the complementary sensitivity function of the whole cascade control system for the primary controlled variable y is as follows

$$T(s) = \frac{y(s)}{w(s)} = \frac{R_M(s)R_S(s)G(s)}{1 + R_S(s)G_S(s) + R_M(s)R_S(s)G(s)}. \tag{3}$$

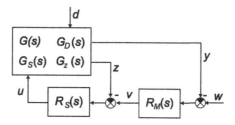

Fig. 1 Cascade control configuration

The cascade control is efficient at rejecting the disturbances on the plant input only, as in Fig. 1, where the secondary measured variable z is located, in line with the idea of cascade control, *between d and y.* The disturbance sensitivity function $S_D(s) = y(s)/d(s)$ is then of the form

$$S_D(s) = \frac{G_D(s) - R_S(s)\det \mathbf{G}(s)}{1 + R_S(s)G_S(s) + R_M(s)R_S(s)G(s)}. \tag{4}$$

If $\mathbf{G}(s)$ is nonsingular, then the presence of $\det \mathbf{G}(s)$ in the numerator of (4) points out the potential of the cascade configuration to minimize the impact of the disturbance on the system response through the secondary controller $R_S(s)$.

Remark 1. As regards the possible compensation of $G_D(s)$ by designing $R_S(s)$ it is worth noting that if the plant itself is of cascade structure too, i.e., if $y(s) = G_M(s)z(s)$, i.e. $G(s) = G_M(s)G_S(s)$, then the matrix $\mathbf{G}(s)$ is always singular, $\det \mathbf{G}(s) = 0$, and the possibility to put down the $S_D(s)$ numerator is excluded. Though the nature of real processes does not comply generally with this condition, in many works dealing with the cascade control scheme this factorized form of the plant model is often assumed.

2 Meromorphic Quasi-integration Transfer Function

Later on we are going to make use of controller transfer functions which are not rational but contain the delay operator $\exp(-s\tau)$ in both the numerator and denominator. A form specific for further considerations is introduced by the following Lemma.

Lemma 1. *Consider the meromorphic transfer function*

$$I_m(s) = \frac{\exp(-s\tau)}{(\Theta_1 s + 1)f(s) - \exp(-s\tau)} \tag{5}$$

where $f(s)$ is a stable aperiodic polynomial of the form $f(s) = (\Theta_2 s + 1)...(\Theta_v s + 1)$ with $\Theta_i < \Theta_1$, $i = 2,...v$, τ is a time delay and Θ_1 is a time constant such that $\Theta_1 \geq \tau$. The frequency response $|I_m(j\omega)|$ is decreasing throughout the frequency range. In the low frequency range, the frequency response $|I_m(j\omega)|$ is for $\omega \to 0$ approaching the asymptotic form

$$I_{a0}(\omega) = \frac{1}{(\tau + \Theta_1 + \Theta_2 + ... + \Theta_\nu)\,\omega} \tag{6}$$

with the negative slope of one amplitude decade to one decade of frequency. In the high frequency range, for $\omega > \Theta_1^{-1}$, $I_m(j\omega)$ is approaching the asymptotic form

$$I_{a\infty}(\omega) = \frac{1}{\Theta_1 \Theta_2 ... \Theta_\nu \omega^\nu}. \tag{7}$$

The function $I_m(s)$ is referred to as quasi-integration and its crossover frequency is approximately given by the crossover of (6), i.e.

$$\omega_C \cong \frac{1}{\tau + \Theta_1 + \Theta_2 + ... + \Theta_\nu}. \tag{8}$$

Proof. After inserting $s = j\omega$ into (5)

$$I_m(j\omega) = \frac{\cos(\omega\tau) - j\cos(\omega\tau)}{(j\omega\Theta + 1)f(j\omega) - \cos(\omega\tau) + j\cos(\omega\tau)} \tag{9}$$

for small angles $\omega\tau$ it holds that $\sin\omega\tau \cong \omega\tau$ and $\cos\omega\tau \cong 1$. The terms with $(j\omega)^i, i = 2, .., \nu$ are negligible compared to $j\omega\Theta_i, i = 1, ...\nu$. Hence in the lowest frequency band it holds

$$\lim_{\omega \to 0} j\omega I_m(j\omega) = \lim_{\omega \to 0} \frac{j\omega\,(1 - j\omega\tau)}{j\omega(\Theta_1 + \Theta_2 + ...\Theta_\nu + \tau)} = \frac{1}{\Theta_S + \tau} \tag{10}$$

where $\Theta_S = \Theta_1 + \Theta_2 + ...\Theta_\nu$. Hence the asymptote for $\omega \to 0$ is $(\omega(\Theta_S + \tau))^{-1}$. On the other hand for the highest frequency band it holds that $|\exp(-\omega\tau)| = 1$ is negligible compared to $|\Theta_i\,\omega|$, $i = 1, ...\nu$ and the product of the time constants is the dominant term in the denominator of $I_m(j\omega)$. Therefore

$$\lim_{\omega \to \infty} |(j\omega)^\nu I_m(j\omega)| = \frac{1}{\Theta_1 \Theta_2 ... \Theta_\nu} \tag{11}$$

The last point of the proof is the crossover frequency ω_C. Since the time constants $\Theta_2, ..., \Theta_\nu$ are less than Θ_1 and τ the terms with the powers $(j\omega)^i$, $i = 2, ...\nu$ in (5) become negligible in frequencies where ω_C is to be expected, i.e. where $|I_m(j\omega_C)| = 1$. Neglecting them and with regard that $|\exp(-j\omega)| = 1$ the condition of the crossover frequency reduces to the condition

$$|\Theta_S j\omega_C + 1 - \cos\omega_C\tau + j\sin\omega_C\tau| = 1 \tag{12}$$

The angle to be found is $\gamma = \omega_C\tau$ and $\omega_C\Theta_S = \kappa\gamma$ where κ is the ratio of Θ_S and τ. For the usual ratio value $1 \le \kappa \le 1.5$ the solution of (12) is approximately $\gamma \cong (1+\kappa)^{-1}$, e.g., if $\kappa = 1$ the true value is $\gamma = 0.507$. The proof is complete.

3 Parameterization Based Design of the Secondary Control Loop

The meromorphic transfer function (5) will now be used for parameterization design of the slave controller. The affine parameterization of stabilizing controllers has become one of the main tools in the design of linear control systems described not only by rational transfer functions but also for a class of time delay systems and was studied e.g. in [7, 8, 10, 12]. It was already shown [4]; [5] that H_2 and H_∞ optimal controllers have a model-based dead-time compensator structure. The parameterization interpreted as the internal model control of linear time-delay systems was investigated in [11]. Consistently with the practical goals of this paper the plant to be controlled is assumed to be asymptotically stable.

Let us first consider the secondary control loop as a self-contained control circuit. As it holds for the secondary plant output that $z(s) = G_S(s)u(s) + G_Z(s)d(s)$, the actuating variable can be expressed as follows

$$u(s) = \frac{R_S(s)}{1 + R_S(s)G_S(s)}[v(s) - G_Z(s)d(s)] \qquad (13)$$

The affine parameterization of the secondary loop leads to introducing the parameterizing transfer function [1]

$$C_S(s) = \frac{R_S(s)}{1 + R_S(s)G_S(s)} \qquad (14)$$

highlighting the importance of $G_S(s)$ inversion in the control action. Using $C_S(s)$ the actuating variable results as

$$u(s) = C_S(s)[v(s) - G_Z(s)d(s)] \qquad (15)$$

and the structure of the cascade scheme may be simplified to a single loop circuit. This interpretation explains the most significant effect of involving the secondary feedback loop: With respect to the faster response of secondary output z the compensation for the disturbance impact is much prompter than in the primary loop due to the term $C_S(s)G_Z(s)d(s)$. It makes the actuating variable u begin to compensate for d considerably sooner than the disturbance impact really appears in the response of y. Since the transfer function $G_S(s)$ is supposed stable the Youla parameterization can be based only on the inner-outer factorization $G_S(s) = G_{SO}(s)G_{SI}(s)$ where the inner part $G_{SI}(s)$ (with $G_{SI}(0) = 1$) absorbs the *delay* in $G_S(s)$ and in general it also would adsorb the possible right-hand-side zeros of $G_S(s)$. The inner part represents the properties which cannot be considered for inversion owing to the stability and causality requirements. After the inner-outer factorization the control parameterizing function can be considered in the form

$$C_S(s) = [G_{SO}(s)F_S(s)]^{-1} \qquad (16)$$

where $F_S(s)$ is a stable polynomial of order v such that $F_S(0) = 1$ [1]. The purpose of polynomial $F_S(s)$ is to achieve $C_S(s)$ as a lead compensation, i.e. to make the $F_S(s)$ zeros lie to the left from the poles of $G_{SO}(s)$. Using the inner-outer factorization $G_S(s) = G_{SO}(s)G_{SI}(s)$ the slave controller appropriate to the suggested $R_S(s)$ results from the relation (13) in the generic form

$$R_S(s) = \frac{C_S(s)}{1 - C_S(s)G_S(s)} = \frac{1}{G_{SO}(s)[F_S(s) - G_{SI}(s)]} \tag{17}$$

The transfer function of the secondary open loop is then independent of $G_{SO}(s)$

$$L_S(s) = R_S(s)G_S(s) = \frac{G_{SI}(s)}{F_S(s) - G_{SI}(s)} \tag{18}$$

Since it holds $G_{SI}(0) = F_S(0) = 1$, it is apparent that $L_S(s)$ has a *pseudo-integrating* character.

Proposition 1. *Suppose the secondary plant transfer function is in the aperiodic form free of right-hand-side zeros*

$$G_S(s) = \frac{K_S \exp(-s\tau_S)}{A_S(s)} \tag{19}$$

where $A_S(s)$ is a stable vth-order polynomial with negative real zeros admitting the factorization

$$A_S(s) = (1 + T_{S1}s)...(1 + T_{Sv}s) \tag{20}$$

so that the inner-outer factorization of $G_S(s)$ results in exponential inner part $G_{SI}(s) = \exp(-s\tau_S)$. Let T_{S1} be the dominant time constant, i.e. $T_{S1} > T_{Si}$, $i = 2,...,v$. For this type of secondary output properties let the parameterizing function be the following lead compensator

$$C_S(s) = \frac{(T_{S1}s + 1)(T_{S2}s + 1)...}{K_S(\alpha T_{S1}s + 1)(\alpha T_{S2}s + 1)...} \tag{21}$$

where α is a coefficient less than one: $0 < \alpha < 1$. Then the transfer function of the secondary open loop is of the form independent of the outer part of the plant model

$$L_S(s) = \frac{\exp(-s\tau_S)}{(\alpha T_{S1}s + 1)(\alpha T_{S2}s + 1)... - \exp(-s\tau_S)} \tag{22}$$

identical with the quasi-integration (5).

Proof. Using this form of $C_S(s)$, according to (17) the slave controller results in the form

$$R_S(s) = \frac{(T_{S1}s + 1)(T_{S2}s + 1)...}{K_S[(\alpha T_{S1}s + 1)(\alpha T_{S2}s + 1).... - \exp(-s\tau_S)]} \tag{23}$$

and the product of this function with $G_S(s)$ leads to (22).

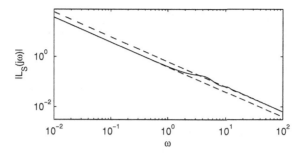

Fig. 2 Frequency response of the slave open loop (29) (solid) and both asymptotes (30) (dashed).

It is apparent that the obtained $L_S(s)$ is of the same form as $I_m(s)$ in (5) and has therefore the quasi-integrating character with the crossover frequency ω_C. As long as the delay term appears in $R_S(s)$ this controller is not a rational but an anisochronic function with delays, partially compensating for the time delay of the plant.

Example 1. Consider a time delay process with the primary transfer function

$$G(s) = \frac{K \exp(-\tau s)}{(1+T_1 s)(1+T_2 s)}, \quad T_1 > T_2. \tag{24}$$

Let an auxiliary output z be available with faster response to both the load d and the actuating variable u so that it is well suited for closing the secondary feedback. Suppose the secondary first order transfer function

$$G_S(s) = \frac{K_S \exp(-\tau_s s)}{1 + T_S s} \tag{25}$$

with considerably shorter T_S and τ_S than $T_{1,2}$ and τ respectively. To achieve a lead compensation by $C_S(s)$ let the polynomial $F_S(s)$ be chosen as $F_S(s) = 1 + \alpha T_S s$ with $\alpha < 1$. According to (21) the parameterizing function is obtained

$$C_S(s) = \frac{1 + T_S s}{K_S(1 + \alpha T_S s)} \tag{26}$$

and the slave controller is then given by (17) as follows

$$R_S(s) = \frac{1 + s T_s}{K_S\left(1 + \alpha s T_S - \exp(-s \tau_S)\right)} \tag{27}$$

The asymptotes of the Bode diagram of $R_S(s)$ for $\omega \to 0$ and $\omega \to \infty$ are as follows

$$\lim_{\omega \to 0} |j\omega R_S(j\omega)| = \frac{1}{K_S(\tau_S + \alpha T_S)}, \qquad \lim_{\omega \to \infty} |j\omega R_S(j\omega)| = \frac{1}{K_S \alpha} \tag{28}$$

The minimum of amplitude ratio is near the cross-over frequency $\omega_C = (\tau + \alpha T_S)^{-1}$. The obtained slave controller provides an quasi-integrating frequency response for the secondary open loop

$$R_S(j\omega)G_S(j\omega) = L_S(j\omega) = \frac{\exp(-j\omega\tau_S)}{1 - \exp(-j\omega\tau_S) + j\omega\alpha T_S} \tag{29}$$

as a first order form of $I_m(j\omega)$. Considering the parameters $T_S = 2, K_S = 1, \tau_S = 1$ and $\alpha = 0.8$, the frequency response of the open loop transfer function is shown in Fig. 2. The asymptotes in Fig. 2 are given by the limits

$$\lim_{\omega \to 0} |j\omega L_S(j\omega)| = \frac{1}{(\tau_S + \alpha T_S)}, \quad \lim_{\omega \to \infty} |j\omega L_S(j\omega)| = \frac{1}{\alpha T_S \omega}. \tag{30}$$

It is easy to see that the transition part between the asymptotes occurs in frequencies higher than the crossover one.

4   Master Controller Design

By means of expressing the actuating variable by (15), the cascade control scheme may be reconsidered as a single loop feedback system with the plant pre-compensated by the secondary loop. The secondary controller is already considered as determined and the only part to be designed is the master controller $R_M(s)$. As soon as the one-loop interpretation of the master controller linkage is introduced it is possible to perform the parameterization in the same way as in the secondary loop. Also for the primary plant transfer function the inner-outer factorization $G(s) = G_O(s)G_I(s)$ may be introduced, where the delays as well as possible right-hand side zeros of $G(s)$ are absorbed in $G_I(s)$. For the outer part $G_O(s)$ the inversion can be assumed to exist. The transfer function of pre-compensated plant is now given by the product $\tilde{G}(s) = C_S(s)G_O(s)G_I(s)$ where only $G_I(s)$ is non-invertible. The parameterizing function $C_M(s)$ for the master controller design may now be selected as the following inversion

$$C_M(s) = [C_S(s)G_O(s)F_M(s)]^{-1} \tag{31}$$

where $F_M(s)$ is a stable polynomial of order n. Again the purpose of this polynomial is not only to make the inversion in (31) feasible but also to adjust a more favourable dynamics in the primary loop. With respect to the expected higher order of $G(s)$ than $G_S(s)$ it is admissible to consider a factorization of this polynomial $F_M(s) = F_S(s)\Phi(s)$ containing the polynomial $F_S(s)$. Using the generic relationship as in (17) the master controller then results as follows

$$R_M(s) = \frac{C_M(s)}{1 - C_M(s)\tilde{G}(s)} = \frac{G_{SO}(s)F_S(s)}{G_O(s)[F_M(s) - G_I(s)]} \tag{32}$$

The transfer function of the pre-compensated primary open loop is then independent of $G_O(s)$

$$L_M(s) = R_M(s)G(s) = \frac{G_I(s)}{F_M(s) - G_I(s)} \qquad (33)$$

Since it holds that $G_I(0) = F_M(0) = 1$, it is apparent that $L_M(s)$ is of quasi-integrating character again.

Proposition 2. *Suppose the primary plant transfer function be again in the aperiodic delayed form, free of right-hand-side zeros*

$$G(s) = \frac{K \exp(-s\tau)}{A(s)} \qquad (34)$$

where $A(s)$ is a stable nth-order polynomial with real negative zeros and with the factorization

$$A(s) = (1 + T_1 s)....(1 + T_n s) \qquad (35)$$

The inner-outer factorization of $G(s)$ results again in exponential inner part $G_I(s) = \exp(-s\tau)$. Let T_1 be the dominant time constant, i.e. $T_1 > T_i$, $i = 2,...n$. For this type of primary plant dynamics let the master parameterizing function be the following lead compensator

$$C_M(s) = \frac{(T_1 s + 1)(T_2 s + 1)...}{K(\beta T_1 s + 1)(\beta T_2 s + 1)...} \qquad (36)$$

where β is a coefficient less than one, $0 < \beta < 1$, again. Then the primary pre-compensated open-loop transfer function $L_M(s) = R_M(s)C_S(s)G(s)$ is of the following form independent of the outer parts of both the primary and secondary plant transfer functions

$$L_M(s) = \frac{\exp(-s\tau)}{(\beta T_1 s + 1)...(\beta T_n s + 1) - \exp(-s\tau)} \qquad (37)$$

and identical with the quasi-integration (5) again.

Proof. Using the form (36) of $C_M(s)$, according to (32) the master controller results as

$$R_M(s) = \frac{K_S F_S(s) A(s)}{K A_S(s) [(\beta T_1 s + 1)(\beta T_2 s + 1).... - \exp(-s\tau)]} \qquad (38)$$

This formula is rather complicated but the product $R_M(s)C_S(s)G(s)$ is of the form

$$L_M(s) = \frac{G_I(s)}{F_M(s) - G_I(s)} = \frac{\exp(-s\tau)}{F_M(s) - \exp(-s\tau)} \qquad (39)$$

apparently identical with (37).

Since the limit of $j\omega L_M(j\omega)$ for $\omega \to 0$ exists and is equal to $(\beta(T_1 + T_2 + ...) + \tau)^{-1}$ the primary open loop obtains an integrating character in low frequency band again, excluding a steady state error. Similarly as in the secondary

loop the crossover frequency of the open loop transfer function is approximately:
$\omega_{CM} = [\tau + \beta(T_1 + T_2 + ..)]^{-1}$.

Example 2. Recall the plant considered in Example 1 with the primary plant transfer function (24). Provided that the slave controller $R_S(s)$ has been taken as in (27) we are to design a master controller $R_M(s)$. According to (36) the master parameterizing function is

$$C_M(s) = \frac{(T_1 s + 1)(T_2 s + 1)}{K(\beta T_1 s + 1)(\beta T_2 s + 1)} \tag{40}$$

and the master controller is then given according to (38) as

$$R_M(s) = \frac{K_S(\alpha T_S s + 1)(T_1 s + 1)(T_2 s + 1)}{K(T_S s + 1)[(\beta T_1 s + 1)(\beta T_2 s + 1) - \exp(-s\tau)]} \tag{41}$$

This transfer function may seem to be rather complicated but the primary open loop transfer function is considerably simpler

$$L_M(s) = \frac{\exp(-s\tau)}{(\beta T_1 s + 1)(\beta T_2 s + 1) - \exp(-s\tau)} \tag{42}$$

and represents a quasi-integration again. In the case of the option $\beta(T_1 + T_2) > \tau$ the primary open loop achieves a favourable phase margin as demonstrated in Section 2. The crossover frequency of the primary control loop is approximately $\omega_{CM} \cong (\tau + \beta(T_1 + T_2))^{-1}$. Considering the following parameters of the plant and controller, $T_1 = 12, T_2 = 5, \tau = 10, K = 1$, and $\beta = 0.6$, the frequency response is shaped as in Fig. 3.

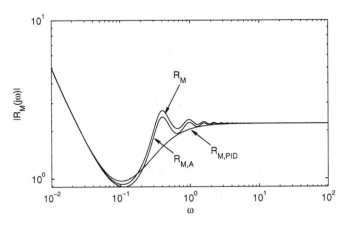

Fig. 3 Frequency response of the master controller (41) and of its approximations (43) and (44).

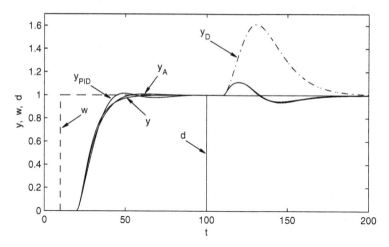

Fig. 4 Setpoint response and disturbance rejection with the master controller (41) - y, (43) - y_A and (44) - y_{PID}. y_D - response with (41) but with no disturbance rejection by the slave control loop.

Remark 2. The master controller (38) turns out to be unpleasantly complicated. A simplified form of the controller can be considered as follows

$$R_{M,A}(s) = \frac{K_S (T_1 s + 1)(a T_2 s + 1)}{K[(\beta T_1 s + 1)(\beta T_2 s + 1) - \exp(-s\tau)]} \quad (43)$$

As it is demonstrated in Fig. 3, the frequency response of (43) has the same asymptotes at low and high frequencies as the response of (41). Nevertheless, the transfer function (41) may also be approximated by PID transfer function with a first-order filtering

$$R_{M,PID}(s) = \frac{r_d s^2 + r_0 s + r_i}{(T_f s + 1)s} \quad (44)$$

where the parameters fulfilling the identity conditions for both the asymptotes are as follows $r_0 = 0.99$, $r_i = 0.049$, $r_d = 5$, $T_f = 2.25$.

In Fig. 4, the set-point responses and disturbance rejections are shown for the closed loops with original master controller (41) and both its approximations (43), (44). For the sake of brevity, the transfer functions from the disturbance d are considered identical with those from the control action u, i.e. $G_D(s) = G(s)$ and $G_Z(s) = G_S(s)$. Besides, the disturbance rejection is also shown in Fig. 4 for the case if $G_Z(s) = 0$, i.e., the disturbance signal is not processed by the slave control loop. As can be seen, processing the disturbance by the slave control loop results in considerable more efficient and faster disturbance rejection. This viewpoint is to be taken into account in selecting the secondary output z of the plant.

5 Concluding Remarks

Unlike the usual cascade control implementation based on the PID controller technique the present method leads to a scheme where both the slave and master controllers are anisochronic, i.e. operating with time shifted (delayed) control errors. As in the standard cascade control, the faster is the response of the secondary measured variable z to the disturbance to be rejected, the better is the final effect of the cascade scheme. The anisochronic character of both the controllers results from the general principle of affine parameterization and therefore the results really achieved in the implementation essentially depend on the validity of the plant models, i.e. on the accordance between the model and the real plant behaviour.

Acknowledgements. The research is supported by the Ministry of Education of the Czech Republic under **Project 1M0567** and by the Grant Agency of the Czech Republic, contract **GA102/08/0186**.

References

1. Goodwin, G.C., Graebe, S.F., Salgado, M.E.: Control System Design. Prentice Hall, Englewood Cliffs (2001)
2. Kaya, I., Atherton, D.P.: Use of Smith Predictor in the outer loop for Cascaded Control of Unstable and Integrating Processes. Industrial and Engineering Chemistry Research 47(6), 1981–1987 (2008)
3. Leva, A., Donida, F.: Autotuning in cascaded systems based on a single relay experiment. Journal of Process Control 19(5), 896–905 (2009)
4. Mirkin, L.: On the extraction of dead-time controllers and estimators from delay-free parameterizations. IEEE Trans. Automatic Control 48(5), 543–553 (2003)
5. Mirkin, L., Raskin, N.: Every stabilizing dead-time controller has an observer-predictor based structure. Automatica 39, 1747–1754 (2003)
6. Morari, M., Zafiriou, E.: Robust Process Control. Prentice Hall, Englewood Cliffs (1989)
7. Pekař, L., Prokop, R., Prokopová: Design of controllers for delayed integration processes using RMS ring. In: Mediterranean Conference on Control and Automation - Conference Proceedings, MED 2008, art. no. 4602062, pp. 146–151 (2008)
8. Pekař, L., Prokop, R., Matušu, R.: Algebraic control of unstable delayed first order systems using RQ-meromorphic functions. Mediterranean Conference on Control and Automation, MED, art. no. 4433754 (2007)
9. Shinskey, F.G.: Process Control Systems: Application, Design and Adjustment. McGraw-Hill, New York (1998)
10. Zhang, W., Allgower, F., Liu, T.: Controller parameterization for SISO and MIMO plants with time delay. Systems and Control Letters 55, 794–802 (2006)
11. Zítek, P., Hlava, J.: Algebraic design of anisochronic internal model control of time-delay systems. Control Engineering Practice 9, 501–516 (2001)
12. Zítek, P., Kučera, V.: Algebraic design of anisochronic controllers for time delay systems. Int. J. Control 76, 1654–1665 (2003)
13. Zítek, P., Kučera, V., Vyhlídal, T.: Affine Parameterization of Cascade Control for Time Delay Plants. In: Proc. of 9th IFAC Workshop on Time Delay Systems, Prague, IFAC-PapersOnline, Time Delay Systems, vol. 9 (2010)

Design of Terminal Cost Functionals and Terminal Regions for Model Predictive Control of Nonlinear Time-Delay Systems

Marcus Reble and Frank Allgöwer

Abstract. In this work, we present results on model predictive control (MPC) for nonlinear time-delay systems. MPC is one of the few control methods which can deal effectively with constrained nonlinear time-delay systems. In order to guarantee stability of the closed-loop, a local control Lyapunov functional in a region around the origin is in general utilized as terminal cost. It is well-known for delay-free systems that a control Lyapunov function calculated for the Jacobi linearization about the origin can also be used as a terminal cost for the nonlinear system for an appropriately chosen terminal region. However, the infinite-dimensional nature of time-delay systems circumvents a straight-forward extension of those schemes to time-delay systems. We present two schemes for calculating stabilizing design parameters based on the Jacobi linearization of the nonlinear time-delay system. The two schemes are based on different assumptions and yield different types of terminal regions. We compare the properties and discuss advantages and disadvantages of both schemes.

1 Introduction

Nonlinear time-delay systems naturally arise in the modelling of many technical, biological and social systems. Examples can be found in mechanics, physics, biology and social sciences and include transportation of material, communication and computational delays in control loops, population dynamics models, disease transmission models, glucose-insulin regulation, and price fluctuations. A comprehensive list of examples is given in [6].

Marcus Reble
Institute for Systems Theory and Automatic Control, University of Stuttgart
e-mail: `reble@ist.uni-stuttgart.de`

Frank Allgöwer
Institute for Systems Theory and Automatic Control, University of Stuttgart
e-mail: `allgower@ist.uni-stuttgart.de`

R. Sipahi et al. (Eds.): Time Delay Sys.: Methods, Appli. and New Trends, LNCIS 423, pp. 355–366.
springerlink.com   © Springer-Verlag Berlin Heidelberg 2012

Although there exist several approaches for the control of such systems, see e.g. [4, 12, 14], only few methods are able to handle constraints. For this reason, we investigate the applicability of model predictive control (MPC) to nonlinear time-delay systems.

MPC, also known as receding horizon control, is one of the key advanced control strategies. One of its major advantages is the explicit consideration of a performance criterion and constraints. However, stability is not generally guaranteed, see e.g. [17] for a practical example. Several MPC schemes have been proposed which guarantee closed-loop stability for delay-free systems by using an approriately chosen terminal cost and a terminal state constraint [1, 13, 18]. The general results on stability have been recently extended to nonlinear time-delay systems [11, 19, 20, 22].

There exist several schemes that allow the determination of stabilizing design parameters such as terminal cost and terminal region . MPC schemes without terminal constraints based on suitably defined terminal cost functionals are presented in [7, 8, 10]. However, a *global* control Lyapunov-Krasovskii functional is necessary in order to guarantee stability whereas schemes with terminal constraint only require *local* control Lyapunov-Krasovskii functionals. The first scheme using a terminal constraint has been proposed in [16]. An expanded zero terminal state constraint is used to ensure stability which is unattractive from a computational point of view. There are several schemes using finite terminal regions for nonlinear time-delay systems [11, 19, 20, 22]. In all cited results, the stabilizing design parameters are calculated based on the Jacobi linearization about the origin using a linear local control law. The first scheme [19, 20, 22] only poses Lyapunov-Krasovskii conditions on the local control law and ensures positive invariance of the terminal region by using an additional bound on the norm. The second scheme [11, 22] uses a definition of the terminal region based on Lyapunov-Razumikhin arguments and requires a combination of Lyapunov-Krasovskii and Lyapunov-Razumikhin conditions for the local control law. In this work, generalizations and new results on the design based on the Jacobi linearization are presented.

In the first part, we extend the results of [20, 22] with a preliminary version of this extension presented in [19]. It is shown that if the linearization is stabilizable by a general linear feedback $u = k(x_t)$, there exist a quadratic terminal cost functional and a finite terminal region such that closed-loop stability can be guaranteed by MPC. Thus, the result allows the use of any known linear feedback design methodology for linear time-delay systems in order to obtain stabilizing design parameters for MPC of the nonlinear system. The terminal region is defined as intersection of a sublevel set of the terminal cost functional and a sphere in the infinite-dimensional space similar to the result in.

In the second part, a generalization of the result in [11] is given. A linear control law based on Lyapunov-Razumikhin arguments is used to derive an appropriate terminal cost functional. The terminal region is of considerably simpler form than the terminal region obtained in the first part. However, the Lyapunov-Razumikhin conditions necessary for this approach are more restrictive than the requirement of stabilizability of the linearization.

The remainder of this paper is organized as follows. In Section 2 the problem setup considered in this work is presented. In Section 3 the MPC setup is given and conditions for asymptotic stability are recalled. In Section 4 it is shown that it is possible to use quadratic terminal cost functionals and particular chosen terminal regions for a general class of systems with stabilizable linearization. In Section 5 a more simple terminal region is designed using additional Lyapunov-Razumikhin arguments. In Section 6 both schemes are compared and advantages and disadvantages are discussed. Concluding remarks are given in Section 7.

Notation: Let $\mathbb{R}$ and $\mathbb{R}^+$ denote the field of real numbers and the set of non-negative real numbers, respectively. Let $\mathbb{R}^n$ denote the n-dimensional Euclidean space with the standard norm $|\cdot|$. $\|P\|$ is the induced 2-norm of matrix P. Given $\tau > 0$, let $\mathscr{C}_\tau = \mathscr{C}([-\tau, 0], \mathbb{R}^n)$ denote the Banach space of continuous functions mapping the interval $[-\tau, 0] \subset \mathbb{R}$ into $\mathbb{R}^n$ with the topology of uniform convergence. A segment $x_t \in \mathscr{C}_\tau$ is defined by $x_t(s) = x(t+s), s \in [-\tau, 0]$. The norm on $\mathscr{C}_\tau$ is defined as $\|x_t\|_\tau = \sup_{\theta \in [-\tau, 0]} |x(t+\theta)|$. $\lambda_{\max}(P)$ and $\lambda_{\min}(P)$ refer to the maximal and minimal eigenvalue of matrix P, respectively. A function $f : \mathbb{R}^+ \to \mathbb{R}^+$ is said to belong to class $\mathscr{K}_\infty$ if it is continuous, strictly increasing, $f(0) = 0$ and $f(s) \to \infty$ as $s \to \infty$.

2 Problem Setup

Consider the nonlinear time-delay system

$$\dot{x}(t) = f(x(t), x(t-\tau), u(t)), \tag{1a}$$
$$x(\theta) = \varphi(\theta), \qquad \forall \theta \in [-\tau, 0], \tag{1b}$$

in which $x(t) \in \mathbb{R}^n$ is the instantaneous state at time t subject to state constraints $x(t) \in \mathscr{X}$, $u(t) \in \mathbb{R}^m$ is the control input subject to input constraints $u(t) \in \mathscr{U}$ and $\varphi \in \mathscr{C}_\tau$ is the initial function. The time-delay $\tau > 0$ is constant and assumed to be known. The function $f : \mathbb{R}^n \times \mathbb{R}^n \times \mathbb{R}^m \to \mathbb{R}^n$ is continuously differentiable. The constraint sets $\mathscr{X} \subset \mathbb{R}^n$ and $\mathscr{U} \subset \mathbb{R}^m$ are compact and contain the origin in their interior. Without loss of generality, $x_t = 0$ is assumed to be an equilibrium of system (1) for $u = 0$, i.e. $f(0,0,0) = 0$. The problem of interest is to stabilize the steady state $x_t = 0$ via model predictive control.

In order to design a locally stabilizing control law the Jacobi linearization of system (1)

$$\dot{\tilde{x}}(t) = A\tilde{x}(t) + A_\tau \tilde{x}(t-\tau) + Bu(t), \tag{2}$$

is used. The matrices are given by

$$A = \left.\frac{\partial f}{\partial x(t)}\right|_{x_t=0, u=0}, \quad A_\tau = \left.\frac{\partial f}{\partial x(t-\tau)}\right|_{x_t=0, u=0}, \quad B = \left.\frac{\partial f}{\partial u(t)}\right|_{x_t=0, u=0}.$$

Define Φ as the difference between the nonlinear system (1) and its Jacobi linearization (2), i.e.

$$\Phi(x_t, u(t)) = f(x(t), x(t-\tau), u(t)) - Ax(t) - A_\tau x(t-\tau) - Bu(t). \tag{3}$$

Since f is continuously differentiable and Φ only consists of higher order terms, i.e. it contains no linear terms, for any $\gamma > 0$ there exists a $\delta = \delta(\gamma) > 0$ such that

$$|\Phi(x_t, u(t))| < \gamma(|x(t)| + |x(t-\tau)| + |u(t)|) \tag{4}$$

for all $\|x_t\|_\tau \le \delta$ and $|u(t)| < \delta$.

3 Model Predictive Control for Nonlinear Time-Delay Systems

Model predictive control (MPC) is formulated as solving online a finite horizon optimal control problem. Based on measurements obtained at time t, the controller predicts the future behavior of the system over a finite prediction horizon T and determines the control input such that a cost functional J is minimized. In order to incorporate a feedback mechanism, the obtained open loop solution to this optimal control problem will be implemented only until the next measurement becomes available. Based on the new measurement, the solution of the optimal control problem is repeated for a now shifted horizon and again implemented until the next sampling instant.

It is well known that an inappropriate definition of the finite horizon optimal control problem may cause instability especially if the horizon is too short, see e.g. the practical example in [17]. In general, appropriately chosen terminal cost and terminal constraint are used to guarantee closed-loop stability [1, 13, 18]. In the following, the general model predictive control setup for nonlinear time-delay systems of [11] is presented and the conditions for asymptotic stability are recalled.

The open loop finite horizon optimal control problem $\mathscr{P}(x_t; T, \Omega)$ at time t with prediction horizon T is formulated as

$$\min_{\bar{u}(\cdot)} J(x_t, u) = \int_t^{t+T} F(\bar{x}(t'), \bar{u}(t')) \, dt' + E(\bar{x}_{t+T}) \tag{5a}$$

subject to

$$\begin{aligned}
\dot{\bar{x}}(t') &= f(\bar{x}(t'), \bar{x}(t'-\tau), \bar{u}(t')), & t' &\in [t, t+T] \\
\bar{u}(t') &\in \mathscr{U}, \quad \bar{x}(t') \in \mathscr{X}, & t' &\in [t, t+T] \\
\bar{x}_{t+T} &\in \Omega \subseteq \mathscr{C}_\tau,
\end{aligned} \tag{5b}$$

in which $\bar{x}(t')$ is the predicted trajectory starting from initial condition $\bar{x}_t(\theta) = x(t+\theta), -\tau \le \theta \le 0$, and driven by $\bar{u}(t')$ for $t' \in [t, t+T]$. The terminal region Ω is a closed set, contains the steady state $0 \in \mathscr{C}_\tau$ in its interior, and is defined such that $x_t \in \Omega$ implies $x(t+\theta) \in \mathscr{X}$ for all $\theta \in [-\tau, 0]$. The terminal cost $E : \mathscr{C}_\tau \to \mathbb{R}^+$

is a suitably defined positive definite terminal cost functional for which a class $\mathcal{K}_\infty$ function $\underline{E} : \mathbb{R}^+ \to \mathbb{R}^+$ exists such that $E(x_t) \geq \underline{E}(|x(t)|)$. The stage cost $F : \mathbb{R}^n \times \mathbb{R}^m \to \mathbb{R}^+$ is continuous, $F(0,0) = 0$ and there is a class $\mathcal{K}_\infty$ function $\underline{F} : \mathbb{R}^+ \to \mathbb{R}^+$ such that

$$F(x,u) \geq \underline{F}(|x|) \quad \text{for all } x \in \mathbb{R}^n, u \in \mathbb{R}^m.$$

The optimal solution of $\mathscr{P}(x_t;T,\Omega)$ is denoted by $u^*(\cdot;x_t,t)$ with associated cost $J^*(x_t;T,\Omega)$. The control input to the system is defined by the optimal solution $u^*(\cdot;x_t,t)$ of problem $\mathscr{P}(x_t;T,\Omega)$ in (5) at sampling instants $t_i = i\Delta$, $i \in \mathbb{N}$, in the usual receding horizon fashion

$$u(t) = u^*(t;x_{t_i},t_i), \quad t_i \leq t \leq t_i + \Delta. \tag{6}$$

Here the sampling time $\Delta > 0$ has to be chosen smaller than the prediction horizon T. The implicit feedback controller resulting from application of (6) is asymptotically stabilizing provided the following conditions are satisfied:

Property 1. The open loop finite horizon problem $\mathscr{P}(x_t;T,\Omega)$ in (5) admits a feasible solution at the initial time $t = 0$.

Property 2. For the nonlinear time-delay system (1), there exists a locally asymptotically stabilizing controller $u(t) = k(x_t) \in \mathscr{U}$, such that the terminal region Ω is controlled positively invariant and

$$\forall x_t \in \Omega : \dot{E}(x_t) \leq -F(x(t),k(x_t)). \tag{7}$$

The main result regarding asymptotic stability of the closed-loop system can be summarized as follows.

Theorem 1 (Stability of closed-loop [11]). *Assume that in the finite horizon optimal control problem $\mathscr{P}(x_t;T,\Omega)$ in (5) the design parameters – the stage cost F, the terminal cost functional E, the terminal region Ω, and the prediction horizon T – are selected such that Properties 1 and 2 are satisfied. Then, the closed-loop system resulting from the application of the model predictive controller (6) to system (1) is asymptotically stable.*

Proof. The proof is given in [11, 22]. □

In the following two sections, we present design schemes in order to obtain suitable a suitable terminal cost functional E and terminal region Ω using the Jacobi linearization about the origin. In particular, these results extend previous results based on certain linear matrix inequality conditions [22]. The first scheme only requires a locally stabilizing control law whereas the second scheme uses a Razumikhin condition, which allows the calculation of a more simple terminal region.

4   General Linearization-Based Design

In this section, previous results on stabilizing design parameters based on the Jacobi linearization [11, 20, 22] are generalized. A preliminary, slightly different version of this result has been presented in [19]. Similar to the results for delay-free systems, it is shown that each system with a stabilizable Jacobi linearization about the origin can be stabilized by MPC with a quadratic terminal cost functional and a finite terminal region. In contrast to the delay-free case, the terminal region is not defined as sublevel set of the terminal cost, but instead as the intersection of such a sublevel set with a sphere in the infinite-dimensional space $\mathscr{C}_\tau$ defined by the norm $\|\cdot\|_\tau$.

To this end, consider a general linear local control law

$$u(t) = k(x_t) = Kx(t) + \int_{-\tau}^{0} K_\tau(\theta)x(t+\theta)d\theta \tag{8}$$

with $K \in \mathbb{R}^{m \times n}$ and matrix function $K_\tau(\theta) \in \mathbb{R}^{m \times n}$, and a quadratic stage cost

$$F(x(t), u(t)) = x(t)^T Q x(t) + u(t)^T R u(t) \tag{9}$$

with symmetric positive definite matrices Q and R.

Using these definitions the main result can be summarized as follows.

Theorem 2 (Systems with stabilizable linearization). *Consider a nonlinear time-delay system of form* (1) *and stage cost* (9). *Assume there exists a linear local control law* (8) *which asymptotically stabilizes the linearized system* (2). *Then there exists a terminal cost functional*

$$E(x_t) = x(t)^T P_0 x(t) + \int_{-\tau}^{0}\int_{-\tau}^{0} x(t+\theta_1)^T P_3(\theta_1, \theta_2)x(t+\theta_2)d\theta_1 d\theta_2 \tag{10}$$

$$+ 2x(t)^T \int_{-\tau}^{0} P_1(\theta)x(t+\theta)d\theta + \int_{-\tau}^{0} x(t+\theta)^T P_2(\theta)x(t+\theta)d\theta$$

with $P_0 = P_0^T \in \mathbb{R}^{n \times n}$ and matrix functions $P_1(\theta) \in \mathbb{R}^{n \times n}$, $P_2(\theta) = P_2(\theta)^T \in \mathbb{R}^{n \times n}$, $P_3(\theta_1, \theta_2) = P_3(\theta_2, \theta_1)^T \in \mathbb{R}^{n \times n}$, and there exists a terminal region

$$\Omega = \left\{ x_t : E(x_t) \le \mu \frac{\alpha^2}{4}, \quad \|x_t\|_\tau \le \frac{\alpha}{2} \right\} \tag{11}$$

in which $\alpha > 0$ and $\mu > 0$, such that the closed-loop with the MPC control law (6) *is asymptotically stable.*

Proof. Since the Jacobi linearization (2) with the linear control law $u(t) = k(x_t)$, i.e.

$$\dot{\tilde{x}}(t) = \tilde{A}\tilde{x}(t) + A_\tau \tilde{x}(t-\tau) + Bk(\tilde{x}_t), \tag{12}$$

is asymptotically stable, there exists a quadratic Lyapunov-Krasovskii functional $\tilde{E}(x_t)$ of form (10) with derivative along trajectories of (12)

$$\dot{E}(x_t) \leq -\varepsilon_1 |x(t)|^2 - \varepsilon_2 |x(t-\tau)|^2 - \varepsilon_3 \int_{-\tau}^{0} |x(t+\theta)|^2 d\theta$$

with $\varepsilon_1, \varepsilon_2, \varepsilon_3 > 0$, see [5]. The derivative of $\tilde{E}$ along trajectories of the nonlinear system $\dot{x}(t) = f(x(t), x(t-\tau), u(t))$ is

$$\dot{\tilde{E}}(x_t) \leq -\varepsilon_1 |x(t)|^2 - \varepsilon_2 |x(t-\tau)|^2 - \varepsilon_3 \int_{-\tau}^{0} |x(t+\theta)|^2 d\theta$$
$$+ 2\Phi(x_t, u(t))^T \left(P_0 x(t) + \int_{-\tau}^{0} P_1(\theta) x(t+\theta) d\theta \right).$$

Due to Inequality (4) and since $2ab \leq a^2 + b^2$ for all $a, b \in \mathbb{R}$,

$$2\Phi(x_t, u(t))^T P_0 x(t) \leq \gamma \|P_0\| \left(4|x(t)|^2 + |x(t-\tau)|^2 + |u(t)|^2 \right)$$

holds for all $\|x_t\|_\tau \leq \delta(\gamma)$ and $|u(t)| \leq \delta(\gamma)$. Similarly, using the Cauchy–Schwarz inequality the following inequalities

$$2\Phi(x_t)^T \int_{-\tau}^{0} P_1(\theta) x(t+\theta) d\theta$$
$$\leq \gamma \left(|x(t)|^2 + |x(t-\tau)|^2 + |u(t)|^2 \right) + 3\gamma \left| \int_{-\tau}^{0} P_1(\theta) x(t+\theta) d\theta \right|^2$$
$$\leq \gamma \left(|x(t)|^2 + |x(t-\tau)|^2 + |u(t)|^2 \right) + 3\gamma\tau \|P_1\|_\tau^2 \int_{-\tau}^{0} |x(t+\theta)|^2 d\theta$$

with $\|P_1\|_\tau = \sup_{\theta \in [-\tau, 0]} \|P_1(\theta)\|$ hold for $\|x_t\|_\tau < \delta(\gamma)$ and $|u(t)| < \delta(\gamma)$. Combining these findings yields

$$\dot{\tilde{E}}(x_t) \leq -(\varepsilon_1 - 4\gamma\|P_0\| - \gamma)|x(t)|^2$$
$$- (\varepsilon_2 - \gamma\|P_0\| - \gamma)|x(t-\tau)|^2$$
$$- (\varepsilon_3 - 3\gamma\tau\|P_1\|_\tau^2) \int_{-\tau}^{0} |x(t+\theta)|^2 d\theta$$
$$+ \gamma(\|P_0\| + 1)|u(t)|^2.$$

Furthermore, we obtain for $u(t) = k(x_t)$

$$|u(t)|^2 = |k(x_t)|^2 \leq 2\|K^T K\| |x(t)|^2 + 2\tau \|K_\tau\|_\tau^2 \int_{-\tau}^{0} |x(t+\theta)|^2 d\theta \qquad (13)$$

with $\|K_\tau\|_\tau = \sup_{\theta \in [-\tau, 0]} \|K_\tau(\theta)\|$ and hence

$$\dot{\tilde{E}}(x_t) \leq -(\varepsilon_1 - 4\gamma\|P_0\| - \gamma - 2\gamma(\|P_0\| + 1)\|K^T K\|)|x(t)|^2$$

$$- (\varepsilon_2 - \gamma \|P_0\| - \gamma) |x(t-\tau)|^2$$

$$- (\varepsilon_3 - 3\gamma\tau \|P_1\|_\tau^2 - 2\tau\gamma(\|P_0\| + 1)\|K_\tau\|_\tau^2) \int_{-\tau}^0 |x(t+\theta)|^2 d\theta$$

for $|k(x_t)|^2 < \delta(\gamma)$. It is clearly possible to choose $\gamma, \beta > 0$ such that

$$\beta \left(\varepsilon_1 - 4\gamma\|P_0\| - \gamma - 2\gamma(\|P_0\| + 1)\|K^T K\|\right) > \|Q + 2K^T RK\|,$$

$$(\varepsilon_2 - \gamma\|P_0\| - \gamma) > 0,$$

$$\beta \left(\varepsilon_3 - 3\gamma\tau \|P_1\|_\tau^2 - 2\tau\gamma(\|P_0\| + 1)\|K_\tau\|_\tau^2\right) > 2\tau\|R\| \|K_\tau\|_\tau^2.$$

Now define the terminal region Ω as in (11) with $\alpha > 0$ chosen such that $\alpha \leq 2\delta(\gamma)$ and such that $x_t \in \Omega$ implies $k(x_t) \in \mathcal{U}$, $|k(x_t)|^2 < \delta(\gamma)$ and $x(t+\theta) \in \mathcal{X}$ for all $\theta \in [-\tau, 0]$. The satisfaction of all three condition is always possible for some small enough $\alpha > 0$. Moreover, define the terminal cost functional $E(x_t) = \beta \tilde{E}(x_t)$. Then clearly for all $x_t \in \Omega$

$$\dot{E}(x_t) \leq -x(t)^T Qx(t) - u(t)^T Ru(t) \tag{14}$$

by using the local control law $u(t) = k(x_t)$. Furthermore, the terminal region Ω is positively invariant for the choice of $\mu = \beta \lambda_{\min}(P_0)$. This can be shown by contradiction similar to the proofs in [15, 20] using $\forall x_t \in \Omega : \dot{E}(x_t) \leq 0$ and by noting that

$$E(x_t) \geq \beta \lambda_{\min}(P_0)|x(t)|^2.$$

Combining (14) and the positive invariance shows that Property 2 is satisfied. Asymptotic stability follows directly by use of Theorem 1.   □

The standard converse Lyapunov Theorem for linear time-delay systems [2, Proposition 7.4] only guarantees the existence of a complete quadratic functional $E(x_t)$ with $\dot{E}(x_t) \leq -\varepsilon_1 |x(t)|^2$, which is not sufficient to ensure $\dot{E}(x_t) \leq -F(x(t), k(x_t))$ for a general linear control law of form (8). Thus, the more general result of [5] is needed in the proof.

The result of Theorem 2 guarantees the existence of a terminal region defined similar to the result in [20] under mild assumptions. In the following section, a similar extension is provided for terminal regions as considered in [11].

5   Design by Combination of Lyapunov-Krasovskii and Lyapunov-Razumikhin Arguments

In this section, we extend results from [11, 22]. Similar to the previous section, we consider a quadratic stage cost as in (9) and calculate the stabilizing design parameters based on the Jacobi linearization about the origin. In contrast to the result in Section 4, we do not only require the existence a locally stabilizing linear control law, but need this local control law to satisfy a Lyapunov-Razumikhin condition as stated in the following assumption.

Property 3. There exists a linear local control law $u(t) = k(x_t) = Kx(t) + \int_{-\tau}^{0} K_\tau(\theta)x(t+\theta)d\theta$ such that the derivative of the Lyapunov-Razumikhin function $V(x(t)) = x(t)^T Px(t)$ along trajectories of the linearized system (2) satisfies $\dot{V}(x(t)) \leq -\varepsilon|x(t)|^2$ whenever

$$V(x(t+\theta)) \leq \rho V(x(t)), \qquad \forall \theta \in [-\tau, 0] \tag{15}$$

for some constants $\varepsilon > 0$ and $\rho > 1$.

Using this additional property, it is possible to use a simpler terminal region in the MPC control law as shown in the following theorem.

Theorem 3 (Combination of Lyapunov-Krasovskii and Lyapunov-Razumikhin arguments). *Consider a nonlinear time-delay system of form (1) and stage cost (9). If Property 3 is satisfied, then there exists a terminal cost functional of form (10) with $\beta > 0$, and there exists a terminal region*

$$\Omega = \left\{ x_t : \max_{\theta \in [-\tau, 0]} V(x(t+\theta)) \leq \alpha \right\} \tag{16}$$

in which $\alpha > 0$, such that the closed-loop with the MPC control law (6) is asymptotically stable.

Proof. In the first part of the proof, it is shown that there is a sufficiently small $\alpha > 0$ such that for all $x_t \in \Omega$ the derivative of V along trajectories of the nonlinear system (1) satisfies $\dot{V}(x(t)) \leq -\varepsilon/2|x(t)|^2$ whenever Condition (15) holds. To this end, note that

$$\dot{V}(x(t)) \leq -\varepsilon|x(t)|^2 + 2x(t)^T P\Phi(x_t, u(t))$$
$$\leq -\varepsilon|x(t)|^2 + 2\gamma|x(t)| \|P\| (|x(t)| + |x(t-\tau)| + |u(t)|)$$
$$\leq -(\varepsilon - 4\gamma\|P\|)|x(t)|^2 + \gamma\|P\| |x(t-\tau)|^2 + \gamma\|P\| |u(t)|^2$$
$$\overset{(13)}{\leq} -(\varepsilon - 4\gamma\|P\| - 2\gamma\|P\| \|K^T K\|)|x(t)|^2$$
$$\qquad + \gamma\|P\| |x(t-\tau)|^2 + 2\tau\|K_\tau\|_\tau^2 \int_{-\tau}^{0} |x(t+\theta)|^2 d\theta$$
$$\overset{(15)}{\leq} \left(-\varepsilon + \gamma\|P\| \left(4 + 2\|K^T K\| + (1 + 2\tau^2\|K_\tau\|_\tau^2)\rho \frac{\lambda_{\max}(P)}{\lambda_{\min}(P)} \right) \right)$$
$$\qquad \times |x(t)|^2.$$

Now choose $\alpha > 0$ such that $\alpha < \lambda_{\min}(P)\delta(\gamma)$ and such that $x_t \in \Omega$ implies $k(x_t) \in \mathscr{U}$, $|k(x_t)|^2 < \delta(\gamma)$ and $x(t+\theta) \in \mathscr{X}$ for all $\theta \in [-\tau, 0]$, which is always possible for small enough α. With $\gamma > 0$ chosen such that

$$\gamma \|P\| \left(4 + 2\|K^T K\| + (1 + 2\tau^2 \|K_\tau\|_\tau^2) \rho \frac{\lambda_{\max}(P)}{\lambda_{\min}(P)} \right) \leq \frac{\varepsilon}{2},$$

the condition $\dot{V}(x(t)) \leq -\varepsilon/2 |x(t)|^2$ holds for all $x_t \in \Omega$ satisfying (15). Furthermore, the terminal region Ω is positively invariant when applying the local control law $k(x_t)$.

Property 3 directly implies that V is a Lyapunov-Razumikhin function for the linearized system with control law $u(t) = k(x_t)$, hence the linearized system is asymptotically stable when using this local control law. Consequently, as shown in the proof of Theorem 2, there exists a quadratic cost functional of form (10) and $\alpha > 0$ small enough such that $\dot{E}(x_t) \leq -x(t)^T Q x(t) - u(t)^T R u(t)$ for all $x_t \in \Omega$ by using the local control law $u(t) = k(x_t)$, see (14).

Therefore, the local control law $k(x_t)$ satisfies Condition (7) in Property 2. Furthermore, the terminal region Ω is positively invariant when applying $k(x_t)$ as shown in the first part of the proof. Thus, the assumptions of Theorem 1 are satisfied and asymptotic stability directly follows.   □

6   Brief Discussion of Both Results

In contrast to the delay-free case, the terminal region is not defined as a sublevel set of the quadratic terminal cost functional in both schemes presented in this work. Indeed, such a set might not be positively invariant along trajectories of the nonlinear system. However, several results from nonlinear delay-free systems heavily rely on the definition of the terminal region as sublevel set of the terminal cost, such as unconstrained MPC schemes [3, 9]. Other Razumikhin-based schemes presented in [19, 21] allow such a definition under additional assumptions.

The first scheme presented in Section 4 can be considered as the most general scheme because it only requires a stabilizing linear local control law. However, the terminal region obtained in this general scheme rather complicated. The second scheme presented in Section 5 is more restrictive due to the requirement of a Lyapunov-Razumikhin function, because there exist no converse theorems regarding Lyapunov-Razumikhin functions [2]. However, it yields a much simpler formulation for the terminal region.

Due to these results and the results in [19, 21], there are now three distinct ways to define an approriate terminal region and terminal cost functional based on the Jacobi linearization and a linear local control law. Each scheme contains the results for delay-free systems as special case. However, each scheme possesses some drawbacks which are not encountered in the delay-free case. On the one hand, it is well known that Razumikhin conditions are restrictive in the sense that even if the system is stabilizable, there might not exist an approriate Lyapunov-Razumikhin function. On the other hand, the scheme using only standard quadratic Krasovskii functionals gives a more complicated terminal region which is not a sublevel set of the terminal cost.

7 Conclusions

This paper presents two schemes for calculating stabilizing design parameters based on the Jacobi linearization of the system. In the first part, an arbitrary stabilizing linear local control laws is considered. It is shown that a stabilizable Jacobi linearization implies the existence of a suitable quadratic terminal cost functional and terminal region In the second part, a simpler terminal region is derived using additional Lyapunov-Razumikhin conditions.

Acknowledgements. This work was supported by the DFG Priority Programme 1305 "Control Theory of Digitally Networked Dynamical Systems" and the Cluster of Excellence in Simulation Technology (EXC 310/1) at the University of Stuttgart.

References

1. Chen, H., Allgöwer, F.: A quasi-infinite horizon nonlinear model predictive control scheme with guaranteed stability. Automatica 34(10), 1205–1218 (1998)
2. Gu, K., Kharitonov, V.L., Chen, J.: Stability of Time-Delay Systems. Birkhäuser, Boston (2003)
3. Hu, B., Linnemann, A.: Toward infinite-horizon optimality in nonlinear model predictive control. IEEE Trans. Autom. Control 47(4), 679–682 (2002)
4. Jankovic, M.: Control of nonlinear systems with time-delay. In: Proc. 42nd IEEE Conf. Decision Contr., Maui, HI, USA, pp. 4545–4550 (2003)
5. Kharitonov, V.L., Zhabko, A.P.: Lyapunov-Krasovskii approach to the robust stability analysis of time-delay systems. Automatica 39(1), 15–20 (2003)
6. Kolmanovskii, V., Myshkis, A.: Introduction to the Theory and Applications of Functional Differential Equations. Kluwer Academic Publishers, Dordrecht (1999)
7. Kwon, W.H., Lee, Y.S., Han, S.H.: Receding horizon predictive control for non-linear time-delay systems. In: International Conference on Control, Automation and Systems, Cheju National Univ. Jeju, Korea, pp. 107–111 (2001)
8. Kwon, W.H., Lee, Y.S., Han, S.H.: Receding horizon predictive control for nonlinear time-delay systems with and without input constraints. In: Proc. 6th IFAC Symposium on Dynamics and Control of Process Systems (DYCOPS-6), Jejudo Island, Korea, pp. 277–282 (2001)
9. Limon, D., Alamo, T., Salas, F., Camacho, E.F.: On the stability of constrained MPC without terminal constraint. IEEE Trans. Autom. Control 51(5), 832–836 (2006)
10. Mahboobi Esfanjani, R., Nikravesh, S.K.Y.: Stabilising predictive control of non-linear time-delay systems using control Lyapunov-Krasovskii functionals. IET Control Theory & Applications 3(10), 1395–1400 (2009)
11. Mahboobi Esfanjani, R., Reble, M., Münz, U., Nikravesh, S.K.Y., Allgöwer, F.: Model predictive control of constrained nonlinear time-delay systems. In: Proc. 48th IEEE Conf. Decision Contr. and 28th Chin. Contr. Conf., Shanghai, China, pp. 1324–1329 (2009)
12. Márquez-Martínez, L.A., Moog, C.H.: Input-output feedback linearization of time-delay systems. IEEE Trans. Autom. Control 49(5), 781–785 (2004)
13. Mayne, D.Q., Rawlings, J.B., Rao, C.V., Scokaert, P.O.M.: Constrained model predictive control: stability and optimality. Automatica 26(6), 789–814 (2000)

14. Mazenc, F., Bliman, P.A.: Backstepping design for time-delay nonlinear systems. IEEE Trans. Autom. Control 51(1), 149–154 (2006)
15. Melchor-Aguilar, D., Niculescu, S.I.: Estimates of the attraction region for a class of nonlinear time-delay systems. IMA Journal of Mathematical Control and Information 24, 523–550 (2007)
16. Raff, T., Angrick, C., Findeisen, R., Kim, J.S., Allgöwer, F.: Model predictive control for nonlinear time-delay systems. In: Proc. 7th IFAC Symposium on Nonlinear Systems, Pretoria, South Africa (2007)
17. Raff, T., Huber, S., Nagy, Z.K., Allgöwer, F.: Nonlinear model predictive control of a four tank system: An experimental stability study. In: Proc. IEEE Conf. Control Applications, Munich, Germany, pp. 237–242 (2006)
18. Rawlings, J.B., Mayne, D.Q.: Model Predictive Control: Theory and Design. Nob Hill Publishing, Madison (2009)
19. Reble, M., Allgöwer, F.: General design parameters of model predictive control for non-linear time-delay systems. In: Proc. 49th IEEE Conf. Decision Contr., Atlanta, GA, USA, pp. 176–181 (2010)
20. Reble, M., Allgöwer, F.: Stabilizing design parameters for model predictive control of constrained nonlinear time-delay systems. In: Proc. IFAC Workshop on Time-Delay Systems, Czech Republic, Prague (2010)
21. Reble, M., Brunner, F., Allgöwer, F.: Model predictive control for nonlinear time-delay systems without terminal constraint. Accepted for publication at IFAC World Congress (2011)
22. Reble, M., Mahboobi Esfanjani, R., Nikravesh, S.K.Y., Allgöwer, F.: Model predictive control of constrained nonlinear time-delay systems. IMA Journal of Mathematical Control and Information (2010), doi:10.1093/imamci/dnq029

Part V
Networked Control Systems and Multi-agent Systems

Networked Control under
Time-Synchronization Errors

Alexandre Seuret* and Karl H. Johansson

Abstract. A robust controller is derived for networked control systems with uncertain plant dynamics. The link between the nodes is disturbed by time-varying communication delays, samplings and time-synchronization. A stability criterion for a robust control is presented in terms of LMIs based on Lyapunov-Krasovskii techniques. A second-order system example is considered and the relation between the admissive bounds of the synchronization error and the size of the uncertainties is computed.

1   Introduction

Internet technology appears as a natural and cheap way to ensure the communication link in remotely controlled systems [1, 8, 16]. Today, the available Quality of Service is often good enough for that kind of applications. However, such a communication link constitutes an additional dynamical system, which great influence on stability was already mentioned in the 60's [4]. Indeed, several dynamics and perturbations (communication delay, real-time sampling, packet dropout and synchronization errors) are unavoidably introduced and have to be taken into account during the design of the control/observation loop.

In the literature, many authors assume that the nodes of the NCS are synchronized [8]. However the synchronization is an fundamental issue of NCS since

Alexandre Seuret · Karl H. Johansson
NeCS Team, Automatic Control Department
GIPSA-Lab, UMR CNRS 5216,
Grenoble.
ACCESS Linnaeus Center, Royal Institute of Technology,
Stockholm, Sweden

* This work was supported by the European project FeedNetBack
   (http://www.feednetback.eu/)

R. Sipahi et al. (Eds.): Time Delay Sys.: Methods, Appli. and New Trends, LNCIS 423, pp. 369–381.
springerlink.com   © Springer-Verlag Berlin Heidelberg 2012

ensuring several nodes are synchronized is not easy and some error in it may reduce the performances of the controller [5]. The article focusses on the lake of time-synchronization and provides a robust controller for continuous networked control systems with synchronization error and to parameter uncertainties. A time-delay representation which takes into account the transmission delays, the sampling and the synchronization errors.

Several works on networked controlled systems introduced the question of transmission delays [2]. It is well known that delays generally lead to unstable behavior [10][11]. Moreover in networked control situations, the delays are basically variable (jitter phenomenon) and unpredictable. This is a source of problem when the classical predictor-based controllers are intended to be applied. These techniques generally need the constant delay, $i.e.$ $h_i(t) = h_i$. In the case of variable delays, some researches have used independent-of-delay conditions. Because such i.o.d. conditions may be conservative in general, particular cases such as constant or $symmetric$ delays were considered [3]. These assumptions refers to the case where the transmission delays are equal, $i.e.$ $h_1(t) = h_2(t) = R(t)/2$, where $R(t)$ denotes the round trip time (RTT). Another interesting approach was recently given in [14], which generalized the predictor techniques to the case of variable delays.

Considering unknown time-varying delays and samplings, some stability and stabilization results, [15] have been provided known introducing bounds of the delays and of the sampling interval (h_m, h_M and T such that $0 \le h_m \le h(t) \le h_M$ and such that the difference between two successive sampling instants is less than T), which is not that restrictive. In this paper, the same assumptions are done to ensure the stability of the NCS using a observer-based controller which extends the controller from [9] to the case of time varying delays, synchronization errors and parameter uncertainties.

The present article is organized as follows. Section 2 concerns the problem formulation providing a presentation of the plant and of the communication. Section 3 exposes the control strategy. Section 4 deals with the stability of the controller. An example is provided in Section 5.

2 Preliminaries

The network control problem is described in Fig.1. The plant and the controller are connected through a network which induces additional dynamics. It is assumed that the time synchronization of the process and controller clocks is not achieved. Then the time t_p given by the plant's clock and the time t_c delivered by the controller's clock do not have the same sense. The reference time is given by the plant clock. It means that $t_c = t_p + \varepsilon(t)$ where ε corresponds to a time-varying error of synchronization. The features of the plant and the assumptions on the network are described in the following.

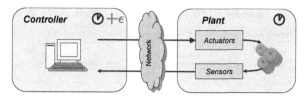

Fig. 1 Plant controller through a network

2.1 Definition of the Plant

Consider the uncertain systems:

$$\dot{x}(t) = (A + \Delta_\gamma A)x(t) + (B + \Delta_\gamma B)u(t),$$
$$y(t) = (C + \Delta_\gamma C)x(t). \tag{1}$$

where $x \in R^n$, $u \in R^m$ and $y \in R^p$ are, respectively, the state, input and output vectors. The constant and known matrices A, B and C correspond to the nominal behavior of the plant. The (time-varying) uncertainties are given in a polytopic representation:

$$\Delta_\gamma A = \gamma \sum_{i=1}^{N} \lambda_i(t)A_i, \ \Delta_\gamma B = \gamma \sum_{i=1}^{N} \lambda_i(t)B_i \ \Delta_\gamma C = \gamma \sum_{i=1}^{N} \lambda_i(t)C_i$$

where N corresponds to the numbers of vertices. The matrices A_i, B_i and C_i are constant and known. The scalar $\gamma \in R$ characterizes the size of the uncertainties. Note that when $\gamma = 0$, no parameter uncertainty is disturbing the system. However the greater the γ, the greater the disturbances. The functions $\lambda_i(.)$ are weighted scalar functions which follow a convexity property, ie. for all $i = 1,..,N$ and for all $t \geq 0$: $\lambda_i(t) \geq 0$ and $\sum_{i=1}^{N} \lambda_i(t) = 1$. It is also assumed that the computation power is low on the plant and its functions are limited to receive control packets, to apply control and to send output measurement data. The computation thus is removed in a centralized controller.

2.2 Synchronization and Delays Models

In addition to parameter uncertainties, the stability of the closed-loop system must be ensured whatever the delays, the possible aperiodicity of the real-time sampling processes and synchronization error. Concerning the transmission delays, the delays are assumed to be non-symmetric but have known minimal and maximal bounds h_m and h_M, so that:

$$\textbf{A1} \ (\textit{maximal allowed delay}): \ h_m \leq h_i(t) \leq h_M. \tag{2}$$

Since we aim at limiting the value of h_m, the use of the User Datagram Protocol (UDP) is preferred to Transmission Control Protocol (TCP), the reliability mechanisms of which may needlessly slow down the feedback loop. Another feature of UDP is that the packets do not always arrive in their chronological emission

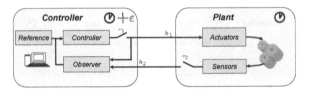

Fig. 2 Architecture of the networked controller

order. The reception function will be added a re-ordering mechanism thanks to some "time-stamps" added in packets. This can be expressed as:

$$\textbf{A2} \; (\textit{packet reordering}): \; \dot{h}_i(t) < 1. \tag{3}$$

Another disturbance implied by the network comes from the samplers and zero-holders. Following the lines of [6], we consider they produce an additional variable delay $t - t_k$, where t_k is the kth sampling instant. Moreover, because of the operating system, the sampling is generally not periodic. So, we only assume there exists a known maximum sampling interval T so that:

$$\textbf{A3} \; (\textit{max. sampling interval}): \; 0 \leq t_{k+1} - t_k \leq T. \tag{4}$$

Assume the function ε is time-varying and there exists a known constant $\bar{\varepsilon}$ such that:

$$\textbf{A4}: \; |\varepsilon(t)| \leq \bar{\varepsilon} \tag{5}$$

3 Observer-Based Networked Control

The system architecture is exposed in Fig. 2. The controller has to estimate present state of plant, using output measurements, and to compute the control value which will be sent to the plant.

D1. **The Control Law:** The controller computes a control law which considers some set-values to be reached. The static state feedback control $u(t) = K\hat{x}(t)$ is defined considering the state estimate $\hat{x}$ given by the observer. The difficulty is to determine a gain K guaranteeing stability despite the delay $\delta_1(t)$.

D2. **Transmission of the Control u:** The kth packets sent by the controller to the process includes the designed control $u(t_{1,k})$ and a sampling time $t_{1,k}$ when it was produced. The plant receives this information at time $t_{1,k}^r$. This time does not have the same meaning for both parts. The term $t_{1,k}^r - t_{1,k}$, corresponding to the transmission delay, corrupted by ε, is estimated by the plant once the packet has reached it.

D3. **Receipt and Processing of the Control Data:** The control, sent at time $t_{1,k}$, is received by the process at time $t_{1,k}^r \geq t_{1,k} + h_m$. There is no raison that the controller also knows the time $t_{1,k}^r$ when the control $u(t_{1,k})$ will be injected into the plant input. Finally, there exists k such that $h_m \leq t_{1,k} \leq h_M + T$ and the process is governed by:

$$\dot{x}(t) = (A + \Delta_\gamma A)x(t) + (B + \Delta_\gamma B)u(t_{1,k}) \tag{6}$$

D4 Transmission of the Output Information: The process have access to its output y only in discrete-time. A packet contains the output $y(t_{2,k'})$ and the sampling time $t_{2,k'}$. The controller receives the output packet at time $t^r_{2,k'}$.

D5 Observation of the Process: For a given $\hat{k}$ and any $t \in [t_{1,\hat{k}} + (h_M - h_m)/2, \ t_{1,\hat{k}+1} + (h_M - h_m)/2[$, there exists a k' such that:

$$\begin{aligned}
\dot{\hat{x}}(t) &= A\hat{x}(t) + Bu(t_{1,\hat{k}} + \varepsilon) - L(y(t_{2,k'}) - \hat{y}(t_{2,k'} - \varepsilon)), \\
\hat{y}(t) &= C\hat{x}(t).
\end{aligned} \tag{7}$$

The design of an observer gain L ensuring stability is not straightforward.

Note that the observation is based on the nominal values of the system definition. No assumption is introduced to estimate the uncertainties and the λ_i functions. The time stamp $t_{1,\hat{k}}$ corresponds to the time where the control input is assumed to be implemented into the plant input. The index k' corresponds to the most recent output information the controller has received. The time $t^r_{1,k}$ and the control $u(t_{1,k})$ (see **D2**) are not known from the observer.

An improve with respect to [13] is that no buffers are required in the controller. This allows considering the input packets as soon as they arrive.

4 Stabilization under Synchronization Error

This section focusses on developing asymptotic stability of the networked control architecture detailed in Fig. 2.

4.1 Closed-Loop System

The input delay approach to sampled-data signals allows a homogenized definition of the delays $\delta_1(t) \triangleq t - t_{1,k}$ where k corresponds to the real sampling implemented in the plant, $\hat{\delta}_1(t) \triangleq t - t_{1,\hat{k}}$ and $\delta_2(t) \triangleq t - t_{2,k'}$ to be considered. The observer dynamics are then driven by:

$$\begin{aligned}
\dot{\hat{x}}(t) &= A\hat{x}(t) + Bu(t - \hat{\delta}_1(t) + \varepsilon) - L(y(t - \delta_2(t)) - \hat{y}(t - \delta_2(t) - \varepsilon)), \\
\hat{y}(t) &= C\hat{x}(t),
\end{aligned} \tag{8}$$

where the features of the system lead to $h_m \leq \delta_i(t) \leq h_M + T$ for $i = 1, 2$. Equivalently, if the average delay $\delta(h_m, h_M, T) = (h_M + T + h_m)/2$ and the maximum delay amplitude $\mu(h_m, h_M, T) = (h_M + T - h_m)/2$ is used, then:

$$\delta - \mu \leq \delta_i(t) \leq \delta + \mu, \quad \forall i = 1, 2. \tag{9}$$

According to (6) and (7) and for given k and any $t \in [t^r_{1,k} + h_m, \ t^r_{1,k+1} + h_m[$, there exist $\hat{k}$ and k' such that the global remote system is governed by:

$$\dot{x}(t) = (A + \Delta_\gamma A)x(t) + (B + \Delta_\gamma B)K\hat{x}(t_{1,k}),$$
$$\hat{x}(t) = A\hat{x}(t) + BK\hat{x}(t_{1,\hat{k}} - \varepsilon) - \Delta_\gamma LCx(t_{2,k'}) - LC(x(t_{2,k'}) - \hat{x}(t_{2,k'} + \varepsilon)). \tag{10}$$

Rewriting the equations with the error $e(t) = x(t) - \hat{x}(t)$, the dynamics become:

$$\dot{x}(t) = (A + \Delta_\gamma A)x(t) + (B + \Delta_\gamma B)K(x(t_{1,k}) - e(t_{1,k}))$$
$$\dot{e}(t) = Ae(t) + LCe(t_{2,k'}) + \Delta Ax(t) + \Delta BK(x(t_{1,k}) - e(t_{1,k})) + \Delta_\gamma LCx(t_{2,k'})$$
$$-BK \int_{t_{1,k}}^{t_{1,\hat{k}}+\varepsilon} [\dot{x}(s) - \dot{e}(s)]ds + LC \int_{t_{2,k'}-\varepsilon}^{t_{2,k'}} [\dot{x}(s) - \dot{e}(s)]ds.$$

Applying the input delay representation [6] yields:

$$\dot{x}(t) = (A + \Delta_\gamma A)x(t) + (B + \Delta_\gamma B)Kx(t - \delta_1) - \Delta_\gamma BKe(t - \delta_1)$$
$$\dot{e}(t) = Ae(t) + \Delta_\gamma Ax(t) + \Delta_\gamma BK(x(t - \delta_1) - e(t - \delta_1)) + L\Delta_\gamma Cx(t - \delta_2) \tag{11}$$
$$+LCe(t - \delta_2) - BK \int_{t-\delta_1}^{t-\hat{\delta}_1+\varepsilon} [\dot{x}(s) - \dot{e}(s)]ds + LC \int_{t-\delta_2-\varepsilon}^{t-\delta_2} [\dot{x}(s) - \dot{e}(s)]ds.$$

with $\delta_1(t) = t - t_{1,k}$ and $\delta_2(t) = t - t_{2,k'}$. From the fact that the communication delays belong to the interval $[h_m, h_M]$ where h_m and h_M are given by the network properties. Then the condition (9) on the delays still holds.

In an ideal case, ie. $\varepsilon = 0$ (from **A2**, synchronized case), the C2P delays are assumed to be well known, ie. $\delta_1(t) = \hat{\delta}_1(t)$ (see [13]) and the model is assumed to be perfectly known and constant ($\gamma = 0$). For this ideal case, Theorem 2 and 3 from [13] deliver controller and observer gains, since the global system is rewritten using the error vector $e(t) = x(t) - \hat{x}(t)$ as:

$$\dot{x}(t) = Ax(t) + BKx(t - \delta_1(t)) - BKe(t - \delta_1(t))$$
$$\dot{e}(t) = Ae(t) + LCe(t - \delta_2(t))$$

4.2 Stability Criteria

It is now accepted that $\delta_1(t) \neq \hat{\delta}_1(t)$ and $\varepsilon \neq 0$. The stability of the controller and of the observer is not ensured anymore by Theorem 2 and 3 in [13], as $\varepsilon \neq 0$ leads error in the delay measurement. As in equation (11), there are interconnection terms between the two variables x and e, a separation principle is no longer applicable to prove the global stabilization. The stability proof requires to consider now both variables simultaneously.

Theorem 1. *For given K and L, suppose that, there exists for q representing the subscript x or e, positive definite matrices : P_{q1}, S_q, R_{qa}, $R_{q\varepsilon}$, S_{xe}, Q_{xe} and R_b and matrices of size $n \times n$: P_{q2}, P_{q3}, Z_{ql} for $l = 1,2,3$, $Y_{ql'}$ for $l' = 1,2$ such that the following LMI's hold :*

$$\begin{bmatrix} \Theta_x^i & \Theta_{x12}^i & \mu P_x^T A_K^i & P_x^T A_K^i & \mu P_x^T A_K^i \\ * & -S_x + 2R_b & 0 & 0 & 0 \\ * & * & -\mu R_{xa} & 0 & 0 \\ * & * & * & -S_{xe} & 0 \\ * & * & * & * & -\mu R_b \end{bmatrix} < 0, \tag{12}$$

$$\begin{bmatrix} \Pi^i & P_e^T \begin{bmatrix} 0 \\ \gamma A_i \\ 0 \end{bmatrix} & \alpha P_e^T \begin{bmatrix} 0 \\ \gamma B_i K \\ 0 \end{bmatrix} & (1+\mu)P_e^T \begin{bmatrix} 0 \\ \gamma L C_i \\ 0 \end{bmatrix} \\ * & -Q_{xe} & 0 & 0 \\ * & * & -\alpha R_b & 0 \\ * & * & * & -(1+\mu)R_b \end{bmatrix} < 0, \qquad (13)$$

$$\begin{bmatrix} R_q & Y_{q1} & Y_{q2} \\ * & Z_{q1} & Z_{q2} \\ * & * & Z_{q3} \end{bmatrix} \geq 0, \quad q \in \{x,e\}, \qquad (14)$$

where $\alpha = (1+2\mu)$, $\beta = 2(\mu + \bar{\varepsilon})$, $P_q = \begin{bmatrix} P_{q1} & 0 \\ P_{q2} & P_{q3} \end{bmatrix}$ and

$$\Pi^i = \begin{bmatrix} \Theta_e & \Theta_{e12}^i & \mu P_e^T A_L & \bar{\varepsilon} P_e^T A_L & \bar{\varepsilon} P_e^T A_L & \beta P_e^T A_K & \beta P_e^T A_K \\ * & -S_e + S_{xe} & 0 & 0 & 0 & 0 & 0 \\ * & * & -\mu R_{ea} & 0 & 0 & 0 & 0 \\ * & * & * & -\bar{\varepsilon} R_{ee} & 0 & 0 & 0 \\ * & * & * & * & -\bar{\varepsilon} R_{xe} & 0 & 0 \\ * & * & * & * & * & -\beta R_{ee} & 0 \\ * & * & * & * & * & * & -\beta R_{xe} \end{bmatrix}$$

$$\Theta_{x12} = P_x^T A_K^i - \begin{bmatrix} Y_{x1}^T \\ Y_{x2}^T \end{bmatrix}, \qquad \Theta_{e12} = P_e^T \begin{bmatrix} 0 \\ LC - \gamma B_i K \end{bmatrix} - \begin{bmatrix} Y_{e1}^T \\ Y_{e2}^T \end{bmatrix},$$

$$\Theta_x^i = \Theta_x^{ni} + \begin{bmatrix} Q_{xe} & 0 \\ 0 & 2\beta R_{xe} + 4\mu R_b \end{bmatrix}, \qquad \Theta_e = \Theta_e^n + \begin{bmatrix} 0 & 0 \\ 0 & 2\beta R_{ee} + 4\mu R_b \end{bmatrix},$$

$$\Theta_x^{ni} = P_x^T \begin{bmatrix} 0 & I \\ \bar{A}_i & -I \end{bmatrix} + \begin{bmatrix} 0 & I \\ \bar{A}_i & -I \end{bmatrix}^T P_x + \begin{bmatrix} S_x + Y_{x1} + Y_{x1}^T + \delta Z_{x1} & Y_{x2} + \delta Z_{x2} \\ * & \delta R_x + 2\mu R_{xa} + \delta Z_{x3} \end{bmatrix},$$

$$\Theta_e^n = P_e^T \begin{bmatrix} 0 & I \\ A & -I \end{bmatrix} + \begin{bmatrix} 0 & I \\ A & -I \end{bmatrix}^T P_e + \begin{bmatrix} S_e + Y_{e1} + Y_{e1}^T + \delta Z_{e1} & Y_{e2} + \delta Z_{e2} \\ * & \delta R_e + 2\mu R_{ea} + \delta Z_{e3} \end{bmatrix},$$

and where $A_K = \begin{bmatrix} 0 \\ BK \end{bmatrix}$, $A_K^i = \begin{bmatrix} 0 \\ B_i K \end{bmatrix}$ and $A_L = \begin{bmatrix} 0 \\ LC \end{bmatrix}$.
Then, the NCS (10) is asymptotic stable.

The proof of Theorem 3 is given in the appendix.

Remark 1. Theorem 1 guarantees the robust stability of the global remote to be guaranteed system with respect to the synchronization error and for observer and controller gains given in [13]. Since the problems of designing observer and controller gains are dual, to develop constructive LMI's is not straightforward.

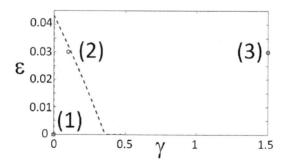

Fig. 3 Maximal synchronization error with respect to the disturbances

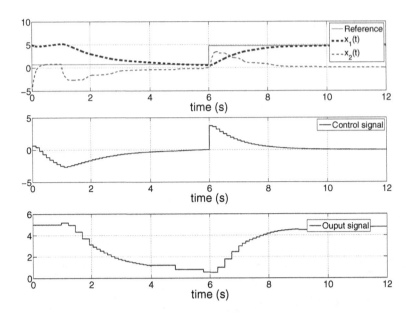

(a) Simulation results for $\gamma = 0.1$ and $\varepsilon = 0.1$ (2)

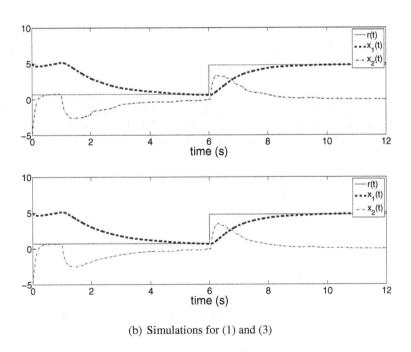

(b) Simulations for (1) and (3)

Fig. 4 Simulation results

5 Application to a Mobile Robot

This study is illustrated on the model of a mobile robot (Slave) which can move in one direction. The identification phase gives the following dynamics:

$$\begin{cases} \dot{x} = \begin{bmatrix} 0 & 1 \\ 0 & -11,32 - \zeta\gamma \end{bmatrix} x + \begin{bmatrix} 0 \\ -11,32 + \zeta\gamma \end{bmatrix} u(t - \delta_1), \\ y = \begin{bmatrix} 1 + \zeta\gamma/10 & 0 \end{bmatrix} x, \end{cases} \tag{15}$$

where the scalar function $\zeta(t)$ lies in $[-1, 1]$ and is taken as $\zeta(t) = sin(6t)$. The characteristics of transmission delays in a classical network (between Lens and Lille in France (50km)) allows $h_m = 0,1s$ and $h_M = 0.4s$. Consider now that the bandwidth of the network allows the sampling period as $T = 0.1s$ to be defined. For these values, Theorems 2 and 3 in [13] produce the following gains $L = \begin{bmatrix} -0.9119 & -0.0726 \end{bmatrix}^T$ and $K = \begin{bmatrix} -0.9125 & -0.0801 \end{bmatrix}$. This gains ensures that, in the ideal case the remote system is α-stable for $\alpha_x = \alpha_e = 1.05$. Theorem 1 ensures that, with these features, the global system is asymptotically stable and robust without any time-varying synchronization error less than $\bar{\varepsilon} = 0.04s$ in (5) for $\gamma = 0$. Figure 3 shows the the maximal admissive $\bar{\varepsilon}$ for greater values of γ. Moreover it guarantees asymptotic stability of the global system without the introduction of a buffer in the controller.

Figure 4(a) shows the simulation results for $\gamma = 0.1$ and $\varepsilon = 0.03$ (point (2) in Figure 3). The state of the process and the sampled input and output are provided. It can be seen that the state convergence to the reference. The stability of the system despite the synchronization error and the parameters uncertainties is ensured.

Figure 4(b) present simulations for $\gamma = 0$ and $\varepsilon = 0$ (point (1)) and for $\gamma = 1.5$ and $\varepsilon = 0.03$ (point (3)). In comparison to Figure 4(a), the results for (1) are closed to the ones obtained for (2). Concerning (3), Theorem 1 does not ensure the stability. However the controller still stabilize the system. It means that the conditions from Theorem 1 are conservative. Further results would investigate in reducing the conservativeness of the stability conditions.

6 Concluding Remarks

This paper presents a strategy for an observer-based control of a networked controlled systems under synchronization erros. No buffering technique was involved, which allows using the available information as soon as received. Various perturbations were dealt with jittery, non-symmetric and unpredictable delays, synchronization error, aperiodic sampling (real-time) and uncertainties in the model. A remaining assumption in [13] which is that the clocks have to be synchronized is not required anymore.

A characteristic feature of this control strategy is to consider that the observer based controller runs in continuous time (i.e., with small computation step) whereas the process provides discrete-time measurements. Thus, the observer keeps on providing a continuous estimation of the current state, even if the data are not sent continuously.

The proposed conditions are conservative. New and less conservative results which guarantee stability of system with sampled-data control recently appears and might help in reducing the conservativeness. It would be interesting to apply these new technics on the present system.

7 Proof of Theorem 1

To analyze the asymptotic stability property of such a system, equations (11) are rewritten by using the descriptor representation [7] with $\bar{x}(t) = col\{x(t), \dot{x}(t)\}$, $\bar{e}(t) = col\{e(t), \dot{e}(t)\}$. In this section, when there is no confusion, any function considered at time 't' will be written without '(t)'. Consider the Lyapunov-Krasovskii (LK) functional:

$$V = V_{xn} + V_{xa} + V_{x\varepsilon} + V_{en} + V_{ea} + V_{e\varepsilon} + V_{xe} \tag{16}$$

where the sub-LK functionals are, for q representing the subscript of the variables 'x' and 'e':

$$\begin{aligned}
V_{qn} &= \bar{q}^T E P_q \bar{q} + \int_{-\delta}^{0} \int_{t+\theta}^{t} \dot{q}^T(s) R_q \dot{q}(s) ds d\theta \\
&\quad + \int_{t-\delta}^{t} q^T(s) S_q q(s) ds, \\
V_{qa} &= \int_{-\mu}^{\mu} \int_{t+\theta-\delta}^{t} \dot{q}^T(s) R_{qa} \dot{q}(s) ds d\theta, \\
V_{q\varepsilon} &= 2 \int_{-\mu-\bar{\varepsilon}}^{\mu+\bar{\varepsilon}} \int_{t+\theta-\delta}^{t} \dot{q}^T(s) R_{q\varepsilon} \dot{q}(s) ds d\theta \\
V_{qb} &= 2 \int_{-\mu}^{\mu} \int_{t+\theta-\delta}^{t} \dot{q}^T(s) R_b \dot{q}(s) ds d\theta
\end{aligned}$$

with $E = diag\{I_n, 0\}$ and P_x, P_e defined in Theorem 1.

The signification of each sub-LK functional has to be explain. The first functionals V_{xn} and V_{en} deal with the stability of the Slave and the observer systems subject to the constant delay δ while V_{xa} and V_{ea} refer to the disturbances due to the delay variations. Even if the functionals do not explicitly depend on each time varying delay, it will be considered both different delays δ_1 and δ_2. The functionals $V_{q\varepsilon}$ are concerned with synchronization errors. The last functionals V_{qb} deals with the interconnection between the variables x and e. Consider as a first step, the polytopic representation of the dynamics in x:

$$\dot{x} = \sum_{i=1}^{N} \lambda_i \{\bar{A}_i x + \bar{B}_i K(x(t-\delta_1) - e(t-\delta_1))\} \tag{17}$$

where $\bar{A}_i = A + \gamma A_i$ and $\bar{B}_i = B + \gamma B_i$. According to Theorem 2 in [12], if LMI (14) holds for $'q = x'$ and for all vertices of the polytopic system, the following inequality holds:

$$\dot{V}_{xn} + \dot{V}_{xa} \leq \sum_{i=1}^{N} \lambda_i \left\{ \xi_x^T \begin{bmatrix} \Psi_{x1}^i & \Theta_{x12}^i \\ * & -S_x \end{bmatrix} \xi_x + \eta_x^i \right\} \tag{18}$$

where $\xi_x = col\{x,\dot{x},x(t-\delta)\}$ and:

$$\eta_x^i = -2\bar{x}^T P_x^T A_K^i e(t-\delta_1), \quad \Psi_{x1}^i = \Theta_x^{ni} + \mu P_x^T A_K^i R_{xa}^{-1} A_K^{iT} P_x.$$

Using the Leibnitz formula and a classical LMI bounding, it yields, for $i = 1,2$:

$$\begin{aligned}\eta_x^i &\leq \bar{x}^T P_x^T A_K^i (S_{xe}^{-1} + \mu R_b^{-1}) A_K^{iT} P_x \bar{x} \\ &+ e^T(t-\delta) S_{xe} e(t-\delta) + |\int_{t-\delta_1}^{t-\delta} \dot{e}^T(s) R_b \dot{e}(s) ds|\end{aligned} \tag{19}$$

where S_{xe} and R_b are positive definite matrices which represent the presence of the error vector in the state equation. Then, the following inequality holds:

$$\begin{aligned}\dot{V}_{xn} + \dot{V}_{xa} &\leq \sum_{i=1}^N \lambda_i \left\{ \xi_x^T \begin{bmatrix} \Psi_{x2}^{ni} & \Theta_{x12}^i \\ * & -S_x \end{bmatrix} \xi_x \right\} \\ &+ e^T(t-\delta) S_{xe} e(t-\delta) + |\int_{t-\delta_1}^{t-\delta} \dot{e}^T(s) R_b \dot{e}(s) ds|,\end{aligned} \tag{20}$$

where $\Psi_{x2}^{ni} = \Theta_x^{ni} + P_x^T A_K^i (S_{xe}^{-1} + \mu R_{xa}^{-1} + \mu R_b^{-1}) A_K^{iT} P_x$. Concerning the errors dynamics, differentiating $V_{en} + V_{ea}$ along the trajectory of (11) and assuming that LMI (14) holds with $q = e$ yields:

$$\begin{aligned}\dot{V}_{en} + \dot{V}_{ea} &\leq \sum_{i=1}^N \lambda_i \{ \xi_e^T \begin{bmatrix} \Psi_{e1} & P_e^T A_L - Y_e^T \\ * & -S_e \end{bmatrix} \xi_e - \eta_{e1}^x \\ &+ \eta_{e1}^e - \eta_{e2}^x + \eta_{e2}^e + \eta_{\Delta A}^{xi} + \eta_{\Delta B}^{xi} + \eta_{\Delta B}^{ei} + \eta_{\Delta C}^{xi} \},\end{aligned} \tag{21}$$

where $\xi_e = col\{e,\dot{e},e(t-\delta)\}$ and where

$$\begin{aligned}\Psi_{e1} &= \Theta_e^n + \mu P_e^T A_L R_{ea}^{-1} A_L^{iT} P_e, & \eta_{e1}^q &= 2\bar{e}^T P_e^T A_K \int_{t1,k}^{t1,\hat{k}+\varepsilon} \dot{q}(s) ds, \\ \eta_{e2}^q &= -2\bar{e}^T P_e^T A_L \int_{t2,k'-\varepsilon}^{t2,k'} \dot{q}(s) ds, & \eta_{\Delta A}^{xi} &= 2\bar{e}^T P_e^T \begin{bmatrix} 0 & \gamma A_i^T \end{bmatrix}^T x, \\ \eta_{\Delta B}^{xi} &= 2\bar{e}^T P_e^T \begin{bmatrix} 0 & \gamma (B_i K)^T \end{bmatrix}^T x(t-\delta_1), & \eta_{\Delta B}^{ei} &= -2\bar{e}^T P_e^T \begin{bmatrix} 0 & \gamma (B_i K)^T \end{bmatrix}^T e(t-\delta_1), \\ \eta_{\Delta C}^{xi} &= 2\bar{e}^T P_e^T \begin{bmatrix} 0 & \gamma (L C_i)^T \end{bmatrix}^T x(t-\delta_1),\end{aligned}$$

with q representing either x or e. Note that the functions η_{ei}^q, for $q = {}$'x','e' and $i = 1,2$ correspond to the disturbance due to the synchronization error. Consider $i = 1$: Noting that from assumption **A4**, inequality $t_{1,\hat{k}} + \varepsilon - t_{1,k} \leq \bar{\varepsilon} + 2\mu$ holds, then a classical bounding leads to:

$$\eta_{q1}^x \leq (\bar{\varepsilon} + 2\mu) \bar{e}^T P_e^T A_K R_{q\varepsilon}^{-1} A_K^T P_e \bar{e} + \int_{t1,k}^{t1,\hat{k}+\varepsilon} \dot{q}^T(s) R_{q\varepsilon} \dot{q}(s) ds. \tag{22}$$

By the same way, the following inequalities hold:

$$\eta_{e2}^q \leq \bar{\varepsilon} \bar{e}^T P_e^T A_L R_{q\varepsilon}^{-1} A_L^T P_e \bar{e} + \int_{t2,k'-\varepsilon}^{t2,k'} \dot{q}^T(s) R_{q\varepsilon} \dot{q}(s) ds. \tag{23}$$

Following the same method as in (19), the following inequalities hold:

$$
\begin{aligned}
\eta_{\Delta A}^{xi} &\le \bar{e}^T P_e^T \begin{bmatrix} 0 \\ \gamma A_i \end{bmatrix} Q_{xe}^{-1} \begin{bmatrix} 0 \\ \gamma A_i \end{bmatrix}^T P_e \bar{e} + x^T Q_{xe} x, \\
\eta_{\Delta B}^{xi} &\le (1+\mu) \bar{e}^T P_e^T \begin{bmatrix} 0 \\ \gamma B_i K \end{bmatrix} R_b^{-1} \begin{bmatrix} 0 \\ \gamma B_i K \end{bmatrix}^T P_e \bar{e} \\
&\quad + x^T(t-\delta) R_b x(t-\delta) + |\int_{t-\delta_1}^{t-\delta} \dot{x}^T(s) R_b \dot{x}(s) ds|, \\
\eta_{\Delta B}^{ei} &\le \mu \bar{e}^T P_e^T \begin{bmatrix} 0 \\ \gamma B_i K \end{bmatrix} R_b^{-1} \begin{bmatrix} 0 \\ \gamma B_i K \end{bmatrix}^T P_e \bar{e} \\
&\quad - 2\bar{e}^T P_e^T \begin{bmatrix} 0 \\ \gamma B_i K \end{bmatrix} e(t-\delta) + |\int_{t-\delta_1}^{t-\delta} \dot{e}^T(s) R_b \dot{e}(s) ds|, \\
\eta_{\Delta C}^{xi} &\le (1+\mu) \bar{e}^T P_e^T \begin{bmatrix} 0 \\ \gamma L C_i \end{bmatrix} R_b^{-1} \begin{bmatrix} 0 \\ \gamma L C_i \end{bmatrix}^T P_e \bar{e} \\
&\quad + x^T(t-\delta) R_b x(t-\delta) + |\int_{t-\delta_2}^{t-\delta} \dot{x}^T(s) R_b \dot{x}(s) ds|.
\end{aligned}
\tag{24}
$$

Finally, the following inequality holds:

$$
\begin{aligned}
\dot{V}_{en} + \dot{V}_{ea} &\le \xi_e^T \begin{bmatrix} \Psi_{e2}^n & \Theta_{12}^{ei} \\ * & -S_e + R_b \end{bmatrix} \xi_e + x^T Q_{xe} x \\
&\quad + 2x^T(t-\delta) R_b x(t-\delta) - 2\bar{e}^T P_e^T \begin{bmatrix} 0 \\ \gamma B_i K \end{bmatrix} e(t-\delta) \\
&\quad + |\int_{t-\delta_2}^{t-\delta} \dot{x}^T(s) R_b \dot{x}(s) ds| + \sum_{q=x,e} \Big\{ |\int_{t-\delta_1}^{t-\delta} \dot{q}^T(s) R_b \dot{q}(s) ds| \\
&\quad + \int_{t_{1,k}}^{t_{1,k}+\varepsilon} \dot{q}^T(s) R_{qp} \dot{q}(s) ds + \int_{t_{2,k'}-\varepsilon}^{t_{2,k'}} \dot{q}^T(s) R_{qp} \dot{q}(s) ds \Big\},
\end{aligned}
\tag{25}
$$

where

$$
\begin{aligned}
\Psi_{e2}^n &= \Theta_e^n + P_e^T A_L (\mu R_{ea} + \bar{e} R_{xe}^{-1} + \bar{e} R_{ee}^{-1})^{-1} A_L^T P_e \\
&\quad + \beta P_e^T A_K (R_{xe}^{-1} + R_{ee}^{-1}) A_K^T P_e + P_e^T \begin{bmatrix} 0 \\ \gamma A_i \end{bmatrix} Q_{xe}^{-1} \begin{bmatrix} 0 \\ \gamma A_i \end{bmatrix}^T P_e \\
&\quad + \alpha P_e^T \begin{bmatrix} 0 \\ \gamma B_i K \end{bmatrix} R_b^{-1} \begin{bmatrix} 0 \\ \gamma B_i K \end{bmatrix}^T P_e + (1+\mu) P_e^T \begin{bmatrix} 0 \\ \gamma L C_i \end{bmatrix} R_b^{-1} \begin{bmatrix} 0 \\ \gamma L C_i \end{bmatrix}^T P_e.
\end{aligned}
$$

Differentiating $V_{x\varepsilon}, V_{e\varepsilon}, V_{xb}$ and V_{eb} leads to:

$$
\begin{aligned}
\dot{V}_{q\varepsilon} &= 2\beta \dot{q}^T R_{q\varepsilon} \dot{q} - 2 \int_{t-\delta-\mu-\bar{e}}^{t-\delta+\mu+\bar{e}} \dot{q}^T(s) R_{x\varepsilon} \dot{q}(s) ds \\
\dot{V}_{qb} &= 4\mu \dot{q}^T R_b \dot{q} - 2 \int_{t-\delta-\mu}^{t-\delta+\mu} \dot{q}^T(s) R_b \dot{q}(s) ds,
\end{aligned}
\tag{26}
$$

Combining (20), (25) and (26) and noting that the sum of the negative integrals in (26) with the integrals from (23) is negative, the following inequality holds:

$$
\dot{V} \le \sum_{i=1}^N \lambda_i \Big\{ \xi_x^T \begin{bmatrix} \Psi_x^i & \Theta_{12}^{xi} \\ * & -S_x + Rex \end{bmatrix} \xi_x + \xi_e^T \begin{bmatrix} \Psi_e & \Theta_{12}^{ei} \\ * & -S_e + S_{xe} \end{bmatrix} \xi_e \Big\}
$$

where $\Psi_x^i = \Psi_{x2}^{ni} + \begin{bmatrix} 0 & 0 \\ 0 & 2\beta R_{xe} + 4\mu R_b \end{bmatrix}$, and $\Psi_e = \Psi_e^n + \begin{bmatrix} 0 & 0 \\ 0 & 2\beta R_{ee} + 4\mu R_b \end{bmatrix}$.

Then the Schur complement leads to the LMI's given in (12) and (13). Then LMI's from Theorem 1 are satisfied, the system (11) is asymptotically stable.

References

1. Abdallah, C.T.: Delay effect in the networked control of mobile robot. In: Chiasson, J., Loiseau, J.-J. (eds.) Application in Time-Delay Systems. LNCIS, vol. 352, Springer, Heidelberg (2007)
2. Azorin, J.M., Reinoso, O., Sabater, J.M., Neco, R.P., Aracil, R.: Dynamic analysis for a teleoparation system with time delay. In: Proceeding of Conference on Control Applications, June 2003, pp. 1170–1175 (2003)
3. Eusebi, A., Melchiorri, C.: Force-Reflecting telemanipulators with Time-delay: Stability Analysis and control design. IEEE trans. on Robotics and Automation 14(4), 635–640 (1998)
4. Ferrel, W.R.: Remote manipulation with transmission delay. IEEE Trans. on Human Factors in Electronics HFE-6, 24–32 (1965)
5. Freris, N.M., Kumar, P.R.: Fundamental Limits on Synchronization of Affine Clocks in Networks. In: Proceedings of the 46th IEEE Conference on Decision and Control, New Orleans, LA, USA, December 12-14 (2007)
6. Fridman, E., Seuret, A., Richard, J.-P.: Robust Sampled-Data Stabilization of Linear Systems: An Input Delay Approach. Automatica 40(8), 1141–1146 (2004)
7. Fridman, E., Shaked, U.: A descriptor system approach to H^∞ control of linear time-delay systems. IEEE Trans. on Automatic Control 47(2), 253–270 (2002)
8. Hespanha, J.P., Naghshtabrizi, P., Xu, Y.: A survey of recent results in networked control systems. Proceedings of the IEEE 95(1), 138–162 (2007)
9. Motestruque, L.A., Antsaklis, P.J.: Stability of model-based networked control system with time-varying transmission time. IEEE Tans. on Automatic Control 49(9), 1562–1572 (2004)
10. Niculescu, S.-I.: Delay Effects on Stability. A Robust Control Approach. Springer, Heidelberg (2001)
11. Richard, J.-P.: Time delay systems: an overview of some recent advances and open problems. Automatica 39(10), 1667–1694 (2003)
12. Seuret, A., Fridman, E., Richard, J.-P.: Sampled-data exponential stabilization of neutral systems with input and state delays. In: IEEE MED, 13th Mediterranean Conference on Control and Automation, Cyprus (June 2005)
13. Seuret, A., Michaut, F., Richard, J.-P., Divoux, T.: Networked Control using GPS Synchronization. In: Minneapolis, U.S. (ed.) American Control Conference, Minneapolis, US (June 2006)
14. Witrant, E., Canudas de Wit, C., Georges, D., Alamir, M.: Remote output stabilization via communication networks with a distributed control law. IEEE Trans. on Automatic Control 52(8), 1480–1485 (2007)
15. Yue, D., Han, Q.-L., Peng, C.: State feedback controller design for networked control systems. IEEE Trans. on Automatic Control 51(11), 640–644 (2004)
16. Zampieri, S.: A survey of recent results in Networked Control Systems. In: Proc. of the 17th IFAC World Congress, Seoul, Korea, July 2008, pp. 2886–2894 (2008)

Modelling and Predictive Congestion Control of TCP Protocols

Rafael C. Melo, Jean-Marie Farines, and Julio E. Normey-Rico

Abstract. This chapter presents the modelling and congestion control of TCP protocols. A nonlinear and a simple linear model are presented to represent TCP including comparative results with NS-2 [1] network'simulator. The simple first order plus dead-time model is selected to design two linear controllers: a generalized predictive controller (GPC) and a PI plus an Smith Predictor controller (SPPI) which are compared to a nonlinear based predictive controller (NBPC). To evaluate the TCP congestion models and controllers, different situations are simulated using NS-2 considering the presence of UDP connections. The obtained results demonstrate that the SPPI offers the best trade-off between complexity and performance to cope with the process dead time and network disturbances.

1 Introduction

The Transmission Control Protocol (TCP) is the most widely used Internet protocol, owing to its reliability skills. Expansion and growth of traffic are the causes of Internet congestion which must be controlled to guarantee the stability of Internet [2].

The TCP connection considers usually a single bottleneck link with a capacity which can depend on time; a buffer is associated to this link and its usual model is

Rafael C. Melo
Department of Systems and Automation (DAS),
Federal University of Santa Catarina (UFSC), Florianopolis/SC - Brazil
e-mail: `rafaelcm@das.ufsc.br`

Jean-Marie Farines
Department of Systems and Automation (DAS),
Federal University of Santa Catarina (UFSC), Florianopolis/SC - Brazil
e-mail: `farines@das.ufsc.br`

Julio E. Normey-Rico
Department of Systems and Automation (DAS),
Federal University of Santa Catarina (UFSC), Florianopolis/SC - Brazil
e-mail: `julio@das.ufsc.br`

R. Sipahi et al. (Eds.): Time Delay Sys.: Methods, Appli. and New Trends, LNCIS 423, pp. 383–393.
springerlink.com   © Springer-Verlag Berlin Heidelberg 2012

based on queue with an integrator representation in order to predict the changes of the congestion window in the sender [3, 5].

In accordance with the classic TCP congestion control defined initially in [4], the sending rate is "clocked" by the received acknowledgments and consequently influenced by the variations on links (forward and reverse) of the TCP connection. Because this "self-clocking" mechanism, the queue length variations due to changes of the window are faster than the integrator can model [5].

In the model representing the relation between the window size and the buffer size, the sending rate is usually assumed as proportional to the window size divided by the round trip delay RTT and the queue is represented by an integrator of the exceeding rate at the link [6]. But this model doesn't take in account the "self-clocking" phenomena where sending rate is controlled by the rate of received acknowledgements. An alternative model, to deal with "self-clocking" is based on an algebraic relation between window and buffer size, reflecting proportionally any window change on buffer queue, with a transmission delay; in this static model, the variations are more abrupt than in the former model [7]. A joint link model, combining both integrator and static model [5] captures in a more accurate way, the queue dynamics when congestion windows changes. This general model deals with N window-based TCP sources, a single bottleneck link and cross-traffic through it, however it does not consider the load in the bottleneck link in an appropriate manner in our point of view. In this chapter we are interested in obtaining a simple model for control purposes that takes into account both, the load and congestion window effect. We consider first the ideas proposed in [5] and we derive a linearized model composed of two first order plus dead time transfer functions, which performs well in all simulated situations.

The classic TCP congestion control is based on a additive-increase and multiplicative - decrease (AIMD) mechanism in order to regulate the size of the congestion window in source. The window size is increased while the network congestion is not reached; as soon as congestion is detected, the mechanism reduces the window, decreasing so the sending rate [4]. This type of control algorithm causes high oscillations of the queue length in the bottleneck link [14]. To improve this behavior, several congestion control strategies based on control theory have been recently presented in literature [5, 8, 9, 10]. In [8] a control strategy is presented based in the Smith Predictor. [9] proposes a control strategy also based in the Smith predictor however including a queue length feedback. In [5] a different control strategy is presented focused on the stability analysis. In [10] the control strategy is based on active queue management (AQM) principle where the control is not performed in the source but in the router.

In this paper we propose a simple PI plus a Smith Predictor controller (SPPI) as the most appropriate solution for the congestion control, as it offers a good trade-off between performance and simplicity. To show that, SPPI is compared to a nonlinear predictor based controller (NPBC) and also to generalized predictive controller (GPC) through simulation comparative results with NS-2 simulator.

This chapter is organized as follows. Section 2 is devoted to the modelling aspects while section 3 presents the proposed control scheme and some comparative simulations. The chapter ends with the conclusions and future work presented in section 4.

2   Modeling TCP Connections

The TCP connection can be seen as a virtual link between the sender and the receiver, including its respective buffers. The following variables, used in the TCP flow control mechanism are available in the sender size: *LBSent* and *LBAcked* which respectively represent the last byte sent and acknowledged and *RWnd* which indicates the free space in the receptor buffer [2].

The congestion control mechanism uses two additional variables: the congestion window (*CWnd*) and a *threshold* value in order to regulate the *CWnd* incremental speed: either exponential or linear. The number of data in transit *LBSent* − *LBAcked* shall be smaller than the minimum between the congestion window *CWnd* and the reception window *RWnd*. The effective window is calculated as following:

$$W = MIN(CWnd, RWnd) - (LBSent - LBAcked) \tag{1}$$

In the present work we assume that the reception buffer is large enough in order to consider the limitation only due to the congestion window (*CWnd*).

Fig. 1 TCP - Dumb-bell Network Topology

To model the dynamics of the TCP connection, for simplicity, the focus here are TCP networks with a single bottleneck link topology (Dumb-bell topology) like the one presented in figure 1 also used in almost all papers. The queue in the bottleneck link is controlled using the *CWnd* of the TCP source and other uncontrolled flow (disturbances) can arrive to this link. The obtained results can be extended to other more complex topologies, however, because of the limited space they are not analyzed here.

2.1   Static Model

On TCP connections, the packets flow from the sender to the receiver and, according to the transmission rate, queues can be observed in the bottleneck link. The queue length depends on the available bandwidth (A_b), the round trip time (*RTT*), the packet size (P_s). The link capacity D is given as $D = A_b \times RTT$ (in Mbits). The queue length present in the bottleneck link depends on the difference between the congestion window *CWnd*, which defines the number of packets that are "on traffic", and the number of packets corresponding to the link capacity (D/P_s).

Considering the topology in figure 1, it will be observed that, when A_b and P_s do not vary, incrementing $CWnd$ of a value of δ packets over D/P_s, the queue length will be increased by δ.

Thus, the static model is given by:

$$P_s \frac{CWnd}{RTT} = A_b \qquad RTT = p + RTT^* \qquad p > 0 \qquad (2)$$

where RTT^* is the minimal value of the RTT and p is the queue time delay.

This is a very simple static model that assumes that the load in the link does not change, as the available bandwidth is given by $A_b = B - x$, where B is the bottleneck link bandwidth and x is the traffic of other flows (which are not being controlled by $CWnd$) in the bottleneck link.

2.2 Dynamical Model

The static model cannot be used for control purposes. To model the dynamic of the queue, some approaches on window based congestion control assumes that the sending rate is proportional to the window size divided by the round trip delay and that the queue behaves like a simple integrator, as in [3]. However, this integrator model does not take into account the self-clocking characteristic of window based schemes, where the sending rate is regulated by the rate of the received acknowledges. As pointed out in [5] this phenomenon can be easily modeled by a static plus dead time model that considers the variations in the queue delay being proportional to the variations in the delayed value of the congestion window. Considering these two aspects a complete model that gives the behavior of the queue $p(t)$ is considered here.

Assuming uniform movements of the packets in the queue, the delay of the queue at time t, $p(t)$, is related to the length of the queue $q(t)$ using $q(t) = Bp(t)$ (in Mbits). If P_s is constant, $q(t) = (B/P_s)p(t)$ (in packets). In this model it is also assumed that the dynamics associated to these movements is negligible. Thus, the dynamical model, that is similar to the one proposed in [5] is:

$$B\frac{dp(t)}{dt} = P_s \frac{CWnd(t-L)}{p(t)+RTT^*} + P_s \frac{dCWnd(t-L)}{dt} - [B - x(t)] \qquad (3)$$

where L is the time spent by the packet to reach the bottleneck link. This model can capture the static and dynamic behavior even when heterogeneous $RTTs$ or non-window based connections like UDP are present. This model can be generalized for N homogeneous and controlled TCP flows.

There are some important points to be considered in this model: (i) it assumes that both L and RTT^* are known; (ii) it is only valid for the situation where the $CWnd$ generates a queue $q(t) > 0$, that is, it is assumed that all the available bandwidth A_b is being used at any time; (iii) the model is open-loop stable. In [5] some open-loop simulations studies using NS-2 were presented to validate this model in the following situations: (a) when $x(t) = 0$ and (b) when $x(t)$ is generated by UDP traffic.

From a control point of view this model has to be considered as a input-output model with $CWnd(t)$ as the manipulated variable, $p(t)$ as the controlled variable and $x(t)$ as the disturbance caused by the flow load. It is important to note that $p(t)$ is not available as the length queue information arrives to the sender L_b times latter. L_b is the time spent between the instant when the packet arrives to the queue link and the instant when the correspondent acknowledge arrives to the source. Thus, a new equation is necessary to define the measured queue delay $p_m(t)$:

$$p_m(t) = p(t - L_b) \tag{4}$$

Note also that $L + L_b = RTT$ and $L_b = RTT^* - L + p(t)$.

2.3   Simple Model for Control Purposes

A linearized model near an operating point will be obtained to be used in the predictor models of the proposed GPC and SPPI controllers. The model considers an operating point given by: $CWnd_0$, p_0 and x_0, respectively, the desired congestion window, queue delay and expected load at the operating point. These values are related by:

$$p_0 = \frac{P_s CWnd_0}{B - x_0} - RTT^* \tag{5}$$

In the Laplace domain the linearized model is given by:

$$\Delta p_m(s) = G_w(s)e^{-RTTs}\Delta CWnd(s) + G_x(s)e^{-L_bs}\Delta x(s)$$

where $\Delta p_m(s), \Delta CWnd(s), \Delta x(s)$ are the incremental variables in the Laplace domain. Using $\tau_1 = RTT_0 = p_0 + RTT^*, \tau_2 = \frac{B}{B-x_0}RTT_0$ follows:

$$G_w(s) = \frac{P_s}{B - x_0}\frac{\tau_1 s + 1}{\tau_2 s + 1} \qquad G_x(s) = \frac{RTT_0}{B - x_0}\frac{1}{\tau_2 s + 1} \tag{6}$$

Note that the disturbance to output dead time (L_b) is smaller that the control to output dead time (RTT).

Moreover, some other hypotheses must be considered for control purposes: (i) in practice L cannot be measured at every sampling time because of cost increasing, as special packets must be sent in order to determine when the packet arrives to the bottleneck link; (ii) the time spent by the packet in the queue is normally smaller than RTT $(RTT^* >> p_0)$, thus, $RTT_0 \cong RTT^*$; (iii) with no load $x_0 = 0$ $G_w(s)$ is a simple gain plus a dead-time model; (iv) $\tau_2 \geq \tau_1 = RTT_0$, that is, the model is dead-time dominant for low values of x_0 and lag-dominant for high values of x_0; (v) the dead-time of the model is almost constant if the packet path doesn't change.

Considering all these model assumptions and some simulation results in NS-2 a simple first order plus dead time model (FOPDT), similar to the one presented on [14], is proposed here to relate the control action to the measured output:

$$\Delta p_m(s) = \frac{K_m}{1 + sT_m} e^{-L_m s} \Delta CW nd(s) \tag{7}$$

where $L_m = RTT_0$, $K_m = \frac{P_s}{B-x_0}$ and $T_m = \tau_2 - 0.5\tau_1$ are chosen using average values of the available bandwidth $A_b = B - x_0$. Note that T_m captures the effect of the "zero" from G_w. As will be shown in next section the controller is able to cope with these model uncertainties. Note that if the packet size does not change and both the queue length and the congestion window are given in packets, the model has unitary static gain.

2.4   Comparative Dynamical Models Simulation Results

Based on the topology presented in figure 1 the presented simple model is compared to the complete non-linear model and to a network simulation using NS-2 [1]. The topology under consideration presents a bottleneck link between the nodes n_1 and n_2 where the system has the lowest bandwidth. In this case the RTT^* is the sum of all transmission times among the links, thus, $RTT^* = 2X(35 + 1 + 1)ms = 74ms$. Assuming that the packet size is 1400 bytes, the link capacity is calculated as follows: $LinkCapacity = (BottleneckBandwidth * RTT^*)/(PacketSize)$, in this case $LinkCapacity \approx 34$ packets. Figure 2 shows a simulation case study when a square-wave signal $CWnd$ is applied to the models and also to NS-2.

To compare the results of both models with the NS-2 simulations, the integral square error (ISE) in percentage was calculated: $ISE_{NonlinearModel} = 0.5883\%$,

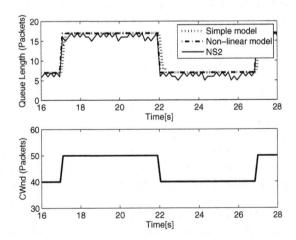

Fig. 2 Comparison between models for packet size equal to 1400bytes

$ISE_{SimpleModel} = 0.8419\%$, which shows that both models capture the network dynamics with similar accuracy. As the simple model approximate quite well the behavior obtained in the NS-2 simulations it can be adopted for controller use in simple controllers, as shown in next section (other simulation tests are given in [13] to show that the model is valid for different packet sizes).

3   Congestion Control

The structure of the proposed controller (SPPI, [11]) is shown in figure 3 and is composed by a Smith predictor with a PI primary controller $C(s) = \frac{K_c(1+sT_i)}{sT_i}$ where the fast model block is given by $G_m(s) = \frac{K_m}{1+sT_m}$. For comparative purposes two other controllers are analyzed. The second controller, NBPC, uses the same predictor structure and PI as the SPPI but a different fast model, which considers the nonlinear model (3) without the delay L_m. The third controller is a simple GPC with unitary control horizon [12] which minimizes the following cost function:

$$J = \sum_{j=1}^{N} [\hat{q}(t+j\,|\,t) - q_r(t+j)]^2 + \lambda\,[\triangle CWnd(t)]^2 \qquad (8)$$

where $\hat{q}(t+j\,|\,t)$ is an optimum j-step ahead prediction of $q(t)$ in data up to time t, N is the prediction horizon, λ is the control weight and $q_r(t+j)$ is the future set-point. The objective of GPC is to compute the incremental control action $\triangle CWnd(t)$ in such a way that the future plant output $q(t+j)$ is driven close to $q_r(t+j)$. The optimal predictions are computed using a discretization of model (7) and a integrated white noise [12].

For the comparisons of the sections the controllers are implemented in the discrete domain (using C code in NS-2 simulator), thus the process models and PI are discretized using a sample time T_s. In addition, the process model is considered adaptive in the sense its parameters are adjusted automatically according the network changes in terms of RTT and bandwidth.

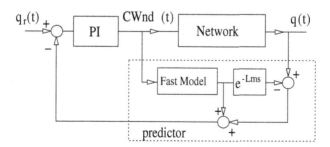

Fig. 3 Controller structure

3.1 Case Study 1: Controller Performance Comparison

The first case study compares the performance of the three controllers with the network topology shown in figure 1 simulated in NS-2 and using a simulation test where at $t = 5s$ a UDP connection starts with a constant bit rate (CBR) of $1Mbps$ (the packet size $P_s = 1400$bytes is constant and both the queue length and the congestion window will be represented in packets). The controllers were tuned with $T_s = 5$ms and using the following parameters: for the PI, $K_c = 0.65$ and $T_i = T_m$; for the GPC, $N = 3$ and $\lambda = 0.8$. This choice was determined using several simulations trying to reduce control complexity without loosing performance.

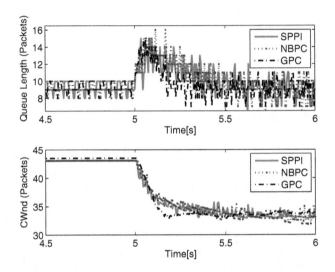

Fig. 4 Case study 1: comparing SPPI, NBPC and GPC with $T_s = 5ms$

The results in figure 4 shows that all the controllers offer similar performance, they are able to compensate the bandwidth variations, caused by the UDP connection, avoiding buffer overflow bringing the queue length back to the desired setpoint. For this simulation, the following results were obtained: $ISE_{SPPI} = 1.38\%$, $ISE_{NBPC} = 1.81\%$ and $ISE_{GPC} = 1.21\%$ calculated in the interval $4.5s \le t \le 6.0s$. All controllers show similar performance on compensating the bandwidth variations, nevertheless the SPPI has better compromise between performance and complexity.

3.2 Case Study 2: Sample Time Effects

The next simulation is related to the influence of the controller sample time. In [15] the authors suggest an approach for selecting the sample time according the round trip time, it means $T_s = RTT^*$, which avoids the need for a dead time compensation, as the obtained model has no delay. In this case a simple PI is enough to control the queue. To compare the performance of the SPPI using $T_s = 5ms$ with the PI using $T_s = RTT^* = 74ms$, a first simulation is presented in figure 5. The set-point is set to 10 packets at $t = 0s$ and at $t = 3s$ an UDP connection starts with $CBR = 1Mbps$. The ISE gives for these cases: $ISE_{SPPI_5} = 1.17\%$ and $ISE_{PI_{74}} = 2.51\%$ calculated in the interval $2.5s \leq t \leq 4.0s$. As can be seen with $T_s = 5ms$ the controller reacts a bit faster to the disturbances although with $T_s = RTT^*$ the controller is also able to compensate the variation caused by the UDP connection. However the overshoot obtained with the PI controller is approximately 8 packets while with the SPPI is approximately 4 packets.

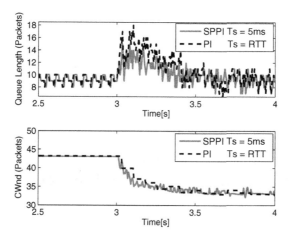

Fig. 5 Effect of T_s on performance

As with $T_s = RTT$ the dynamic response exhibits higher overshoot, under worse operational conditions, the controller can drive the system to a packet loss. This effect can be observed on figure 6 where the bottleneck buffer in this case is limited to 17 packets. In this case once the buffer overflows after the UDP starts, packets are lost and the transmission stops approximately at $t \approx 3.2s$ and restarts $100ms$ later. Note that the PI closed-loop system takes approximately $1s$ to reach again the set-point, causing loss of performance.

3.3 Case Study 3: RTT Variation Effects

As SPPI controller can be considered the more appropriate solution considering its simplicity and performance, a simulation is performed to test the closed-loop system under RTT variations. RTT variations simulates a soft-handoff in a mobility management scheme [16], where the receiver node moves from one network to another one and the packets are tunneled to another queue location, therefore changing the delay of the route. In this simulation the bottleneck bandwidth is not modified and

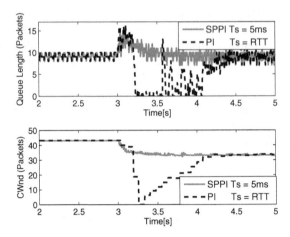

Fig. 6 Packet loss effect caused by the use $T_s = RTT^* = 74ms$ in the PI controller with bottleneck buffer limited to 17 packets

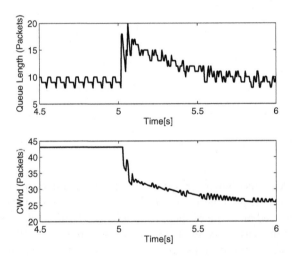

Fig. 7 Effect of RTT variation on the proposed SPPI with $T_s = 5ms$

only the bottleneck delay is changed from $35ms$ to $15ms$ at $t = 5s$. The results presented on figure 7 show the qualities of the proposed simple SPPI controller, which is also able to compensate RTT network variations.

4 Conclusion

This work has analyzed the modelling and predictive congestion control of TCP protocols. Using NS-2 network simulations it was shown that a first order plus dead time model (FOPDT) can be used with a good compromise between simplicity and accuracy to represent the network behavior. This model was used in a simple SPPI controller to solve the congestion control problem. The obtained closed-loop time responses are compatible with the network dynamics, avoiding buffer overflow and keeping the bandwidth usage near the maximum available limit. When compared to more complex solutions, GPC and NBPC, the proposed SPPI showed similar performance. Therefore, considering SPPI simplicity, it is a promising strategy to be used in real time congestion control applications. Among the good results obtained in this case the proposed approach has the additional advantage that the model and controller can be generalized for the case of multiple senders and receivers, and also deals with mobility case.

References

1. NS-2, Network Simulator (2008), http://www.isi.edu/nsnam/ns/
2. Kurose, J.F., Ross, K.W.: Computer Networking - A Top-Down Approach Featuring the Internet, 3rd edn. Addison-Wesley (2000)
3. Baiocchi, A., Vacirca, F.: TCP fluid modeling with a variable capacity bottleneck link. In: IEEE Infocom., vol. 1 (2007)
4. Jacobson, V., Karels, M.: Congestion Avoidance and Control. ACM Computer Communications Review 18(4), 314–329 (1988)
5. Tang, A., Jacobsson, K., Andrew, L.L.H., Low, S.H.: Accurate Link Model and Its Application to Stability Analysis of FAST TCP. In: IEEE Infocom., vol. 1 (2007)
6. Hollot, C., Misra, V., Towsley, D., Gong, W.: A control theoretic analysis of RED. In: IEEE Infocom., vol. 1 (2001)
7. Wang, J., Wei, D.X., Low, S.H.: Modeling and stability of FAST TCP. In: Wireless Communications. IMA Volumes in Mathematics and its Applications, vol. 143, Springer Science (2006)
8. Mascolo, S.: Modeling the Internet Congestion Control Using a Smith Controller with Input Shaping. In: IFAC (2003)
9. Gerla, M.: Generalized Window Advertising for TCP Congestion Control, number 990012, 21 p (1999)
10. Hollot, C., Misra, V., Towsley, D., Gong, W.: Analysis and Design of Controllers for AQM Routers Supporting TCP Flows. IEEE/ACM Trans. Automatic Control 47(6), 945–959 (2002)
11. Normey-Rico, J.E., Bordons, C., Camacho, E.F.: Improving the Robustness of Dead-Time Compensating PI Controllers. Control Engineering Practice 5(6), 801–810 (1997)
12. Camacho, E.F., Bordons, C.: Model Predictive Control. Springer, Berlin (2004)
13. Melo, R.C., Normey-Rico, J.E., Farines, J.M.: TCP Modelling and Predictive Congestion Control. In: IFAC (2009)
14. Melo, R.C., Normey-Rico, J.E., Farines, J.M.: Modelagem e controle de congestionamento em conexões TCP. In: Brazilian Control Conference (2008)
15. Haeri, M., Rad, A.H.M.: Adaptive model predictive TCP delay-based congestion control. Computer Communications 29 (2006)
16. Eddy, W.M., Swami, Y.P.: Adapting End Host Congestion Control for Mobility. NASA/CR-2005-213838 (2005)

Dependence of Delay Margin on Network Topology: Single Delay Case

Wei Qiao and Rifat Sipahi

Abstract. This article studies the indirect relationship between the delay margin τ^* of coupled systems and different graphs $\mathscr{G}$ these systems form via their different topologies. A four-agent linear time invariant (LTI) consensus dynamics is taken as a benchmark problem with a single delay τ and second-order agent dynamics. In this problem, six different topologies exist without disconnecting an agent from all others. We first start with CTCR to reveal the delay margin τ^*. We next investigate how τ^* is affected as one graph transitions to another when some links between the agents weaken and vanish. This line of research is recently growing and new results along these lines promise delay-tolerant topology design for coupled dynamical systems with delays.

1 Introduction

Studying the dynamics of coupled systems has impacts on multi-agent problems in formation control, sensor networks, mobile robots, autonomous vehicles, automated highway systems and vehicular traffic follow, [1, 9, 17, 19, 22, 24, 25, 26, 34]. One of the crucial problems in these applications is the group agreement (consensus), which can be achieved with the awareness of the group members about the states of the other members, combined with an appropriate control law. A well-known

Wei Qiao

Northeastern University, Department of Mechanical and Industrial Engineering,
Boston, MA 02115 USA
e-mail: qiao.w@husky.neu.edu

Rifat Sipahi

Northeastern University, Department of Mechanical and Industrial Engineering,
Boston, MA 02115 USA
e-mail: rifat@coe.neu.edu

R. Sipahi et al. (Eds.): Time Delay Sys.: Methods, Appli. and New Trends, LNCIS 423, pp. 395–405.
springerlink.com   © Springer-Verlag Berlin Heidelberg 2012

consensus problem arises, for instance, in follow-the-leader traffic flow, where each driver makes decisions to maintain a constant headway and to match the vehicle speed to that of the preceding vehicle, [11, 23].

When multiple dynamics couple each other, the scale of the problem grows, but more importantly coupling pattern (network topology) also plays a role in determining the overall dynamic behavior. This coupling pattern, that is, the network topology with graph $\mathscr{G}$ is the first focus in this paper. Often, when dynamics couple via a network, information sharing across the network becomes *delayed* due to many reasons, including communication lines, actuation, and decision-making [14]. In such a scenario, instantaneous information is not available due to delays τ. The presence of delays, which is a well-known source of instability, forms the second focus of this paper.

The problem studied here is on the stability of coupled systems with $\mathscr{G}$ in the presence of delays in the communication lines among the systems. More specifically, we are interested in finding the maximum amount of delay τ^* that a particular network with graph $\mathscr{G}$ is tolerant without losing stability. Delay τ^* is expected to be different for different $\mathscr{G}$, and we seek to reveal how τ^* relates to different network topologies $\tau^* = \tau^*(\mathscr{G})$. To demonstrate the results, linear time-invariant (LTI) consensus dynamics with a single delay is taken as a benchmark problem.

What is unique in this paper is that we explore the relationship between two indirectly related concepts; the network topology $\mathscr{G}$ and stability margin τ^*. There exist many studies that focus on addressing τ^* [10, 12, 15, 30], however only few work attempt to connect $\mathscr{G}$ and τ^* [26, 27, 31, 33]. This article complements the existing work by considering higher order dynamics, damping effects, and topology transitions. To achieve these non-trivial tasks without overwhelming the capacities of computational tools, we start with a LTI four-agent consensus problem, which can exhibit six different topologies $\mathscr{G}_1, \ldots, \mathscr{G}_6$ without disconnecting any agent from the remaining ones. We then revisit CTCR [30] to reveal the τ^* for each $\mathscr{G}$. Next, we study the effects of damping to τ^* and report the results with comparisons. Based on the results, we also interpret how τ^* is affected as one transitions from one graph to another.

The text is organized as follows. We first present the preliminaries related to mathematical modeling of coupled dynamics with delays. Next, the stability analysis and arising results are presented. Future research directions and conclusions end the paper.

Notations are standard. We use $\mathbb{C}_+$, $\mathbb{C}_-$, $j\mathbb{R}$ for right half, left half and the imaginary axis of the complex plane, respectively. $\mathbb{R}$ represents the set of real numbers and j is the complex number. Matrices, vectors and sets are denoted by bold face, while the scalar entities are with normal font. The vector $\mathbf{v} \in \mathbb{R}_+^L$ defines an L-dimensional vector with positive real entries.

2 Preliminaries

2.1 Linear-Time Invariant Model with Delays

A commonly studied continuous-time LTI consensus dynamics is given by

$$\ddot{x}_k(t) = \sum_{\mu=1,\mu\neq k}^{4} \alpha_{\mu k}\left(x_\mu(t-\tau)-x_k(t)\right) - \beta\dot{x}_k(t), \tag{1}$$

where, $x_k(t)$ and $\dot{x}_k(t)$ are the states of agent k, $k=1\ldots4$, $\tau>0$ is the delay, the first term on the right-hand side of (1) is the controller with $\alpha_{\mu k}\in\mathbb{R}_+$ and μ is the subscript representing the agents sharing their state information with agent k based on the topology ($\mathscr{G}$) that determines which $\alpha_{\mu k}$ is zero or non-zero. Eq.(1) represents the dynamics of four agents, where $\beta\neq0$ represents the damping effects, and each agent is aware of its state instantaneously, but only delayed states of the remaining agents.

It is straightforward to re-write Eq. (1) in matrix form as

$$\ddot{\mathbf{x}}(t) = \mathbf{A}\mathbf{x}(t) + \mathbf{B}\mathbf{x}(t)(t-\tau) - \mathbf{C}\dot{\mathbf{x}}(t), \tag{2}$$

where $\mathbf{x}(t) = (x_1(t)\ldots,x_n(t))^T \in \mathbb{R}^{4\times1}$, $n=4$ is the number of states, and $\mathbf{A},\mathbf{B},\mathbf{C} \in \mathbb{R}^{4\times4}$ are matrices with constant real entries. If the states of the agents converge to each other as time elapses, synchronization is said to be achieved [19, 24].

The delay differential equation (1) is not new, and has been studied extensively in various forms including string stability in vehicular traffic flow, time-variant effects, switching topology, single delay case and topology dependency [2, 19, 20, 21, 23, 24, 26, 29]. On the other hand, the non-conservative stability features of (2) in connection with $\mathscr{G}$ are open problems.

Remark 1. Note that the system in (2) has an invariant pole $s=0$ in $\mathbb{C}$ for any delay τ. As we confirmed, this pole, which creates the rigid body dynamics of the agents, does not invite any additional $s=0$ poles for some $\tau\geq0$, and thus it does not play any role on the stability of (2) for $\tau>0$. Consequently, we focus only on the stability of the remaining infinitely many dynamic modes of (2). If these modes are asymptotically stable, then we call Eq.(2) asymptotically stable disregarding the rigid body dynamics.

2.2 Stability Analysis

The characteristic equation of (2) is given by

$$f(s,\tau,\alpha_{\mu k}) = \det[s^2\mathbf{I} + s\mathbf{C} - \mathbf{A} - \mathbf{B}\,e^{-\tau s}] = 0, \tag{3}$$

where $s \in \mathbb{C}$ is the Laplace variable and $\mathbf{I}$ is the identity matrix. In order to study the asymptotic stability of (2) with respect $\alpha_{\mu k}$ and τ, one needs to investigate the

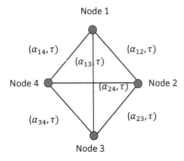

Fig. 1 Four-agents forming six undirected links.

location of the roots of (3), $s = s(\alpha_{\mu k}, \tau)$, on the complex plane $\mathbb{C}$. As per stability condition, for any given τ and $\alpha_{\mu k}$, the dynamics in (2) is asymptotically stable if and only if all the infinitely many roots of (3) are in $\mathbb{C}_-$ [10]. Clearly, checking the stability condition in the entire s domain is a prohibitively difficult task.

In this study, we deploy CTCR [30] in order to solve for the delay margin τ^*, that is, the largest delay value that the system in (2) can withstand without loosing stability. For problems with multiple delays, we also refer the readers to ACFS. For conciseness, ACFS and CTCR are not discussed here, and the readers are referred to the citations.

3 Numerical Results

We now consider a four-agent problem where each agent is a node and exhibits the dynamics defined in (1), and each node is coupled with undirected communication lines to the other nodes. The fully connected graph $\mathscr{G}_1$ is the topology we start with, where each link between the nodes are represented by a coupling strength $\alpha_{\mu k}$ (or a controller gain) and a communication delay τ.

Notice that six different topologies with graphs $\mathscr{G}_p$, $p = 1, \ldots, 6$ arise by connecting the four nodes in Figure 1 in different ways, without disconnecting a node from all others, Figure 2. We now quest what the delay margin τ^* is for each one of the topologies. This can be done by setting some of the $\alpha_{\mu k}$ coupling strengths to zero accordingly, as explained next.

3.1 Dependence of Delay Margin on Topology - Weak Damping Case

In order to reveal the relationship between τ and $\mathscr{G}$ without the interference of other parameters, we assume that agents have identical coupling strengths, $\alpha_{\mu k} = \alpha$, and $\beta = 0.01$. For the graph $\mathscr{G}_1$ in Figure 1, the system matrices in (2) are given by

$$\mathbf{A}_1 = -3\alpha\mathbf{I}, \ \mathbf{B}_1 = \alpha(\mathbf{1} - \mathbf{I}), \ \mathbf{C}_1 = \beta\mathbf{I},$$

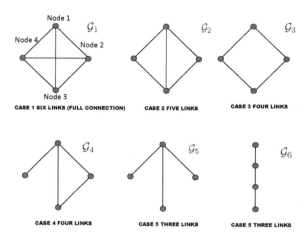

Fig. 2 Six different configurations of topology.

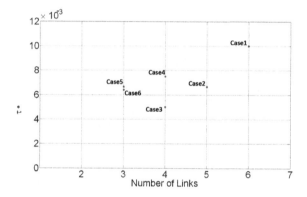

Fig. 3 Delay margin τ^* with respect to Case p for topology graph $\mathscr{G}_p$ in Figure 2. $\beta = 0.01$

with compatible dimensions, and where **1** is a matrix with all entries unity. Matrices corresponding to other graphs can be similarly obtained and are thus suppressed.

We next compute τ^* for each topology in Figure 2. The results are shown in Figure 3. Inspecting Figure 3 results that the fully connected $\mathscr{G}_1$ exhibits the largest delay margin among all. Interestingly, $\mathscr{G}_3$ with four links has smaller τ^* than $\mathscr{G}_6$ that has three links. This indicates that a network with missing connections does not always mean that the dynamics it couples with will exhibit smaller τ^*, but in the contrary, in some cases, the topology induces larger stability margins by losing a connection. This observation is valid for both transitions $\mathscr{G}_2 \rightarrow \mathscr{G}_4$ and $\mathscr{G}_3 \rightarrow \mathscr{G}_6$.

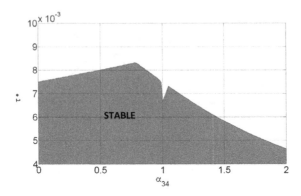

Fig. 4 Delay margin τ^* with respect to α_{34} in $\mathscr{G}_2 \to \mathscr{G}_4$ as $\alpha_{34} \to 0$. Shaded region below τ^* corresponds to stable dynamics. $\beta = 0.01$

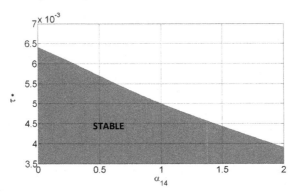

Fig. 5 Delay margin τ^* with respect to α_{14} in $\mathscr{G}_3 \to \mathscr{G}_6$ as $\alpha_{14} \to 0$. Shaded region below τ^* corresponds to stable dynamics. $\beta = 0.01$

3.2 Effects of Topology Transition on Stability

3.2.1 Stability Transition When $\mathscr{G}_2 \to \mathscr{G}_4$

Discovered from Figure 3, the topology transition from $\mathscr{G}_2$ to $\mathscr{G}_4$ gains larger stability margin. The coupling strength α_{34} is responsible for this transition, and we investigate its effects to the stability margin, see Figure 4 for results.

Furthermore, we notice that transition from $\mathscr{G}_2$ to $\mathscr{G}_4$ (by losing one connection) enhances stability margin, contrary to the transition from $\mathscr{G}_1$ to $\mathscr{G}_2$. In both transitions, we reveal significantly high sensitivity in τ^* when the responsible coupling strength becomes unity, see also Figure 6.

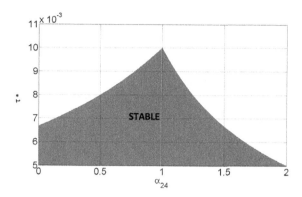

Fig. 6 Delay margin τ^* with respect to α_{24} in $\mathcal{G}_1 \rightarrow \mathcal{G}_2$ as α_{24}. Shaded region below τ^* corresponds to stable dynamics. $\beta = 0.01$

3.2.2 Stability Transition When $\mathcal{G}_3 \rightarrow \mathcal{G}_6$

Topology transition from $\mathcal{G}_3$ to $\mathcal{G}_6$ while losing one link also improves τ^*. The stability transition with α_{14} sweeping from 0 to 2 is captured in Figure 5. Similar to the transition from $\mathcal{G}_2$ to $\mathcal{G}_4$, losing one link yields larger τ^* and ultimately larger stable regions. Different from $\mathcal{G}_2 \rightarrow \mathcal{G}_4$ transition, we do not observe any high sensitivity with respect to the responsible coupling strength α_{14}, and the system monotonically loses stability region as α_{14} increases from zero to two.

3.2.3 Stability Transition When $\mathcal{G}_1 \rightarrow \mathcal{G}_2$

The transition from $\mathcal{G}_1$ to $\mathcal{G}_2$ occurs when α_{24} sweeps from $\alpha_{24} = 1$ to $\alpha_{24} = 0$, see Figure 2. The stability margin during this transition is shown in Figure 6. In Figure 6, as α_{24} approaches unity (thus graph $\mathcal{G}_1$ is recovered), τ^* becomes sensitive to changes in α_{24}, and a small change in α_{24} could result in a significant change of the shaded stability region. This interesting feature also occurs in the transition from $\mathcal{G}_2$ to $\mathcal{G}_4$ as α_{34} approaches unity, see Figure 4.

3.3 Stability Analysis with Respect to Damping

In the previous sections, damping constant is chosen as $\beta = 0.01$. In this section, the effects of β to τ^* are inspected by choosing β as 0.1, 0.2, 0.3. We then repeat the same analysis as in the previous section to investigate the damping effects.

3.3.1 Damping Effects to Transition in $\mathcal{G}_2$

Figure 7 shows the transition of $\mathcal{G}_2$ with respect to α_{34} for different β values, we see in this figure that sensitivity around $\alpha_{34} \approx 1$ is decreasing as β gets larger. Moreover,

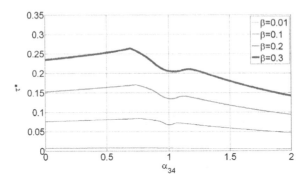

Fig. 7 Damping effect to delay margin as α_{34} changes in $\mathscr{G}_2$.

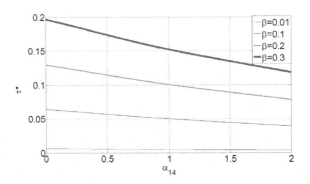

Fig. 8 Damping effects to delay margin as α_{14} changes in $\mathscr{G}_3$.

we see that stable region below the respective curves enlarges as β increases, and the singularity point remains around $\alpha_{34} \approx 0.78$.

3.3.2 Damping Effects in $\mathscr{G}_1$ and $\mathscr{G}_3$

Similar to $\mathscr{G}_2$, stable region enlarges with increasing β in $\mathscr{G}_1$ and $\mathscr{G}_3$, see Figure 8 and Figure 9, respectively.

Remark 2. In many examples in the literature, it is shown that damping enhances the stability and enlarges the stability regions in delay parameter space. In our calculations for the three specific topologies investigated, we have revealed similar results that damping increases the delay margin. We finally note that the stability analysis results presented in this article are also confirmed by TRACE-DDE software [3].

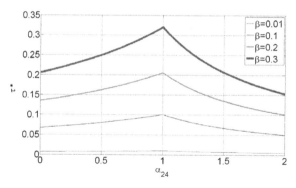

Fig. 9 Damping effect to transition with respect to α_{24} in $\mathscr{G}_1$.

4 Conclusion

The effects of different topologies of coupled systems to the largest delay, that is, the delay margin τ^*, these systems can withstand is studied via CTCR. We present the results on a linear time-invariant four-agent consensus dynamics in which agents become aware of each other's states only after a delay τ. Some interesting and counter-intuitive results are also captured. The most interesting result is that in some cases topologies with less number of links among the agents have larger delay margins compared to topologies with more links. We find that this claim holds true even when a topology transits to another while a responsible link weakens and eventually vanishes. We finally note that the stability margin of some topologies become extremely sensitive with respect to the coupling strengths of some links between the agents. Future work along these lines include the consideration of larger scale problems with multiple delays, as well as the design of delay-independent topologies using algebraic tools.

Acknowledgements. This research has been supported in part by the award from the National Science Foundation ECCS 0901442 and Dr. R. Sipahi's start-up funds available at Northeastern University. A part of this study was presented at IFAC workshop on Time Delay System held in Prague, Czech Republic, 2010.

References

1. Beard, W.B., McLain, T.W., Nelson, D.B.: Decentralized Cooperative Aerial Surveillance Using Fixed-Wing Miniature UAVs. Proc. IEEE 94, 1306–1324 (2006)
2. Bose, A., Ioannou, P.A.: Analysis of Traffic Flow with Mixed Manual and Semiautomated Vehicles. IEEE Tran. Intell. Transp. Sys. 4, 173–188 (2003)
3. Breda, D., Maset, S., Vermiglio, R.: Pseudospectral Differencing Methods for Characteristic Roots of Delay Differential Equations. SIAM Journal of Scientific Computing 27, 482–495 (2006)

4. Chen, J., Latchman, H.A.: Frequency sweeping tests for asymptotic stability independent of delay. IEEE Transactions on Automatic Control 40, 1640–1645 (1995)
5. Chen, J., Gu, G., Nett, C.N.: A new method for computing delay margins for stability of linear delay systems. Systems and Control Letters 26, 107–117 (1995)
6. Cooke, K.L., Driessche, P.V.D.: On zeros of some transcendental equations. Funkcialaj Ekvacioj 29, 77–90 (1986)
7. Collins, G.E.: The calculation of multivariate polynomial resultants. Journal of the Association for Computing Machinery 18, 515–532 (1971)
8. Datko, R.: A procedure for determination of the exponential stability of certain differential-difference equations. Quart. Appl. Math. 36, 279–292 (1978)
9. Delice, I.I., Sipahi, R.: Advanced Clustering with Frequency Sweeping(ACFS) Methodology for the Stability. IEEE Transactions on Automatic Control 56, 467–472 (2011)
10. Gu, K., Niculescu, S.-I., Chen, J.: On Stability Crossing Curves for General Systems with Two Delays. J. Math. Anal. Appl. 311, 231–253 (2005)
11. Helbing, D.: Traffic and Related Self-Driven Many-Particle Systems. Rev Mod Phy 73, 1067–1141 (2001)
12. Jarlebring, E.: Computing the Stability Region in Delay-Space of a TDS Using Polynomial Eigenproblems. In: 6th IFAC Workshop on Time-Delay Systems, L'Aquila, Italy (2006)
13. Louisell, J.: A matrix method for determining the imaginary axis eigenvalues of a delay system. IEEE Transactions on Automatic Control 46, 2008–2012 (2001)
14. Loiseau, J.J., Michiels, W., Niculescu, S.-I., Sipahi, R.: Topics in Time Delay Systems Analysis Algorithms and Control. Springer (2009)
15. Michiels, W., Niculescu, S.-I.: Stability and Stabilization of Time-Delay Systems: An Eigenvalue-Based Approach. SIAM Advances in Design and Control 12 (2007)
16. Moreau, L., Ghent, S.: Stability of Continuous-Time Distributed Consensus Algorithms. In: 43rd IEEE Conference on Decision and Control, Atlantis Paradise Island, Bahamas (2004)
17. Münz, U., Papachristodoulou, A., Allgöwer, F.: Consensus Reaching in Multi-Agent Packet-Switched Networks with Nonlinear Coupling. International Journal of Control 82, 953–969 (2009)
18. Olfati-Saber, R.: Flocking for Multi-Agent Dynamic Systems: Algorithms and Theory. IEEE Trans. Aut. Cont. 51, 401–420 (2006)
19. Olfati-Saber, R., Murray, R.M.: Consensus Problems in Networks of Agents with Switching Topology and Time-Delays. IEEE Trans. Aut. Cont. 49, 1520–1533 (2004)
20. Olfati-Saber, R., Murray, R.M.: Consensus Protocols for Networks of Dynamic Agents. In: American Control Conference Denver, Colarado (2003)
21. Olfati-Saber, R., Murray, R.M.: Agreement Problems in Networks with Directed Graphs and Switching Topology. In: Conference on Decision and Control Maui, Hawaii (2003)
22. Olgac, N., Ergenc, A.F., Sipahi, R.: Delay Scheduling: A New Concept for Stabilization in Multiple Delay Systems. Journal of Vibration and Control 11, 1159–1172 (2005)
23. Orosz, G., Krauskopf, B., Wilson, R.E.: Bifurcations and multiple traffic jams in a car-following model with reaction time delay. Phys. D 211, 277–293 (2005)
24. Ren, W., Beard, R.W., Atkins, E.M.: A Survey of Consensus Problems in Multi-agent Coordination. In: American Control Conference Portland, OR (2005)
25. Ren, W., Moore, K.L., Chen, Y.: High-order and Model Reference Consensus Algorithms in Cooperative Control of MultiVehicle Systems. ASME J. Dyn. Sys. Meas. Cont. 129, 678–688 (2007)

26. Schöllig, A., Münz, U., Allgöwer, F.: Topology-Dependent Stability of a Network of Dynamical Systems with Communication Delays. In: Proceedings of the European Control Conference Kos, Greece (2007)

27. Sipahi, R., Acar, A.: Stability Analysis of Three-Agent Consensus Dynamics with Fixed Topology and Three Non-Identical Delays. In: ASME Dynamic Systems and Control Conference Ann (October 2008)

28. Sipahi, R., Olgac, N.: A Comparison of Methods Solving the Imaginary Characteristic Roots of LTI Time Delayed Systems. SIAM J. Cont. Optim. 45, 1680–1696 (2006)

29. Sipahi, R., Niculescu, S.-I., Atay, F.: Effects of Short-Term Memory of Drivers on Stability Interpretations of Traffic Flow Dynamics. In: American Control Conference New York (2007)

30. Sipahi, R., Olgac, N.: Complete Stability Map of Third Order LTI, Multiple Time-Delay Systems. Automatica 41, 1413–1422 (2005)

31. Sipahi, R., Qiao, W.: Responsible Eigenvalue Concept for the Stability of a Class of Single-Delay Consensus Dynamics with Fixed Topology. IET Control Theory and Applications (accepted and in print, 2011)

32. Stepan, G.: Retarded Dynamical Systems: Stability and Characteristic Functions. ser. Pitman Research Notes in Mathematics Series. Longman Scientific and Technical, copublisher John Wiley and Sons, Inc., New York (1989)

33. Qiao, W., Sipahi, R.: Responsible Eigenvalue Approach For Stability Analysis And Control Design Of A Single-Delay Large-Scale System With Random Coupling Strengths. In: ASME 3rd Dynamic Systems and Control Conference, Cambridge, MA (2010)

34. Zou, Y., Pagilla, P.R., Misawa, E.: Formation of a Group of Vehicles with Full Information Using Constraint Forces. ASME J. Dyn. Sys. Meas. Cont. 129, 654–661 (2007)

Consensus in Networks under Transmission Delays and the Normalized Laplacian

Fatihcan M. Atay

Abstract. We study the consensus problem on directed and weighted networks in the presence of time delays. We focus on information transmission delays, as opposed to information processing delays, so that each node of the network compares its current state to the past states of its neighbors. The connection structure of the network is described by a normalized Laplacian matrix. We show that consensus is achieved if and only if the underlying graph contains a spanning tree. Furthermore, this statement holds independently of the value of the delay, in contrast to the case of processing delays. We also calculate the consensus value and show that, unlike the case of processing delays, the consensus value is determined not just by the initial states of the nodes at time zero, but also on their past history over an interval of time.

1 Introduction

We consider the consensus problem for multi-agent systems in continuous time

$$\dot{x}_i(t) = u_i(t) \tag{1}$$

under the consensus protocol

$$u_i(t) = \frac{1}{d_i} \sum_{j=1}^{n} a_{ij}(x_j(t - \tau) - x_i(t)). \tag{2}$$

Here, $x_i(t) \in \mathbb{R}$ is the state, or "opinion", of the agent i at time t, $i = 1, \ldots, n$, which changes under the interaction with the other agents in a manner described by the "consensus protocol" u_i. The quantity $\tau \geq 0$ denotes the time delay in information

Fatihcan M. Atay
Max Planck Institute for Mathematics in the Sciences, Inselstrasse 22,
04103 Leipzig, Germany
e-mail: atay@member.ams.org

R. Sipahi et al. (Eds.): Time Delay Sys.: Methods, Appli. and New Trends, LNCIS 423, pp. 407–416.
springerlink.com

transmission between the agents, $a_{ij} \geq 0$ is the strength of the influence of agent j on i, and $d_i = \sum_{j=1}^{n} a_{ij}$. We assume $d_i > 0$ for all i, that is, each unit has some input from the others. The numbers a_{ij} define a directed and weighted graph G on n vertices, where there is an arc from vertex j to i if and only if $a_{ij} \neq 0$. Special cases of an unweighted graph is obtained if a_{ij} are restricted to binary values $\{0, 1\}$, and an undirected graph is obtained if $a_{ij} = a_{ji}\ \forall i, j$. The quantity d_i is then the (in-) degree of vertex i in a (directed) graph. The interpretation of (2) is that the state of each agent is modified according to the average difference between his state and of those who influence him. We say that the system (1) reaches consensus if for any set of initial conditions there exists some $c \in \mathbb{R}$ (depending on the initial conditions) such that $\lim_{t \to \infty} x_i(t) = c$ for all i. The number c is then called the consensus value.

In previous works, delayed consensus problems have often been studied with the protocol

$$u_i(t) = \sum_{j=1}^{n} a_{ij}(x_j(t - \tau) - x_i(t - \tau)), \tag{3}$$

see e.g. [14] for the continuous-time case in undirected networks and [12] for a distributed-delay version. The main difference with (2) is that the term x_i is also delayed in (3). Hence, (3) represents *delayed information processing*, where the difference of the states $x_j - x_i$ influences the dynamics after some time delay τ. By contrast, (2) represents *delayed information transmission*, where agent i compares its state to the information coming from its neighbor j after some time delay τ. Information transmission delays have so far not attracted much attention in consensus problems, although they arise naturally in many dynamical processes on networks. Among the rare examples on transmission delays, one can mention [13], who studied the stability of the continuous-time consensus algorithm in a network with time-varying connections with the protocol

$$u_i(t) = \sum_{j=1}^{n} a_{ij}(t)(x_j(t - \tau) - x_i(t)) \tag{4}$$

although it did not obtain the consensus value, as well as [15] who studied a time-invariant version with constant vertex degrees and obtained complicated conditions involving matrix inequalities. Here, we derive the exact consensus condition and calculate the consensus value.

Another difference between (2) and (3) is the normalization in (2) of the summation term by the degree d_i. This gives rise to a normalized version of the graph Laplacian. Recall that the Laplacian matrix of a graph is given by $D - A$, where $A = [a_{ij}]$ is the weighted adjacency matrix, and D is the diagonal matrix of vertex degrees

$$D = \mathrm{diag}\{d_1, \ldots, d_n\}.$$

The Laplacian $D - A$ arises naturally if one writes the system in vector form under the protocol (3) or (4). By contrast, for (2) the relevant connection matrix is the normalized Laplacian $I - D^{-1}A$. In the normalized version, each agent acts on the average difference between its state and those of its neighbors, regardless of how many neighbors it has. This is the natural coupling model for several important network dynamics, for instance for opinion formation in social networks or models of synchronization in neural systems. Dynamics on networks under the normalized Laplacian and has been studied, for instance, in the context of delay-induced stability [1, 5] and synchronization in delayed [4] and undelayed networks [7]. Notice that, if $d_i = k$ for all i, then the two versions of the Laplacian matrix differ simply by a factor of k.

In the following, we study the consensus problem (1)–(2) under delayed information transmission in weighted and directed networks. We show that consensus is independent of the transmission delays and can be achieved if and only if the network has a spanning tree. This result is in contrast to the case of processing delays (3), where consensus is not possible for large values of the delay [12, 14]. Furthermore, we calculate the consensus value explicitly, and show that it depends on the initial history of the agents over a time interval as well as the details of the network structure. For undirected networks, the latter involves the degree sequence of the underlying graph.

2   The Normalized Laplacian and Consensus in the Absence of Delays

With $\mathbf{x} = (x_1, \ldots, x_n)$, the consensus problem (1)–(2) can be written in vector form as

$$\dot{\mathbf{x}}(t) = -\mathbf{x}(t) + D^{-1}A\mathbf{x}(t - \tau). \tag{5}$$

Let $\mathbf{1} = (1, 1, \ldots, 1)^\top \in \mathbb{R}^n$. Reaching a consensus is equivalent to the convergence $\lim_{t \to \infty} \mathbf{x}(t) = c\mathbf{1}$ for some scalar c, which ultimately is the consensus value. The form of the right hand side of (5) motivates the definition of the normalized Laplacian $L = I - D^{-1}A$. This matrix has similar properties to the combinatorial Laplacian $D - A$, and its spectral properties play a significant role in the dynamics of (5). Let $\{\lambda_1, \ldots, \lambda_n\}$ denote the set of eigenvalues of L. By an application of Gershgorin's theorem [10] it can be seen that

$$|1 - \lambda_i| \le 1, \quad \forall i. \tag{6}$$

Note that $L\mathbf{1} = 0$ since L has zero row sums. Hence, zero is always an eigenvalue of L, which we assign to λ_1, with the corresponding eigenvector $\mathbf{v}_1 = \mathbf{1}$. We assume in the following that L has a complete set of eigenvectors[1] $\{\mathbf{v}_1, \ldots, \mathbf{v}_n\}$ corresponding to the eigenvalues λ_i: $L\mathbf{v}_i = \lambda_i\mathbf{v}_i$, $i = 1, \ldots, n$. Let $\{\mathbf{u}^1, \ldots, \mathbf{u}^n\}$ be the dual basis corresponding to $\{\mathbf{v}_1, \ldots, \mathbf{v}_n\}$. That is, the $\mathbf{u}^i$ are linearly independent vectors such that $\langle \mathbf{u}^i, \mathbf{v}_j \rangle = \delta^i_j$. It is easy to see that $\mathbf{u}^i$ are the left eigenvectors of L, $\mathbf{u}^i L = \lambda_i \mathbf{u}^i$. Furthermore, for $\mathbf{x} = \sum_{i=1}^n \alpha_i \mathbf{v}_i \in \mathbb{R}^n$ one has $\alpha_i = \langle \mathbf{u}^i, \mathbf{x} \rangle$.

---

[1] Although this assumption is not strictly necessary, it is generically satisfied for matrices in $\mathbb{R}^{n \times n}$ and makes the subsequent analysis and notation easier to follow.

We briefly consider the undelayed case for motivation and comparison. When the delays are absent in (5) (i.e., when $\tau = 0$), the system simplifies to

$$\dot{\mathbf{x}}(t) = -L\mathbf{x}(t) \tag{7}$$

whose solution is

$$\mathbf{x}(t) = e^{-Lt}\mathbf{x}(0). \tag{8}$$

Expressing the initial condition $\mathbf{x}(0) = \mathbf{x}_0$ in the eigenbasis of L as $\mathbf{x}_0 = \sum_{i=1}^{n} \langle \mathbf{u}^i, \mathbf{x}_0 \rangle \mathbf{v}_i$, (8) yields

$$\mathbf{x}(t) = \sum_{i=1}^{n} \langle \mathbf{u}^i, \mathbf{x}_0 \rangle e^{-\lambda_i t} \mathbf{v}_i. \tag{9}$$

Consider now the condition

$$\min_{i \geq 2} \mathrm{Re}(\lambda_i) > 0. \tag{10}$$

If (10) is satisfied, (9) implies

$$\mathbf{x}(t) \to \langle \mathbf{u}^1, \mathbf{x}_0 \rangle \mathbf{v}_1,$$

since $\lambda_1 = 0$. Hence, the system reaches consensus, since $\mathbf{v}_1 = \mathbf{1}$. On the other hand, if $\mathrm{Re}(\lambda_i) \leq 0$ for some $i \geq 2$, then (9) gives

$$|\langle \mathbf{u}^i, \mathbf{x}(t) \rangle| \geq |\langle \mathbf{u}^i, \mathbf{x}_0 \rangle| \text{ for all } t \geq 0,$$

so that $\mathbf{x}(t)$ always has a component along an eigenvector different from $\mathbf{1}$, provided $\mathbf{x}(0)$ has a nonzero component along that eigenvector. Hence, (10) is also a necessary condition to reach consensus from arbitrary initial conditions.

To summarize, system (7) reaches consensus from arbitrary initial conditions if and only if (10) is satisfied. In view of (6), the condition (10) holds if and only if zero is a simple eigenvalue of L. Furthermore, if (10) holds, then the consensus value is

$$c = \langle \mathbf{u}^1, \mathbf{x}_0 \rangle, \tag{11}$$

i.e. a weighted average of initial values, where the weights are given by the components of the left eigenvector $\mathbf{u}^1$ of L corresponding to the zero eigenvalue.

3 Consensus under Transmission Delays

To analyze the dynamics of the delayed consensus problem (5), we express $\mathbf{x}$ in the eigenbasis of L, similar to the development in Section 2,

$$\mathbf{x}(t) = \sum_{i=1}^{n} \alpha_i(t) \mathbf{v}_i, \quad \text{where } \alpha_i(t) = \langle \mathbf{u}^i, \mathbf{x}(t) \rangle. \tag{12}$$

Operating on both sides of (5) with $\mathbf{u}_i$ shows that the evolution of α_i is governed by the equation

$$\dot{\alpha}_i(t) = -\alpha_i(t) + (1 - \lambda_i)\alpha_i(t - \tau) \tag{13}$$

which has the characteristic equation

$$\chi_i(s) := s + 1 - (1 - \lambda_i)e^{-s\tau} = 0, \tag{14}$$

We have the following result on the characteristic roots of (14).

Lemma 1. *If $\lambda_i = 0$, then the characteristic equation (14) has a simple root at zero and all other roots have negative real parts. If $\lambda_i \neq 0$, then all roots have negative real parts.*

Proof. Rearranging (14), s is a root of χ_i if and only if

$$s + 1 = (1 - \lambda_i)e^{-s\tau}. \tag{15}$$

We first claim that if χ_i has a root s with $\mathrm{Re}(s) \geq 0$, then s must be zero. To prove the claim, let $s = \sigma + i\omega$ with real part $\sigma \geq 0$. Then,

$$\begin{aligned}
|s + 1| = |\sigma + 1 + i\omega| &\geq 1 \tag{16} \\
&\geq |(1 - \lambda_i)| \\
&\geq |(1 - \lambda_i)||e^{-s\tau}|,
\end{aligned}$$

where the second line follows by (6). Now the inequality (16) is strict whenever $\sigma > 0$ or $\omega \neq 0$, in which case (15) fails to hold, contradicting the assumption that s is a characteristic root. Thus, we must have $\sigma = \omega = 0$, which proves the claim. Therefore, all roots of χ_i have negative real parts, except possibly for a root at zero. Now from (14), $s = 0$ is a root of χ_i if and only if $\lambda_i = 0$, in which case,

$$\chi_i'(0) = 1 + \tau \neq 0,$$

which shows that zero is a simple root. This completes the proof.   □

We can now state the main consensus result.

Theorem 1. *The system (5) reaches consensus if and only if zero is a simple eigenvalue of the Laplacian L. Furthermore, the consensus value is given by*

$$c = \frac{1}{1 + \tau} \langle \mathbf{u}^1, \mathbf{x}(0) + \int_{-\tau}^0 \mathbf{x}(\xi)\, d\xi \rangle. \tag{17}$$

where $\mathbf{u}^1$ is the left eigenvector of L corresponding to the zero eigenvalue.

Proof. Suppose that zero is a simple eigenvalue of the Laplacian L; that is, $\lambda_i \neq 0$ for $i \geq 2$. Then using Lemma 1, we conclude that solutions of (13) satisfy $\lim_{t \to \infty} \alpha_i(t) = 0$ for $i \geq 2$. Consequently, by (12),

$$\lim_{t \to \infty} \|\mathbf{x}(t) - \alpha_1(t)\mathbf{1}\| = 0. \tag{18}$$

It remains to study the dynamics of α_1, which is governed by

$$\dot{\alpha}_1(t) = -\alpha_1(t) + \alpha_1(t - \tau), \tag{19}$$

which follows from (13) using the fact that $\lambda_1 = 0$. To this end, we let $\mathscr{C} = C([-\tau, 0], \mathbb{R})$ denote the Banach space of real-valued continuous functions on $[-\tau, 0]$, equipped with the supremum norm. Similarly, let $\mathscr{C}^* = C([0, \tau], \mathbb{R})$, and define the bilinear form (ψ, ϕ) for $\phi \in \mathscr{C}$ and $\psi \in \mathscr{C}^*$ by

$$(\psi, \phi) := \psi(0)\phi(0) + \int_{-\tau}^{0} \psi(\xi + \tau)\phi(\xi)\,d\xi. \tag{20}$$

Let $\mathscr{C}_0$ and $\mathscr{C}_0^*$ be the one-dimensional subspace of constant functions in $\mathscr{C}$ and $\mathscr{C}^*$, respectively. We choose the constant function $\Phi(\theta) \equiv 1$ as a basis for $\mathscr{C}_0$ and the constant function $\Psi(\theta) \equiv (1 + \tau)^{-1}$ as a basis for $\mathscr{C}_0^*$, so that $(\Phi, \Psi) = 1$. Now let $a_t \in \mathscr{C}$ be defined by

$$a_t(\theta) = \alpha_1(t + \theta), \quad \theta \in [-\tau, 0]. \tag{21}$$

By Lemma 1, the characteristic equation corresponding to (19) has a simple root at zero and all other roots have negative real parts. Note that $\mathscr{C}_0$ is the eigenspace corresponding to the characteristic value zero. By the theory of functional differential equations [9], the space $\mathscr{C}$ can be decomposed using the invariant subspace $\mathscr{C}_0$ and its complement. Hence, we can write a_t as the sum $a_t = a_t^0 + b_t$, where $a_t^0 \in \mathscr{C}_0$ for all $t > 0$, and $\lim_{t \to \infty} b_t = 0$ since all eigenvalues other than zero have negative real parts. Further, a_t^0 is given by the constant function $a_t^0 = \Phi(\Psi, a_0)$. Using (20) and (21), we conclude that a_t (and hence $\alpha_1(t)$) approaches the constant function

$$a_t^0 = \Phi(\Psi, a_0) = \frac{1}{1 + \tau}\left(\alpha_1(0) + \int_{-\tau}^{0} \alpha_1(\xi)\,d\xi\right).$$

Hence, by (18) and (12), consensus is achieved, with consensus value given by (17).

Finally, to prove the necessity of a simple zero eigenvalue of the Laplacian, suppose $\lambda_i = 0$ for some $i \geq 2$. Then by Lemma 1, zero is a characteristic root of equation (13) and all other characteristic roots have negative real parts. Thus, $\alpha_i(t)$ does not approach zero for general initial conditions. Consequently, (12) implies that $\mathbf{x}(t)$ has a component along the eigenvector $\mathbf{v}_i$, which is not a scalar multiple of $\mathbf{1}$; hence, consensus is not achieved from arbitrary initial conditions. $\square$

4   Remarks

4.1   *Zero Eigenvalue and Spanning Trees*

The main result of this paper, as stated by Theorem 1, indicates that the stability of the consensus algorithm does not depend on the delays but on the network structure.

Namely, it is required that zero is a simple eigenvalue of the Laplacian L. This condition is equivalent to the more geometric condition that the network has a spanning tree; that is, there exists a vertex from which all other vertices can be reached along directed paths. This is a known fact for the matrix $D - A$ (see e.g. [11]), and also holds for the normalized Laplacian $I - D^{-1}A$ with a similar proof [6]. Clearly, an undirected graph has a spanning tree if and only if it is connected.

4.2 Undirected Networks

Results from graph theory allow more detailed information for undirected networks (see e.g. [8]). Indeed, if the underlying network G is undirected, then the Laplacian L has real eigenvalues, which can be ordered as $0 = \lambda_1 \le \lambda_2 \le \cdots \le \lambda_n \le 2$, counting multiplicities. The smallest eigenvalue λ_1 is zero, and $\lambda_2 > 0$ if and only if the graph is connected. Hence, by the above results, the network reaches consensus if and only if it is connected, regardless of the delay. Furthermore, for undirected graphs one can calculate the left eigenvector $\mathbf{u}^1$ of the normalized Laplacian corresponding to the zero eigenvalue to be

$$\mathbf{u}^1 = \frac{1}{\text{vol}(G)}(d_1,\ldots,d_n), \tag{22}$$

where $\text{vol}(G) = \sum_{i=1}^n d_i$ is the *volume* of the graph. Hence $\mathbf{u}^1$ is simply the degree sequence of the graph up to a normalization. Thus, when the system reaches consensus, the consensus value is given by a weighted average of the initial opinions where the weights are determined by the vertex degrees. The final consensus value is influenced more by those units having a larger number of neighbors. For instance, in scale-free networks the consensus value is essentially determined by the hubs. For the relation between the degree sequence (22) of the graph and the spectral properties of the normalized Laplacian, the reader is referred to [3].

4.3 Normalized versus Non-normalized Laplacian

In this work we have used the normalized Laplacian $L = I - D^{-1}A$, as opposed to the existing literature that mainly uses $D - A$. If the in-degree is the same for all vertices, i.e. if $\sum_{j=1}^n a_{ij} = k$ for all i, then the two Laplacians differ by the constant multiple k, and in particular $A = k(I - L)$. In this case, the results in this paper also apply to the protocol

$$u_i(t) = \sum_{j=1}^n a_{ij}(x_j(t - \tau) - x_i(t)),$$

(i.e., without the normalization by d_i) under which the system (1) becomes

$$\dot{\mathbf{x}}(t) = -k\mathbf{x}(t) + A\mathbf{x}(t - \tau) \tag{23}$$
$$= -k\mathbf{x}(t) + k(I - L)\mathbf{x}(t - \tau)\,ds.$$

This equation can be put into the form (5) after scaling time by the factor k, which does not affect consensus. Hence, we conclude that the system (23) reaches consensus if and only if zero is a simple eigenvalue of the Laplacian (in either definition). Ref. [15] studied (23) but the conditions they derive are convoluted expressions involving matrix inequalities and limits. As shown in Section 3 and 4.1, the actual condition only involves the existence of a spanning tree, and the consensus value is given by (17).

Although both Laplacians have similar properties, an essential difference regarding consensus problems is the difference in their left eigenvectors. One can see this more clearly in undirected graphs: Since $D - A$ is then a symmetric matrix, the left eigenvector $\mathbf{u}^1$ corresponding to the zero eigenvalue is obtained by transposing the right eigenvector $\mathbf{v}_1 = \mathbf{1}$ with the normalization $\langle \mathbf{u}^1, \mathbf{v}_1 \rangle = 1$, i.e.,

$$\mathbf{u}_1 = \frac{1}{n}(1, 1, \ldots, 1).$$

Hence, the consensus value, say (11), in the case of the non-normalized Laplacian is based on the average of the agents' initial conditions, whereas the normalized Laplacian, in view of (22), weights the initial conditions by the vertex degrees. This property of the normalized Laplacian makes it a more appropriate choice for modeling opinion dynamics, where one naturally expects that the highly-connected hubs have a more pronounced influence on the final outcome as compared to the poorly-connected members of the network.

4.4   Transmission versus Processing Delays

When one considers processing delays, modeled by (3), the consensus problem takes the form

$$\dot{\mathbf{x}}(t) = L\mathbf{x}(t - \tau),$$

and the expansion (12) gives that the perturbations α_i along the eigendirections are governed by

$$\dot{\alpha}_i(t) = -\lambda_i \alpha_i(t - \tau).$$

Although these are delay differential equations for $\lambda_i \neq 0$, they become ordinary differential equations for $\lambda_i = 0$. In particular, since $\lambda_1 = 0$ corresponds to the eigenvector $\mathbf{1}$, one obtains an ordinary differential equation for the mode that eventually determines the consensus value. Hence, if the system reaches consensus, the consensus value depends only on the initial value of the state $\mathbf{x}$ at a single time point. By contrast, under signal transmission delays modeled by the protocol (2), the consensus value depends on the initial history of $\mathbf{x}$ over some time interval, as given by (17). This property can imply additional robustness for the consensus value. For instance, if the initial states of agents are constant, but their measurements are subject to small noise, then the consensus value calculations based on $\mathbf{x}(0)$ may be subject to some error. However, if the noise terms have zero temporal mean, then their effect is reduced by averaging over some time interval such as (17). Since Theorem 1 implies that reaching consensus is independent of the delay, a suitably large τ can facilitate noise reduction in calculation of the consensus value, without impeding consensus itself. More details in this direction can be found in an extended paper [2].

Finally we note the difference between transmission and processing delays with respect to delay magnitude. As we have shown in this study, stability of consensus is independent of the signal transmission delays for the protocol (2). This is in contrast to signal processing delays of the protocol (3), where large delays are known to prohibit consensus [12, 14].

4.5 Distributed Delays, Discrete-Time Systems

In the foregoing, we have considered a fixed transmission delay τ in the consensus protocol. More generally, in place of (2) one can study protocols with distributed delays

$$u_i(t) = \frac{1}{d_i} \sum_{j=1}^{n} a_{ij} \left(\int_0^{\tau} f(s) x_j(t-s)\, ds - x_i(t) \right),$$

where $f(s)$ denotes a distribution of delays on an interval $[-\tau, 0]$. One can similarly define a consensus problem in discrete time with a protocol with distributed delays

$$u_i(t) = \frac{1}{d_i} \sum_{j=1}^{n} a_{ij} \left(\left(\sum_{s=0}^{\tau} f_s x_j(t-s) \right) - x_i(t) \right),$$

where f_s is a discrete probability distribution. We refer the reader to the paper [2] for extended results in these general settings.

References

1. Atay, F.M.: Oscillator death in coupled functional differential equations near Hopf bifurcation. J. Differential Equations 221(1), 190–209 (2006), doi:10.1016/j.jde.2005.01.007
2. Atay, F.M.: Consensus in networks with transmission delays (forthcoming)
3. Atay, F.M., Bıyıkoğlu, T., Jost, J.: Synchronization of networks with prescribed degree distributions. IEEE Trans. Circuits and Systems I 53(1), 92–98 (2006), doi:10.1109/TCSI.2005.854604
4. Atay, F.M., Jost, J., Wende, A.: Delays, connection topology, and synchronization of coupled chaotic maps. Phys. Rev. Lett. 92(14), 144101 (2004), doi:10.1103/PhysRevLett.92.144101
5. Atay, F.M., Karabacak, Ö.: Stability of coupled map networks with delays. SIAM J. Applied Dyn. Syst. 5(3), 508–527 (2006), doi:10.1137/060652531
6. Bauer, F.: Spectral graph theory of directed graphs (forthcoming)
7. Bauer, F., Atay, F.M., Jost, J.: Synchronization in discrete-time networks with general pairwise coupling. Nonlinearity 22(1), 2333–2351 (2009), doi: 10.1088/0951-7715/22/10/001

8. Chung, F.R.K.: Spectral Graph Theory. American Mathematical Society, Providence (1997)
9. Hale, J.K.: Theory of Functional Differential Equations. Springer (1977)
10. Horn, R.A., Johnson, C.R.: Matrix Analysis. Cambridge University Press, Cambridge (1985)
11. Merris, R.: Laplacian matrices of graphs: a survey. Linear Algebra and its Applications 197-198, 143–176 (1994), doi:10.1016/0024-3795(94)90486-3
12. Michiels, W., Morărescu, C.I., Niculescu, S.I.: Consensus problems with distributed delays, with application to traffic flow models. SIAM Journal on Control and Optimization 48(1), 77–101 (2009), doi:10.1137/060671425
13. Moreau, L.: Stability of continuous-time distributed consensus algorithms. In: Proc. 43rd IEEE Conference on Decision and Control, pp. 3998–4003 (2004)
14. Olfati-Saber, R., Murray, R.M.: Consensus problems in networks of agents with switching topology and time-delays. IEEE Transactions on Automatic Control 49(9), 1520–1533 (2004), doi:10.1109/TAC.2004.834113
15. Seuret, A., Dimarogonas, D.V., Johansson, K.H.: Consensus under communication delays. In: Proc. 47th IEEE Conference on Decision and Control, CDC 2008, Cancun, Mexique (2008)

Consensus with Constrained Convergence Rate and Time-Delays

Irinel-Constantin Morărescu, Silviu-Iulian Niculescu, and Antoine Girard

Abstract. In this paper we discuss consensus problems for networks of dynamic agents with fixed and switching topologies in presence of delay in the communication channels. The study provides sufficient agreement conditions in terms of delay and the second largest eigenvalue of the Perron matrices defining the collective dynamics. We found an exact delay bound assuring the initial network topology preservation. We also present an analysis of the agreement speed when the asymptotic consensus is achieved. Some numerical examples complete the presentation.

1 Introduction

The analysis of multi-agent systems has various applications in many areas encompassing cooperative control of vehicles [3, 5], congestion control in communication networks, flocking [12], distributed sensor networks [2]. In many of these applications, all the agents need to agree with respect to a priori fixed criteria and the agreement might be subject to some speed constraints. Furthermore, the communication channels between the agents are not ideal and may introduce time-delays into dynamics.

The consensus problem for directed graphs with a fixed topology was treated in the case where no convergence speed is imposed (see [11]). We also note that

Irinel-Constantin Morărescu
CRAN (UMR-CNRS 7039), Nancy-Université, 2 avenue de la Forêt de Haye,
54516, Vandoeuvre-lès-Nancy
e-mail: `constantin.morarescu@ensem.inpl-nancy.fr`

Silviu-Iulian Niculescu
L2S (UMR CNRS 8506), CNRS-Supélec, 3, rue Joliot Curie, 91192, Gif-sur-Yvette
e-mail: `Silviu.Niculescu@lss.supelec.fr`

Antoine Girard
Joseph Fourier University, Jean Kuntzmann Laboratory, Tour IRMA,
51 rue des Mathématiques, 38400, Saint Martin d'Hères
e-mail: `Antoine.Girard@imag.fr`

R. Sipahi et al. (Eds.): Time Delay Sys.: Methods, Appli. and New Trends, LNCIS 423, pp. 417–428.
springerlink.com   © Springer-Verlag Berlin Heidelberg 2012

the consensus problem for undirected graphs with a switching topology generated by a bounded confidence was considered in [5, 6]. The characterizations given in these works use the notions of *periodically linked together* (the agents are periodically linked together) or *finally linked together* (i.e. the agents are linked in $\bigcup_{t \geq s} G(t)$, $\forall s \geq 0$, where $G(t)$ is the graph describing the network topology at instant t). A frequency domain approach for the consensus problem of a fixed topology network of continuous-time integrators agents communicating trough delayed channels was developed in [10]. Precisely when the time-delays in all communication channels is given by the same constant, the time-delay bound guaranteing the consensus is given in terms of the smallest eigenvalue of the matrix defining the collective dynamics (called the Perron matrix).

The updating rule used in this paper matches in the free of delay case with those considered by [1] (see also [9]). The time-delay introduced in our model is constant and is the same for all the communication channels. This delay value can be seen either as a computational time or as a communication latency, depending on the application under consideration. Our aim is to find the delay bound that guarantees the agreement with at least an a priori given speed, in terms of the second largest eigenvalue of the Perron matrix.

The remainder of this paper is organized as follows. In Section 2 we introduce some basic notions related to graph theory and we formulate the model. In Section 3 we present the convergence result and derive the exact delay margin assuring an agreement speed which preserves the network topology. Section 4 provide an analysis of the agreement speed when the consensus for a switching topology case is assured. Some numerical examples are presented in Section 5 and Section 6 ends the paper with some concluding remarks.

2 Preliminaries

2.1 Algebraic Graph Theory Elements

Let $G = (\mathcal{V}, E)$ denotes a directed graph with the set of vertices $\mathcal{V}$ and the set of edges E. Each vertex is labeled by $v_i \in \mathcal{V}$, $i = 1, \ldots, n$ and one says that $(i, j) \in E$ if there exists an edge between v_i and v_j. We consider $N_i = \{ v_j \in \mathcal{V} \mid (i, j) \in E \}$. Given two graphs with the same set of vertices, $G_1 = (\mathcal{V}, E_1)$ and $G_2 = (\mathcal{V}, E_2)$ we say that $G_1 \subset G_2$ if $E_1 \subset E_2$.

Definition 1. A **path** in a given graph $G = (\mathcal{V}, E)$ is a union of edges $\bigcup_{k=1}^{p} (i_k, j_k)$ such that $i_{k+1} = j_k, \forall k \in \{1, \ldots, p-1\}$.

  Two nodes v_i, v_j are **connected** in a graph $G = (\mathcal{V}, E)$ if there exists at least a path in G joining v_i and v_j (i.e. $i_1 = i$ and $j_p = j$).

  A **connected graph** has all the nodes connected.

The adjacency matrix $A = [a_{ij}] \in \mathbb{R}^{n \times n}$ of the graph G is the integer matrix with the ij-entry equals to the number of arcs from i to j which is usually 0 or 1. The graph Laplacian of G is than defined as

$$L = D - A \tag{1}$$

where $D = diag(d_1, \ldots, d_n)$ is the degree matrix of G with $d_i = \sum_{j \neq i} a_{ij}$. Therefore, L has the right eigenvector $\mathbb{1} = (1, 1, \ldots, 1)$ associated with the eigenvalue 0 ($L\mathbb{1} = 0$).

2.2 Consensus Protocol

In the following we consider that each vertex v_i represents a dynamic agent and the state of the network will be given by $x(\cdot) = (x_1(\cdot), x_2(\cdot), \ldots, x_n(\cdot))^\top \in \mathbb{R}^n$ where $x_i(t)$ is a scalar real value assigned to v_i at the moment t. The value $x_i(t)$ will be called the opinion of the agent v_i at the moment t.

In order to motivate the updating rule proposed in the sequel, we consider a synchronized sensors network (i.e. all the sensors clocks are synchronized and each sensor knows at which time was sent a specific information even if it does not receive this information instantaneously). This allows us to ignore the communication delay. On the other hand some encoding-decoding delays are associated to each sensor. One considers that these delays are constant and all are equal τ. Therefore, the sensors update their opinion/decision at the time-step $t + 1$ considering the information available at time $t + 1 - \tau$ via the network configuration available at the time-step t. Thus, the discrete-time collective dynamics is defined by

$$x(t + 1) = P(t)x(t + 1 - \tau) \tag{2}$$

where $P(t)$ is a matrix considered to be doubly stochastic (called the Perron matrix of the system). In the sequel one denotes by $p_{ij}(t)$ the entries of the Perron matrix $P(t)$. This model presented above can be also interpreted in terms of interactions between a set of agents and a virtual environment. Considering τ the round-trip delay associated to each agent, the actions at instant $t + 1$ are determined by the reactions received via the network configuration at time t of the action done at the time-step $t + 1 - \tau$. In other words, any opinion at the instant $t + 1$ is a weighted average of the opinion values at the instant $t + 1 - \tau$:

$$x_i(t + 1) = \sum_{j=1}^{n} p_{ij}(t)x_j(t + 1 - \tau), \quad \forall i \in \{1, \ldots, n\}, t \geq 0 \tag{3}$$

In the sequel we assume that the Perron matrix $P(t)$ satisfies the following properties:

Assumption 1 *For $t \in \mathbb{N}$, the coefficients $p_{ij}(t)$ satisfy*

1. $p_{ij}(t) \in [0, 1]$, *for all $v_i, v_j \in \mathcal{V}$.*
2. $\sum_{j=1}^{n} p_{ij}(t) = 1 = \sum_{i=1}^{n} p_{ij}(t)$, *for all $v_i \in \mathcal{V}$.*

Precisely, $p_{ij}(t) > 0$ if agent j communicates at instant t its current value to agent i and $p_{ij}(t) = 0$ otherwise. Since $P(t)$ is supposed doubly stochastic we actually consider only balanced graphs. Let us denote by

$$1 = \lambda_1(t) \geq \lambda_2(t) \geq \ldots \geq \lambda_n(t) \geq 0$$

the eigenvalues of the symmetric positive semi-definite matrix $P(t)^\top P(t)$.

Remark 1. An extensively used example of matrix P satisfying Assumption 1 is $I - \alpha L$ (see for instance [5, 6, 10]) where I is the identity matrix and L is the graph Laplacian matrix associated to the graph G. When G is undirected L is symmetric. Moreover L and P obtained like this are positive semi-definite and $\mu_2 = \sqrt{\lambda_2} = 1 - \lambda_2(L)$, where $\lambda_2(L)$ is the smallest nonzero eigenvalue of L (one considers G is connected so the multiplicity of the eigenvalue zero of L is one). It is noteworthy that $\lambda_2(L)$ is called the connectivity of G (see [4] for details on $\lambda_2(L)$ and graph theory).

In this paper we consider a set of agents that updates their opinions using the algorithm (2) and $G(t) = (\mathcal{V}, E(t))$ is the graph representation of the corresponding *dynamic network* with a *switching topology* that is time-dependent.

Letting E the set of edges of the graph $G(0)$, we consider that the evolution of the network topology is given by

$$E(t) = \{(i,j) \in E \mid |x_i(t) - x_j(t)| \leq M\rho^t\} \tag{4}$$

where $M \in \mathbb{R}_+$ and $\rho \in (0,1)$ are some parameters fixed by the designer. It means that, at each time-step the agents become more confident in their own opinion and they take into account only the neighbors whose opinion approaches enough their own opinion. This procedure lead either to a fast convergent consensus algorithm or to a partition of agents in several groups where the agreement is reached.

Remark 2. It is clear that a smaller ρ leads to a higher convergence speed. We also note that $\rho = 1$, corresponding to the bounded confidence model, imposes no convergence speed constraints.

A natural assumption that is made in this paper states that $P(t_1) = P(t_2)$ when $E(t_1) = E(t_2)$ (for similar network configurations the same weighted average is used to determine the opinion values of the next step). It is worth noting that

$$\sum_{i=1}^n x_i(t+1) = \sum_{i=1}^n \sum_{j=1}^n p_{ij}(t)x_j(t-\tau+1) = \sum_{j=1}^n \sum_{i=1}^n p_{ij}(t)x_j(t-\tau+1)$$

$$= \sum_{j=1}^n x_j(t-\tau+1), \quad \forall t \leq 0 \tag{5}$$

Considering an initial condition $x(t) = x^0, t \in [-\tau, 0]$, one easily obtains that $S = \sum_{j=1}^n x_i(t)$ and $Ave(x(t)) = \frac{S}{n}$ are invariant quantities.

Definition 2 (agreement). We say nodes v_i and v_j *asymptotically agree* if and only if $\lim_{t\to\infty} x_i(t) = \lim_{t\to\infty} x_j(t)$. Two nodes *asymptotically disagree* if $\lim_{t\to\infty} x_i(t) \neq \lim_{t\to\infty} x_j(t)$. An algorithm guarantees *asymptotic consensus* if: for every initial condition $x(t) = x^0, t \in [-\tau, 0]$ and for every sequence $\{P(t)\}$ allowed by (4) and Assumption 1, all the nodes asymptotically agree.

Remark 3. Since $Ave(x(t))$ is an invariant quantity, the algorithm (2) may guaranty only the asymptotic *average-consensus* $(\lim_{t\to\infty} x_i(t) = Ave(x(t)), \forall i)$.

3 Consensus Problem for Networks with Fixed Topology

Let us consider the positive function

$$V(t) = ||x(t) - x^*||$$

where $x^* = Ave(x(t))\mathbb{1}$ where $\mathbb{1}$ represents the column vector whose elements are all equal to one. For all $(i, j) \in E$ the following holds:

$$V(0)^2 \geq |x_i(0) - x_i^*|^2 + |x_j(0) - x_j^*|^2 \geq \frac{1}{2}(|x_i(0) - x_i^*| + |x_j(0) - x_j^*|)^2$$

$$\geq \frac{1}{2}|x_i(0) - x_j(0)|^2$$

Therefore, in the sequel we set $M = \sqrt{2}V(0)$ in order to guaranty that all the possible initial transmission lines appear in the initial configuration of the network.

3.1 Convergence Result

In this paragraph we prove that the algorithm (2) assures the asymptotic *average-consensus* when the network topology does not evolve according to (4) but is fixed. Obviously, in order to reach the consensus we have to assume that the network topology is given by a strongly connected graph (the opinion of each agent can be accessed after a given delay by all the other agents). Precisely, the discrete-time collective dynamics is defined by

$$x(t) = Px(t - \tau) \tag{6}$$

where P is a doubly stochastic matrix. The eigenvalues of the symmetric stochastic matrix $P^\top P$ are denoted by

$$0 \leq \lambda_n \leq \lambda_{n-1} \leq \ldots \leq \lambda_2 \leq \lambda_1 = 1$$

Since the graph is connected, 1 is a simple eigenvalue so $\lambda_2 < 1$ and the following result holds.

Proposition 1. *The algorithm* (6) *guarantees the asymptotic average-consensus. Moreover if* $\mu_2 = \sqrt{\lambda_2}$ *one has*

$$V(t + k\tau) \leq \mu_2^k V(0), \forall t \in [-\tau, 0], k \in \mathbb{N}$$

Proof. Let us recall that a doubly stochastic matrix P (and $P^\top$) has always the left and right eigenvector $\mathbb{1}$ corresponding to the eigenvalue $\lambda_1 = 1$. If $\{u_i\}$ is the set of orthonormal eigenvectors of the symmetric stochastic matrix $P^\top P$, then

$$x(t) = \sum_{i=1}^{n} a_i(t)u_i, \quad a_i(t) \in \mathbb{R}, \forall i, \forall t \geq -\tau$$

and using the linearity of P we get

$$P^\top x(t) = P^\top P x(t - \tau) = \sum_{i=1}^{n} \lambda_i a_i(t - \tau)u_i$$

Since $\lambda_1 = 1$ and $u_1 = \mathbb{1}$ one obtains that

$$P^\top x(t + k\tau) = P^\top P x(t + (k-1)\tau) = a_1(0) \cdot \mathbb{1}$$
$$+ \sum_{i=2}^{n} \lambda_i^k a_i(0)u_i, \forall t \in [-\tau, 0]$$

which leads to

$$P^\top x(t) \xrightarrow[t \to \infty]{} a_1(0) \cdot \mathbb{1}$$

Multiplying by $\mathbb{1}^\top$ at the left one arrives at

$$\sum_{j=1}^{n} x_i(t) = \mathbb{1}^\top x(t) \xrightarrow[t \to \infty]{} n a_1(0)$$

so using the invariance of S (see (5)) one has $x^* = a_1(0) \cdot \mathbb{1}$ which means

$$x(t) \xrightarrow[t \to \infty]{} a_1(0) \cdot \mathbb{1} \tag{7}$$

Let $t \in [-\tau, 0]$ and $k \in \mathbb{N}$, it is straightforward that

$$V^2(t + k\tau) = (x(t + k\tau) - x^*)^\top (x(t + k\tau) - x^*)$$
$$= (x(t + (k-1)\tau) - x^*)^\top P^\top P (x(t + (k-1)\tau) - x^*)$$
$$= \left(\sum_{i=2}^{n} a_i(t + (k-1)\tau)u_i \right)^\top \left(\sum_{i=2}^{n} \lambda_i a_i(t + (k-1)\tau)u_i \right)$$
$$= \sum_{i=2}^{n} \lambda_i a_i(t + (k-1)\tau)^2 \leq \lambda_2 V^2(t + (k-1)\tau)$$

Repeating the procedure k times and taking into account that $V(t) = V(0), \forall t \in [-\tau, 0]$ the proof is finished.   □

Remark 4. A version of Proposition 1 in the free of delays case can be found in [10]. Even if the presence of delay is not crucial, for the sake of completeness we preferred to provide a proof in the delayed case. Moreover, our result emphasizes both the convergence speed of the consensus protocol (6) and the consensus value.

3.2 Delay Margin Assuring a Fixed Network Topology

Let us consider now the network topology evolution (4) and derive the relation between ρ (in (4)), $\lambda_2(0)$ and τ assuring that $E(t) = E, \forall t \geq 0$. In other words, given an initial configuration (which determines $P(0)$ and consequently $\lambda_2(0)$) we derive a delay margin that assures an agreement speed at least equal with the convergence speed of the sequence $\{\rho^t\}$.

Proposition 2. *The algorithm* (2) *guarantees the asymptotic average-consensus of a network of agents whose topology evolution is given by* (4) *and* $E(t) = E, \forall t \geq 0$ *if* $\tau \leq \log_\rho \mu_2(0)$.

Proof. Let us consider again $t \in [-\tau, 0]$ and denote $P = P(0)$, $\lambda_2 = \lambda_2(0)$. Like in the proof of Proposition 1 we deduce that

$$V(1) = ||P(x(1-\tau) - x^*)|| \leq \mu_2 V(0) \tag{8}$$

For $\rho \in (0,1)$ one has $\tau \leq \log_\rho \mu_2 \Leftrightarrow \mu_2 \leq \rho^\tau \leq \rho$. On the other hand $|x_i(1) - x_j(1)| \leq \sqrt{2}V(1)$. Therefore, (8) leads to

$$|x_i(1) - x_j(1)| \leq \sqrt{2}V(0)\rho = M\rho$$

So, if $(i, j) \in E$ one gets $(i, j) \in E(1)$ which means $E(1) = E$ (see (4)).
   Supposing that $E(T) = E, \forall T \leq \tau - 1$ (and consequently $P(T) = P, \forall T \leq \tau - 1$) one obtains

$$V(T+1) = ||P(x(T - \tau + 1) - x^*)|| \leq \mu_2 V(0)$$

Therefore

$$V(T+1) \leq \rho^\tau V(0) \leq \rho^{T+1}V(0)$$

and as above

$$|x_i(T+1) - x_j(T+1)| \leq \sqrt{2}V(0)\rho^{T+1} = M\rho^{T+1}$$

implying $E(T+1) = E$. Resuming, we have proved that

$$V(T) \leq \rho^T V(0) \quad \& \quad E(T) = E, \quad \forall T \in [0, \tau] \tag{9}$$

Let us now suppose that $E(T) = E, \forall T \leq T^*$ (and consequently $P(T) = P, \forall T \leq T^*$) and consider $T^* = t^* + k\tau, k \in \mathbb{N}, t^* \in [0, \tau - 1]$. The procedure used in the proof of Proposition 1 and (9) yield

$$V(T^* + 1) = V(t^* + k\tau + 1) \leq \mu_2^k V(t^* + 1) \leq \rho^{k\tau}\rho^{t^*+1}V(0)$$

which immediately leads to

$$|x_i(T^*+1) - x_j(T^*+1)| \le \sqrt{2}V(0)\rho^{T^*+1} = M\rho^{T^*+1}$$

implying $E(T^*+1) = E$. In other words we have proved the statement by induction.

$\square$

Definition 3. Considering $u_i(0), i = 1,\ldots,n$ the orthonormal eigenvectors of the symmetric matrix $P(0)^\top P(0)$, any initial condition can be written as $x(0) = \sum_{i=1}^n a_i(0)u_i(0)$. The set of *admissible initial conditions* is then defined by $H = \{x(0) \in \mathbb{R}^n \mid a_2(0) \ne 0\}$.

The Lebesgue measure of the set $\mathbb{R}^n \setminus H$ is zero and this set will be neglected in the further development. In other words we always consider that $x(0) \in H$.

The next result shows that for all admissible initial conditions $x(0) \in H$, $\tau \le \log_\rho \mu_2(0)$ is a minimal requirement assuring that the network topology is fixed.

Proposition 3. *Let consider a network of agents whose topology evolution is given by (4) and the discrete-time collective dynamics given by (2). If $\tau > \log_\rho \mu_2(0)$ then for all $x(0) \in H$ there exists $t > 0$ such that $E(t) \subsetneq E$.*

Proof. Let us suppose that the statement is false. Thus $E(t) = E, \forall t \ge 0$ which implies $P(t) = P, \lambda_i(t) = \lambda_i, u_i(t) = u_i, \forall i, \forall t \ge 0$ reducing (2) to (6). Since the network topology evolution is given by (4) we deduce that

$$|x_i(t) - x_j(t)| \le M\rho^t, \quad \forall(i,j) \in E, t \ge 0 \qquad (10)$$

and as in the proof of Proposition 2

$$V(t) \le \rho^t V(0), \quad \forall t \ge 0 \qquad (11)$$

On the other hand (7) leads to $x(t) - x^* = \sum_{i=2}^n a_i(t)u_i, \quad \forall t$. Therefore, taking into account the orthogonality of $\{u_i\}$ one obtains

$$
\begin{aligned}
V(t+k\tau)^2 &= (x(t+k\tau) - x^*)^\top (x(t+k\tau) - x^*) \\
&= \sum_{i=2}^n \lambda_i^k a_i(0)^2 \ge \lambda_2^k a_2(0)^2, \forall t \in [-\tau, 0]
\end{aligned}
\qquad (12)
$$

Combining (11) and (12) one arrives at

$$\lambda_2^k a_2(0)^2 \le V(0)^2 \rho^{2k\tau} \rho^{2t} \Leftrightarrow \left(\frac{\rho^\tau}{\mu_2}\right)^k \ge \frac{|a_2(0)|}{V(0)\rho^t}, \quad \forall t \in [-\tau, 0], \forall k \in \mathbb{N}$$

But $0 < \rho^\tau < \mu_2$ means $\left(\frac{\rho}{\mu_2}\right)^t \xrightarrow[t\to\infty]{} 0$ and $x(0) \in H$ means $\frac{|a_2(0)|}{V(0)\rho^t} > 0$ which leads to a contradiction.

$\square$

4 Agreement Speed in Networks with Dynamic Topology

In the sequel we consider a network of agents with the discrete-time collective dynamics given by (2) whose topology evolution is given by (4). In the previous sections we have shown that $\tau \leq \log_\rho \mu_2(0)$ guarantees a fixed network topology, the consensus and an agreement speed higher than the convergence speed of $(\rho^t)_{t \geq 0}$. We have also noticed that $\tau > \log_\rho \mu_2(0)$ is equivalent with a time-dependent network topology. In this section we analyze the agreement speed for $\tau > \log_\rho \mu_2(0)$. It is worth noting that under this hypothesis the global consensus may not be achieved and the agents organize themselves into several communities where a local agreement is reached (see Section 5). However in this paper we analyze only the case when the global consensus can be achieved, thus the agents are at least *finally linked together*.

Assumption 2 *There exist $M^* > 0$, $\gamma < \rho$ such that $|x_i(t) - x_i^*| \leq M^* \gamma^t, \forall i, t$.*

Proposition 4. *Consider a network of agents with the discrete-time collective dynamics given by (2) whose topology evolution is given by (4). Suppose the agents asymptotically agree and Assumption 2 holds, then*

$$\exists T^* > 0 \text{ such that } |x_i(t) - x_j(t)| \leq M\rho^t, \forall t \geq T^* \tag{13}$$

Therefore, the network topology is fix for all $t \geq T^*$ i.e. $E(t) = E(T^*) = E(0)$, $P(t) = P(T^*) = P(0), \forall t \geq T^*$. Moreover $\tau < \log_\rho \mu_2(T^*)$.

Proof. Assumption 2 yields

$$|x_i(t) - x_j(t)| \leq |x_i(t) - x_i^*| + |x_j(t) - x_j^*| \leq 2M^* \gamma^t$$

On the other hand, it is clear that $\gamma < \rho$ implies the existence of $T^* > 0$ such that $2M^* \gamma^t \leq M\rho^t, \forall t \geq T^*$. Thus, (13) holds and $E(t) = E(T^*) = E(0)$, $P(t) = P(T^*) = P(0), \forall t \geq T^*$.

The last part of the statement will be proven by contradiction. Let us consider that $\mu_2(T^*) > \rho^\tau$. It is noteworthy that we can suppose $x(T^*) \in H$ (see [8] for details). Then Proposition 3 implies the existence of a time step $t > T^*$ such that $E(t) \neq E(T^*)$ contradicting the first part of the statement. In other words the following sequence of inequality must hold $\mu_2(T^*) \leq \rho^\tau$ and the proof is finished. $\square$

Since $E(0)$ has a finite number of elements and $E(t) \subset E(0)$ we deduce that $E(t)$ belongs to a finite set of possible configurations. Therefore, $P(t)$ belongs to a finite set of doubly stochastic matrices $\mathcal{M} = \{P_1, \ldots, P_m\}$. The subspace of dimension $n-1$ orthogonal to $span\{\mathbb{1}\}$ will be denoted by $\mathcal{F}$.

Lemma 1. *The subspace $\mathcal{F}$ of $\mathbb{R}^n$ is an P_k - invariant subspace for all $k \in 1, \ldots, m$.*

Proof. Let $x \in \mathcal{F}$ and $P \in \mathcal{M}$. From the definition of $\mathcal{F}$ one has that $\mathbb{1}^\top x = 0$. Since P is doubly stochastic it follows that $\mathbb{1}^\top Px = \mathbb{1}^\top x = 0$. Thus, $Px \perp \mathbb{1}$ which means $Px \in \mathcal{F}$. $\square$

In the sequel we consider $P_k : \mathscr{F} \mapsto \mathscr{F}$. Thus $P_k \in \mathbb{R}^{(n-1)\times(n-1)}$ and the spectrum of the initial P_k is obtained by adding $\{1\}$ to the spectrum of the new P_k (see [5] for details). In the proof of Proposition 1 we have noticed that $x(0) - x^* \in \mathscr{F}$, so, $x(t) - x^* \in \mathscr{F}, \forall t \geq 0$.

Definition 4. The joint spectral radius of a set of matrices $\mathscr{M} = \{P_1, \ldots, P_m\}$ is defined by:

$$\delta(\mathscr{M}) = \limsup_{k \to \infty} \delta_k(\mathscr{M})$$

where

$$\delta_k(\mathscr{M}) = \max_{s(i) \in \{1, \ldots, m\}, \forall i \in 1, \ldots, k} ||P_{s(k)} \ldots P_{s(1)}||^{1/k}, \forall k \geq 1$$

Proposition 5. *Consider a network of agents with the discrete-time collective dynamics given by (2) whose topology evolution is given by (4).*

a) If $\tau > \log_\rho \mu_2(0)$ the consensus cannot be achieved faster than $O(\rho^t)$.
b) If the agreement speed is not faster than $O(\rho^t)$ then

$$\tau \geq \log_\rho \delta(\mathscr{M}).$$

Proof. a) If the consensus is achieved with an agreement speed higher than the convergence speed of $(\rho^t)_{t \geq 0}$, Proposition 4 assures us that $\tau \leq \log_\rho \mu_2(0)$. Thus, when $\tau > \log_\rho \mu_2(0)$ the consensus cannot be achieved with an agreement speed higher than the convergence speed of $(\rho^t)_{t \geq 0}$.

b) The agreement speed is at most the convergence speed of $(\rho^t)_{t \geq 0}$ if there exists $M^* > 0$ and a subsequence $(x(t_p))_{p \geq 0}$ such that

$$M^* \rho^{t_p} \leq ||x(t_p) - x^*||, \quad \forall p \geq 0$$

For $t = t^* + k\tau, t^* \in [-\tau + 1, 0], k \geq 1$ one has

$$||x(t) - x^*|| = ||A_k \ldots A_1(x(t) - x^*)|| \leq ||A_k \ldots A_1|| \cdot V(0)$$

where $A_i \in \mathscr{M}, \forall i = 1, \ldots, k$. Therefore,

$$M^* \rho^{t_p^* + k_p \tau} \leq ||A_{k_p} \ldots A_1|| \cdot V(0), \forall t_p^* \in [-\tau + 1, 0], p \geq 0 \qquad (14)$$

where t_p^* and k_p are chosen such that $t_p^* + k_p \tau = t_p, \forall p \geq 0$. Relation (14) can be further rewritten as

$$M^* \leq M^* \rho^{t_p^*} \leq \left(\frac{||A_{k_p} \ldots A_1||^{1/k_p}}{\rho^\tau} \right)^{k_p} \cdot V(0)$$

$$\leq \left(\frac{\delta(\mathscr{M})}{\rho^\tau} \right)^{k_p} \cdot V(0), \forall t_p^* \in [-\tau + 1, 0], p \geq 0 \qquad (15)$$

Inequality (15) may hold only if $\frac{\delta(\mathscr{M})}{\rho^\tau} > 1$. Thus, we conclude that $\tau \geq \log_\rho \delta(\mathscr{M})$. $\qquad \square$

5 Numerical Examples

We illustrate our results using the karate club network initially studied by Zachary in [13]. This is a social network with 34 agents shown in Figure 1. The Perron matrix defining the collective dynamics is given by $P(t) = I - \alpha L(t)$ where I is the identity matrix, $L(t)$ is the graph-Laplacian matrix at the moment t, and α is set to 0.05. The parameters of the model where chosen as follows: $M = \sqrt{2}V(0)$ and $\rho = 0.99$. The model was simulated for an initial condition chosen randomly in $[0, 1]^{34}$. The computations show that $\mu_2(0) = 0.9766$ and $\log_\rho \mu_2(0) = 2.356$. Simulations were performed as long as enabled by floating point arithmetics for $\tau = 2$, $\tau = 3$ and $\tau = 5$ (see (2)). The experimental results have proven that the network topology remains fix for $\tau = 2 < \log_\rho \mu_2(0)$ (Figure 1 left). For $\tau = 3 > \log_\rho \mu_2(0)$ the topology changes. Furthermore, only local consensus can be reached in this case as can be seen in Figures 1 right.

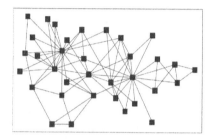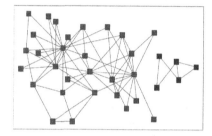

Fig. 1 Graph of the Zachary karate club network. Left: For $\tau = 2$ the initial network configuration is kept constant and the agents asymptotically reach the consensus. Right: For $\tau = 3$ some links are canceled during the simulation leading to two disconnected communities inside the graph.

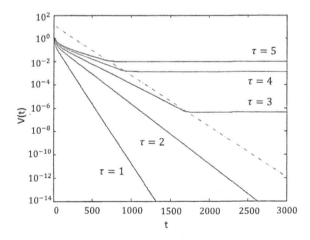

Fig. 2 The Lyapunov functions using a logarithmic scale for $\tau \in \{1,2,3,4,5\}$. Dashed line represents $\sqrt{n}M\rho^t$ using the same logarithmic scale.

The study of the agreement speed is summarized in Figure 2. The agreement speed decreases when τ increases. The agents try to reach a global agreement as far as the Lyapunov function $V(t) \leq \sqrt{n} M \rho^t$. When this condition is violated the agents organize themselves in communities where a local agreement is reached. It is noteworthy that the condition holds inside each group replacing n by the number of agents in the corresponding community.

6 Conclusions

In this paper we have analyzed a model of multi-agent system with decaying confidence and time-invariant delay in the communication channels. The study provide an exact delay bound that assure the preservation of the network topology. We have also shown that increasing the delay, the agreement speed decreases and the global consensus may not be reached. A sufficient condition for the global agreement of the agents in our model can be expressed in terms of the eigenvalues of the matrix defining the collective dynamics. Numerical simulations show us the necessity of this condition.

References

1. Blondel, V.D., Hendrickx, J.M., Olshevsky, A., Tsitsiklis, N.: Convergence in multiagent coordination, consensus, and flocking. In: Proceedings of the 44th IEEE CDC, pp. 2996–3000 (2005)
2. Cortes, J., Bullo, F.: Coordination and geometric optimization via distributed dynamical systems. SIAM J. Control Optim. 44(5), 1543–1574 (2005)
3. Fax, A., Murray, R.M.: Information flow and cooperative control of vehicle formations. IEEE Trans. Automat. Contr. 49, 1465–1476 (2004)
4. Godsil, C., Royle, G.: Algebraic Graph Theory. Springer, New York (2001)
5. Jadbabaie, A., Lin, J., Morse, A.S.: Coordination of groups of mobile autonomous agents using nearest neighbor rules. IEEE Trans. Automat. Contr. 48(6), 988–1001 (2003)
6. Li, S., Wang, H.: Multi-agent coordination using nearest-neighbor rules: revisiting the Vicsek model. arXiv:cs/0407021 (2004)
7. Merris, R.: Laplacian Matrices of Graphs: A Survey. Linear Algebra and its Applications 197, 143–176 (1994)
8. Morărescu, I.-C., Girard, A.: Opinion dynamics with decaying confidence: application to community detection in graphs. IEEE Trans. Automat. Contr. (2009) (to appear, 2011)
9. Moreau, L.: Stability of multiagent systems with time-dependent communication links. IEEE Trans. Automat. Contr. 50(2), 169–182 (2005)
10. Olfati-Saber, R., Murray, R.M.: Consensus problems in networks of agents with switching topology and time-delays. IEEE Trans. Automat. Contr. 49(9), 1520–1833 (2004)
11. Olfati-Saber, R., Murray, R.M.: Consensus protocol for networks of dynamic agents. In: Proc. of American Control Conference (2003)
12. Vicsek, T., Czirók, A., Ben-Jacob, E., Cohen, O., Shochet, I.: Novel type of phase transition in a system of self-deriven particles. Phys. Rev. Lett. 75(6), 1226–1229 (1995)
13. Zachary, W.W.: An information flow model for conflict and fission in small groups. Journal of Anthropological Research 33(4), 452–473 (1977)

H_∞ Control of Networked Control Systems via Discontinuous Lyapunov Functionals

Kun Liu and Emilia Fridman

Abstract. This chapter presents a new stability and L_2-gain analysis of linear Networked Control Systems (NCS). The new method is inspired by discontinuous Lyapunov functions that were introduced in [13] and [12] in the framework of impulsive system representation. Most of the existing works on the stability of NCS (in the framework of time delay approach) are reduced to some Lyapunov-based analysis of systems with uncertain and bounded time-varying delays. This analysis via time-independent Lyapunov functionals does not take advantage of the sawtooth evolution of the delays induced by sample-and-hold. The latter drawback was removed in [4], where *time-dependent* Lyapunov functionals for sampled-data systems were introduced. This led to essentially less conservative results. The objective of the present chapter is to extend the time-dependent Lyapunov functional approach to NCS, where variable sampling intervals, data packet dropouts and *variable network-induced delays* are taken into account. The new analysis is applied to a novel network-based static output-feedback H_∞ control problem. Numerical examples show that the novel discontinuous terms in Lyapunov functionals essentially improve the results.

1 Introduction

Three main approaches have been used to the sampled-data control and later to the NCS, where the plant is controlled via communication network. The first one is based on discrete-time models [22]. This approach is not applicable to the performance analysis (e.g. to the exponential decay rate) of the continuous-time

Kun Liu
Department of Electrical Engineering-Systems, Tel-Aviv University
e-mail: liukun@eng.tau.ac.il

Emilia Fridman
Department of Electrical Engineering-Systems, Tel-Aviv University
e-mail: emilia@eng.tau.ac.il

R. Sipahi et al. (Eds.): Time Delay Sys.: Methods, Appli. and New Trends, LNCIS 423, pp. 429–440.
springerlink.com   © Springer-Verlag Berlin Heidelberg 2012

closed-loop systems. The second one is a *time delay approach*, where the system is modeled as a continuous-time system with a time-varying *sawtooth delay* in the control input [6]. The time delay approach via *time-independent* Lyapunov-Krasovskii functionals or Lyapunov-Razumikhin functions lead to Linear Matrix Inequalities (LMIs) [2] for analysis and design of linear uncertain NCS [3, 8, 20, 21]. The third approach is based on the representation of the sampled-data system in the form of *impulsive model* [1]. Recently the impulsive model approach was extended to the case of uncertain sampling intervals [13] and to NCS [12]. In [13] and [12] a discontinuous Lyapunov function method was introduced, that improved the existing Lyapunov-based results.

For systems with time-varying delays, stability conditions via time-independent Lyapunov functionals guarantee also the stability of the corresponding systems with constant delay. However, it is well-known (see examples in [10] and discussions on *quenching* in [14], as well as Example 1 below) that in many particular systems the upper bound on the sawtooth delay that preserves the stability may be higher than the corresponding bound for the constant delay. In the recent paper [4] *time-dependent Lyapunov functionals* have been introduced for the analysis of sampled-data systems in the framework of time delay approach. The introduced time-dependent terms of Lyapunov functionals lead to qualitatively new results, taking into account the sawtooth evolution of the delays induced by sample-and-hold. In some well-studied numerical examples, the results of [4] approach the analytical values of minimum L_2-gain and of maximum sampling interval, preserving the stability.

The objective of the present chapter is to extend the *discontinuous Lyapunov functional method* (in the framework of time delay approach) *to network-based H_∞ control*, where data packet dropouts and *variable network-induced delays* are taken into account. Our Lyapunov functional depends on time and on the upper bound of the network-induced delay and it does not grow along the input update times. We apply our analysis results to a novel *static output-feedback H_∞ control*. We note that the observer-based control via network is usually encountered with some waiting strategy and buffers [17]. The implementation of the network-based static output-feedback controller is simple. Sufficient conditions for the stabilization via the continuous static output-feedback can be found in the survey [19]. Similar to the sampled-data H_∞ control [18], we consider an H_∞ performance index that takes into account the updating rates of the measurement. This index is related to the energy of the measurement noise. Numerical examples show that the novel discontinuous terms in the Lyapunov functional essentially reduce the conservatism.

Notation: Throughout the chapter the superscript 'T' stands for matrix transposition, $\mathscr{R}^n$ denotes the n dimensional Euclidean space with vector norm $\| \cdot \|$, $\mathscr{R}^{n \times m}$ is the set of all $n \times m$ real matrices, and the notation $P > 0$, for $P \in \mathscr{R}^{n \times n}$ means that P is symmetric and positive definite. The symmetric elements of the symmetric matrix will be denoted by $*$. Given a positive number $\tau_M > 0$, the space of functions $\phi : [-\tau_M, 0] \to \mathscr{R}^n$, which are absolutely continuous on $[-\tau_M, 0)$, have a finite $\lim_{\theta \to 0-} \phi(\theta)$ and have square integrable first order derivatives is denoted by W

with the norm $\|\phi\|_W = \max_{\theta \in [-\tau_M, 0]} |\phi(\theta)| + \left[\int_{-\tau_M}^0 |\dot{\phi}(s)|^2 ds \right]^{\frac{1}{2}}$. We also denote $x_t(\theta) = x(t+\theta), \dot{x}_t(\theta) = \dot{x}(t+\theta), (\theta \in [-\tau_M, 0])$.

2 Problem Formulation

Consider the system

$$\dot{x}(t) = Ax(t) + B_2 u(t) + B_1 w(t),$$
$$z(t) = C_1 x(t) + D_{12} u(t), \tag{1}$$

where $x(t) \in \mathscr{R}^n$ is the state vector, $w(t) \in \mathscr{R}^q$ is the disturbance, $u(t) \in \mathscr{R}^m$ is the control input and $z(t) \in \mathscr{R}^r$ is the signal to be controlled or estimated, A, B_1, B_2, C_1 and D_{12} are system matrices with appropriate dimensions. We will consider exponential stabilization (for $w = 0$) and H_∞ control of (1) via static output feedback.

Consider the static output-feedback control of NCS shown in Figure 1. The sampler is time-driven, whereas the controller and the Zero-Order Hold (ZOH) are event-driven (in the sense that the controller and the ZOH update their outputs as soon as they receive a new sample). We assume that the measurement output $y(s_k) \in \mathscr{R}^p$ is available at discrete sampling instants $0 = s_0 < s_1 < \cdots < s_k < \cdots$, $\lim_{k \to \infty} s_k = \infty$ and it may be corrupted by a measurement noise signal $v(s_k)$ (see Figure 1):

$$y(s_k) = C_2 x(s_k) + D_{21} v(s_k). \tag{2}$$

We take into account data packet dropouts by allowing the sampling to be nonuniform. Thus, in our formulation $y(s_k)$, $k = 0, 1, 2...$ correspond to the measurements that are not lost.

Denote by t_k the updating instant time of the ZOH, and suppose that the updating signal at the instant t_k has experienced a signal transmission delay η_k, so we have $s_k = t_k - \eta_k$, which corresponds to the sampling time of the data that has not been lost. Following [12], we allow the delays η_k to grow larger than $s_{k+1} - s_k$, provided that the sequence of input update times t_k remains strictly increasing. This means that if an old sample gets to the destination after the most recent one, it should be dropped.

The static output-feedback controller has a form $u(t_k) = Ky(t_k - \eta_k)$, where K is the controller gain. Thus, considering the behavior of the ZOH, we have

$$u(t) = Ky(t_k - \eta_k), \quad t_k \le t < t_{k+1}, \ k = 0, 1, 2, \ldots \tag{3}$$

with t_{k+1} being the next updating instant time of the ZOH after t_k.

As in [8, 12, 21], we assume that

$$t_{k+1} - t_k + \eta_k \le \tau_M, \ 0 \le \eta_k \le \eta_M, \ k = 0, 1, 2, \ldots \tag{4}$$

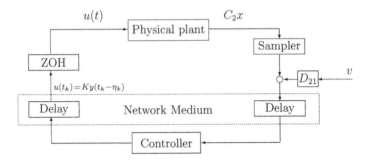

Fig. 1 Networked static output-feedback control system.

where η_M is a known upper bound on the network-induced delays η_k and τ_M denotes the maximum time span between the time $s_k = t_k - \eta_k$ at which the state is sampled and the time t_{k+1} at which next update arrives at the ZOH.

As a Corollary from the main result, we will formulate sufficient conditions for stabilization of NCS with constant delay $\eta_k \equiv \eta_M$.

Remark 1. The assumption (4) is equivalent to $s_{k+1} - s_k + \eta_{k+1} \leq \tau_M$, $0 \leq \eta_{k+1} \leq \eta_M$, $k = 0,1,2,...$ The latter implies that $s_{k+1} - s_k \leq \tau_M$, i.e. that the sampling intervals and the numbers of successive packet-dropouts are uniformly bounded.

Defining

$$\tau(t) = t - t_k + \eta_k, \ t_k \leq t < t_{k+1}, \tag{5}$$

we obtain the following closed-loop system (1), (3):

$$\dot{x}(t) = Ax(t) + A_1 x(t - \tau(t)) + A_2 v(t - \tau(t)) + B_1 w(t),$$
$$z(t) = C_1 x(t) + D_1 x(t - \tau(t)) + D_2 v(t - \tau(t)), \tag{6}$$

where

$$A_1 = B_2 K C_2, \quad A_2 = B_2 K D_{21},$$
$$D_1 = D_{12} K C_2, \quad D_2 = D_{12} K D_{21}. \tag{7}$$

Under (4) and (5), we have $0 \leq \tau(t) \leq t_{k+1} - t_k + \eta_k \leq \tau_M$ and $\dot{\tau}(t) = 1$ for $t \neq t_k$.

Denote $\bar{v}(t) = v(t - \tau(t))$ $(t \geq t_0)$. Then (6) has two disturbances $\bar{v} \in L_2[t_0, \infty)$ and $w \in L_2[t_0, \infty)$, where

$$\|\bar{v}\|_{L_2}^2 = \int_{t_0}^{\infty} v^T(t - \tau(t)) v(t - \tau(t)) dt = \sum_{k=0}^{\infty} (t_{k+1} - t_k) v^T(t_k - \eta_k) v(t_k - \eta_k). \tag{8}$$

For a given scalar $\gamma > 0$, we thus define the following performance index [18]:

$$J = \|z\|_{L_2}^2 - \gamma^2 (\|\bar{v}\|_{L_2}^2 + \|w\|_{L_2}^2)$$

$$= \int_{t_0}^{\infty} [z^T(s)z(s) - \gamma^2 w^T(s)w(s)]ds - \gamma^2 \sum_{k=0}^{\infty} (t_{k+1} - t_k)v^T(t_k - \eta_k)v(t_k - \eta_k). \quad (9)$$

Our objective is to find a controller of (3) that internally exponentially stabilizes the system and that leads to L_2-gain of (6) less than γ. The latter means that along (6) $J < 0$ for the zero initial function and for all non-zero $w \in L_2, v \in l_2$ and for all allowable sampling intervals, data packet dropouts and network-induced delays, satisfying (4).

We note that *the last term* of the performance index J *takes into account the updating rates of the measurement* and is thus related to the *energy of the measure-ment noise* [18]. For the sampled-data control under uniform sampling intervals, a conventional performance index has a form [16]:

$$J_{samp} = \int_{t_0}^{\infty}[z^T(s)z(s) - \gamma^2 w^T(s)w(s)]ds - \gamma^2 \sum_{k=0}^{\infty} v^T(t_k)v(t_k). \quad (10)$$

Index J_{samp} has a little physical sense for NCS since it does not take the updat-ing rates into account. For the sake of brevity, in the remainder of this chapter the notation τ stands for the time-varying delay $\tau(t)$.

3   Main Results

3.1   *Exponential Stability and L_2-Gain Analysis*

In this section, we analyze the closed-loop systems (6). Exponential stability of (6) with $w = v = 0$, i.e. of

$$\dot{x}(t) = Ax(t) + A_1 x(t - \tau), \ \tau = t - t_k + \eta_k, \ t_k \leq t < t_{k+1}, \quad (11)$$

as well as the L_2-gain analysis of (6) will be based on the following

Lemma 1. *Let there exist positive numbers* α, β, δ *and a functional* $V : \mathscr{R} \times W \times L_2[-\tau_M, 0] \to [t_0, \infty)$ *such that* $\beta|\phi(0)|^2 \leq V(t, \phi, \dot{\phi}) \leq \delta\|\phi\|_W^2$. *Let the function* $\bar{V}(t) = V(t, x_t, \dot{x}_t)$ *be continuous from the right for* $x(t)$ *satisfying (6), absolutely continuous for* $t \neq t_k$ *and satisfies* $\lim_{t \to t_k^-} \bar{V}(t) \geq \bar{V}(t_k)$.
(i) *If along (11)*

$$\dot{\bar{V}}(t) + 2\alpha\bar{V}(t) \leq 0, \quad almost \ for \ all \ t, \quad (12)$$

then $\bar{V}(t) \leq e^{-2\alpha t}\bar{V}(t_0)$, *i.e.* $|x(t)|^2 \leq e^{-2\alpha t}\frac{\delta}{\beta}\|x_{t_0}\|_W^2$ *for* $x_{t_0} \in W$ *and thus (11) is exponentially stable with the decay rate* α.
(ii) *For a given* $\gamma > 0$, *if along (6)*

$$\dot{\bar{V}}(t) + z^T(t)z(t) - \gamma^2 w^T(t)w(t) - \gamma^2 v^T(t_k - \eta_k)v(t_k - \eta_k) < 0, \quad (13)$$
$$\forall w \neq 0, \ v \neq 0, \ t_k \leq t \leq t_{k+1},$$

then the performance index (9) achieves $J < 0$ for all nonzero $w \in L_2, v \in l_2$ and for the zero initial function.

Proof. For (i) see [4].

(ii) Given $N \gg 1$, we integrate the first inequality (13) from t_0 till t_N. We have

$$\bar{V}(t_N) - \bar{V}(t_{N-1}) + \bar{V}(t_{N-1}^-) - \bar{V}(t_{N-2}) \ldots + \bar{V}(t_1^-) - \bar{V}(t_0)$$
$$+ \int_{t_0}^{t_N} [z^T(t)z(t) - \gamma^2 w^T(t)w(t)]dt - \gamma^2 \sum_{k=0}^{N-1} (t_{k+1} - t_k)v^T(t_k - \eta_k)v(t_k - \eta_k) < 0.$$

Since $\bar{V}(t_N) \geq 0$, $\bar{V}(t_{k-1}^-) - \bar{V}(t_{k-1}) \geq 0$ for $k = 2, \ldots, N$ and $\bar{V}(t_0) = 0$, we find

$$\int_{t_0}^{t_N} [z^T(t)z(t) - \gamma^2 w^T(t)w(t)]dt - \gamma^2 \sum_{k=0}^{N-1} (t_{k+1} - t_k)v^T(t_k - \eta_k)v(t_k - \eta_k) < 0.$$

Thus, for $N \to \infty$ we arrive to $J < 0$.

A standard time-independent functional for delay-dependent stability of (11) with fast varying delay $\tau \in [0, \tau_M]$ has a form (see [5, 15])

$$
\begin{aligned}
V_0(x_t, \dot{x}_t) &= x^T(t)Px(t) + \int_{t-\tau_M}^{t} e^{2\alpha(s-t)}x^T(s)Sx(s)ds \\
&+ \frac{1}{\tau_M} \int_{-\tau_M}^{0} \int_{t+\theta}^{t} e^{2\alpha(s-t)}\dot{x}^T(s)R\dot{x}(s)dsd\theta, \quad P > 0, \ S > 0, \ R > 0,
\end{aligned}
\tag{14}
$$

where $\alpha > 0$ corresponds to exponential stability with the decay rate $\alpha > 0$. In the existing papers [2, 3, 6, 8, 21] in the framework of input delay approach, time-independent Lyapunov functionals are usually involved.

For the case of $\eta_k \equiv 0$ (when there are no network-induced delays) and $\tau = t - t_k$, the following time-dependent functional has been introduced in [4]

$$V_s(t, x_t, \dot{x}_t) = \bar{V}_s(t) = x^T(t)Px(t) + \sum_{i=1}^{2} V_{is}(t, x_t, \dot{x}_t),$$

where the discontinuous terms V_{1s} and V_{2s} have the form

$$
\begin{aligned}
V_{1s}(t, x_t, \dot{x}_t) &= \frac{\tau_M - \tau}{\tau_M} [x(t) - x(t - \tau)]^T X [x(t) - x(t - \tau)], \quad X > 0, \\
V_{2s}(t, x_t, \dot{x}_t) &= \frac{\tau_M - \tau}{\tau_M} \int_{t-\tau}^{t} e^{2\alpha(s-t)}\dot{x}^T(s)U\dot{x}(s)ds, \quad U > 0, \ \alpha > 0.
\end{aligned}
\tag{15}
$$

For $\eta_k \equiv 0$, V_{1s} and V_{2s} do not increase along the jumps, since these terms are nonnegative before the jumps at t_k and become zero just after the jumps (because $t_{|t=t_k} = (t - \tau)_{|t=t_k}$). Thus, $\bar{V}_s(t)$ does not increase along the jumps and the condition $\lim_{t \to t_k^-} \bar{V}_s(t) \geq \bar{V}_s(t_k)$ holds.

In the case of $\eta_k \neq 0$ and $\tau = t - t_k + \eta_k$, the discontinuous terms (15) cannot be used, because $t_{|t=t_k} \neq (t - \tau)_{|t=t_k} = t_k - \eta_k$. In this case, the upper bound τ_M, that preserves the stability, is between the corresponding upper bounds on the (arbitrary) fast varying delay and on the sampling in the absence of network-induced delays (with $\eta_k \equiv 0$). Since the biggest bound is the one for the case of $\eta_k \equiv 0$, we construct the discontinuous terms of Lyapunov functional that correspond to the "worst case", where we have the maximum network-induced delay η_M. Defining

$$\tau_1 = max\{0,\ \tau - \eta_M\} = max\{0,\ t - t_k - \eta_M + \eta_k\},\ t_k \leq t < t_{k+1}, \tag{16}$$

we note that $\tau_1|_{t=t_k} = 0$ and that $\tau_1 \leq \tau_M - \eta_M$. We consider the functional of the form

$$V(t,x_t,\dot{x}_t) = \bar{V}(t) = V_0(x_t,\dot{x}_t) + \Sigma_{i=1}^2 V_i(t,x_t,\dot{x}_t), \tag{17}$$

where V_0 is defined by (14) and

$$
\begin{aligned}
V_1(t,x_t,\dot{x}_t) &= \tfrac{\tau_M-\tau}{\tau_M-\eta_M}[x(t)-x(t-\tau_1)]^T X[x(t)-x(t-\tau_1)],\ X > 0,\\
V_2(t,x_t,\dot{x}_t) &= \tfrac{\tau_M-\tau}{\tau_M-\eta_M}\int_{t-\tau_1}^t e^{2\alpha(s-t)}\dot{x}^T(s)U\dot{x}(s)ds,\quad U > 0,\ \alpha > 0.
\end{aligned}
\tag{18}
$$

Along the input update times $t = t_k$, V_1 and V_2 do not increase since these terms are nonnegative before t_k and become zero just after t_k (because $t|_{t=t_k} = (t - \tau_1)|_{t=t_k}$). Thus, $\bar{V}$ does not increase along the input update times and the condition $\lim_{t\to t_k^-}\bar{V}(t) \geq \bar{V}(t_k)$ holds.

By using the discontinuous Lyapunov functional (17), we obtain the following sufficient conditions:

Theorem 1. (i) Given $\alpha > 0$, let there exist $n \times n$-matrices $P > 0, R > 0, U > 0, X > 0, S > 0, T_{11}, P_{2i}, P_{3i}, T_{2i}, M_{2i}, Y_{ij}$ and $Z_{ij}(i,j = 1,2)$ such that the following four LMIs

$$\Psi_{11} = \begin{bmatrix} \Phi_{11} & \Phi_{12} & \tau_M Z_{11}^T & Z_{11}^T & \Phi_{16} \\ * & \Phi_{22} & \tau_M Z_{12}^T & Z_{12}^T & \Phi_{26} \\ * & * & -Re^{-2\alpha\tau_M} & 0 & 0 \\ * & * & * & -Se^{-2\alpha\tau_M} & 0 \\ * & * & * & * & T_{11}+T_{11}^T \end{bmatrix} < 0, \tag{19}$$

$$\Psi_{12} = \begin{bmatrix} \Phi_{11} & \Phi_{12} & \eta_M Y_{11}^T & (\tau_M-\eta_M)Z_{11}^T & Z_{11}^T & \Phi_{16} \\ * & \Phi_{22} & \eta_M Y_{12}^T & (\tau_M-\eta_M)Z_{12}^T & Z_{12}^T & \Phi_{26} \\ * & * & -\tfrac{\eta_M}{\tau_M}Re^{-2\alpha\tau_M} & 0 & 0 & \eta_M T_{11} \\ * & * & * & -\tfrac{\tau_M-\eta_M}{\tau_M}Re^{-2\alpha\tau_M} & 0 & 0 \\ * & * & * & * & -Se^{-2\alpha\tau_M} & 0 \\ * & * & * & * & * & T_{11}+T_{11}^T \end{bmatrix} < 0, \tag{20}$$

$$\Psi_{21} = \begin{bmatrix} \Omega_{11}-\tfrac{1-2\alpha(\tau_M-\eta_M)}{\tau_M-\eta_M}X & \Omega_{12}+X & Z_{21}^T & (\tau_M-\eta_M)Z_{21}^T & \eta_M M_{21}^T & \Omega_{17}+\tfrac{1-2\alpha(\tau_M-\eta_M)}{\tau_M-\eta_M}X & \Omega_{18} \\ * & \Omega_{22}+U & Z_{22}^T & (\tau_M-\eta_M)Z_{22}^T & \eta_M M_{22}^T & \Omega_{27}-X & \Omega_{28} \\ * & * & -Se^{-2\alpha\tau_M} & 0 & 0 & 0 & T_{22} \\ * & * & * & \Omega_{55}|_{\tau=\eta_M} & 0 & 0 & (\tau_M-\eta_M)T_{22} \\ * & * & * & * & \Omega_{66} & 0 & 0 \\ * & * & * & * & * & \Omega_{77}-\tfrac{1-2\alpha(\tau_M-\eta_M)}{\tau_M-\eta_M}X & 0 \\ * & * & * & * & * & * & -T_{22}-T_{22}^T \end{bmatrix} < 0, \tag{21}$$

$$\Psi_{22} = \begin{bmatrix} \Omega_{11}-X & \Omega_{12} & Z_{21}^T & (\tau_M-\eta_M)Y_{21}^T & \eta_M M_{21}^T & \Omega_{17}+\tfrac{1}{\tau_M-\eta_M}X & \Omega_{18} \\ * & \Omega_{22} & Z_{22}^T & (\tau_M!-\eta_M)Y_{22}^T & \eta_M M_{22}^T & \Omega_{27} & \Omega_{28} \\ * & * & -Se^{-2\alpha\tau_M} & 0 & 0 & 0 & T_{22} \\ * & * & * & \Omega_{44}|_{\tau=\tau_M} & 0 & (\tau_M-\eta_M)T_{21} & 0 \\ * & * & * & * & \Omega_{66} & 0 & 0 \\ * & * & * & * & * & \Omega_{77}-\tfrac{1}{\tau_M-\eta_M}X & 0 \\ * & * & * & * & * & * & -T_{22}-T_{22}^T \end{bmatrix} < 0 \tag{22}$$

are feasible, where

$$\Phi_{11} = A^T P_{21} + P_{21}^T A + 2\alpha P + S - Y_{11} - Y_{11}^T, \quad \Phi_{12} = P - P_{21}^T + A^T P_{31} - Y_{12},$$
$$\Phi_{22} = -P_{31} - P_{31}^T + R, \quad \Phi_{16} = Y_{11}^T - Z_{11}^T + P_{21}^T A_1 - T_{11}, \quad \Phi_{26} = Y_{12}^T - Z_{12}^T + P_{31}^T A_1,$$
$$\Omega_{11} = A^T P_{22} + P_{22}^T A + 2\alpha P + S - Y_{21} - Y_{21}^T, \quad \Omega_{12} = P - P_{22}^T + A^T P_{32} - Y_{22},$$
$$\Omega_{17} = Y_{21}^T - M_{21}^T - T_{21}, \quad \Omega_{18} = M_{21}^T - Z_{21}^T + P_{22}^T A_1, \quad \Omega_{22} = -P_{32} - P_{32}^T + R,$$
$$\Omega_{27} = Y_{22}^T - M_{22}^T, \quad \Omega_{28} = M_{22}^T - Z_{22}^T + P_{32}^T A_1,$$
$$\Omega_{44} = -(\tau - \eta_M)[\frac{1}{\tau_M} e^{-2\alpha\tau_M} R + \frac{1}{\tau_M - \eta_M} e^{-2\alpha(\tau_M - \eta_M)} U],$$
$$\Omega_{55} = -\frac{\tau_M - \tau}{\tau_M} R e^{-2\alpha\tau_M}, \quad \Omega_{66} = -\frac{\eta_M}{\tau_M} R e^{-2\alpha\tau_M}, \quad \Omega_{77} = T_{21} + T_{21}^T.$$

$$(23)$$

Then system (11) is exponentially stable with the decay rate α for all delays (5) satisfying (4). If the above LMIs hold with $\alpha = 0$, then they are feasible for a small enough $\alpha_0 > 0$, i.e. (11) is exponentially stable with the decay rate α_0.

(ii) Given $\gamma > 0$, if the following LMIs

$$
\left[
\begin{array}{c|ccc}
 & P_{2i}^T A_2 & P_{2i}^T B_1 & C_1^T \\
 & P_{3i}^T A_2 & P_{3i}^T B_1 & 0 \\
\Psi_{ij}|_{\alpha=0} & 0 & 0 & 0 \\
 & 0 & 0 & D_1^T \\
\hline
* & -\gamma^2 I & 0 & D_2^T \\
* & * & -\gamma^2 I & 0 \\
* & * & * & -I
\end{array}
\right] < 0, \quad i,j = 1,2
$$

$$(24)$$

with notations given in (23) are feasible. Then (6) is internally exponentially stable and has L_2-gain less than γ.

Proof. (i) see [9].

(ii) Consider Lyapunov functional of (17) with $\alpha = 0$. By using arguments similar to the proof of (i), we find that (13) holds if LMIs (24) are feasible.

When the network-induced delay is constant, i.e. $\eta_k \equiv \eta_M$, we have only one case of $\tau \in [\eta_M, \tau_M]$. So the following result is obtained from Theorem 1:

Corollary 1. *(i) Consider (6) with $\eta_k \equiv \eta_M$. Given $\alpha > 0$, let there exist $n \times n$-matrices $P > 0$, $R > 0, U > 0, X > 0, S > 0, P_{22}, P_{32}, T_{2i}, M_{2i}, Y_{2i}$ and $Z_{2i}(i = 1,2)$ such that two LMIs $\Psi_{2i} < 0 (i = 1,2)$ with notations given in (23) are feasible. Then (11) is exponentially stable with the decay rate α. If LMIs $\Psi_{2i} < 0 (i = 1,2)$ hold with $\alpha = 0$, then (11) is exponentially stable with a small enough decay rate.*

(ii) Given $\gamma > 0$, if two LMIs (24), where $i = 2, j = 1,2$ are feasible, then (6) is internally exponentially stable and the cost function (9) achieves $J < 0$ for all nonzero $w \in L_2, v \in l_2$ and for the zero initial condition.

3.2 Application to Network-Based Static Output-Feedback Design

We apply the results of the previous section to *static output-feedback design* problem. It is well-known that static output-feedback stabilization is a non-convex problem. We suggest here some solution to this problem (which may be conservative). Assume that B_2 is of full rank. Then there exists a mapping $x \mapsto \tilde{T}x$ with nonsingular $n \times n$-matrix $\tilde{T}$, such that B_2 has the following partitioned form $B_2^T = [0 \ B^T]$, where $B \in \mathscr{R}^{m \times m}$ is non-singular. Hence, without loss of generality, we take B_2 in the above form.

Corollary 2. *Given $\gamma > 0$ and tuning scalar parameters $\varepsilon_i(i = 2,3)$ and a constant matrix $G \in \mathcal{R}^{m \times (n-m)}$, let there exist $n \times n$-matrices $P > 0$, $R > 0, U > 0, X > 0, S > 0, T_{11}, T_{2i}, M_{2i}, Y_{ij}, Z_{ij}(i,j = 1,2)$ and matrices $K \in \mathcal{R}^{m \times p}, G_{k1} \in \mathcal{R}^{(n-m) \times (n-m)}, G_{k2} \in \mathcal{R}^{(n-m) \times m} (k = 2,3)$, such that four LMIs (24), where the slack variables $P_{2i}, P_{3i}, i = 1,2$ are chosen of the following form:*

$$P_{2i} = \begin{bmatrix} G_{21} & G_{22} \\ G & \varepsilon_2 I_m \end{bmatrix}, P_{3i} = \begin{bmatrix} G_{31} & G_{32} \\ G & \varepsilon_3 I_m \end{bmatrix}, i = 1,2 \qquad (25)$$

with notations given in (7), (23), are feasible. Then (6) is internally exponentially stable and the cost function (9) achieves $J < 0$ for all nonzero $w \in L_2, v \in l_2$ and for the zero initial condition. If in the above conditions only two LMIs, corresponding to $i = 2, j = 1,2$, are feasible, then the results are valid for (6) with constant delay $\eta_k \equiv \eta_M$.

Proof. Taking into account (25), we have $P_{ji}^T B_2 K C_2 = \text{col}\{G^T BKC_2, \varepsilon_j BKC_2\}$, $i = 1,2$, $j = 2,3$. Substitution of (7) and (25) into matrix inequalities of Theorem 1 and Corollary 1 completes the proof.

4 Examples

Example 1. Consider the following system from [21, 22] for exponential stability and L_2-gain analysis.

$$\dot{x}(t) = \begin{bmatrix} 0 & 1 \\ 0 & -0.1 \end{bmatrix} x(t) + \begin{bmatrix} 0 \\ 0.1 \end{bmatrix} u(t) + \begin{bmatrix} 0.1 \\ 0.1 \end{bmatrix} w(t),$$
$$z(t) = [0 \ 1]x(t) + 0.1u(t), \qquad (26)$$

where $u(t) = -[3.75 \ 11.5]x(t_k - \eta_k)$, $t_k \leq t < t_{k+1}$. The closed-loop system with $w = 0$ and with constant delay τ

$$\dot{x}(t) = \begin{bmatrix} 0 & 1 \\ 0 & -0.1 \end{bmatrix} x(t) + \begin{bmatrix} 0 & 0 \\ -0.375 & -1.15 \end{bmatrix} x(t - \tau) \qquad (27)$$

is asymptotically stable for $\tau \leq 1.16$ and becomes unstable for $\tau > 1.17$. The latter means that all the existing methods via time-independent Lyapunov functionals cannot guarantee the stability of (27) for the sampling intervals that may be greater than 1.17. When there is no network-induced delay, i.e. $\eta_k \equiv 0$, the resulting τ_M determines an upper bound on the variable sampling intervals $t_{k+1} - t_k$. The results (obtained by various methods in the literature and by Theorem 1 with $\alpha = 0$) for the admissible upper-bounds on the sampling intervals, which preserve the stability, are listed in Table 1. From Table 1, we can see that the result by Theorem 1 almost coincides with the result of [4] and is close to the exact bound 1.72 for the constant sampling.

For the values of η_M given in Table 2, by applying various methods in the literature and by Theorem 1 with $\alpha = 0$, we obtain the maximum values of τ_M that

Table 1 Example 1: Maximum upper bound on the variable sampling

method	[15]	[13]	[11]	[7]	[4]	Th 1
τ_M	1.04	1.11	1.36	1.36	1.69	1.68

Table 2 Example 1: Maximum value of τ_M for different η_M

$\tau_M \setminus \eta_M$	0	0.2	0.4	0.6	0.8
[15]	1.04	1.04	1.04	1.04	1.04
[12]	1.11	1.01	0.95	0.90	0.88
Th 1	1.68	1.26	1.18	1.14	1.10

preserve the stability (see Table 2). We note that in this example the results of Corollary 1 for the constant $\eta_k \equiv \eta_M$ coincide with the results of Theorem 1 for the variable $0 \leq \eta_k \leq \eta_M$.

Consider next the static output-feedback controller $u(t) = -0.1122y(t_k - \eta_k)$, $t_k \leq t < t_{k+1}$, where

$$y(t_k - \eta_k) = [1\ 0]x(t_k - \eta_k) + 0.2v(t_k - \eta_k),\qquad(28)$$

and where $\eta_M = 0.1$, $\tau_M = 1.5$. Applying LMIs of Theorem 1 (with the zero and with the non-zero X and U), we find that the resulting closed-loop system has an L_2-gain less than $\gamma = 1.27$ (for $X = U = 0$) and less than $\gamma = 1.17$ (for $X > 0$ and $U > 0$). Hence, the discontinuous terms of Lyapunov functional improve the performance (the exponential decay rate and the I_2-gain).

Example 2. We consider the following simple and much-studied problem (see [14] and the references therein):

$$\dot{x}(t) = -x(t_k - \eta_k),\quad t_k \leq t < t_{k+1}.\qquad(29)$$

It is well-known that the equation $\dot{x}(t) = -x(t - \tau)$ with constant delay τ is asymptotically stable for $\tau \leq \pi/2$ and unstable for $\tau > \pi/2$, whereas for the fast varying delay it is stable for $\tau < 1.5$ and there exists a destabilizing delay with an upper bound greater than 1.5. It is easy to check, that in the case of pure (uniform) sampling, the system remains stable for all constant samplings less than 2 and becomes unstable for samplings greater than 2. Conditions of [15] guarantee asymptotic stability for all fast varying delays from the interval $[0, 1.339]$. By applying Theorem 1 with $\alpha = 0$ and $\eta_M = 0$, we find that for all variable samplings from the interval $[0, 1.99]$ the system remains stable. The latter coincides with the result of [4].

By applying Theorem 1 and Corollary 1 in the cases of variable and constant delay η_k respectively, we obtain the maximum values of τ_M listed in Table 3.

Table 3 Example 2: Max. value of τ_M for variable and constant delay η_k

$\tau_M \backslash \eta_M$	0	0.1	0.2	0.4	0.6	0.8	1	1.2
$Th1$	1.99	1.64	1.57	1.50	1.46	1.43	1.40	1.36
$Cor1$	1.99	1.64	1.57	1.50	1.46	1.44	1.42	1.41

Example 3. static output-feedback H_∞ control.

We consider (26) with the measurement given by (28). It is assumed that the network-induced delay η_k satisfies $0 \le \eta_k \le \eta_M = 0.1$ and that $0 \le t_{k+1} - s_k \le \tau_M = 1.5$. Choosing $\varepsilon_2 = \varepsilon_3 = 10, G = 0.5$ and applying Corollary 2, we obtain a minimum performance level of $\gamma = 1.51$. The corresponding static output feedback control law is $u(t) = -0.1122y(t)$. As we have seen in Example 1, the above controller, in fact, leads to a smaller performance level of $\gamma = 1.17$ (which follows from the application of Theorem 1 to the resulting closed-loop system). The latter improvement of γ illustrates the conservatism of the design method.

5 Conclusions

A piecewise-continuous in time Lyapunov functional method has been presented for analysis and design of linear networked control systems, where variable sampling intervals, data packet dropouts and variable network-induced delays are taken into account. This method has been developed in the framework of time delay approach. The presented results depend on the upper bound η_M of the network-induced delays. The new analysis has been applied to a novel static output-feedback H_∞ control. Differently from the observer-based control, the static one is easy for implementation.

The presented method essentially reduces the conservatism. It gives insight for new constructions of Lyapunov functionals for systems with time-varying delays. The method can be applied to different networked control *design* problems.

Acknowledgements. This work was partially supported by Israel Science Foundation (grant no. 754/10) and by China Scholarship Council.

References

1. Basar, T., Bernard, P.: H_∞ Optimal Control and Related Minimax Design Problems. A Dynamic Game Approach. In: Systems and Control: Foundation and Applications. Birkhauser, Boston (1995)
2. Boyd, S., El Ghaoui, L., Feron, E., Balakrishnan, V.: Linear Matrix Inequalities in Systems and Control Theory. SIAM Studies in Applied Mathematics. SIAM, Philadelphia (1994)
3. Fridman, E.: Discussion on:"stabilization of networked control systems with data packet dropout and transmission delays: continuous-time case". European Journal of Control 11(1), 53–55 (2005)

4. Fridman, E.: A refined input delay approach to sampled-data control. Automatica 46, 421–427 (2010)
5. Fridman, E., Shaked, U.: Delay-dependent stability and H_∞ control: constant and time-varying delays. International Journal of Control 76(1), 48–60 (2003)
6. Fridman, E., Seuret, A., Richard, J.P.: Robust sampled-data stabilization of linear systems: an input delay approach. Automatica 40, 1441–1446 (2004)
7. Fujioka, H.: Stability analysis of systems with aperiodic sample-and-hold devices. Automatica 45, 771–775 (2009)
8. Gao, H., Chen, T., Lam, J.: A new system approach to network-based control. Automatica 44(1), 39–52 (2008)
9. Liu, K., Fridman, E.: Stability analysis of networked control systems: a discontiuous Lyapunov functional approach. In: Proceedings of the 48th IEEE Conference on Decision and Control, Shanghai, China (December 2009)
10. Louisell, J.: New examples of quenching in delay differential equations having time-varying delay. In: Proceedings of the 5th European Control Conference, Karlsruhe, Germany (September 1999)
11. Mirkin, L.: Some remarks on the use of time-varying delay to model sample-and-hold circuits. IEEE Transactions on Automatic Control 52(6), 1109–1112 (2007)
12. Naghshtabrizi, P., Hespanha, J., Teel, A.: Stability of delay impulsive systems with application to networked control systems. In: Proceedings of the 26th American Control Conference, New York, USA (July 2007)
13. Naghshtabrizi, P., Hespanha, J., Teel, A.: Exponential stability of impulsive systems with application to uncertain sampled-data systems. Systems & Control Letters 57, 378–385 (2008)
14. Papachristodoulou, A., Peet, M., Niculescu, S.: Stability analysis of linear systems with time-varying delays: delay uncertainty and quenching. In: Proceedings of the 46th IEEE Conference on Decision and Control, New Orleans, USA (December 2007)
15. Park, P.G., Ko, J.W.: Stability and robust stability for systems with a time-varying delay. Automatica 43(10), 1855–1858 (2007)
16. Sagfors, M., Toivonen, H.: H_∞ and LQG control of asynchronous sample-data systems. Automatica 33, 1663–1668 (1997)
17. Seuret, A., Richard, J.P.: Control of a remote system over network including delays and packet dropout. In: 17th IFAC World Congress, Seoul, Korea (July 2008)
18. Suplin, V., Fridman, E., Shaked, U.: Sampled-data H_∞ control and filtering: nonuniform uncertain sampling. Automatica 43, 1072–1083 (2007)
19. Syrmos, V., Abdallah, C., Dorato, P., Grigoriadis, K.: Satic output feedback-a survey. Automatica 33, 125–137 (1997)
20. Yu, M., Wang, L., Chu, T., Hao, F.: Stabilization of networked control systems with data packet dropout and transmission delays: continuous Time Case. European Journal of Control 11(1), 41–49 (2005)
21. Yue, D., Han, Q.L., Lam, J.: Networked-based robust H_∞ control of systems with uncertainty. Automatica 41(6), 999–1007 (2005)
22. Zhang, W., Branicky, M., Phillips, S.: Stability of networked control systems. IEEE Control System Magazine 21(1), 84–99 (2001)

Index

Lecture Notes in Control and Information Sciences

Edited by M. Thoma, F. Allgöwer, M. Morari

Further volumes of this series can be found on our homepage:
springer.com